9 780898 715590 90000

Applied Mathematics Entering the 21st Century

Applied Mathematics Entering the 21st Century

Invited Talks from the ICIAM 2003 Congress

Edited by
James M. Hill
Ross Moore

Applied Mathematics Entering the 21st Century
Invited Talks from the ICIAM 2003 Congress

Royalties from the sale of this book go to ANZIAM (Australian and New Zealand Industrial and Applied Mathematics), a Division of the Australian Mathematical Society, organizers of the ICIAM 2003 congress.

10 9 8 7 6 5 4 3 2 1

Library of Congress Control Number: 2004102328
ISBN 0-89871-559-8

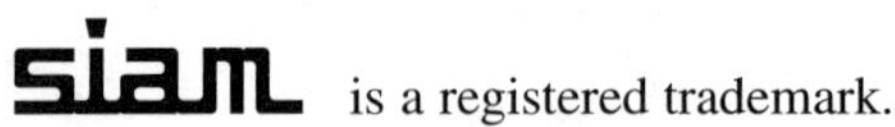

Foreword

Papers appearing in this volume are the Invited Talks given at ICIAM 2003, the 5th International Congress of Industrial and Applied Mathematics, held in Sydney over the period July 7 to 11, 2003. This event was organised by ANZIAM (Australian–New Zealand Industrial and Applied Mathematics) a division of the Australian Mathematical Society (AustMS), on behalf of the International Council for Industrial and Applied Mathematics (ICIAM). The mathematical significance of the Congress, these papers and their authors, is discussed briefly in the Preface, written by Professor Ian Sloan, who was chair of the International Programming Committee for the Congress, and who is currently President of ICIAM.

That the Congress took place at all, and proved to be such a great success for the 1900+ delegates who attended at least some of the daily sessions, was the result of much planning, and work by a great many people, spanning up to seven years prior to July 2003. It could not have happened without help from many sponsors, and the dedication of many academics from at least eleven different Australian universities and research organisations. Foremost among this was the commitment from the Australian Mathematical Society to provide the bridging finance needed to effectively underwrite the whole Congress. The value of the contribution of the University of Technology Sydney, partly in terms of sponsorship but mainly for use of their facilities, as well as time devoted by academic staff, is incalculable. Participating individuals include the ICIAM 2003 Organising Committee: Noel Barton (Congress Director; CSIRO Mathematical and Information Sciences), Ian Sloan (International Program Committee; University of NSW), Lindsay Botten (Local Program Committee; University of Technology Sydney), Neville de Mestre (Chair of ANZIAM; Bond University), David Hunt (Representative of the Australian Mathematical Society; University of NSW), Bill Summerfield (Treasurer; University of Newcastle), Ross Moore (Information Technology; Macquarie University), Jan Thomas (Social and Cultural; Australian Mathematical Sciences Institute), Nalini Joshi (Abstract Reviews; University of Sydney). Also on this Committee, working professionally through ICMS Australasia Pty Ltd were: Bryan Holliday (Director of ICMSAust), Shonna Peasley (Congress Manager), Sally Hobbs (Congress Co-ordinator), Brooke Cartwright, Felicity Brailey, who handled most of the non-academic aspects of the Congress.

Major sponsors, such as the CSIRO Division of Mathematical and Information Sciences, the University of New South Wales, Axiss Australia, and the Department of Education Science and Training of the Government of NSW, provided substantial support in various ways, including sponsoring the visits of some of the invited speakers whose papers appear within this volume. Mathematics departments from universities around Australia, and other research oranisations, similarly made contributions to the support of invited speakers. A complete list of sponsors appears in other publications prepared for the ICIAM 2003 congress and distributed to delegates.

The main Industrial and Applied Mathematics theme of the ICIAM 2003 congress was further enriched by five 'embedded' submeetings. This included the 6th Australia–New Zealand Mathematics Convention, organised by David Hunt

(University of NSW), combining the annual meetings of the Australian and New Zealand mathematical societies, as well as four specialist groups within the applied mathematics community. These played a significant role in making the congress a success.

ASOR 17th Conference of the Australian Society for Operations Research
Simon Perkins (PriceWaterhouse–Coopers) and Layna Groen (University of Technology Sydney)

CTAC 11th Computational Techniques and Applications Conference
Jerard Barry (Australian Nuclear Science and Technology Organisation)

EMAC 6th Engineering Mathematics and Applications Conference
Leigh Wood (University of Technology Sydney)

NSFM 2nd National Symposium on Financial Mathematics
Eckhard Platen (University of Technology Sydney)

Other participating individuals include Phil Broadbridge (now at University of Delaware), Jim Hill (University of Wollongong), Catherine Ross (formerly ICMSAust PL), Ernie Tuck (University of Adelaide), and Alf van der Poorten (Macquarie University). All did much significant early work, either for the Congress itself or for the initial bid. Also many other mathematicians from Universities in the Sydney region helped with the proof-reading and classification of abstracts, needed in the preparation of the academic program; they have been listed by name elsewhere.

Concerning production of this volume, all the authors submitted manuscripts, written in LaTeX or related TeX variants, along with images in PostScript or other image formats. These were combined with a SIAM book format and merged into a single volume using pdfTeX programming techniques developed by one of the editors (RM), specifically for handling volumes of contributed papers. Much editing, especially of the mathematics and figures, was done to unify the presentation style and appearance across all of the papers. Corrections and reviewing were handled using PDF format versions of individual papers, made available to the authors and reviewers via the internet from the ICIAM website.

Finally, we thank the ANZIAM Review team, of Grant Cox, Timothy Marchant, Scott McCue, Mark Nelson, and Song-Ping Zhu all from the University of Wollongong, for their proof-reading and other review comments on the contents of the papers appearing in this volume. Their work helped to locate numerous errors and mis-spellings that might otherwise have gone uncorrected.

James M Hill
School of Mathematics and Applied Statistics
University of Wollongong
New South Wales, Australia.

Ross Moore
Department of Mathematics
Macquarie University
Sydney, Australia.

Opening Address, by Her Excellency Professor Marie Bashir AC, Governor of New South Wales

Dr Noel Barton, Professor Olavi Nevanlinna, Professor Ian Sloan, distinguished guests — International and Australian:

It is a pleasure to join you all this brilliant winter morning for the 5th International Congress on Industrial and Applied Mathematics, and to extend a warm welcome to those delegates who have travelled far distances from across our nation and across the world to participate.

Thank you Uncle Max Eulo for your 'welcome to country', and to our didgeridoo player with his ancient and stirring music. It is a privilege also to record my respect for the traditional owners of this land, and their descendants, and indeed all aboriginal Australians who have nurtured this wonderful continent for tens of thousands of years. Despite your busy schedule, I hope that many of you will have time to see something of our hospitable and cosmopolitan city of Sydney and beyond, including some of the unique art and culture of our indigenous people.

This is certainly a most significant gathering, drawing together for a mathematics Olympiad, some of the most outstanding mathematical scholars of our age. Indeed, it is awesome to contemplate the immeasurable contribution which you have made, pushing forward the frontiers of science since time immemorial, you, whose thought processes, intellectual creativity, and capacity to convert high abstract concepts into reality which can be tested and re-tested. As noted in the program's message: "mathematics is a supreme creation of the human spirit, and a vital contributor in collaboration with many fields of human endeavour."

The fact that this 5th congress is being held in Australia now, not only engenders a great sense of pride and appreciation in the considerable efforts of Professor Sloan and his committee. Most importantly, this gathering of the world's most eminent mathematicians comes at a critical time, (and just one week after the International Congress of Genetics), when those responsible for charting new pathways in the educational curriculum for our young Australians, are re-emphasising the need for quality teaching and competencies in mathematics and technology, particularly in the years of preadolescence and junior high school.

I want to add for those concerned for other aspects of cultural enrichment, that Shakespeare will also be mandatory. I note that among the special feature days in your program, that an Education Day is scheduled which will illuminate to high school teachers and the community the importance and diverse relevance of mathematics, including its indispensability in other scientific disciplines.

For Australia, which has produced many outstanding men and women scientists across many fields, such renewed encouragement from a source such as this is critical, for we have a challenging issue in regard to our human capital in that many of our finest mathematical minds are choosing to work overseas. The importance, the effectiveness and the contribution of mathematics — of mathematicians — to our nation's progress and prosperity are not adequately appreciated. The outstanding range of speakers and the wide-ranging material to be presented will provide a focus for these considerations: ship hydrodynamics, statistical physics, mathematical modelling, turbulence, dynamics of the nervous system and mathematical biology.

Whenever we see the frontiers of industry and science expanding, we know that mathematics is fuelling and enabling these directions. And as one, who before my appointment as Governor, had spent a professional lifetime in medicine, my gratitude to my mathematical colleagues is boundless.

In recent years, enlightened physicians have progressed to demand evidence — statistically sound evidence for diagnostic categories and treatment applications and for outcomes. We have awaited excitedly for what mathematicians have enabled to unfold from the Human Genome Project, and subsequently the immense applications from analysis of the DNA genetic components, as we track down factors which confer risk or actual presence of some of the most tragic and hitherto incurable disorders of mankind. And there is more to come with trials and evaluation of new interventions. The challenges which face you are thrilling and endless.

I would like to leave you with the exhortation of the way ahead in the message from Australia's Chief Scientist, Dr Robin Batterham:

> "to sustain the flow of mathematical inventions, to capture the benefits of this research for society and to ensure that mathematics gets the recognition it deserves."

I know that you will have a most enriching and inspiring gathering, and I have great pleasure in declaring your 5th Congress *open.*

Her Excellency Professor Marie Bashir AC
Governor of New South Wales

Speech on the occasion of the official opening of the 5th International
Congress on Industrial and Applied Mathematics (ICIAM 2003)
Sydney Convention & Exhibition Centre, Darling Harbour
Monday 7th July 2003.

Preface

As we enter the new century, the rôle of mathematics in science, technology and society has become even more important in terms of human comprehension of ever-increasing complex processes, often involving massive data systems. The discipline of mathematics provides the essential tool necessary to discern order, trends and themes emanating from such complexity. The distinguished mathematician Hermann Weyl (1885–1955) who held the Chairs of Mathematics at Zürich Technische Hochschule and Göttingen, and worked at the Institute for Advanced Study at Princeton from 1933 until he retired in 1952, has said:

> "Mathematics sets the standard of objective truth for all intellectual endeavours, and science and technology bear witness to its practical usefulness. It is one of the primary manifestations of the free creative power of the human mind, and it is the universal organ for world-understanding through theoretical construction."

The Čzechoslovakian-born mathematician Igor Kluvánek (1931–1993), at one time Professor of Mathematics at the Flinders University of South Australia, has said:

> "A person understands some information available to him or her only if he or she grasps the connections, the relationships between phenomena, concepts and ideas to which the information refers. It can be said that the understanding of information consists precisely in the grasping of such relations."

It is in this arena where mathematics plays such a unique and fundamental rôle.

This volume presents an overview of contemporary applications of mathematics, with the coverage ranging from the rhythms of the nervous system (Nancy Kopell), to optimal transportation (Yann Brenier), elasto-plasticity (Alexandre Mielke), computational drug design (Peter Deuflhard), hydrodynamic and meteorological modelling (Ernie Tuck, Rupert Klein), and valuation in financial markets (Mark Davis). One wonders how many of the titles could have been understood by readers a century earlier, given that so many are direct products of the computer revolution: grid generation (Marsha Berger), multiscale modelling (Thomas Hou), high-dimensional numerical integration (Harald Niederreiter), nonlinear optimization (Philippe Toint), accurate floating-point computations (James Demmel) and advanced iterative methods (Henk van der Vorst) could hardly have developed if computers had not driven their development.

Many of the papers demonstrate the close dependence on developments in mathematics itself, (Yoshikazu Giga, Jonathan Keating, Peter Markowich, Philippe Toint) and the increasing importance of statistics (Brian Anderson, Mark Davis, Peter Deuflhard, Alice Guionnet, Thomas Hou, Jonathan Keating, Nancy Kopell,

Harald Niederreiter). Others relate to the study of properties of fluids and fluid-flows (Berger, Brenier, Hou, Klein, Tuck, Ying), or add to our understanding of Partial Differential Equations (Brenier, Giga, Markowich, Toint, Vorst, Ying).

The papers in this volume represent the majority[1] of the invited talks at ICIAM 2003, the 5th International Congress of Industrial and Applied Mathematics, held in Sydney during 7–11 July, 2003. The authors of these papers are all acknowledged masters of their fields, having been chosen through a rigorous selection process by a distinguished International Program Committee (listed below). Their contributions affirm that mathematics in application is in good health!

Like its predecessors in Paris, Washington, Hamburg and Edinburgh, the Sydney ICIAM was the major applied and industrial mathematics event worldwide in the four-year cycle, attracting some 1,700 delegates from all parts of the world, notwithstanding SARS and war in Iraq. It was characterised by intense scientific activity, with 250 minisymposia, as many as 43 minisymposia and contributed paper sessions held in parallel, including five embedded meetings. The scientific level was high.

As well as presenting the latest ideas and work in Applied Mathematics, the ICIAM congresses celebrate outstanding achievements over the years that have lain the foundations for new and future discoveries. Four ICIAM prizes were announced at the congress, each with particular specifications for the type of achievement being honoured. The recipients for 2003 are listed in the final chapter of this book, along with a description of their achievements, and naming the members of the distinguished panel which made the selection.

Ian H Sloan
Chair, International Program Committee, ICIAM 2003
President, International Council for Industrial and Applied Mathematics

International Program Committee:

- Ian Sloan (Chair; University of NSW, Australia)
- Heinz Engl (Johannes Kepler Universität, Linz)
- Philip Holmes (Princeton University, USA)
- Sam Howison (OCIAM, University of Oxford)
- Li Daqian (Fudan University, Shanghai)
- Takuzo Iwatsubo (Kobe University, Japan)
- Masayasu Mimura (Hiroshima University, Japan)
- Stefan Müller (Max Planck Institute, Leipzig)
- Robert O'Malley (University of Washington, Seattle)
- Linda Petzold (University of California, Santa Barbara)
- Olivier Pironneau (Université Pierre et Marie Curie, Paris)
- Bernard Prum (Université d'Évry, France)
- Alfio Quarteroni (École Polytechnique Fédérale de Lausanne, Switzerland)
- Nick Trefethen (Oxford University Computing Laboratory)
- Margaret Wright (Courant Institute, New York University)

[1]Other invited talks were given by Franco Brezzi, Jennifer Chayes, David Donoho, Tom Leighton, Michael Ortiz, George Papanicolaou, Vladimir Zakharov.

List of Contributors

Brian D. O. Anderson
National ICT Australia; and Research School of Information Sciences and Engineering, The Australian National University, Canberra, Australia.
email: brian.anderson@anu.edu.au

Marsha Berger
Courant Institute of Mathematical Sciences, New York University, NY 10012, USA.
email: berger@cims.nyu.edu

Yann Brenier
CNRS, Laboratoire Dieudonné, Nice; and Université Paris 6, France.
email: brenier@math.unice.fr

Mark Davis
Imperial College, London, UK.
email: mark.davis@imperial.ac.uk

James Demmel
Mathematics Department and Computer Science Division, University of California, Berkeley, USA.
email: demmel@cs.berkeley.edu

Peter Deuflhard
Zuse Institute Berlin (ZIB); and Faculty for Mathematics and Computer Science, Freie Universität Berlin, Germany.
email: deuflhard@zib.de

Yoshikazu Giga
Department of Mathematics, Hokkaido University, Sapporo, Japan.
email: gjr@math.sci.hokudai.ac.jp

Nick Gould
Computational Science and Engineering Department, Rutherford Appleton Laboratory, Chilton, UK.
email: gould@rl.ac.uk

Alice Guionnet
UMPA, Ecole Normale Supérieure de Lyon, France.
email: Alice.GUIONNET@umpa.ens-lyon.fr

Thomas Y. Hou
Applied & Computational Mathematics, California Institute of Technology, Pasadena, USA.
email: hou@ama.caltech.edu

Jonathan P. Keating
School of Mathematics, University of Bristol, UK.
email: J.P.Keating@bristol.ac.uk

Rupert Klein
FB Mathematik und Informatik, Freie Universität Berlin; and Potsdam Institut für Klimafolgenforschung (PIK), Potsdam, Germany.
email: rupert.klein@pik-potsdam.de

Plamen Koev
Department of Mathematics, Massachusetts Institute of Technology, Cambridge, Mass., USA.
email: plamen@math.mit.edu

Nancy Kopell
Center for BioDynamics, Boston University, USA.
email: nk@bu.edu

Peter A. Markowich
Wolfgang Pauli Institute Vienna; and Department of Mathematics, University of Vienna, Austria.
email: peter.markowich@univie.ac.at

Alexander Mielke
Institut für Analysis, Dynamik und Modellierung, Universität Stuttgart, Germany.
email: mielke@mathematik.uni-stuttgart.de

Olavi Nevanlinna
Institute of Mathematics, Helsinki University of Technology, Finland.
email: olavi.nevanlinna@hut.fi

Harald Niederreiter
Department of Mathematics, National University of Singapore, Republic of Singapore.
email: nied@math.nus.edu.sg

Christof Schütte
Department of Mathematics and Computer Science, Freie Universität Berlin, Germany.
email: schuette@math.fu-berlin.de

Ian Sloan
School of Mathematics, University of New South Wales, Australia.
email: I.Sloan@unsw.edu.au

Christof Sparber
Wolfgang Pauli Institute Vienna; and Department of Mathematics, University of Vienna, Austria.
email: christof.sparber@univie.ac.at

Philippe L. Toint
Department of Mathematics, University of Namur, Belgium.
email: philippe.toint@fundp.ac.be

Ernest O. Tuck
The School of Applied Mathematics, Adelaide University, South Australia.
email: etuck@maths.adelaide.edu.au

Henk A. van der Vorst
Mathematical Institute, Utrecht, The Netherlands.
email: vorst@math.uu.nl

Lung-an Ying
Key Laboratory of Pure and Applied Mathematics, School of Mathematical Sciences, Peking University, People's Republic of China.
email: yingla@math.pku.edu.cn

Editors

James M Hill
School of Mathematics and Applied Statistics, University of Wollongong, Australia.
email: jhill@uow.edu.au

Ross Moore
Department of Mathematics, Macquarie University, Sydney, Australia.
email: ross@maths.mq.edu.au

Contents

Smoke ceremony performed by Uncle Max Eulo at the Opening Ceremony of ICIAM 2003. Seated from left: Ian Sloan (Chair, International Program Committee), Olavi Nevanlinna (President, International Council for Industrial and Applied Mathematics), Noel Barton (Director, ICIAM 2003), Her Excellency Professor Marie Bashir (Governor of the State of New South Wales). photography by *Happy Medium Photo Co.*

Brian Anderson has just stepped down as President and Chief Executive Officer of the National Information and Communications Technology Centre of Excellence, NICTA. He is Professor of Systems Engineering at The Australian National University, and from 1994 until 2002 he was Director of the Research School of Information Sciences and Engineering, Canberra. His 37 years as a researcher have focussed on work in electrical networks, communication systems, control systems and signal processing. A major current focus of interest is the theory of adaptive control systems, where intelligent controllers are able to change their control strategy on-line, through learning something about what they are controlling.

Professor Anderson was born in Sydney, took his undergraduate degrees in mathematics and electrical engineering at Sydney University, and his doctoral degree in electrical engineering at Stanford University in 1966. He is a Fellow of the Royal Society of London, and of the Australian Academy of Science in which he held the office of President between 1998 and 2002. He has received many prestigious awards, the most recent being the 2001 IEEE James H Mulligan, Jr Education medal, and he holds a number of honorary doctorates.

Chapter 1

Pulling the Information Out of the Clutter

Brian D. O. Anderson[*][†]

Abstract: Fixed and wireless telecommunications systems, sonar systems, navigation devices, image processing algorithms — these are all examples of where signal processing is used. Much signal processing is based on statistical models of processes generating the signals and the contaminating noise. This paper traces the development of statistical processing theories, beginning with Wiener filtering, continuing through Kalman filtering, and ending with Hidden Markov Models. Different assumptions underpin these theories, and also very different mathematics. Yet a number of common features remain.

Contents

*Brian D. O. Anderson is Chief Scientist at National ICT Australia, and Professor at Research School of Information Sciences and Engineering, The Australian National University, Canberra.

†National ICT Australia is funded by the Australian Government's Department of Communications, Information Technology and the Arts and the Australian Research Council through Backing Australia's Ability and the ICT Centre of Excellence program.

1 Introduction

Much of signal processing is concerned with extracting relevant information from measurements in which that relevant information is contained in some way, but generally buried in noise. Determining the presence of genuine target reflections in a signal picked up by a radar receiver; figuring out what the message is in a fax which is blurred; minimising the effect of extraneous signals for a user of a hearing aid trying to talk to another individual; these, and a whole host of other examples, constitute situations where signal processing is a must.

In the following sections, we will discuss three different approaches to signal processing using statistical ideas, those associated with Wiener filtering [1], Kalman filtering [2], [3], and Hidden Markov Model filtering [4], [5]. A further section is provided on what is termed as smoothing, and this attempts to further illustrate the very real similarities among the different approaches to filtering, despite the huge differences in mathematical tools. The final section summarises key conclusions.

We conclude this section by recording an example drawn from a recent project.

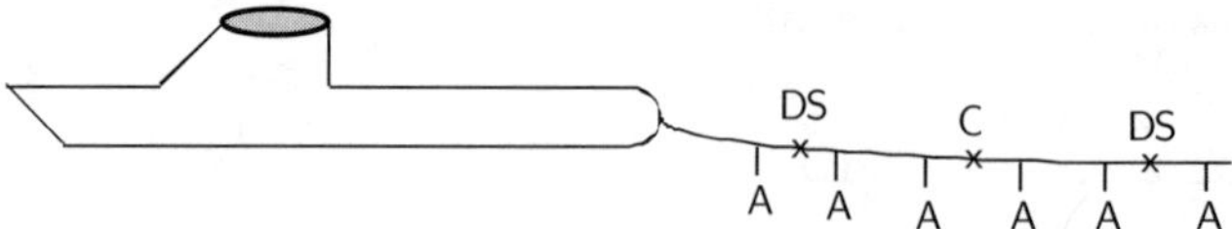

Figure 1. *Submarine with towed array.*

Figure 1 is a diagrammatic representation of a submarine trailing a towed array. A towed array is a cable on which are located a large number of acoustic sensors (labelled with the letter A), and the purpose of the acoustic sensors is to listen for other vessels. Location of acoustic sensors on the towed array rather than location of acoustic sensors on board the submarine allows use of an effectively much bigger and thus more effective acoustic antenna, and lessens difficulties associated with self-noise generated by the submarine. Satisfactory use of the collection of acoustic sensor signals, however, requires knowledge of the shape of the array. The known motion of the submarine together with equations of motion of a towed cable (generally modelled with a nonlinear partial differential equation) allow the generation of an estimate of the shape of the array, but this will be deficient for at least two reasons. First, the equations of motion of the towed cable are only approximations of reality (i.e., there is modelling error); and second, there may well be currents

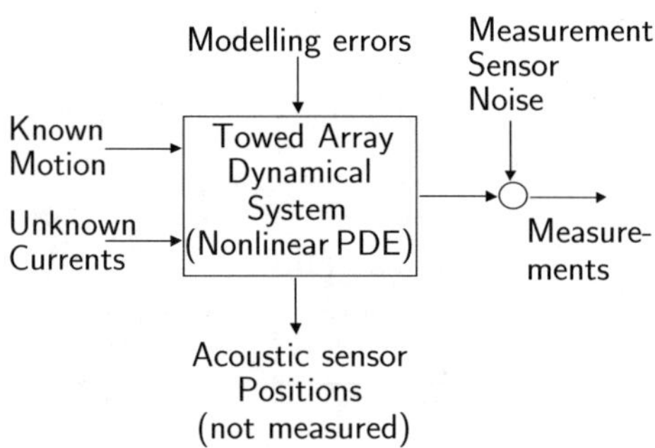

Figure 2. *Abstract representation of array shape estimation problem.*

giving rise to further forces on the array, and thus distortions of it.

For this reason, it is desirable to find techniques for improving the determination of array shape. To this end, one can contemplate including along the array some depth sensors and compasses (labelled DS and C), which provide some sort of (noisy) measurement information relevant to determining the shape of the array.

At this stage, one can represent the situation abstractly via Figure 2. The inputs on the left include the known motion, which is the submarine motion. The rest of the diagram is self-evident.

In order to obtain the acoustic sensor positions, a filter is needed. The signals driving the filter are the measurements from the sensors as well as the motion of the submarine, and, of course, the equations defining the filter in some way will depend on the model of the towed array. Filtering theory in the sense of Wiener [1] and Kalman [2], [3] attempts to provide a technique for determination of a filter, and for predicting the performance of the filter (as measured, for example, by the mean-square error in the estimates produced by the filter).

A description of the towed-array problem can be found in [6].

2 Wiener Filtering

Wiener filtering theory [1] is probably the first attempt to provide optimal filters in situations where signals are characterised by random processes. Indeed, the preface to [1] states: "Largely because of the impetus gained during World War II, communication and control engineering have reached a very high level of development today .... The point of departure may well be the recasting and verifying of the theories of control and communication ... *on a statistical basis.*" Actually, the preface also notes that the work had its origins in ideas of Kolmogorov and Kosulajeff which was published between 1939 and 1941.

A basic situation handled by Wiener filtering is depicted in Figure 3. Figure 3(a) denotes the signal model. The designation 'measurement' is just that: it is what is available to an observer, and can be used for processing. The measurement itself is to be regarded as a sum of signal and noise. The noise, more precisely termed 'measurement noise', is assumed to be a stationary process with

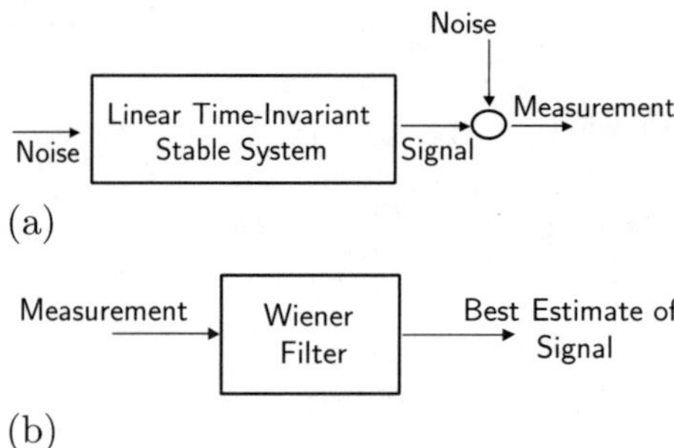

Figure 3. (a) *Signal model;* (b) *basic set-up for Wiener filtering.*

zero mean and known spectrum. The signal is also assumed to be a stationary process. Its generation is modelled through the use of a so-called 'input-noise process', assumed to be zero mean stationary and white[1], passing into a linear time-invariant stable system. Virtually any smooth spectrum can be achieved in this way. The input-noise process and the measurement-noise process are normally assumed to be independent (which implies that the signal and the measurement-noise process are independent). Note that there is no real loss of generality in assuming that the input-noise process is a white process, because if it is not, a 'shaping filter' driven by white noise can be regarded as generating the input process, and then the shaping filter can be combined with the linear, time-invariant stable system between the input-noise process and the signal. The measurement process is often taken as white, but not always.

The Wiener filter, exhibited in Figure 3(b), is the device which reconstructs in real time a 'best estimate' of the signal from the measurement process.

One can either assume that the two noise processes are Gaussian, in which case one can show that the Wiener filter is necessarily linear. Or one can postulate that one is seeking a best estimate among those achievable by linear filters, and not then require that the noise processes are Gaussian. Either way, the Wiener filter is itself linear.

What is meant in this context by the words "constructs in real time a 'best estimate' of the signal from the measurement process"?

Denote the signal by $s(\,.\,)$ the measurement by $z(\,.\,)$ and the signal estimate by $\hat{s}(\,.\,)$ The mean square error at time t associated with the estimate $\hat{s}(t)$ of $s(t)$ is $E[s(t)-\hat{s}(t)]^2$ and it is this quantity which is minimised for the particular choice of filter (a Wiener filter) linking $z(\,.\,)$ to $\hat{s}(\,.\,)$ The words "in real time" connote that t is a running variable, $\hat{s}(t)$ is available at time t, and necessarily only depends on values of $z(\tau)$ for $\tau < t$ or possibly $\tau \leq t$.

There is of course a theory explaining how the Wiener filter can be calculated from problem data. The central component of the calculation is termed 'spectral factorisation'. With some simplification at the edges, the spectral factorisation

[1] A white process is one with constant spectrum. A gaussian white process is obtainable as the derivative of a Wiener process.

problem looks like the following:

> Suppose $S(\omega)$ is the spectrum of the signal process, and suppose $N(\omega)$ is the spectrum of the measurement-noise process. This means, given independence of the signal and the noise processes, that the measurement process has spectrum $S(\omega)+N(\omega)$. Spectral factorisation involves finding a function $H(s)$ of the complex variable s with the following properties:
>
> (i) $H(s)$ and $H^{-1}(s)$ are analytic in $\mathrm{Re}[s] \geq 0$
>
> (ii) $\left|H(j\omega)\right|^2 = S(\omega) + N(\omega)$ for all real ω.

Much of Wiener's work was involved with explaining how to compute the function H. Reference [1] restricts attention, as we have done above, to the case where the signal and noise are scalar processes. In this situation there is in formal terms a formula involving integrals for obtaining H from the spectral information; however, use of approximate integration methods in order to obtain numerical results may be perilous. There is an extension to the vector process case, but not of the formula itself. However, in the event that $S(\omega)+N(\omega)$ is rational in ω, the calculations are hugely simplified; there is even a simple way to deal with the vector process case.

2.1 An Example

By way of illustration, consider Figure 4.

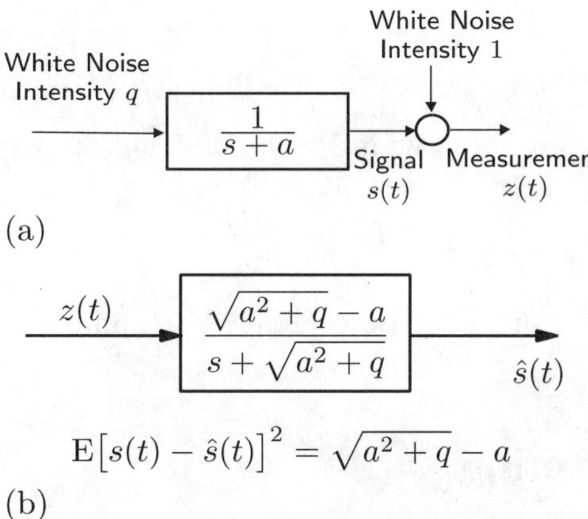

Figure 4. *(a) Signal model; (b) Basic set-up for Wiener filtering.*

To those unfamiliar with the block diagram notation widely used by of electrical engineers, they should understand that Figure 4(a) captures the notion that $\dot{s}+as = w$ where $w(\,.\,)$ connotes the input-noise process, here assumed to be white noise with intensity q. This means that the spectrum of w assumes the value q at all frequencies. The spectrum of the process $s(\,.\,)$ is $S(\omega) = \dfrac{q}{\omega^2 + a^2}$; then

$$S(\omega) + N(\omega) = \frac{q}{\omega^2 + a^2} + 1 = \frac{\omega^2 + a^2 + q}{\omega^2 + a^2} \tag{2.1}$$

and

$$H(s) = \frac{s + \sqrt{a^2 + q}}{s + a} \tag{2.2}$$

in that

$$\left|H(j\omega)\right|^2 = S(\omega) + N(\omega) \tag{2.3}$$

with H and H^{-1} analytic in $\mathrm{Re}[s] \geq 0$.

Also shown in the figure is the Wiener filter and the value of the mean square error. The block diagram notation is shorthand for

$$\dot{\hat{s}} + \left(\sqrt{a^2 + q}\,\right)\hat{s} = \left(\sqrt{a^2 + q} - a\right)z\,. \tag{2.4}$$

It turns out that when the measurement noise is white noise of intensity one, the Wiener filter transformation is always $1 - H^{-1}$.

When the spectra are rational, the calculation of H requires in effect simply the determination of the zeros of two polynomials. The numerator and denominator of $S(\omega) + N(\omega)$ are even in ω^2. With the substitution $\omega^2 = -s^2$, the resulting numerator and denominator polynomials in s, call them $m(s)$ and $n(s)$ respectively, are both factored as

$$m(s) = g(s)\,g(-s)\,, \qquad n(s) = h(s)\,h(-s)$$

where all zeros of $g(\,.\,)$ and $h(\,.\,)$ lie in $\mathrm{Re}[s] < 0$. Then

$$H(s) = \frac{g(s)}{h(s)}\,. \tag{2.5}$$

Obviously, H and H^{-1} are analytic in $\mathrm{Re}[s] \geq 0$, and it is easily checked that $S(\omega) + N(\omega) = \left|H(j\omega)\right|^2$.

3 Kalman Filtering

The Kalman Filter [2], [3] is a variant on the Wiener Filter in several respects. As with the Wiener Filter, we contemplate a signal model and an associated filter. See Figure 5. The key changes between the Wiener Filter and Kalman Filter set-up are as follows:

- The input noise to the signal modal must be white (although this is no real loss of generality, since a device called a 'shaping filter' can be used to cope with a non-white process, at least if it has a stationary, rational spectrum).

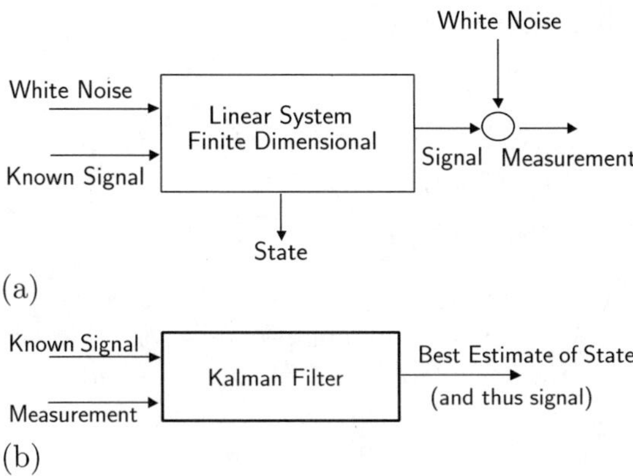

Figure 5. *(a) Signal model; (b) Basic set-up for Kalman filtering.*

- There can be a known, deterministic input in addition to the white-noise process. (The towed array problem is an example where there is a known input signal, the known motion of the submarine.)
- The system linking the input noise to the signal process is necessarily finite-dimensional; i.e., describable by an ordinary differential equation.
- The system linking the input noise to the signal process *is not* assumed to be time-invariant or stable. If however it is unstable, it must be viewed as being switched on at some finite time, rather than the infinitely remote past, to negate the possibility of signals having infinite variance.
- The measurement process is always assumed to be white.
- Quite evidently, the Kalman Filter in order to produce a best estimate of the signal should use both the known input signal and the measurement. But, as it turns out, the Kalman filter seeks not just to estimate 'the signal' but also the whole state of the finite-dimensional linear system sitting between the input process and the signal itself. Obviously, it should be intuitively clear that with a best estimate of the state, one should be able to come up with a best estimate of the signal also.

The great advances in the Kalman Filter are the introduction of possible time-variation in the signal model and non-stationarity in the underlying random processes: the input process and the measurement process need not be stationary; the linear system need not be time-invariant; operations do not have to begin in the infinitely remote past.

The Kalman Filter itself turns out to be a linear finite-dimensional system. It will only be a time-invariant system in the event that all processes in the signal model are stationary.

The calculation of the Kalman Filter is totally different to the calculation of the Wiener Filter. To set it out, we need to formalise some of the above.

3.1 Kalman Filter Equation

The signal model is defined for $t \geq t_0$ by

$$\dot{x} = F(t)\,x + G(t)\,w(t) + \Gamma(t)\,u(t) \tag{3.1}$$

$$s(t) = H(t)\,x \tag{3.2}$$

$$z(t) = H(t)\,x + n(t) \tag{3.3}$$

where $E\big[w(t)\big] = 0$, $E\big[w(t)\,w^{\mathsf{T}}(s)\big] = Q(t)\,\delta(t-s)$, $E\big[n(t)\big] = 0$, $E\big[n(t)\,n^{\mathsf{T}}(s)\big] = R(t)\,\delta(t-s)$ with $w(\,.\,)$, $n(\,.\,)$ independent processes. The initial condition $x(t_0)$ at time t_0 is a random variable independent of $w(\,.\,)$ and $n(\,.\,)$ with mean $\bar{x}(t_0)$ and variance $\bar{P}(t_0)$. The symbol $u(t)$ denotes the deterministic external input. The matrices $F(\,.\,)$, $G(\,.\,)$ etc. are of appropriate dimensions, $Q(t) = Q^{\mathsf{T}}(t) \geq 0$ and $R(t) = R^{\mathsf{T}}(t) > 0$ for all t.

The Kalman filter is defined by

$$\begin{aligned}\dot{\hat{x}}(t) &= \big[F(t) - P(t)\,H(t)R^{-1}(t)\,H^{\mathsf{T}}(t)\big]\,\hat{x}(t)\\ &\qquad + P(t)\,H(t)R^{-1}(t)\,z(t) + \Gamma(t)\,u(t)\end{aligned} \tag{3.4}$$

$$\hat{x}(t_0) = \bar{x}(t_0) \tag{3.5}$$

$$\hat{s}(t) = H(t)\,x(t) \tag{3.6}$$

and $P(t)$ is a symmetric nonnegative definite matrix solving the following matrix Riccati differential equation:

$$\dot{P} = PF^{\mathsf{T}} + FP - PHR^{-1}H^{\mathsf{T}}P + Q\,, \qquad P(t_0) = \bar{P}(t_0)\,. \tag{3.7}$$

Among all possible filters where $\dot{x}(t)$ is constructed using $z(s)$ for $s \leq t$, the Kalman filter ensures

$$E\Big\{\big[x(t) - \hat{x}(t)\big]^{\mathsf{T}}\big[x(t) - \hat{x}(t)\big]\Big\} \tag{3.8}$$

is minimised for all t (minimum variance property).

As intimated above, with the right assumptions on the signal model, the Kalman Filter will be time-invariant. Also, in the event that the deterministic signal at the input of the signal model is zero, the signal model for the Kalman Filter and the signal model for the Wiener Filter have the potential to coincide. It follows therefore that there are some circumstances in which the optimal filter can be calculated either with the Wiener approach or with the Kalman approach. Such situations are necessarily those where there is stationarity and finite-dimensionality of the signal model, white constant intensity input noise and white constant intensity-measurement noise, and a time of commencement in the infinitely remote past, so that all signals are stationary.

3.2 Example

To illustrate this point, we explain how the Kalman Filter can be calculated for the example of Figure 4. The signal model of Figure 4(a) can be written as

$$\dot{x} = -ax + w\,, \qquad s = x\,, \qquad z = x + n$$

i.e., $F = -a$, $G = 1$, $\Gamma = 0$, $H = 1$, $Q = q$, $R = 1$ and initial time is $-\infty$.

The Riccati equation is

$$\dot{P} = -2Pa - P^2 + q\,.$$

It can be verified that for any $\bar{P}(t_0) > 0$, when $t_0 \to -\infty$, the solution of the Riccati equation is independent of the initial conditions, and is constant, *viz.*,

$$P(t) = \sqrt{a^2 + q^2} - a\,.$$

Accordingly, the Kalman filter is

$$\dot{\hat{x}} = -\sqrt{a^2 + q^2}\,\hat{x} + (\sqrt{a^2 + q^2} - a)z\,, \qquad \hat{s} = \hat{x}\,.$$

This is precisely what Figure 4(b) shows.

In situations where either Kalman or Wiener filtering ideas can be used on the same problem, one has stationarity and one has a rational spectrum. It turns out that the Kalman filter approach involves steady-state Riccati equations and there is a deep connection with rational spectral factorisation.

3.3 Time Constants and Exponential Forgetting

There is an important property common to Wiener and Kalman Filters that is actually not universal, but is obtained in most circumstances. Strict theorems are available defining the circumstances under which these properties are obtained; see for example [2], [3], and [7]. There is a notion of a 'time constant' associated with Wiener and Kalman filters. What does this mean?

- Old measurements are forgotten exponentially fast. The best estimate of the state or the signal at a particular time t depends in an exponentially decaying fashion on prior measurements. If measurements sufficiently long ago were inaccurate not just because of the noise but because of sensor failure, this would not affect matters sufficiently far away from the time at which the erroneous measurements were collected.

- Initial state information of the Kalman Filter is forgotten exponentially fast. To understand this statement, recall that at some time (which might be in the infinitely remote past) the Kalman Filter has to be turned on. There is a best initial state for the Kalman Filter, namely, the mean assumed for the initial condition of the signal model. Obviously, one must reckon with the possibility that a highly inappropriate initial state for the Kalman Filter is selected. The point of the observation is that any damage caused by the inappropriate selection will be forgotten exponentially fast.

- Round-off and similar errors can only accumulate to a limited extent. Suppose that the Wiener or the Kalman Filter is implemented on a computer, so that at every step in the calculation some little error is introduced. One should have an *a priori* concern as to whether these errors will accumulate in an ultimately damaging way. The point of the remark is that this will not happen.

The array shape estimation problem introduced in Section 1, see Figure 2, was tackled in [6] using Kalman Filtering ideas. A key step in the modelling is to replace the non-linear partial differential equation model of the towed array by a linear-system model. Also, the replacement model must be finite-dimensional. Thus, as with many practical problems, the initial mathematical problem has to be approximated or modified, to suit the Kalman Filtering framework. The acoustic sensor positions correspond to some components of the state vector in the signal model and their estimates to some components of the state vector in the Kalman Filter. The approximation error in replacing the Figure 2 model is assumed to be swept up in some way by the incorporation of noise signals in the signal model; i.e., the noise signals in the signal model are meant to capture not just genuine noise from sensors, or the uncertainty associated with currents, but also (obviously in a very crude fashion) the inaccuracies associated with the approximation inherent in the modelling process. That these inaccuracies will not overwhelm the calculations as they evolve in time is also a consequence of the time-constant/exponential-forgetting concept described above.

4 Hidden Markov Models

Wiener and Kalman filtering theories are concerned with filtering of signals and linear systems. The theory can be pushed to consider some levels of nonlinearity, typically when linearisation is applicable; but in no sense do the theories provide a general theory for the filtering of nonlinear systems. There is however a theory which can capture many nonlinear filtering problems, and that is based on Hidden Markov Models [4], [5].

As noted in the abstract of [5], Hidden Markov Models (HMMs) were initially introduced in the late 1960s and early 1970s (i.e., about 30 years ago), and their popularity has slowly grown. To quote from the abstract of [5]:

> "There are two strong reasons why this has occurred. First, the models are very rich in mathematical structure and hence can form the theoretical basis for use in a wider range of applications. Second the models, when applied properly, work very well in practice with some important applications."

The reasons might just as well have been advanced for the Wiener filter and Kalman filter. But for HMMs, the mathematical structure is very different, and the successful applications are also very different. The mathematical structure does not include spectral factorisation or Riccati equations. But it does include the theory of positive matrices (to be distinguished from positive-definite matrices), including (as work of this decade has revealed) a form of time-varying extension of Perron–Frobenius theory, [8].

4.1 Examples of Hidden Markov Models

Before defining what a Hidden Markov Model is, let us give several examples. The first is a very old one, the random telegraph wave (see Figure 6).

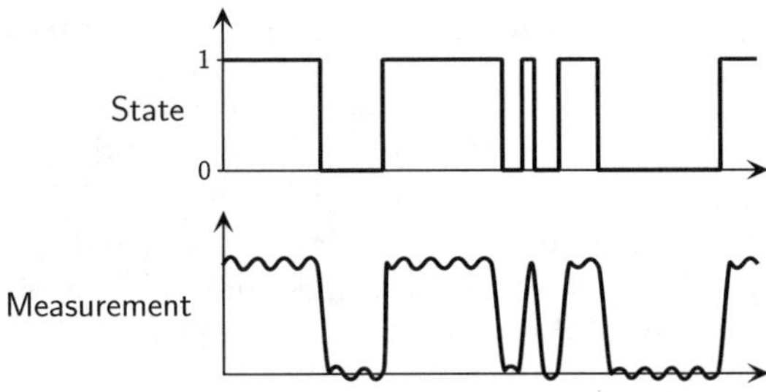

Figure 6. *Noisy measurement of a random telegraph wave.*

One assumes that a signal is transmitted which takes the value zero or one. One has available noisy measurements of that signal, and from the noisy measurements, one is required to reconstruct the original signal. (The common noise model is additive white gaussian noise.) The transitions within the original signal are assumed to occur in a Poisson manner, and the number of levels in the original signal (here two) can be generalised to be finite, but not to be infinite.

For the second example, consider the problem of listening from one submarine for the engine of another submarine. One can postulate that the engine speed of the other submarine has a fundamental frequency lying in one of a finite set of frequency ranges, and the transition probability for movement from one range to another is known. Noisy estimates (in effect noisy measurements) are available of the particular range in which the fundamental frequency of the other submarine's engine lies, and the problem is to properly reconstruct the activity of the other submarine's engine from the estimates.

For a third example, consider a significantly nonlinear variant on the signal model of Figure 4(a). In particular, suppose that replacing the equation

$$\dot{x} = -ax + w \tag{4.1}$$

there appears the equation

$$\dot{x} = -a(x) + w$$

where $a(\,.\,)$ is some nonlinear function of x. One cannot expect Kalman or Wiener filtering to work well, unless perhaps, and then only perhaps, $a(\,.\,)$ is close to being linear. To cope with the nonlinear case, we could imagine partitioning up the real axis on which x lies into a finite set of intervals (two of which would be semi-infinite), and consider transitions in the variable x from one region to another. When x is in the ith region, we could regard that as equivalent to the ith state of a Hidden Markov Model with a finite number of states being the current state of the HMM. The finer the sub-division, the more accurately would the HMM capture the model with continuous x. This is the way in which Hidden Markov Models can be used to cope with problems which are too nonlinear for Kalman filters, where it happens to be true that some linearisation can be acceptable.

4.2 Formal Description with Finite-State Hidden Markov Model

In this sub-section we shall try to capture in a more abstract framework the contents of the previous examples. We shall assume that the underlying time process is a discrete time one rather than continuous time, which eases the exposition very substantially. The use of discrete-time modelling in signal processing has become increasingly common, driven not just by questions of ease of exposition, but the very digital nature of much of the hardware which serves as the implementation platform.

There is an underlying state process $\{X_0, X_1, \ldots, X_k, \ldots\}$ and X_k (the process at time k) can assume one of a finite set of values, for convenience $1, 2, \ldots, N$. The quantity $\Pr\left[\, X_{k+1} = i \mid X_k = j \,\right] = a_{ij}$ is a transition probability, and X_k is a Markov process. We denote by A the matrix $\left[a_{ij}\right]$. There is also an output process $Y_0, Y_1, \ldots$. We shall assume that Y_k takes one of a finite set of values, for convenience labelled $1, 2, \ldots, M$. (There are many examples where Y_k assumes continuous values; for example, when it is equal to gaussian noise plus the state. However, for the sake of this paper we shall assume the simpler case of a finite set of values. This finite set incidentally could arise through quantisation of a continuum.) The link between the state process and the output process is defined by $\Pr\left[\, Y_k = m \mid X_k = n \,\right] = c_{mn}$ and we denote by C the matrix $\left(c_{mn}\right)$.

Evidently, then, two matrices whose entries are all probabilities, A and C, describe the Hidden Markov Model process.

4.3 Hidden Markov Model Filter

An HMM filter is, in quite precise terms, a device for calculating the N-vector whose i^{th} entry is $\Pr\left[\, X_k = i \mid Y_0, Y_1, \ldots, Y_k \,\right]$. This means that the filter uses the measurements up to time k to provide the best possible statement concerning the knowledge of the state at time k. For the simple HMM setup that we have described, it is fairly easy to obtain filtering equations by straightforward application of Bayes' Theorem. The update process involves two steps, incorporating a time update of the state variable with no extra measurements, and then adding in the extra measurement associated with an update. More precisely, let $\Pi_{k|k}$ be the vector with i^{th} entry $\Pr\left[\, X_k = i \mid Y_0, \ldots, Y_k \,\right]$, and $\Pi_{k+1|k}$ the vector with i^{th} entry $\Pr\left[\, X_{k+1} = i \mid Y_0, \ldots, Y_k \,\right]$. Then

$$\Pi_{k+1|k} = A\,\Pi_{k|k} \tag{4.2}$$

$$\Pi_{k+1|k+1} = \frac{1}{[1 \ldots 1]\, C_{Y_{k+1}} \Pi_{k+1|k}}\, C_{Y_{k-1}} \Pi_{k+1|k}\,. \tag{4.3}$$

4.4 The Forgetting Property

At this stage one can ask similar questions to those which can be asked regarding Wiener and Kalman filters:

Are old measurements forgotten, is an inappropriate filter initialisation forgotten, and are round-off and similar errors guaranteed not to overpower the calculation?

As for Kalman and Wiener filtering problems, the answer is, in general, *yes*. The qualification is one which can be expressed in technical terms [9]–[13]; in broad terms it demands that filtering problems be well-posed. The general conclusion is in fact that *there is an exponential forgetting property*, just like that for Kalman and Wiener filtering. Incidentally, obtaining these conclusions for continuous-time Hidden Markov Models is much more difficult technically.

There is a new angle here, which does not arise in Kalman and Wiener filters, and it should be noted. What we have just said is that the calculations leading to the conditional probability associated with the filtering problem are ones in which an exponential forgetting property is found. Suppose that one focuses on the actual production of a state estimate. Thus, one could logically define a filtered state estimate by saying that $\hat{x}_{k|k} = J$ if J maximises $\Pr\big[\, X_k = i \mid Y_0, \ldots, Y_k \,\big]$ over i. Then it turns out that $\hat{x}_{k|k}$ is determined with a *finite memory*; i.e., $\hat{x}_{k|k}$ depends on $Y_k, Y_{k-1}, \ldots, Y_{k-l}$ for some fixed l and all k. At this stage, theory is not available to estimate l easily [14].

4.5 Mathematics of the Exponential Forgetting Property

The exponential forgetting property can be established by an elegant extension of the Perron–Frobenius theory on the eigen-properties of matrices of non-negative or positive entries, [8].

Consider equations 4.2 and 4.3, and the following two equations:

$$\Sigma_{k+1|k} = A\,\Sigma_{k|k} \tag{4.4}$$
$$\Sigma_{k+1|k+1} = C_{Y_{k+1}}\,\Sigma_{k+1|k}\,. \tag{4.5}$$

The effect of the scalar division on the right side in 4.3 and absent in 4.5 is to normalize; i.e., scale the right side of 4.3 so that the vector entries add up to 1. This means that 4.4 and 4.5 together constitute an unnormalized version of 4.2 and 4.3. Thus the behaviour of 4.4 and 4.5 will be able to predict, in many ways, the behaviour of 4.2 and 4.3.

Now observe that 4.4 and 4.5 can be combined together to give

$$\Sigma_{k+1|k+1} = C_{Y_{k+1}} A\,\Sigma_{k|k}\,. \tag{4.6}$$

The matrix $C_{Y_{k+1}}$ can only assume one of M values and so we could rewrite 4.6 as

$$\Sigma_{k+1|k+1} = D_k\,\Sigma_{k|k}\,, \tag{4.7}$$

where D_k is drawn from a finite set of known matrices, call them $A_1, A_2, \ldots, A_M$. The extension of the Perron–Frobenius theorem is found in [8]. Let $\{A_1, \ldots, A_M\}$ denote a finite set of matrices with positive entries, and let $E_N = D_N D_{N-1} \cdots D_1$, where $D_i \in \big[A_1, \ldots, A_M\big]$. Then as $N \to \infty$, $E_N \to \mu_N \nu^\top$ for variable vector μ_N and some fixed vector ν, exponentially fast. (The Perron–Frobenius theorem deals with the case where all the D_i are the same.) There are incidentally extensions of this inhomogeneous product result to cope with non-negative matrices, and such extensions can be useful for the application to HMMs.

The above result implies

$$\Sigma_{N|N} \longrightarrow \mu_N \, \nu^{\top} \, \Sigma_{0|0}$$

exponentially fast. Observe that different values of $\Sigma_{0|0}$ will lead to different values of $\nu^{\top} \Sigma_{0|0}$, which is a scalar; i.e., $\Sigma_{0|0}$ only affects the scaling of $\Sigma_{N|N}$.

If there is normalization, as there is in calculating $\Pi_{N|N}$ from $\Sigma_{N|N}$, $\nu^{\top} \Sigma_{0|0}$ drops out completely. Equivalently, the initial condition is forgotten exponentially fast. Similarly, one argues that old measurements are forgotten exponentially fast.

4.6 Rapprochement between HMMs and Wiener–Kalman Theory

One can pose the question:

> Are there situations in which the Hidden Markov Model approach and the Kalman filter approach or the Wiener approach overlap?

If by the HMM approach, one is talking about finite-state processes, the answer is *no.* However, one can usually regard the type of signal model which appears in a Kalman filter or Wiener filter problem as a limiting version of a Hidden Markov Model signal model. Unpublished work of R. L. Streit has demonstrated that as the limit of the number of states in a certain finite-state Hidden Markov Model is allowed to become infinite, the HMM converges in a certain sense to a Kalman filter signal model, and one can establish convergence of the associated filters as well.

5 Smoothing

In this section, we aim to explain a variant on filtering which applies to each of the three types of filtering we have described.

Any processing of the measurements is often described by the generic term 'filtering'. However, one can particularize the meaning of the word filtering, to distinguish it from a related concept called 'smoothing'. This is the point of this section. Also, we will record some distinctions in the properties of filters and smoothers.

5.1 The Difference between Filtering and Smoothing

Consider Figure 7. For convenience, we shall explain the smoothing concept in terms of a Kalman filtering problem. The explanations carry over to Wiener and HMM problems, and to discrete-time formulations.

This figure depicts one entry of the unknown true state of the system on which the filtering is being performed: it depicts the measurements taken at the output of that system and it depicts one entry of the filtered estimate of the state, obtained at the output of the Kalman filter. The notation $\hat{x}(t \mid t)$ is used to denote a filtered

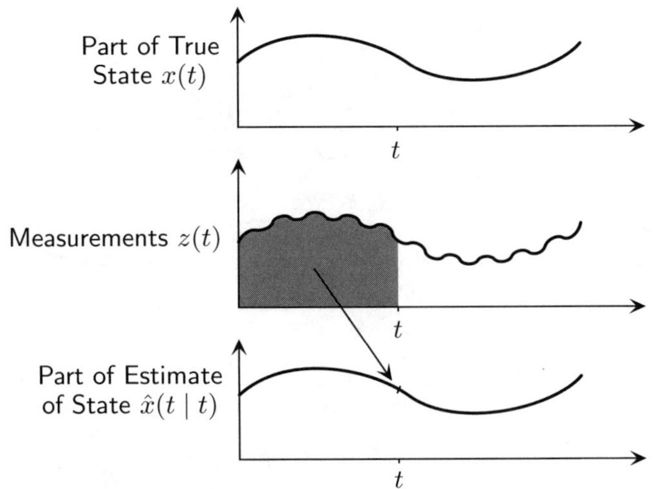

Figure 7. *Representation of filtered-estimate dependence on measurements.*

estimate. The first occurrence of t signifies that we are achieving this estimate of the state at time t and the second occurrence of t signifies we are using measurements occurring up to a time t.

It is intuitively obvious that measurements received after time t must contain some sort of information about $x(t)$. If those measurements are in some way usable, we ought to be able to obtain an improved estimate of $x(t)$; i.e., one with lesser mean-square error. Figure 8 illustrates the distinction between filtering and 'smoothing', where in smoothing we are using measurements not just up to time t but up to some later time T, in order to produce our estimate of $x(t)$. The new estimate is termed $\hat{x}(t \mid T)$.

Leaving aside for the moment the question of how exactly such an estimate might be constructed, it is important to realize that there is one key disadvantage of working with a smoothed estimate; namely, the estimate is not available in real time but only with some delay. For a control application, this may be a fatal disadvantage. However, if one is analyzing what happened in an experiment subsequent to that experiment, there may be no disadvantage at all.

For the sake of completeness, we should mention also the concept of prediction using measurements $z(t)$ up until time t. One seeks to estimate not $x(t)$ but $x(t+\delta)$ for some $\delta > 0$. (The quantity δ may be fixed and t a running variable.) Such an estimation may be relevant in, for example, a rendezvous problem with a moving target with which a rendezvous is sought at a future time. If one can do filtering, it is generally very easy to do prediction; we will devote almost no attention to it.

There are in fact several different types of smoothing which need to be distinguished. These are termed fixed-interval smoothing, fixed-point smoothing, and fixed-lag smoothing.

In fixed-lag smoothing, t is variable and T is set to equal $t + \Delta$, with a fixed

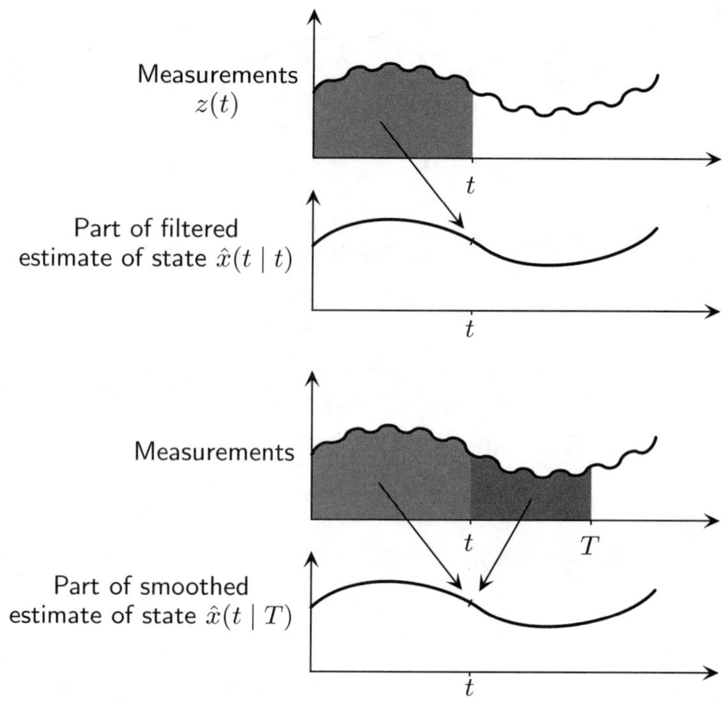

Figure 8. *Contrast of smoothed and filtered estimate dependence on measurements.*

quantity Δ termed the 'lag'. Thus fixed-lag smoothing is like filtering with delay. Figure 9 illustrates how various measurements give rise to a fixed-lag estimate at different time instants.

Fixed-lag smoothing is treated in the Wiener filtering context in [1], in the Kalman filter context in [7], and in the HMM context in [13], with a precursor in [15]. Figure 10 depicts traces of the state of a discrete-time system, a filtered estimate of that state and a fixed-lag estimate if that state together with error performance of the filter and the smoother. An inspection by eye suggests the greater accuracy of the fixed-lag estimate, and this is confirmed by a calculation of the error variance.

In relation to the towed array problem, it is evident that in filtering, measurement information up to a time t would allow estimates of the acoustic-sensor positions at time t and allow listening for other vessels using those sensor estimates. Smoothing would allow a better estimate of acoustic-sensor positions, and allow better listening—but there would be a delay.

5.2 A Mathematical Road Block

Aside from the (admittedly modest) movement in conceptual complexity in passing from filtering to smoothing, there were practical problems associated with the implementation of smoothers. These can best be understood by considering the same

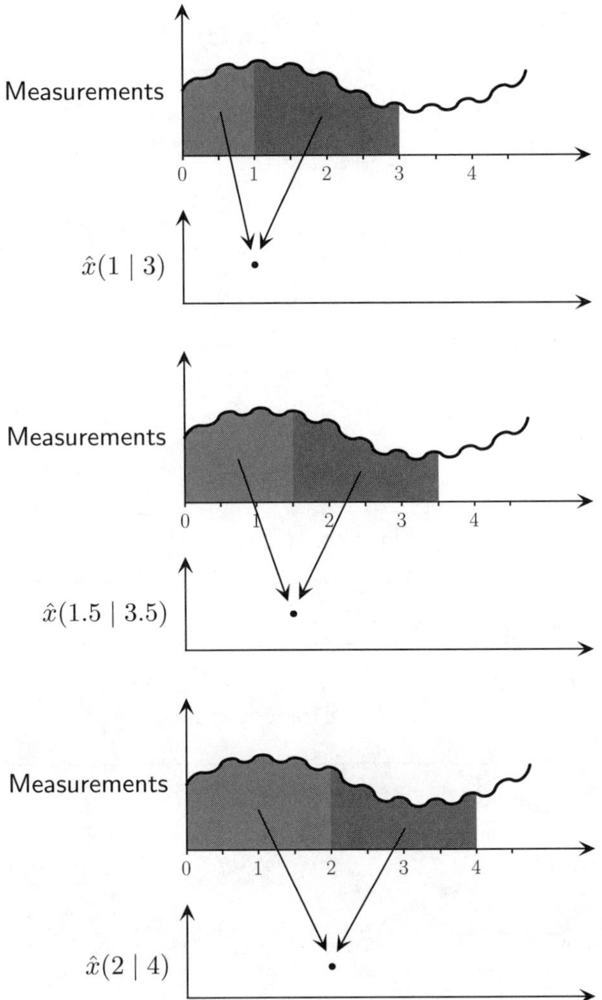

Figure 9. *Representation of generation of a fixed-lag estimate.*

simple example as earlier, illustrated in Figure 4.

With a fixed lag of Δ, the fixed lag estimate $\hat{s}(t-\Delta \mid t)$ can be shown to be given by

$$\begin{aligned}\hat{s}(t-\Delta \mid t) &= \hat{s}(t-\Delta \mid t-\Delta) \\ &\quad + (b-a)\frac{\exp(-s\Delta)-\exp(-b\Delta)}{s-b}\left[z(t)-\hat{s}(t \mid t)\right] \qquad (5.1)\end{aligned}$$

where $b = \sqrt{a^2+q}$ and we have mixed Laplace transform notation with pure time-domain quantities. It turns out that the hardware implementation of a device

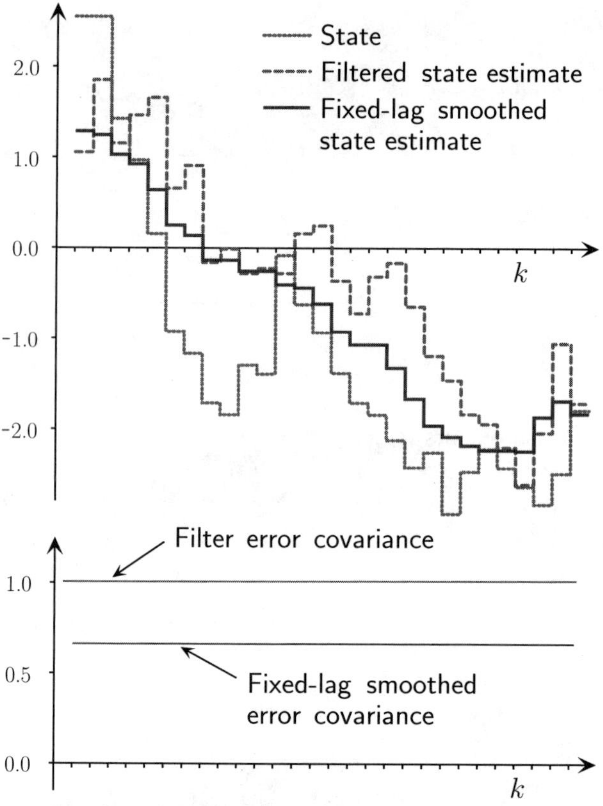

Figure 10. *Simulation data for filtering and fixed-lag smoothing comparison.*

corresponding to

$$\frac{\exp(-s\Delta) - \exp(-b\Delta)}{s - b} \tag{5.2}$$

is apparently straightforward, but this is not actually the case.

More precisely, the "obvious" implementation to an electrical engineer is fatally flawed, due to the inclusion of a guaranteed instability. The "obvious" hardware device has to capture the equivalent of

$$\dot{y} - by = u(t - \Delta) - \exp(-b\Delta)\, u(t) \tag{5.3}$$

[with $u(\,.\,)$ an input and y an output]. An exact solution of this equation results in a bounded mapping from $u \in L_p$ to $y \in L_p$ for any $p \in [1, \infty)$ (with an appropriate initial condition). Nevertheless, the associated homogenous equation

$$\dot{y} - by = 0 \tag{5.4}$$

has exponentially divergent solutions, and any hardware (including software algorithm) based on direct solution of the forced equation will be overpowered by the instability. Fortunately, the problem does not occur in discrete time. (To explain this would lead us too far afield.)

For information on these points, see [16],[17],[18] for two approaches to circumventing the instability above (which is generic, and not particular to this example), [7] for discrete time Kalman filters, and [13] for discrete time HMM filtering.

5.3 Comparative Advantages of Smoothing over Filtering

We have already referred to the key disadvantage of using smoothing as opposed to filtering, including fixed-lag smoothing. This is the delay in obtaining an estimate. The key advantage is the greater accuracy in the estimate. This naturally raises the question: "What improvement can we expect?" A subsidiary question is: "How much lag should one use in fixed-lag smoothing to capture all the significant improvement?"

These questions have been addressed for Wiener filtering problems in a number of papers: see e.g., [19]–[22]. The first key conclusion is that at *high signal-to-noise ratios smoothing gives greater improvement over filtering than at low signal-to-noise ratios.* Denote by P_s and P_f the mean square error in estimating the signal with a smoother and with a filter, respectively. Then [19] provides for a significant family of systems, a bound for the minimum possible value of P_s/P_f in terms of the maximum signal-to-noise ratio. The bound is depicted in Figure 11.

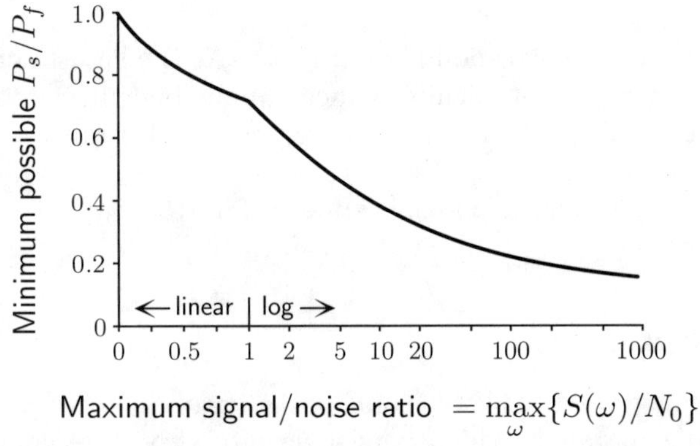

Figure 11. *Smoothing improvement against maximum signal-to-noise ratio.*

At low signal ratios then it is impossible to get much improvement. Note also that the curve does not guarantee that at high signal-to-noise ratios, there has to be a lot of improvement. It simply indicates that there may be a lot of improvement. Nevertheless, examples supported in various references testify to the conclusion that at high signal-to-noise ratios, a significant improvement can be expected.

(Notice that the SNR can go to infinity either as signal power goes to infinity or noise power goes to zero. In the latter case P_s and P_f, both goes to zero and in the limit, the issue of improvement is irrelevant. However for high signal-to-noise ratios, improvement may nevertheless be very desirable: modern digital communications systems after all do seek extremely low error rates.)

The same qualitative conclusions hold for HMM filters. Experimental data was obtained in [15]. The theoretical underpinnings however are not yet complete, as there are no nice formulae for the error measure of an HMM filter. Nevertheless, for the high signal-to-noise ratio case, analytic justification was obtained in [23].

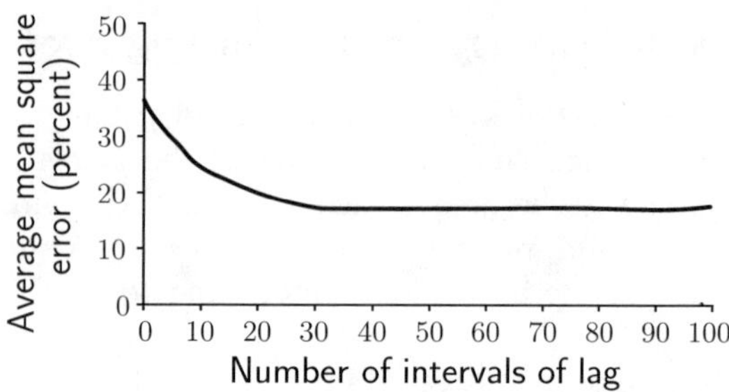

Figure 12. *Variation of smoothing performance with lag.*

A second key conclusion relates to the choice of Δ for fixed-lag smoothing. If Δ is taken to be several times the dominant time-constant of the Wiener or Kalman filter, then one will obtain all the practical improvement that it is possible to obtain using fixed-lag smoothing. A typical curve illustrating the situation is shown in Figure 12.

The case of zero lag corresponds to filtering. As the lag is increased, the mean square error goes down monotonically in fact but the benefit from further increases in lag gradually tails off (in fact it tails off exponentially) until a lag is reached at which further increase of Δ is pointless.

For HMMs, experimental evidence appeared in [15], but it took approximately 25 years before the results could be explained, in [13].

6 Conclusion

In this paper, we have tried to sketch some of the parallels between Wiener, Kalman and HMM filtering, parallels which persist despite very different styles of mathematics. The key results we have chosen to highlight concern:

(1) exponential forgetting properties in filtering, with the relations of the associated time-constant to the understanding of lag adjustment in fixed-lag smoothing;

(2) the benefit of fixed-lag smoothing over filtering being much greater for low-noise situations.

While the details of (1) are well worked out, this is not the case for (2). The absence of error-rate formulae for HMMs is one roadblock.

One would imagine the ideas would be relevant in other domains; e.g., Image Processing.

Bibliography

[1] N Wiener, *Extrapolation, interpolation and smoothing stationary time series*, MIT Press 1949.

[2] R. E. Kalman, R. S. Bucy, *New results in linear filtering and prediction theory*, J. Basic. Eng. Trans. ASME Series D Vol 83 (1961), pp. 95–108.

[3] R. E. Kalman, *A new approach to linear filtering and prediction problems*, J. Basic. Eng. Trans. ASME Series D Vol (1960), pp. 35–45.

[4] R. J. Elliott, L. Aggoun, J. B. Moore, *Hidden Markov Models: estimation and control*, Springer-Verlag 1994.

[5] L. R. Rabiner, *A tutorial on Hidden Markov Models and selected applications in speech recognition.*, Proc. IEEE Vol 77 (1989), pp. 257–285.

[6] A. D. Grey, B. D. O. Anderson, R. R. Bitmead, *Towered array shape estimation using Kalman filters Part I — Theoretical Model*, J. Oceanic Engineering Vol 18 (1993) pp. 543–556.

[7] B. D. O. Anderson, J. B. Moore, *Optimal Filtering*, Prentice–Hall Inc., 1979.

[8] E. Seneta, *Non-negative matrices and markov chains*, 2nd ed. Springer-Verlag, 1981.

[9] B. D. O. Anderson, *New development in the theory of positive systems*, in Systems and Control in the Twenty-First Centrury, C. I. Byrnes, B. N. Datta, C. F. Martin, D. S. Gilliam, eds., Birkhauser, Boston, 1997.

[10] F. LeGland, L. Mevel, *Geometric ergodiety and selected applications in speech recognition*, Proc. IEEE, Vol 77 (1989), pp. 257–285.

[11] A. Arapostathis, S. I. Marcus, *Analysis of an identification algorithm arising in adaptive estimation of Markov chains*, Math. of Control, Signals and Systems, Vol 3 (1990), pp. 1–29

[12] L. E. Baum, T. Petrie, *Statistical inference for probabilistic functions of finaite state Markov chains*, Annals of Math. Stats Vol 37 (1966), pp. 1154–1567.

[13] L. Shue, B. D. O. Anderson, S. Dey, *Exponential stability of filters and smoothers for hidden Markov models*, IEEE Trans. Signal Processing, Vol 46, No. 8, August 1998, pp. 2180–2194.

[14] B. D. O. Anderson, *Forgetting properties for hidden markov models*, in Defenece Applications of Signal Processing, Proc. of the US/Australia Joint Workshop on Defence Applications of Signal Processing, D. Cochran, W. Moran, L. B. White (eds), Elsevier, Amsterdam, The Netherlands, 2001, pp. 26–39.

[15] D. Clements, B. D. O. Anderson, *A nonlinear fixed-lag smoother for finite-state Markov processes*, IEEE Trans. Inform. Theory, Vol IT–21 (1975), pp. 446–452.

[16] S. Chirarattanon, B. D. O. Anderson, *Outline design for stable, continuous-time processes*, Electronic Letters, Vol 8 (1972) pp. 163–264

[17] S. Chirarattanon, B. D. O. Anderson, *Stable fixed-lag smoothing of continuous-time processes*, IEEE Trans. Infor. Theory, Vol IT–19 (1973), pp. 25–36.

[18] P. K. S. Tam, J. B. Moore, *Stable realization of fixed-lag smoothing equations for continuous-time signals*, IEEE Trans. Auto. Control, Vol AC–19 (1974), pp. 84–87.

[19] S. Chirarattanon, B. D. O. Anderson, *Smoothing as an improvement on filtering: a universal bound*, Electronic Letters, Vol 7 (1971), p. 524.

[20] B. D. O. Anderson, *Properties of optimal linear smoothing*, IEEE Trans. Auto. Control, Vol AC–14 (1969), pp. 114–115.

[21] B. D. O. Anderson, S. Chirarattanon, *New linear smoothing formulas*, IEEE Trans. Auto. Control, Vol AC–17 (1972), pp. 160–161.

[22] J. B. Moore, K. L. Teo, *Smoothing as an improvement on filtering in high noise*, Systems and Control Letters, Vol 8 (1986), pp. 51–54.

[23] L. Shue, B. D. O. Anderson, F. De Bruyne, *Asypototic smoothing errors for hidden Markov models*, IEEE Trans. Signal Processing, Vol 28, No. 12, Dec 2000, pp. 3289–3302.

At the Icebreaker Reception: Marsha Berger (USA), Larry Forbes (Australia), Giles Auchmuty (USA). photography by *Happy Medium Photo Co.*

Marsha Berger is a Professor of Computer Science and Mathematics at the Courant Institute of New York University. She works in scientific computing; in particular, in computational fluid dynamics. More specifically her major research efforts have been in developing adaptive mesh refinement methods, developing automatic methods for mesh generation using Cartesian methods with embedded boundaries, and in parallel computing.

One area of Marsha's research involves the time-consuming task of grid generation; needed, for example, to compute the flow around a full aircraft in three dimensions. She also develops adaptive algorithms for use in these applications. Adaptive algorithms concentrate the computational effort where it is most needed (by concentrating additional grid points in these regions, for example). This technique can reduce the turnaround time for flow simulations from months to days or hours.

Chapter 2
Putting Together the Pieces: Grid Generation and Flow Solvers for Complex Geometries

*Marsha Berger**

Abstract: We discuss some of the steps involved in preparing for and carrying out a fluid flow simulation in complicated geometry. Our goal is to automate this process as much as possible to enable high quality inviscid flow calculations. To this end we have developed algorithms based on the use of Cartesian grids with embedded geometry. This work is in collaboration with Michael Aftosmis and Scott Murman, at NASA Ames Research Center.

Contents

1 Introduction

There are many steps in carrying out a fluid flow simulation in realistic engineering geometries. To briefly enumerate for a steady state flow, the steps include:

- geometry specification,
- surface mesh generation,

*This research was supported by AFOSR Grant F49620–00–1–0099, and DOE Grant DE–FG02–88ER25053.

- volume mesh generation,
- flow solution,
- post-processing.

If the computation is time-dependent, the last two steps are repeated at each time-step. If the geometry is moving, the iteration repeats from the mesh-generation steps, and if the geometry itself is changing, for example bodies in relative motion or in an optimization loop modifying a wing shape, all steps are repeated. Especially in these last cases, it is essential to automate the process in an accurate and robust way. This is the goal of our research.

Twenty years ago, the bottleneck step was computing the flow. However, with algorithmic advances, along with advances in the hardware and software for high performance parallel computing, this step has been reduced from weeks to hours or minutes. Twenty years of Moore's law has affected all the automated steps, but left the interactive steps almost completely unchanged. The bottleneck steps now are the first two, with the initial step of geometry acquisition the most time-consuming and least automatic and robust of them all.

In this talk, I will sketch some of the algorithmic advances we have made in some of the above categories, toward the goal of automating the entire simulation process. The central piece in our strategy is the use of Cartesian grids with embedded (non-body-fitted) geometry (see figure 1). This has greatly facilitated and affected all the other steps, as we will show below.

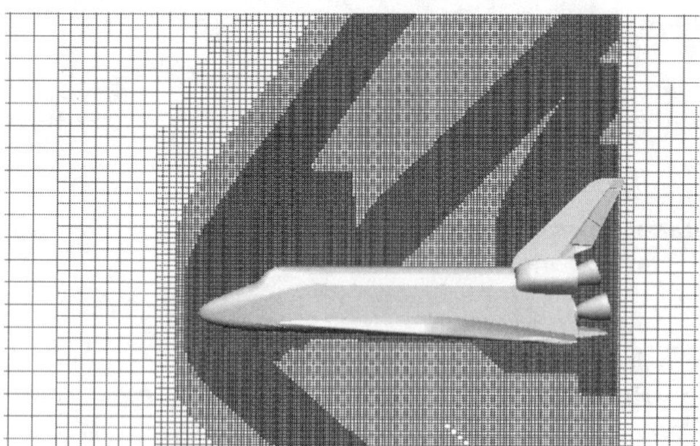

Figure 1. *Illustration of a Cartesian non-body-fitted grid around a complicated geometry. When the object intersects a Cartesian cell, it simply* cuts out *pieces of the cell.*

2 Geometry Representation

Even describing a complicated geometry is difficult, without the added burden of turning it into a surface that is suitable for solving fluid flow equations. The description of a geometry can come in many forms: as a set of cross-sectional curves, or a scatter of points on a surface, such as a laser scanner would produce. Often, geometries come from CAD programs, many of which represent the object using a solid model. This is most often done using NURBS (Non-Uniform Rational B-Splines), along with topological information such as which surface is connected to which other surface. However, when these high order polynomials intersect, they do not match perfectly, leading to gaps or overlapping surfaces. that can cause a program to make topologically inconsistent decisions. For example, a basic step in generating meshes is determining whether a point is inside or outside the geometry. With a closed watertight surface description this can be done using ray-casting: draw a line from the point in question to a known point exterior to the geometry. If it intersects the surface once, then the point was inside. However if there are tiny gaps in the surface, we will no longer robustly be able to tell what is inside and what isn't. Similarly, if the surfaces overlap, we may find two intersection points between the ray and the surface, and conclude the point is exterior to the geometry when it isn't. These kind of topological inconsistencies can cause a program to fail catastrophically.

There are other common sources of errors from NURBS. For example, going from a point x, y, z in physical space to a u, v point on a NURBS surface involves a nonlinear iteration with tolerances. This would be a necessary step for finding matching points across different NURBS patches. NURBS are only valid in regions described using *trimming curves*, which are themselves another potential source of inconsistency as well as floating point error.

In our work we have instead chosen to start with a lower order accurate but completely robust representation of the surface using triangles. This has the drawback that highly curved surfaces need many of them, but the benefit that such a description is unambiguous, and simple to work with. The triangulations will themselves come from NURBS descriptions of objects, but section 3.1 on surface intersections describes how a part of this process can be simplified.

3 Surface Meshing

Given a geometry description, the next step is to produce a nice tessellation for it. The type and quality of the surface mesh that is needed will depend on what kind of volume mesh is used. For structured hexahedral meshing, for example used by chimera methods ([18, 6]), structured quadrilateral patches on the surface are necessary. Tetrahedral volume grids, on the other hand, are usually generated by marching out from the surface triangulation. Both of these *body-fitted* grid types need high quality surface meshs since the surface is a facet of the volume mesh. In contrast, Cartesian meshes are not body-fitted; triangles on the surface are not used for computation, but are used only to cut away part of the Cartesian cell. This decoupling of the surface and volume meshes allows for much greater flexibility in generating the surface mesh. With Cartesian grids, the surface triangulation only has to be an accurate representation of the solid body. The use of Cartesian meshes greatly reduces the difficulty of generating a nice surface-mesh for many types of calculations.

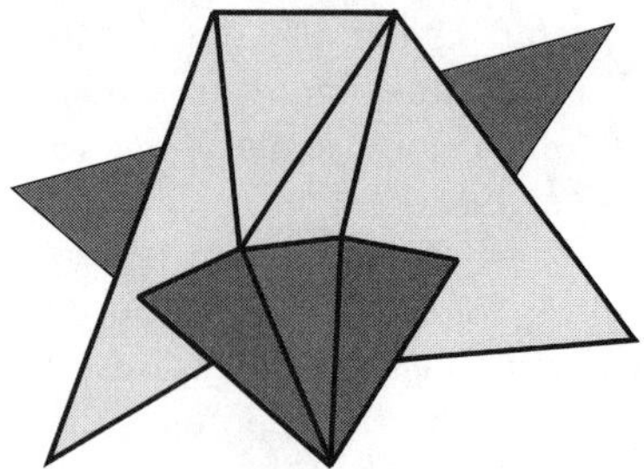

Figure 2. *The colored triangles come from different component triangulations. Three white triangles are shown intersecting a single blue triangle, with their intersection indicated in bold. The middle white triangle has been retriangulated around the intersected segments.*

3.1 Surface Intersections

Most complicated geometries are assemblies of individually described parts, for example, a wing, a nacelle, and a fuselage are different parts in a CAD system describing an airplane. Often, the wing described by the NURBS extends into the fuselage a bit. A trimming curve marks the intersection, indicating that on one side the wing should be used, and on the other side the fuselage.

As mentioned above, this is an error-prone part of the process. Instead, using the flexibility of Cartesian non-body-fitted grids, we can do the following. We describe each component with its own closed, watertight triangulation, ignoring the surface–surface interactions. We have designed a component intersection algorithm using robust arithmetic to intersect the triangulations, (a much more robust process), and extract the *wetted* surface. The triangles that are intersected are retriangulated, (using a constrained Delaunay triangulation, also a robust procedure), and the triangles that end up being interior to the wetted surface are discarded with a robust process for inside–outside detection. Figure 2 schematically illustrates these steps using two components. The details of this intersection algorithm can be found in [2].

Since the intersecting triangles have their intersection points computed in floating point, it suffers from round-off error. (It would be prohibitively expensive, at this point at least, to use extended precision packages for these calculations). However, the *topology* of the intersecting surfaces can be determined exactly, without much cost. The strategy then is to use the robust topological information to drive the algorithm.

For example, suppose we are computing the intersection of a line segment and a triangle. If the intersection point is near an edge of the triangle, it is possible that due to round-off, the point will be outside the first triangle, but the adjacent triangle will also think it doesn't intersect. Somehow the line goes from one side of the triangulated surface to the other without intersecting any triangles! This kind of logical inconsistency can cause a program to bomb.

The determination of whether or not a line segment intersects a triangle can be cast as a series of determinant evaluations. The 4×4 determinant

$$\begin{vmatrix} p_x & p_y & p_z & 1 \\ q_x & q_y & q_z & 1 \\ r_x & r_y & r_z & 1 \\ a_x & a_y & a_z & 1 \end{vmatrix} \tag{3.1}$$

gives six times the signed volume of the tetrahedron whose vertices correspond to the determinant rows. For example, consider figure 3 with line segment ab and triangle pqr. If the tetrahedron $pqra$ and $pqrb$ have volumes of opposite signs, then the line segment crossed the plane of triangle pqr. A few more determinant evaluations, for example the tetrahedron $pqab$, are needed to show that the line segment actually intersects within the triangle boundaries.

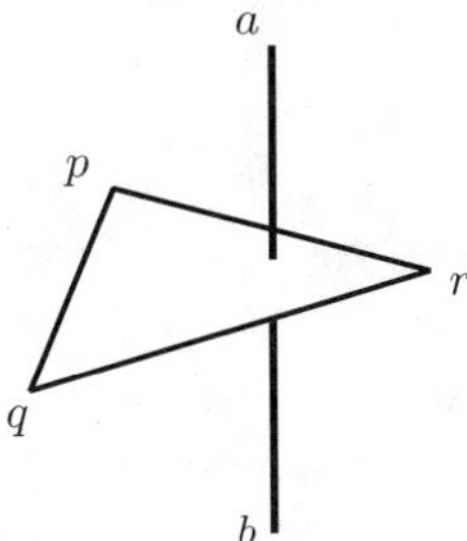

Figure 3. *The determinant in eq. (3.1) is six times the signed volume of tetrahedron pqra.*

Assuming we treat the input data as exact, the determinant involves at most a polynomial in the data to the third power, and so is of bounded complexity. In addition, all that is needed is the sign. An extremely useful tool in this approach is Jonathan Shewchuk's Predicates package [17] using *adaptive* arithmetic. The approximate determinant is computed along with an error bound. If the answer exceeds the error bound, the sign can be trusted. If not, the next more accurate approximation is computed. Not only is his approach extremely fast, but the more expensive computations only need to be done for a small fraction of the intersections (typically under 1%).

This still leaves the question of how to handle the degenerate intersections. For example, suppose a line segment intersects the vertex of a several triangles. The determinant will be exactly zero. Should this be counted as one intersection, none or many? This step is the heart of many ray-casting algorithms, which determine if a point is inside an object or outside? If a line segment starts from the exterior of an object, and intersects it once (or an odd number of times), the ending point of the line segment is inside. If the line segment intersects an even number of times, the last point is exterior again. If the count is wrong, this again leads to a logical inconsistency. To resolve these degenerate cases in an automatic way, we use the Simulation of Simplicity approach of Edelsbruner and Mucke [8] in the intersection, constrained Delaunay re-triangulation and inside/outside testing.

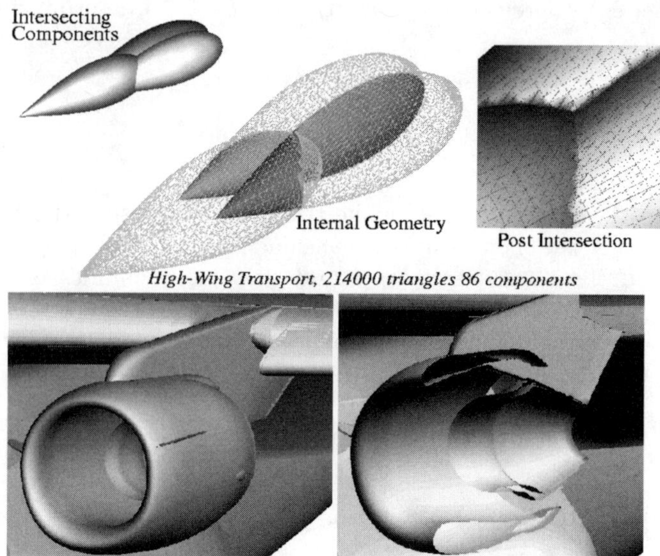

Figure 4. *Several examples of surface intersections involving many input components, resulting in a single watertight geometry.*

Figure 4 shows several intersection results. In the high-wing transport example, the 86 initial components had a total of 214,000 triangles, of which 10,400 were intersecting. The average intersected triangle was divided into 6 smaller ones by the constrained Delaunay procedure, with the largest retriangulation divided into 232 smaller triangles. As you can see from the figure, there are sliver triangles and triangles that are not nicely graded in the resulting surface, but they don't matter.

4 Volume Mesh Generation

A volume mesh fills the domain between the surface geometry and the far field, hopefully in a way that allows for an accurate solution with a minimum number of elements. Cartesian meshes allow for a very accurate solution over most of the domain where the mesh is completely regular. Most of the work of the mesh generator comes from the intersection of the geometry with the mesh.

A Cartesian mesh generation does two things. First, it uses the curvature of the body, approximated by the variation in the surface normals of the triangles, to decide where the elements need to be refined. In other words, multilevel mesh refinement is needed to resolve the geometry, not just for the flow solution. An example of this is shown in figure 5, showing a mesh around the tail end of the space shuttle. Second, the mesh generator determines which cells are intersected, and if so, computes the volume, face areas, and intersected surface areas that will be needed by a second-order accurate finite-volume flow solver.

Figure 6 shows a blow-up of a cut cell intersecting the geometry, where the

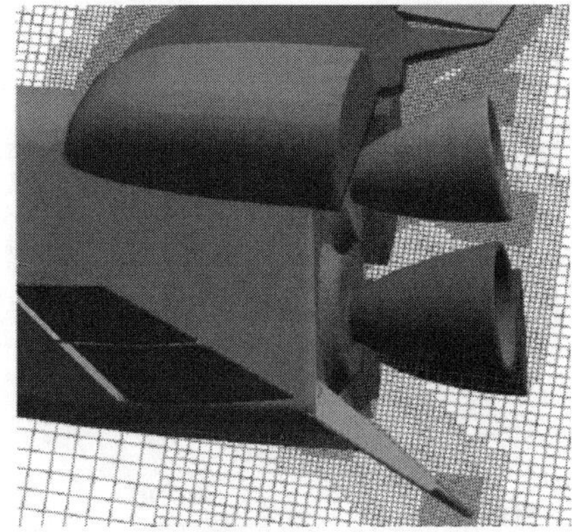

Figure 5. *Mesh around part of space shuttle, showing curvature based refinement in order to improve the resolution of the geometry.*

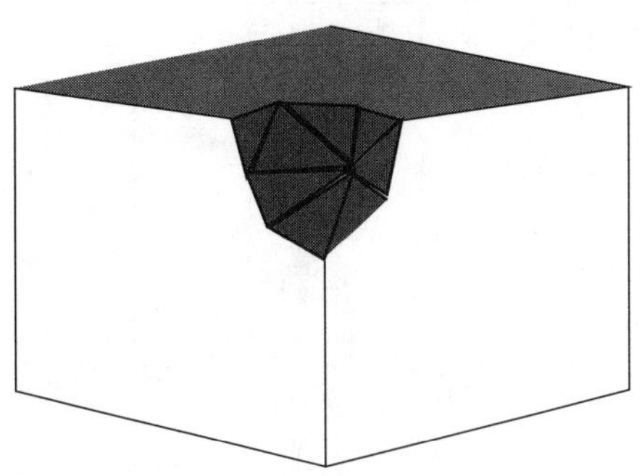

Figure 6. *Illustration of a cut cell. The geometric quantities that need to be computed for it include the face areas, the volume, and the centroids (face, volume, and surface).*

bulk of the computation is done. Note that a cut cell need not be convex. Also, one difficult case that must be accommodated is when a thin piece of geometry bisects a cell into two or more separate polyhedra, creating what we call a *split* cell. This is conceptually simple, but is where most of the program complexity is found. The mesh generator typically runs in minutes on a desktop workstation.

5 Flow Solver

It is easy to implement a finite-volume multigrid-accelerated Cartesian mesh method over most of the flow domain, where the cells are regular. (Parallelization is never easy, at least yet, but Cartesian meshes present no new problems here). The only new difficulty in the numerical discretization comes from the cut cells. These are completely irregular: every aspect of the scheme needs to be re-thought, from gradient calculations to limiter evaluations. For time-dependent problems using explicit time-marching schemes, cut cells have cell volumes that are orders of magnitude smaller than the regular cells, and will be unstable with the usual CFL number without a lot of work. This is a topic of much research in the last few years. See [5, 3, 4, 12, 11] for several approaches to this problem.

What I will talk about here is our approach to parallelizing the flow solver. The typical approach here uses domain decomposition to distribute the computation in a load-balanced way. There are several industrial strength packages that do this, for example [9], but none that take advantage of a Cartesian mesh, an easier problem to solve than the general unstructured graph-based partitioning problem. A natural approach that we have adopted, and one that also gives us on-the-fly partitioning for any number of processors, uses space-filling curves.

A space-filling curve through a 3 dimensional mesh gives a linear ordering of the cells. It has high locality, since the curve enters a cell through one face

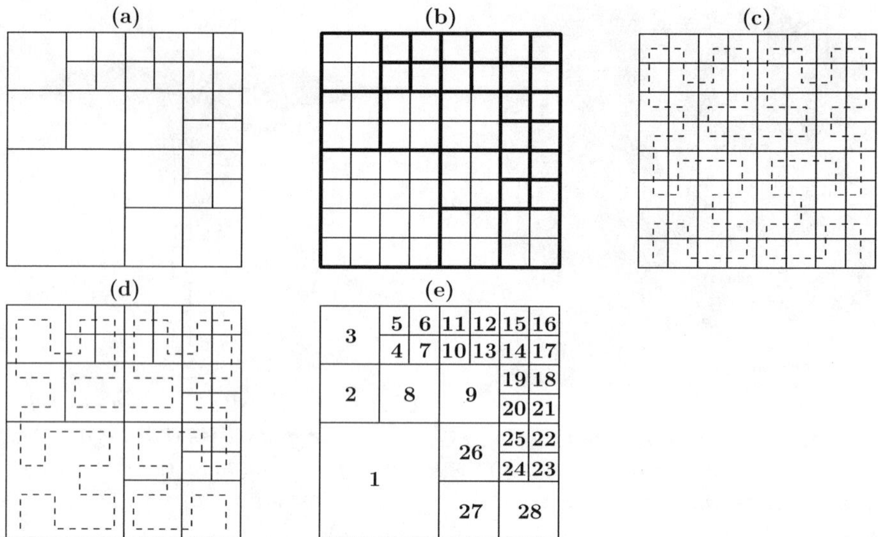

Figure 7. *Illustration of the Peano–Hilbert space-filling curve (sfc) numbering steps for a multilevel Cartesian mesh: (a) 2D multilevel mesh; (b) corresponding uniform mesh at finest level; (c) Peano–Hilbert curve through mesh; (d) curve on original mesh; (e) final ordering of cells using sfc numbering.*

and leaves through another so at least two of a cell's neighboring cells will be consecutively numbered, leading to good cache performance too. There has been a flurry of activity using space-filling curve (sfc) orderings in the last decade, see [16, 13, 14, 1]. Figure 7 illustrates pictorially the Peano–Hilbert ordering for a 2 dimensional Cartesian mesh, its adaptive generalization, and the final ordering of the cells.

Once the mesh is reordered using the sfc-numbering, it can be partitioned dynamically at run-time. The computational work W on the mesh can be easily totaled, accounting for the extra work of integrating a cut cell over a full flow cell (empirically found to be 1.5). With p processors, each processor gets W/p, which can be distributed with only a second pass over the complete mesh. Since Cartesian meshes have large regular blocks, these partitions are largely regular themselves, have low overhead, and communication statistics only slightly worse than completely uniform Cartesian grids. More details of this approach, including parallel speed-ups, can be found in [1]. Figure 8 shows an example of the space shuttle in launch configuration partitioned into 16 domains.

6 Extensions

The flow solver described above has been used for several years to compute steady state inviscid solutions around complex geometry. Our most recent work however has focused on extending these algorithms to the time-dependent case [12, 11]. As

Figure 8. *The domain including the space shuttle has been divided into 16 partitions. Note that most domains are fairly rectangular, although the are not always compact.*

a first pass, we have simply put together all the pieces described above, along with some that haven't been presented here, to compute flow with moving geometry. Figure 9 shows one time-step of a time-dependent computation performed by Michael Aftosmis and Scott Murman, collaborators at NASA Ames. After the devastating breakup of the Columbia, computations were performed to try to determine if the foam could have hit the wing's leading edge, as it was later concluded by the CAIB to have done. These computations helped determine a possible trajectory, as well as the likely range of velocities and mass of the debris that was observed in the launch footage.

7 Open Problems

This talk has hopefully demonstrated many of the successes in performing high quality simulations in complicated geometry using Cartesian meshes with embedded boundaries. However, there are still many open research problems that I hope this talk encourages. Algorithms for moving geometry are in their infancy. The method briefly described above, for example, is implicit and stable, but open issues (for example conservation) still remain for large time-steps. Another unknown is whether Cartesian methods can be used efficiently for highly anisotropic flows. The issue here is whether anisotropy can be folded into the framework of Cartesian methods in a way that takes advantage of their robustness and simplicity. This will be necessary if Cartesian mesh methods are to be extended for viscous flows with high Reynolds numbers where boundary-layer zoning is typically used for efficiency. A first pass on this was in the thesis of [7]. A third avenue to investigate is higher-order discretizations at the cut cells. Since most of the mesh is Cartesian it would be

Figure 9. *Composite of time-steps from a simulation of foam debris falling from the bi-pod ramp of the space shuttle. Simulation by Aftosmis and Murman.*

simple to implement higher than second-order schemes in this region. The question is whether anything more accurate can be done at the cut cells. Even a fully second-order scheme at the cut cells is difficult [10]. Finally, a favorite topic of mine to think about is whether the representation of the geometry itself can be extended in a robust way to higher than first order (planar facets). A solution to any of these problems would find immediate application.

Bibliography

[1] M. J. Aftosmis, M. J. Berger and G. Adomavicius; *A Parallel Multilevel Method for Adaptively Refined Cartesian Grids with Embedded Boundaries.* AIAA Paper 2000–0808, Reno, NV., Jan. 2000.

[2] M. Aftosmis, M. Berger and J. Melton; *Robust and Efficient Cartesian Mesh Generation for Component-Based Geometry.* AIAA J. **36**(6), June, 1998.

[3] S. A. Bayyuk, K. G. Powell and B. van Leer; *A Simulation Technique for 2-D Unsteady Inviscid Flows around Arbitrarily Moving and Deforming Bodies of Arbitrary Geometry.* AIAA–93–3391.

[4] M. Berger, C. Helzel and R. LeVeque; *H-box Methods for the Approximation of Hyperbolic Conservation Laws on Irregular Grids.* SIAM J. Num. Anal. **41**(3), 2003.

[5] I.-L. Chern and P. Colella; *A Conservative Front Tracking Method for Hyperbolic Conservation Laws.* Lawrence Livermore Laboratory Technical Report UCRL–97200, July 1987.

[6] G. Chesshire and W. D. Henshaw; *Composite Overlapping Meshes for the Solution of Partial Differential Equations.* J. Comp. Phys. **90** (1), 1990.

[7] W. Coirier; *An Adaptively Refined Cartesian, Cell-Based Scheme for the Euler and Navier–Stokes Equations.* University of Michigan PhD Thesis, 1994.

[8] H. Edelsbruner and E. Mucke; *Simulation of Simplicity: a Technique to Cope with Degenerate Cases in Geometric Algorithms.* ACM Symp. on Computational Geometry, 1986.

[9] G. Karypis and V. Kumar; *A Fast and High Quality Multilevel Scheme for Partitioning Irregular Graphs.* Technical Report, Dept. of Computer Science TR95–035, University of Minnesota, 1995.

[10] C. Helzel and M. Berger; *A High Resolution Rotated Grid Method for Conservation Laws with Embedded Boundaries.* NYU CMCL Report 03–005, June 2003. Submitted to SIAM J. Sci. Comp.

[11] S. Murman, M. J. Aftosmis and M. J. Berger; *Numerical Simulation of Rolling-Airframes Using a Multi-Level Cartesian Method.* AIAA Paper 2002–2798, Reno, NV., June 2002.

[12] S. Murman, M. J. Aftosmis and M. J. Berger; *Implicit Approaches for Moving Boundaries in a 3-D Cartesian Method.* AIAA Paper 2003–1119, Reno, NV., Jan. 2003.

[13] M. Parashar and J. C. Browne; *Distributed Dynamic Data-Structures for Parallel Adaptive Mesh Refinement.* Proc. Intl. Conf. High Performance Computing, 1995.

[14] J. R. Pilkington and S. B. Baden; *Dynamic Partitioning of Non-uniform Structured Workloads with Space-filling Curves.* IEEE Trans. Parallel and Distrib. Systems **7**(3), March 1996.

[15] J. J. Quirk; *An Alternative to Unstructured Grids for Computing Gas Dynamics Flows around Arbitrarily Complex Two Dimensional Bodies.* ICASE report 92–7, 1992.

[16] J. K. Salmon and M. S. Warren and G. S. Winckelmans; *Fast Parallel Tree Codes for Gravitational and Fluid Dynamical N-body Problems.* Intl. J. Supercomp. Appl. **8**(2), 1994.

[17] J. Shewchuk; *Robust Adaptive Floating-Point Geometric Predicates.* Proc. 12th Annual ACM Symp. Comp. Geometry, May 1996.

[18] J. Steger, Benek and C. Dougherty; *A Flexible Grid Embedding Technique with Application to the Euler Equations.* AIAA–83–1944. 6th CFD conference, Danvers, Mass., July 1983.

Yann Brenier is Professor of Mathematics at the Université de Paris 6, currently on leave of absence as a senior researcher at the Université de Nice. His research interests have included nonlinear partial differential equations arising in fluid mechanics, plasma physics, with a special focus on their numerical and geometrical features.

Professor Brenier was born in 1957 in Saint-Chamond, France, and studied at the Université de Paris 6 before taking his PhD and later his "doctorat d'État" at the Université de Paris Dauphine. He started his career in 1979 as a researcher at INRIA, the French research center for computer sciences. Later he was a Hedrick Assistant Professor at UCLA in 1985 and became a professor at the Université de Paris 6 in 1990, teaching at École Normale Supérieure, from 1990 to 1997. He has been a member of the Institut Universitaire de France since 1996 and was recently invited to be a speaker of the International Congress of Mathematicians in Beijing in 2002.

Chapter 3

Optimal Transportation Theory and Geometric Partial Differential Equations

Yann Brenier

Abstract: The first optimal transportation problem was addressed by Monge in 1781, considering a civil engineering problem where parcels of materials have to be displaced from one site to another one with minimal transportation cost. A treatment of this problem by Kantorovich in 1942 established interesting links between statistics, combinatorial optimization and probability theory. In the late 80s, a new connection was made between optimal transportation theory and non-linear partial differential equations, relating the Monge transportation problem with two of the most interesting and challenging (both analytically and numerically) nonlinear PDEs, namely the (real) Monge–Ampère equations and the Euler equation of incompressible inviscid fluids.

More recently, other important PDEs, in particular the heat equation and several dissipative equations were discovered to be gradient flows of some suitable functionals with respect to the Wasserstein distance. Optimal transportation has turned out to be a powerful concept, giving new insights on both analysis and modelling. The aim of this paper is to give some flavour of this flourishing field of pure and applied mathematics. In particular, recent connections with fluid mechanics and electrodynamics are addressed.

Contents

1 Introduction

The first optimal transportation problem was addressed by Monge in 1781 in his "mémoire sur la théorie des déblais et des remblais", a Civil Engineering problem where parcels of materials have to be displaced from one site to another one with minimal transportation cost. A modern treatment of this problem was initiated by Kantorovich in 1942, based on a probabilistic approach to reduce it to an infinite dimensional linear program. This already established interesting links between Statistics, Combinatorial Optimization and Probability theory, through the so-called Monge–Kantorovich theory and the related concept of Wasserstein distance, as referred, for instance, in the recent book by Rachev and Rüschendorf.

In the late 80s, a new connection was made, this time between Optimal Transportation Theory and non-linear Partial Differential Equations. In particular, a variant of the original Monge transportation problem was related to two of the most interesting and challenging (both analytically and numerically) non-linear PDEs, namely the (real) Monge–Ampère equations and the Euler equation of incompressible inviscid fluids. Both equations have strong geometric features (the Monge–Ampère equation describes surfaces with prescribed Gaussian curvature and the Euler equation describes geodesics of volume preserving maps).

More recently, other important PDEs, in particular the heat equation and several dissipative equations (porous medium equations, lubrication equations, limited flux diffusion equations, granular flow equations,...) were discovered to be gradient flows of some suitable functionals with respect to the Wasserstein distance. Rapidly, optimal transportation has turned out to be a powerful concept, giving new insights on both analysis (functional inequalities, log-Sobolev inequalities, geometry of convex bodies, ...) and modelling (inverse problems in computational cosmology and geophysics, image processing, ...).

The aim of this paper is to give some flavour of this flourishing field of pure and applied mathematics. In particular, connections with fluid mechanics and electrodynamics will be addressed.

2 Optimal transportation and the Monge–Ampère equation

The simplest optimal transportation problem is the following basic *combinatorial optimization* problem (sometimes called *linear assignment* problem):

1) N people are subject to move, ...
   a. from prescribed initial locations: $X_1, ..., X_N$ in the Euclidean space $\mathbb{R}^d$,
   b. to possible destinations: $Y_1, ..., Y_N$.
2) The transportation cost to go from x to y is *given* by a function $c(x, y)$.
3) An assignment is defined as a *permutation*: $\sigma : (1, ..., N) \to (\sigma_1, ..., \sigma_N)$.

An optimal assignment just corresponds to a permutation σ that *minimizes* the total transportation cost

$$c(X_1, Y_{\sigma_1}) + ... + c(X_N, Y_{\sigma_N}) . \tag{2.1}$$

An important example (indeed, the only one considered in this paper) is the *quadratic cost*

$$c(x,y) = \tfrac{1}{2}|x-y|^2, \tag{2.2}$$

where $|\cdot|$ denotes the Euclidean norm.

It is worth noticing that the computational cost to find σ for a general cost scales like $O(N^3)$ (*cf.* [3]), which is quite satisfactory from the combinatorial optimization viewpoint. However, for the quadratic cost above, we make a specific:

Conjecture 2.1. *There is an algorithm to solve the linear assignment problem with quadratic cost whose computational cost scales like* $O(N \log N)$.

(Note that the conjecture is true for $d = 1$.) Finding such an algorithm would be a computational breakthrough, with many possible applications.

For the quadratic cost, the optimality of a permutation σ means precisely that the reordered sequence $Z = (Y_{\sigma_i})_{i=1,N}$ is *cyclically monotone.* This means that, for any cycle in $\{1, ..., N\}$,

$$i_1 \to i_2 \to ... \to i_L \to i_1\,, \tag{2.3}$$

$$Z_{i_1} \cdot (X_{i_1} - X_{i_2}) + ... + Z_{i_k} \cdot (X_{i_k} - X_{i_{k+1}}) + ... + Z_{i_L} \cdot (X_{i_L} - X_{i_1}) \geq 0\,, \tag{2.4}$$

where $\cdot$ denotes the Euclidean inner product. In particular, Z is monotonic:

$$(Z_k - Z_j) \cdot (X_k - X_j) \geq 0\,, \quad \forall\, j,k = 1, ..., N\,. \tag{2.5}$$

A theorem due to Rockaffellar shows that: *A sequence of points* $(X_1, Z_1), ..., (X_N, Z_N)$ *in* $\mathbb{R}^d \times \mathbb{R}^d$ *is* cyclically monotone *if and only if there is a* convex *Lipschitz function* Φ, *defined on the whole Euclidean space, such that, for all* $i = 1, ..., N$,

$$Z_i \in \partial\Phi(X_i)\,, \tag{2.6}$$

which just means

$$\Phi(x) \geq \Phi(X_i) + Z_i \cdot (x - X_i)\,, \quad \forall x \in \mathbb{R}^d\,. \tag{2.7}$$

Notice that Φ is explicitly given by

$$\Phi(x) = \sup\big(Z_{i_1} \cdot (X_{i_2} - X_{i_1}) + ... + Z_{i_k} \cdot (X_{i_{k+1}} - X_{i_k}) + ... + Z_{i_L} \cdot (x - X_{i_L})\big) \tag{2.8}$$

where the supremum is performed over all sequences $i_1, ..., i_L$. (Unfortunately this formula does not seem suitable for computational purposes.)

Thus, we can attach a *continuous* object, namely the *convex* function Φ, to the quadratic transportation problem, whose solution is a purely *discrete* object, namely the optimal permutation σ for which $Z = (Y_{\sigma_i})$, $i = 1, ..., N$ is cyclically monotone. This observation enables us to consider *continuous* versions of the optimal transportation problem by letting N go to ∞. Ironically enough, the earliest work on optimal transportation, *le Mémoire sur les déblais et des remblais*, written by Monge in 1781, was setup in a continuous framework. Monge's theory was completely renewed by Kantorovich in the 1940s, who used probabilistic and duality concepts. The main reference is [27].

Let us now consider the discrete probability measures

$$\mu_N = \frac{1}{N}\big(\delta_{X_1} + \ldots + \delta_{X_N}\big), \qquad \nu_N = \frac{1}{N}\big(\delta_{Y_1} + \ldots + \delta_{Y_N}\big), \tag{2.9}$$

and assume that, as N goes to ∞, they respectively converge to some probability measures $\alpha(x)\,\mathrm{d}x$ and $\beta(y)\,\mathrm{d}y$ where α and β are smooth positive functions with unit integral.

Consider the probability measure associated with the optimal permutation σ and defined on $\mathbb{R}^d \times \mathbb{R}^d$ by :

$$\gamma_N = \frac{1}{N}\big(\delta_{(X_1,Y_{\sigma_1})} + \ldots + \delta_{(X_N,Y_{\sigma_N})}\big) = \frac{1}{N}\big(\delta_{(X_1,Z_1)} + \ldots + \delta_{(X_N,Z_N)}\big), \tag{2.10}$$

which gives the probability for a departure point x to be transported to a destination point y (namely $1/N$ if $x = X_i$ and $y = Y_{\sigma_i}$ for some $i = 1, \ldots, N$). Then, we have

Theorem 2.2. *(see [9, 10, 28, 15, 16].) As $N \to \infty$,*

$$\gamma_N = \frac{1}{N}\big(\delta_{(X_1,Y_{\sigma_1})} + \ldots + \delta_{(X_N,Y_{\sigma_N})}\big) \quad \to \quad \delta\big(y - \nabla\Phi(x)\big)\alpha(x)\,\mathrm{d}x, \tag{2.11}$$

where Φ is a smooth strictly convex *function uniquely defined by α and β through the Monge–Ampère equation:*

$$\beta\big(\nabla\Phi(x)\big)\,\det\big(D^2\Phi(x)\big) = \alpha(x). \tag{2.12}$$

This fully nonlinear elliptic PDE is traditionally related to the Minkowski problem, which amounts to finding hypersurfaces of prescribed Gaussian curvature. Here, we have a completely different interpretation coming from combinatorial optimization.

3 References on optimal transportation

Once the relationship between optimal transportation and the Monge–Ampère equations had been established in the late 80s, a lot of applications were found. Here is a selected list.

1. Textbook and Lecture Notes:
   Villani, *Topics in optimal transportation* [29]; see also reviews and lecture notes by Ambrosio, Brenier, Caffarelli, Gangbo and McCann, Evans, Urbas, ...

2. Dissipative and Parabolic PDEs Revisited:
   Otto [23, 24, 25], Jordan *et al.* [20], Otto, Carrillo, McCann and Villani (preprint), ...

3. Geometric Inequalities: (Brunn–Minkowski, Brascamp–Lieb, log-Sobolev, Bakry–Emery, Gagliardo–Nirenberg, ...)
   McCann [21], Barthe [4], Otto and Villani [26], Bobkov and Ledoux [6], Agueh *et al.* [1], Cordero–Erausquin, Gangbo and Houdré, preprint, ...

4. Computations, Image Processing and Inverse Problems:
   Benamou and Brenier [5], Gangbo and McCann [18], Haker *et al.* [19], Frisch *et al.* [17], Oliker [22], ...

5. Optimal Transportation with Different Cost:
   many recent contributions; in particular by Ambrosio, Bouchitté, Buttazzo, Caffarelli, Evans, Feldman, Gangbo, Kirchheim, McCann, Trudinger, Wang, Xia, ...

4 Modelling fluids and plasmas through optimal transportation rules

The main task of this section is to derive, or approximate, as many as possible of the fluid or plasma equations, from elementary dynamical systems combined with optimal transportation principles.

4.1 Springs with optimal assignment

1) Consider N particles with locations at time t: $X_1(t), ..., X_N(t)$ moving in the Euclidean space $\mathbb{R}^d$.
2) Let there be also N fixed and evenly spaced particles: $Y_1, ..., Y_N$ in the unit cube.
3) Each moving particle X_i is attached to one of the fixed particles Y_{σ_i} by a spring.
4) The assignment: $\sigma : (1, ..., N) \to (\sigma_1, ..., \sigma_N)$ is subject to change in time so that the assignment remains optimal.

The resulting equations are

$$\epsilon \frac{\mathrm{d}^2}{\mathrm{d}t^2} X_i + X_i = Y_{\sigma_i}, \tag{4.1}$$

where $\epsilon > 0$ is fixed, and $\sigma = \sigma(t)$ is updated in order to minimize

$$\frac{1}{2\epsilon}\left(\left|X_1(t) - Y_{\sigma_1}\right|^2 + \cdots + \left|X_N(t) - Y_{\sigma_N}\right|^2\right), \tag{4.2}$$

which is the total potential energy of the springs.

Numerical simulations

The following computations involve $N = 10 \times 10$ particles whose trajectories are drawn in Figure 1 below. In the first picture, σ is *not* updated. (We get independent elliptic trajectories.) In the second one, σ is updated (by optimal assignment).

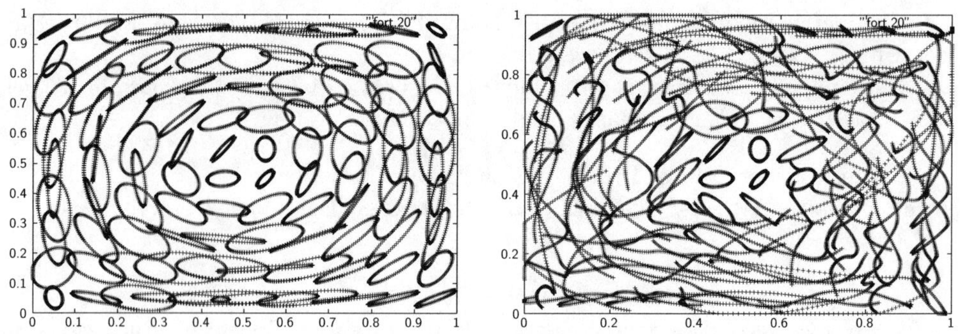

Figure 1. *Trajectories of $N{=}10 \times 10$ particles: (left) not updated, and (right) updated by optimal assignment.*

A theoretical result

Theorem 4.1. *(cf. [13].) Under appropriate assumptions on the initial data, as $N \to \infty$ and $\epsilon \to 0$ (with $N\epsilon^{8d} \to \infty$), the (discrete) velocity field generated by the springs under optimal transportation, namely*

$$\frac{1}{N}\Big(\frac{\mathrm{d}X_1(t)}{\mathrm{d}t}\delta_{X_1(t)} + \ldots + \frac{\mathrm{d}X_N(t)}{\mathrm{d}t}\delta_{X_N(t)}\Big)\,, \tag{4.3}$$

behaves as the velocity field of an inviscid incompressible fluid, *subject to the Euler equations.*

The Euler equations, introduced in 1755, have a very neat geometric interpretation. They describe curves, along the group of volume preserving diffeomorphisms, which are geodesic with respect to the L^2 metric; *cf.* [2].

Let us now briefly explain why the convergence theorem is natural. First, using the optimal transportation theorem, we see that, as $N \to \infty$, with $\epsilon > 0$ being fixed, the discrete density field

$$\frac{1}{N}\big(\delta_{X_1(t)} + \ldots + \delta_{X_N(t)}\big) \tag{4.4}$$

and the discrete velocity field respectively behave like a continuous density field $\rho(t,x)$ and a continuous velocity field $v(t,x)$, subject to the *Euler–Monge–Ampère equations*:

$$(\partial_t + v\cdot\nabla_x)v = \nabla_x\phi\,, \qquad \partial_t\rho + \nabla_x\cdot(\rho v)\,, \qquad \det(I+\epsilon D_x^2\phi) = \rho\,, \tag{4.5}$$

where I is the identity matrix and $\Phi(t,x) = \frac{1}{2}|x|^2 + \epsilon\,\phi(t,x)$ is the convex solution of the Monge–Ampère equation associated to the optimal transportation problem at time t. Next, as ϵ gets very small, the Euler–Monge–Ampère system gets very close to the Euler–Poisson system

$$(\partial_t + v\cdot\nabla_x)v = \nabla_x\phi\,, \qquad \partial_t\rho + \nabla_x\cdot(\rho v)\,, \qquad \epsilon\Delta_x\phi = \rho - 1\,, \tag{4.6}$$

that describes the motion of electrons in an electrically neutralizing background with unit density (ϵ being the electric constant). Indeed

$$\det(I + \epsilon D_x^2 \phi) = 1 + \epsilon \Delta_x \phi + \mathrm{O}(\epsilon^2) \,. \tag{4.7}$$

Notice, at this point, that our model of springs with optimal assignment can be seen as a caricature of charged particles subject to Coulomb interactions.
Finally, letting ϵ go to zero, we find

$$(\partial_t + v \cdot \nabla_x) v = \nabla_x \phi \,, \qquad \nabla_x \cdot v = 0 \,, \qquad \rho = 1 \tag{4.8}$$

which is nothing but the Euler equations for an inviscid incompressible fluid with unit density.

Comment

The Euler equations can be related to the theory of optimal transportation in a different way through the concept of multiphase (or generalized) flow, as developed in [11, 12, 14].

4.2 Vibrating strings and Born–Infeld nonlinear electromagnetism

Our goal is twofold[1]:

1. design a genuinely dynamical approach to optimal transportation (with many applications in mind: motion pictures, dynamical allocation in telecommunications and networks, etc, ...);
2. interpret Electromagnetism as a kind of dynamical optimal transportation theory.

Let us describe our model:

1) N particles with locations : $X_1(t), ..., X_N(t)$ moving in $\mathbb{R}^d$;
2) N other particles with locations : $Y_1(t), ..., Y_N(t)$ also moving in $\mathbb{R}^d$;
3) N chains made of L springs of equal stiffness with nodes located at $Z_{k,i}(t)$, for $k = 0, ..., L$, $i = 1, ..., N$.
4) Each chain must start (for $k = 0$) from one one of the particles X_i and end up (for $k = L$) at one of the particles Y_j, so that there is always a pairing between the X_i and the Y_j through one and only one chain.
5) The chains vibrate but may reconnect as they cross in order to keep minimimal the total potential energy of the springs:

$$\begin{aligned} &|Z_{0,1} - Z_{1,1}|^2 + ... + |Z_{L-1,1} - Z_{L,1}|^2 + \dots \\ &\quad + |Z_{0,N} - Z_{1,N}|^2 + ... + |Z_{L-1,N} - Z_{L,N}|^2 \,. \end{aligned} \tag{4.9}$$

[1]see preprints 2002/2003, `http://www-math.unice.fr/~brenier/`

Numerical simulations

$N = 20$ chains, made of $L = 100$ springs each, are drawn at a given time. Optimal reconnections are performed and plotted in Figure 2.

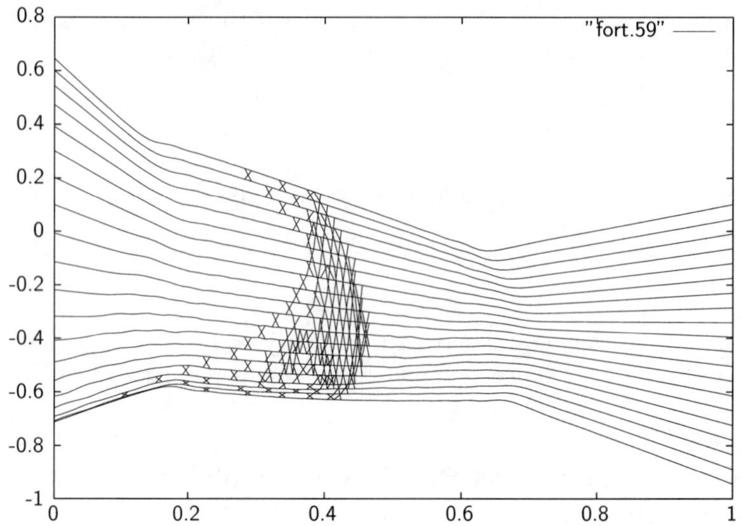

Figure 2. *Plot of optimal reconnections for N=20 chains, each made of L=100 springs.*

Analysis

As $N, L \to \infty$ the model describes a continuum of vibrating strings

$$s \in [0,1] \to Z(t,s;a)\,, \tag{4.10}$$

parameterized by (for instance) their right end-point a (so that $Z(t, s = 1; a) = a$), and vibrating according to the one-dimensional wave equation

$$\partial_{tt} Z = \partial_{ss} Z\,. \tag{4.11}$$

Let us assume the associated density field defined by

$$\rho(t,s,x) = \int \delta\big(x - Z(t,s;a)\big)\,\mathrm{d}a \tag{4.12}$$

to be smooth and bounded from zero. Introduce the related vectorial fields:

$$b(t,s,x) = \frac{1}{\rho(t,s,x)} \int \partial_s Z(t,s;a)\,\delta\big(x - Z(t,s;a)\big)\,\mathrm{d}a\,, \tag{4.13}$$

$$v(t,s,x) = \frac{1}{\rho(t,s,x)} \int \partial_t Z(t,s;a)\,\delta\big(x - Z(t,s;a)\big)\,\mathrm{d}a\,. \tag{4.14}$$

These fields (ρ, b, v) formally satisfy:

$$(\partial_t + v \cdot \nabla_x)b = (\partial_s + b \cdot \nabla_x)v\,, \qquad \partial_t\rho + \nabla_x \cdot (\rho v) = 0\,, \tag{4.15}$$

$$(\partial_t + b \cdot \nabla_x)v = (\partial_s + v \cdot \nabla_x)b\,, \qquad \partial_s\rho + \nabla_x \cdot (\rho b) = 0\,. \tag{4.16}$$

For two space dimensions, these equations can be written in a more symmetric way after introducing notations as follows.

$$X = (X_1, X_2, X_3) = (x_1, x_2, s)\,, \qquad \nabla = (\partial_1, \partial_2, \partial_s)\,, \tag{4.17}$$

$$B(t, X) = \rho(t, s, x_1, x_2)\big(b_1(t, s, x_1, x_2), b_2(t, s, x_1, x_2), 1\big)\,, \tag{4.18}$$

$$V(t, X) = \big(v_1(t, s, x_1, x_2), v_2(t, s, x_1, x_2), 0\big)\,. \tag{4.19}$$

We then get a system that we call the *pressureless MHD equations*:

$$\partial_t(\rho V) + \nabla \cdot \big(\rho V \otimes V - \tfrac{1}{\rho} B \otimes B\big) = 0\,, \qquad \partial_t\rho + \nabla \cdot (\rho V) = 0\,, \tag{4.20}$$

$$\partial_t B + \nabla \times (B \times V) = 0\,, \qquad \nabla \cdot B = 0\,. \tag{4.21}$$

The pressureless MHD equations and the Maxwell equations correspond to two different regimes of the Born–Infeld equations [7, 8]:

$$\partial_t B + \nabla \times \big(B \times V + \tfrac{1}{\rho} D\big) = 0\,, \qquad \nabla \cdot B = 0\,, \tag{4.22}$$

$$\partial_t D + \nabla \times \big(D \times V - \tfrac{1}{\rho} B\big) = 0\,, \qquad \nabla \cdot D = 0\,, \tag{4.23}$$

$$\rho = \sqrt{1 + B^2 + D^2 + |D \times B|^2}\,, \qquad V = \tfrac{1}{\rho} D \times B\,. \tag{4.24}$$

The linear Maxwell equations just describe weak fields $B, D \ll 1$, with $\rho \sim 1$. The Born–Infeld model yields additional conservation laws:

$$\partial_t(\rho V) + \nabla \cdot \big(\rho V \otimes V - \tfrac{1}{\rho}(B \otimes B + D \otimes D)\big) = \nabla\big(\tfrac{1}{\rho}\big)\,, \quad \partial_t\rho + \nabla \cdot (\rho V) = 0\,. \tag{4.25}$$

As we consider the regime in which D and V are of unit size, while B and ρ are both very large of the same order of magnitude, we recover the pressureless MHD equations:

$$\partial_t(\rho V) + \nabla \cdot \big(\rho V \otimes V - \tfrac{1}{\rho} B \otimes B\big) = 0\,, \qquad \partial_t\rho + \nabla \cdot (\rho V) = 0\,, \tag{4.26}$$

$$\partial_t B + \nabla \times (B \times V) = 0\,, \qquad \nabla \cdot B = 0\,. \tag{4.27}$$

Thus, dynamical optimal transportation as described by the pressureless MHD equations, and Electromagnetism as described by the linear Maxwell equations, just correspond to two extreme regimes of the same Born–Infeld theory!

5 Conclusion

Optimal transportation theory has become a successful field of research in nonlinear PDEs, modelling and analysis in the last ten years, mostly by providing a fresh and parallel view on many classical topics. Models based on transportation theory tend to describe a strange combinatorial world relying heavily on the concept of permutations, which somewhat mimics the classical world of continuum mechanics and physics. We hope that many developments will follow from the more recent concepts of dynamical or multiphase transportation problems addressed in this paper.

Bibliography

[1] Agueh M, Ghoussoub N, and X Kang; *Geometric inequalities via a general comparison principle for interacting gases*, submitted to Geom. Funct. Anal., (2003).

[2] Arnold V I, and B Khesin; "Topological methods in hydrodynamics," Applied Mathematical Sciences, 125. Springer-Verlag, New York, 1998.

[3] Balinski M L; *A competitive (dual) simplex method for the assignment problem*, Math. Programming, 34 (1986) pp. 125–141.

[4] Barthe F; *On a reverse form of the Brascamp–Lieb inequality*, Invent. Math., 134 (1998) pp. 335–361.

[5] Benamou J D, and Yann Brenier; *A computational fluid mechanics solution to the Monge–Kantorovich mass transfer problem*, Numer. Math., 84, (2000) pp. 375–393.

[6] Bobkov S G, and M Ledoux; *From Brunn–Minkowski to Brascamp–Lieb and to logarithmic Sobolev inequalities*, Geom. Funct. Anal., 10 (2000) pp. 1028–1052.

[7] Born, Max and Leopold Infeld; *Foundations of the new field theory*, Proc. Roy. Soc. London, A 144 (1934) pp. 425–451.

[8] Born, Max; *Théorie non-lineaire du champ électromagnétique*, Ann. Inst. H. Poincaré, 7 (1937) pp. 155–265.

[9] Brenier, Yann; *Polar decomposition and increasing rearrangement of vector fields*, (in french), C. R. Acad. Sci. Paris Sér. I Math., 305 (1987) pp. 805–808.

[10] Brenier, Yann; *Polar factorization and monotone rearrangement of vector-valued functions*, Comm. Pure Appl. Math., 44 (1991) pp. 375–417.

[11] Brenier, Yann; *A homogenized model for vortex sheets*, Arch. Rational Mech. Anal., 138 (1997) pp. 319–343.

[12] Brenier, Yann; *Minimal geodesics on groups of volume-preserving maps and generalized solutions of the Euler equations*, Comm. Pure Appl. Math., 52 (1999) pp. 411–452.

[13] Brenier, Yann; *Derivation of the Euler equations from a caricature of Coulomb interaction*, Commun. Math. Physics, 212 (2000) pp. 93–104.

[14] Brenier, Yann and Marjolaine Puel; *Optimal multiphase transportation with prescribed momentum*, ESAIM Control Optim. Calc. Var., 8 (2002) pp. 287–343.

[15] Caffarelli, Luis; *The regularity of mappings with a convex potential*, J. Amer. Math. Soc., 5 (1992) pp. 99–104.

[16] Caffarelli, Luis; *Boundary regularity of maps with convex potentials. II*, Ann. Math., (2) 144 (1996) pp. 453–496.

[17] Frisch U, Matarrese S, Mohayaee R, and A Sobolevski; *A reconstruction of the initial conditions of the Universe by optimal mass transportation*, Nature, 417 (2002) pp. 260–262.

[18] Gangbo W, and R J McCann; *Shape recognition via Wasserstein distance*, Quart. Appl. Math., 58 (2000) pp. 705–737.

[19] Haker S, Tannenbaum A, and R Kikinis; "Mass Preserving Mappings and Image Registration," Medical Image Computing and Computer-Assisted Intervention — MICCAI 2001: 4th International Conference. LNCS 2208, Springer-Verlag, Heidelberg, (2002) pp. 120–127.

[20] Jordan R, Kinderlehrer D, and F Otto; *The variational formulation of the Fokker–Planck equation*, SIAM J. Math. Anal., 29 (1998) pp. 1–17.

[21] McCann R J; *A convexity principle for interacting gases*, Adv. Math., 128 (1997) pp. 153–179.

[22] Oliker V; *On the geometry of convex reflectors*, in "PDEs, submanifolds, and affine differential geometry." Banach Center Publ. 57, Polish Acad. Sci., Warsaw (2002) pp. 155–169.

[23] Otto F; *Lubrication approximation with prescribed nonzero contact angle*, Comm. PDE, 23 (1998) pp. 2077–2164.

[24] Otto F; *Evolution of microstructure in unstable porous media flow: a relaxational approach*, Comm. Pure Appl. Math., 52 (1999) pp. 873–915.

[25] Otto F; *The geometry of dissipative evolution equations: the porous medium equation*, Comm. PDE, 26 (2001) pp. 101–174.

[26] Otto F, and C Villani; *Generalization of an inequality by Talagrand and links with the logarithmic Sobolev inequality*, J. Funct. Anal., 173 (2000) pp. 361–400.

[27] Rachev S T, and L Rüschendorf; "Mass transportation problems, Vol. II. Applications." Probability and its Applications, Springer-Verlag, New York, 1998.

[28] Smith C S, and M Knott; *Note on the optimal transportation of distributions*, J. Optim. Theory Appl., 52 (1987) pp. 323–329.

[29] Villani, C; "Topics in optimal transportation,", Graduate Studies in Mathematics, 58, AMS, Providence RI, 2003.

Mark Davis is Professor of Mathematics at Imperial College London, specialising in financial mathematics, credit risk models, pricing in incomplete markets and stochastic volatility. He also acts as a consultant to Hanover Square Capital, a newly-founded capital markets company, and is affiliated with the Financial and Actuarial Mathematics group at TU Wien. From 1995–1999 he was Head of Research and Product Development at Tokyo-Mitsubishi International, leading a front-office group providing pricing models and risk analysis for fixed-income, equity and credit-related products.

Dr Davis holds a PhD from the University of California at Berkeley. He is the author of three books on stochastic analysis and optimisation. In 1983 he was awarded a Doctor of Science (ScD) from Cambridge University. From 1978 to 1995 he was Editor-in-Chief of the journal *Stochastics and Stochastics Reports.* He was a founding co-editor of *Mathematical Finance* (1990–93) and is currently an associate editor of *Quantitative Finance.* In 2002 he was awarded the Naylor Prize by the London Mathematical Society, for "excellent pioneering contributions to stochastic analysis, stochastic control and filtering theory and more recently to mathematical finance. His expository work is of enormous influence in introducing modern mathematical methods into these areas."

Chapter 4

Valuation, Hedging and Investment in Incomplete Financial Markets

Mark H. A. Davis[*]

Abstract: Financial assets are contracts entitling the holder to receive deterministic or stochastic cash flows at various times in the future and to take certain kinds of action. For example, the holder of an ordinary share receives dividends and has the right to sell the share at any time at the prevailing market price. *Valuation* means stating a 'fair price' now for entering a non-exchange-traded contract; *hedging* means offsetting the associated risks by trading other assets in the market, while *investment* involves forming a portfolio of assets with a view to maximising investment returns. Hedging and valuation are very closely related. Indeed, if an asset can be *perfectly hedged* or *replicated* — i.e., its cash flows are equal to those of some portfolio of traded assets — then its value is equal to the value of the replicating portfolio, else there will be an *arbitrage* opportunity (the availability of 'something for nothing').

The standard Black–Scholes option pricing theory is based on this 'arbitrage pricing' idea. However, in today's markets there are many contracts that cannot be replicated. Examples include equity-linked insurance contracts with guarantees, or credit derivatives (insurance against default of a corporation on its debt). In these cases arbitrage arguments give only upper and lower bounds on prices that are typically very weak. We argue that for valuation of these contracts, *hedging* should be replaced by *investment*: the value to an investor is related to the use he/she can make of the asset in forming portfolios with superior returns. Of course this depends on the investor's attitude to risk, but that is inevitable in an environment where all risk cannot be 'hedged away'.

We give specific examples and outline the mathematics, from stochastic analysis and stochastic control theory, needed to turn this approach into reality.

[*]Department of Mathematics, Imperial College, London

Contents

1 Introduction

Financial instruments or 'positions' are legal contracts between two parties providing for exchange of fixed or contingent payments at various times in the future. For example, in a forward foreign exchange contract entered at time t_0, Party A agrees to pay Party B \$1.5 at time $t_1 > t_0$ while Party B pays Party A £1. The value of this contract to Party A at t_1 is $\$(f(t_1) - 1.5)$ where $f(t_1)$ is the £/\$ FX rate at time t_1. Since this is unknown at t_0, the parties are entering an agreement, with no payment at t_0, providing for a net random payment in one direction or the other at time t_1. There may also be optionality: if the contract gives Party A the right *but not the obligation* to effect the above exchange then the value at t_1 is $H = \max(f(t_1) - 1.5), 0)$ since Party A will not exercise her right if $f(t_1) < 1.5$. Since $H \geq 0$, Party B will not agree to this arrangement unless there is some compensating payment to her, which could be made at time t_1 but more conventionally appears as an "upfront" option premium paid by Party A at t_0. Thus Party A is exchanging a fixed payment now for a random payment later. The classic option pricing problem is to determine a 'fair value' for the premium.

Banks – or even individuals – will be holding portfolios containing hundreds or thousands of positions. The requirements for mathematical modelling are three:

1. *Valuation:* determine a value for a portfolio that is consistent with prices of exchange traded assets. This process is known as *marking to market.* The value of any exchange-traded asset such as an ordinary share is simply its

current market price, but a model is needed to value an option on the share or indeed any non-exchange traded position.

2. *Hedging* is the process of mitigating risk by engaging in offsetting transactions. The writer of an option accept a premium payment but has to make a — possibly unbounded — random payment later. How should she trade, using the premium payment, so as to be in a position to meet the later payment of at least reduce the risk of loss? This is the hedging problem.

3. *Investment* is the process of trading in the market with a view to increasing the value of one's assets. Ever since the pioneering work of Markowitz [27] it has been appreciated that there is a trade-off between risk and return: investors can only increase expected return by accepting a higher level of risk (i.e., variance of returns). The portfolio selection problem of Markowitz-style investment theory is to maximise expected returns subject to a bound on the level of risk.

Let X_t be the value of a portfolio obtained by trading in the market starting with initial capital $X_0 = x$ (a more precise description is given in the next section). An *arbitrage opportunity* is the availability of 'something for nothing'; i.e., the existence of a portfolio process X_t and a time $T > 0$ such that

- $x = 0$
- $X_T \geq 0$ a.s
- $P[X_T > 0] > 0$ (equivalently, $E[X_T] > 0$)

This portfolio requires no initial investment, entails no possibility of loss but a positive probability of gain. Real markets may or may not contain (albeit transitory) arbitrage opportunities, but in mathematical modelling we must insist on models that are arbitrage-free since otherwise there will be strategies leading to infinite riskless profits, an obviously unrealistic conclusion.

Suppose we have an option-like contract in which a premium p paid at time zero is exchanged for a random variable H at time $T > 0$, and suppose there is a portfolio process X_t such that $X_T = H$ almost surely. Then $x = X_0$ is the unique arbitrage-free value for the premium p: if $p > x$ then we can simply pocket the difference $p - x$ since x is enough to form the portfolio X_t which perfectly hedges our obligation at time T (this is called a *replicating portfolio*). If $p < x$ a similar arbitrage accrues to the option purchaser. This argument is sometimes known as the *law of one price*: if two positions involve identical cash flows in the future then they have the same value now.

It is surprising how far this process of pricing by absence of arbitrage will go. A market model is *complete* if there is a replicating portfolio for 'any' contingent claim H (a contingent claim is one whose exercise value H is a function of traded asset prices) and it is a remarkable fact that the standard market model of Samuelson–Merton–Black–Scholes in which asset prices are geometric Brownian motion (see Section 2 below), is complete. In a complete market, valuation and hedging are one and the same thing: underlying assets are valued at their market price and contingent claims at the capital value required for replication.

At first sight, hedging and investment appear to be very different things since hedging is aimed at mitigating risk while investment is risk-seeking, as we learn from Markowitz. However on closer inspection the two turn out to be very closely related, the connection being through the duality theory of optimal investment, which is outlined briefly in Section 4 below. This theory gives us very clear hints what to do in the case of incomplete markets, in which there are unhedgeable contingent claims. Generally, markets are incomplete when there are not enough traded assets or when trading restrictions limit the class of portfolios we can construct. In this case *hedging* and *investment* become synonymous: writing an option entails real risk, and therefore its value can only be assessed in relation to some investment objective. We will see below in Section 4 that pricing rules can be obtained this way which are direct extensions of the complete market case; i.e., coincide with the no-arbitrage price for all claims that actually are hedgeable. The corresponding investment strategy — replication in the case of hedgeable claims — becomes the utility-maximising strategy when perfect hedging is impossible. Ideas of this sort have been developed in one form or another by economists for upwards of a century, but it is only recently that the connections with arbitrage pricing theory have been explored.

Study of incomplete markets is far from being a purely academic matter. In the area of conventional derivatives, 'stochastic volatility' models, aiming at a more accurate description than the Black–Scholes model, generally lead to incompleteness (see Davis [11]). In recent years a huge market in *credit derivatives* has developed (see Schönbucher [34]) and few if any of the contracts traded there are exactly hedgeable other than by trivial back-to-back deals. Another important source of incompleteness is a long time horizon. Generally long-term option contracts are unhedgeable becaue of uncertainty about future volatility. There is also a huge category of liabilites held by insurance companies involving long-term guarantees on pension payments, annuity rates and the like. In line with our philosophy that "hedging = investment", we need to study long-term investment to understand valuation and strategies for such contracts. An excellent survey of these problems is given by Campbell and Viceira [4].

This paper gives a survey of some of the above topics, and is laid out as follows. Section 2 introduces the basic framework of price processes, investment strategies and arbitrage-free models. Section 3 describes the standard approach to optimal investment based on maximising expected utility and solution techniques based on dynamic programming. Section 4 covers duality theory and associated pricing concepts.

In the long-term problems mentioned above, parameter uncertainty is a crucial feature. Section 5 discusses some ways in which this can be handled. The paper concludes with some final remarks in Section 6.

Throughout the paper, the intention is not to aim at a maximum level of generality but, on the contrary, to concentrate on specific cases and solved problems which give insight into the nature of optimal strategies for hedging and investment.

2 Market model

We start with a vector of continuous semimartingales $S_0(t), \ldots, S_n(t)$ on a filtered probability space $(\Omega, \mathcal{F}, (\mathcal{F}_t), P)$, representing the prices of $n+1$ traded assets. Suppose $S_0(0) = 1$ and $S_0(t) > 0$. A *trading strategy* is an n-vector locally bounded $\mathcal{F}_t$-predictable process ϕ_t, where $\phi_i(t)$ is the number of units of asset i held at time t. All residual value is invested in asset S_0. The evolution of portfolio value X_t starting with initial capital x is then[1]

$$\mathrm{d}X_t = \phi_t \,\mathrm{d}S(t) + \frac{X_t - \phi_t S(t)}{S_0(t)} \,\mathrm{d}S_0(t), \quad X_0 = x. \tag{2.1}$$

ϕ is *admissible* if $X_t \geq 0$ for all t. Define $\tilde{X}_t = X_t / S_0(t)$ etc. Then applying the Ito formula we obtain the following convenient expression for the evolution of portfolio value in normalised units:

$$\tilde{X}_t^\phi = x + \int_0^t \phi_u \,\mathrm{d}\tilde{S}(u). \tag{2.2}$$

The "first fundamental theorem of asset pricing" (Delbaen and Schachermayer, [12]) states, roughly speaking, that *there is no arbitrage if and only if there exists an* equivalent measure $Q \sim P$ *under which the normalised price processes* $\tilde{S}(t)$ *are local martingales.* We say (S_0, Q) is a *numéraire pair.* Since trading strategies are predictable, the representation (2.2) shows that $\tilde{X}_t^\phi$ is a local martingale under measure Q.

A *contingent claim* H exercised at time T is an integrable $\mathcal{F}_T$-measurable random variable. The classic example is $H = \max(S_k(T) - K, 0)$ a *call option* on asset k. If $\tilde{H} = \tilde{X}_T^\phi$ a.s. for some ϕ then we say ϕ 'replicates' H. If X^ϕ is a Q-martingale, as opposed to just a local martingale, then by the martingale property

$$x = E_Q\big[\tilde{X}_0^\phi\big] = E_Q\big[\tilde{X}_T^\phi\big] = E_Q\big[\tilde{H}\big]$$

so that

$$x = E_Q\Big[\frac{H}{S_0(T)}\Big] \tag{2.3}$$

and x is the unique arbitrage-free value for H at time 0. If a replicating strategy exists for 'any' contingent claim, the market is *complete.* The "second fundamental theorem of asset pricing" states that the market is complete if and only if there is a *unique* martingale measure Q. In this case (2.3) gives the unique arbitrage-free value for arbitrary H.

An *incomplete* market is one in which there are many martingale measures Q. In general, replication is not possible in incomplete markets and there is an interval I of arbitrage-free prices given by

$$I = \Big(\inf_{Q \in \mathcal{M}} E_Q\big[\tilde{H}\big], \sup_{Q \in \mathcal{M}} E_Q\big[\tilde{H}\big]\Big)$$

[1] ϕS denotes the inner product $\sum_1^n \phi_i S_i$, so $X_c - \phi_t S(t)$ is the cash available for investment in the 0th asset

This was shown by Kramkov [25]. It seems clear that nothing should depend on which asset we choose as the numéraire asset $S_0(t)$ (except that we must have $S_0(t) > 0$ for all t), and indeed this is the case: If (S_0, Q) is a numéraire pair and N is another numéraire then (N, Q_N) is a numéraire pair where

$$\frac{dQ_N}{dQ}\bigg|_{\mathcal{F}_T} = \frac{N(T)}{S_0(T)}.$$

Thus the quantity $\zeta_t = (dQ_N/dP)_t/N(t)$ is numéraire invariant and (2.3) is expressed as

$$x = E\big[\zeta_T H\big].$$

The process ζ_t is known as a *deflator* or *state price density.*

2.1 Example: the Black–Scholes world

In the market model of Black and Scholes [2] the price processes $S_i(t)$, $i - 1, \dots, n$ satisfy equations

$$dS_i = \mu_i S_i \, dt + \sigma_i S_i \, dw_t^i$$

where $(w_t^1, \dots, w_t^n)$ is a vector of correlated Brownian motions, while

$$dS_0 = rS_0 \, dt$$

for some constant interest rate r. Thus the asset prices $S_1(t) \dots S_n(t)$ are log-normally distrbuted, while the numéraire asset is a risk-free savings account. Taking $\gamma = (\mu - r)/\sigma$ we find $d\tilde{S} = \sigma\tilde{S}(dw_t + \gamma_d t) =: \sigma\tilde{S}\, d\widehat{w}_t$. $\widehat{w}_t$ is a Brownian motion under an equivalent measure Q given by the Girsanov theorem. The replication requirement is

$$e^{-rT} H = x + \int_0^T \phi\sigma\tilde{S}\, d\widehat{w}. \tag{2.4}$$

Existence of (x, ϕ) satisfying (2.4) follows from the martingale representation theorem for BM; this is a complete market. The value of the contingent claim H at time $t < T$ is then given by

$$v_t = E_Q\big[e^{-r(T-t)} H\big]$$

so that in particular $v_0 = x$.

3 Optimal Investment and Consumption

In this section we see that optimal investment is essentially a *stochastic control* problem. A special role is played by so-called growth-optimal portfolios, and these can be studied in a very direct way as described below. We then move on to look at a general formulation in terms of maximisation of expected utility, and solution of some problems by dynamic programming.

3.1 Growth-optimal portfolios

For a single portfolio trajectory we can define a process $\eta(t)$ by

$$X_t = x\, e^{\eta(t)t}$$

so that $\eta(t) = \log(X_t/x)/t$ is the *growth rate* over the time interval $[0,t]$. Thus choosing the investment strategy to maximise $E[\log X_T]$ maximises the expected growth rate, giving the *growth-optimal portfolio.* If we take the normalised portfolio process $\tilde{X}_t = X_t/S_0(t)$ then $\log \tilde{X}_t = \log X_t - \log S_0(t)$ so *choice of numéraire is irrelevant to the optimisation problem*, a key property of the logarithmic criterion.

In the semimartingale model, suppose price processes take the form

$$S_i(t) = M_i(t) + \int_0^t \alpha_i(u)\, \mathrm{d}\mu(u)$$

and

$$\langle M_i, M_j \rangle_t = \int_0^t \beta_{ij}(u)\, \mathrm{d}\mu(u)$$

where the processes $M_i(t)$ are continuous martingales, μ is a fixed measure $\beta_{ii}(t) > 0$ for all t a.s. and $\langle M_i, M_j \rangle$ denotes the quadratic co-variation of martingales M_i, M_j (see §IV.23 of [33]). If we write the trading strategy ϕ as $\phi_t = \pi_t \tilde{X}_t$ then the portfolio value evolves as

$$\mathrm{d}\tilde{X}_t = \pi_t \tilde{X}_t\, \mathrm{d}\tilde{S}_t$$

so that by the Ito formula

$$\mathrm{d}(\log \tilde{X}_t) = \left(\pi'\alpha - \tfrac{1}{2}\pi'\beta\pi\right) \mathrm{d}\mu + \pi'\, \mathrm{d}M\,. \tag{3.1}$$

We maximise the expectation by maximising the 'drift' term; i.e., taking $\pi = \pi^*$ where

$$\pi^* = \beta^{-1}\alpha\,,$$

giving the maximal value

$$E(\log X_T) = E \log(x S_0(T)) + E \int_0^T \tfrac{1}{2}\alpha'\beta^{-1}\alpha\, \mathrm{d}\mu\,.$$

The first term is the value obtained by simply investing everything in the numéraire asset.

It turns out that there is a close connection between numéraire pairs and the problem of maximising logarithmic utility [1]. Indeed, suppose (Y, Q) is a numéraire pair; then using the inequality $\log x \le x - 1$ we have for an arbitrary portfolio process X_t

$$E_Q \log X_T - E_Q \log Y_T = E_Q\left[\log(X_T/Y_T)\right] \le E_Q\left[X_T/Y_T\right] - 1 = 0\,.$$

Thus Y_T maximises logarithmic utility under Q.

The converse is also true: if Y_T maximises logarithmic utility under a certain measure Q then (Y, Q) is a numéraire pair. In particular if we use the log-optimal portfolio, calculated as above, as numéraire, the 'physical' measure P is a martingale measure. This is intriguing because it shows we do not need to change the measure to do pricing: we can change the numéraire instead.

To show this, assume S_0 is the log-optimal portfolio process under P. Then

$$E\log\tilde{X}_T = E\log X_T - E\log S_0(T) \leq 0 \tag{3.2}$$

for any portfolio process X corresponding to trading stategy ϕ, and using (3.1) and (3.2) we see that for all π

$$\begin{aligned}-E\log\tilde{X}_T &= \tfrac{1}{2}\int_0^T(\pi'\beta\pi - 2\pi'\alpha)\,\mathrm{d}\mu \\ &= \tfrac{1}{2}\int_0^T\left[(\pi-\beta^{-1}\alpha)'\beta(\pi-\beta^{-1}\alpha) - \alpha'\beta^{-1}\alpha\right]\mathrm{d}\mu \geq 0\,.\end{aligned}$$

Thus $\alpha_i(t) = 0$ a.s. $\mathrm{d}\mu$, showing that $\tilde{S}_i(t) = M_i(t)$, a local martingale.

3.2 Utility functions

A utility function is a smooth, concave function $U : R^+ \to R$ satisfying

$$U'(0) = +\infty\,, \quad U'(\infty) = 0\,.$$

The function

$$I(y) = (U')^{-1}(y)$$

is then decreasing and convex. A utility function U defines a preference ordering $\preceq$ on the set of probability measures on R^+, given by

$$\mu_1 \preceq \mu_2 \longleftrightarrow \int_{R^+} U(x)\,\mu_1(\,\mathrm{d}x) \leq \int_{R^+} U(x)\,\mu_1(\,\mathrm{d}x),$$

and the objective of maximising expected utility with a concave utility function corresponds to 'rational behaviour' of a risk-averse agent; see for example Ferguson [14].

We consider various problems of maximising expected utility of the form

$$E\Big\{\int_0^T U_1(c_t)\,\mathrm{d}t + U_2(X_T)\Big\}\,,$$

where c_t is a *consumption rate.* This objective trades off personal gratification (the first term) against leaving something for the next generation (the second term). General methods for solving such problems are

- Dynamic programming [15]: this provides a very direct computational technique but applies to Markovian systems only
- Convex duality [24]: this can be applied to very general price process models but the computational side is less obvious except in very special cases.

3.3 Infinite-horizon dynamic programming

The problem described here was originally studied by Merton [28] and is remarkable in being one of the few nonlinear stochastic control problems that can be solved explictly, as well as having a solution with intuitively satisfying content. See [9] for a clean treatment.

The objective is to maximise

$$E\int_0^\infty e^{-\delta t}U(c_t)\,\mathrm{d}t$$

for a scalar price model

$$\mathrm{d}S_t = \mu S_t\,\mathrm{d}t + \sigma S_t\,\mathrm{d}w_t\,.$$

The wealth equation is

$$\mathrm{d}X_t = X_t(r+(\mu-r)\pi)\,\mathrm{d}t - c_t\,\mathrm{d}t + X_t\pi_t\sigma\,\mathrm{d}w_t \tag{3.3}$$

where π is the fraction of wealth invested in the risky asset. The *Bellman equation* of dynamic programming is

$$\max_{\pi,c}\left\{\frac{\partial v_0}{\partial x}\Big(x\big(r+(\mu-r)\pi\big)-c\Big)+\tfrac{1}{2}\pi^2\sigma^2x^2\frac{\partial^2 v_0}{\partial x^2}+U(c)-\delta v_0\right\}=0\,. \tag{3.4}$$

On performing the maximisation, this becomes (with $v_0' = \partial v_0/\partial x$ etc.)

$$v_0' r x - \tfrac{1}{2}\left(\frac{\mu-r}{\sigma}\right)^2\frac{(v_0')^2}{v_0''} + U\big(I(v_0')\big) - I(v_0')v_0' - \delta v_0 = 0\,. \tag{3.5}$$

If v_0 satisfies (3.5) then $v_0(x)$ is the maximum achievable expected utility starting with capital x, and the optimal investment strategy is given by the maximising values of c,π in (3.4). For example, when $U(c)=c^\gamma$ then

$$\pi^* = \frac{1}{1-\gamma}\,\frac{\mu-r}{\sigma^2},\qquad c_t^* = C_\gamma X_t\,.$$

The optimal strategy is to invest a constant fraction of total wealth in the risky asset and consume at a rate proportional to total wealth. The same form of strategy is optimal for logarithmic utility, with $\pi^* = (\mu-r)/\sigma^2$.

3.4 Optimal investment with income

In the previous section we are simply concerned with investing an initial 'endowment' x. It is perhaps more natural to consider the case in which the investor also has an income, and this was already studied by Merton in the following simple case. Suppose we start with capital x and receive an income at a constant rate a per unit time. The wealth equation becomes

$$\mathrm{d}X_t = X_t\big(r+(\mu-r)\pi\big)\,\mathrm{d}t + (a-c_t)\,\mathrm{d}t + X_t\pi_t\sigma\,\mathrm{d}w_t \tag{3.6}$$

and the Bellman equation is

$$v'r\left(x+\frac{a}{r}\right)-\tfrac{1}{2}\left(\frac{\mu-r}{\sigma}\right)^2\frac{(v')^2}{v''}+U\big(I(v')\big)-I(v')v'-\delta v=0\,. \tag{3.7}$$

The following is easily checked.

Proposition 1. *If v_0 satisfies (3.5) then $v_a(x) := v(x+a/r)$ satisfies (3.7).*

The intuition here is clear: the value at time 0 of the income stream is

$$\frac{a}{r}=\int_0^\infty a\,e^{-rt}\,\mathrm{d}t\,.$$

The optimal strategy is to borrow this amount and apply the the optimal strategy with initial capital $x+a/r$ and no income. The income stream finances the debt. We will see later that this is a special case of a much more general result.

3.5 Optimal hedging with annuity liability

Suppose we start with capital x and have the obligation to *pay* a perpetual annuity at rate a. This is now a hedging problem: we have to manipulate our funds so as to be in a position to meet our obligation to pay the annuity, if it is possible to do so. If $x \geq a/r$ the solution is simple: the liability is perfectly hedged by placing a/r in the riskless account. The remaining funds can be invested optimally as before, achieving expected utility $v(x-a/r)$.

If $0 < x < a/r$ we cannot guarantee to pay the annuity for ever. Let X_t be the wealth process and define stopping times

$$\begin{aligned}\tau_0 &= \inf\{t : X_t = 0\}\\ \tau_1 &= \inf\{t : X_t = \frac{a}{r}\}\\ \tau &= \begin{cases}\tau_0 & \text{if} \quad \tau_0<\tau_1\\ \infty & \text{if} \quad \tau_1<\tau_0\,.\end{cases}\end{aligned}$$

Then τ is the time for which we can pay the annuity and it is reasonable to maximise the expected NPV

$$E\int_0^\tau a\,e^{-rt}\,\mathrm{d}t=\frac{a}{r}\left(1-Ee^{-r\tau}\right).$$

This problem has the following explicit solution, due to Sid Browne [3]: define

$$\begin{aligned}\theta &= \tfrac{1}{2}\left(\frac{\mu-r}{\sigma}\right)^2\\ d &= \theta^2+4r\theta\\ \eta^+ &= \frac{1}{2r}\left(2r+\theta+\sqrt{d}\right).\end{aligned}$$

Then the optimal strategy is $c_t \equiv 0$ and $\pi_t = \pi^*(X_t)$ where

$$\pi^*(x) = \frac{\mu - r}{\sigma^2(\eta^+ - 1)}\Big(\frac{a}{r} - x\Big)$$

The value function is

$$E_x^{\pi^*} \exp(-r\tau) = (a - rx)^{\eta^+}.$$

Again, this solution is intuitively reasonable: the preportion of wealth invested in stock is proportional to the distance from the 'safe' level a/r. When wealth is low one has to gamble to have any chance of survival, whereas near the safe level volatility should be reduced. (In this solution the upper barrier $x = a/r$ is never actually hit.)

4 Duality and Utility-Based Pricing

4.1 Duality

In this section we take the numéraire asset as $S_0(t) \equiv 1$ for ease of exposition. Equivalently, we consider everything in discounted units. Let $\mathcal{X}(x)$ be the set of wealth processes corresponding to admissible trading strategies ϕ with initial capital x and define

$$u(x) = \sup_{X \in \mathcal{X}(x)} E\big[U(X_T)\big]. \tag{4.1}$$

Here U is a utility function as before. This is the problem studied by Kramkov and Schachermayer [26] with semimartingale price processes $S_i(\cdot)$, following earlier work by several authors based on a diffusion process price model (see Karatzas and Shreve [24] and the references there).

The utility function U is said to have *reasonable asymptotic elasticity* if

$$\limsup_{x \to \infty} \frac{xU'(x)}{U(x)} < 1. \tag{4.2}$$

Define the *dual function* V by

$$V(y) = \max_x \big(U(x) - xy\big).$$

This is a convex decreasing function, and the maximum is achieved at $x = I(y) = (U')^{-1}(y)$. The dual optimisation problem is to calculate

$$v(y) = \inf_{Y \in \mathcal{Y}} E\big[V(yY_T)\big], \tag{4.3}$$

where $\mathcal{Y}$ is the set of non-negative processes such that $Y_0 = 1$ and XY is a supermartingale for all $X \in \mathcal{X}(1)$. Note that $L \in \mathcal{Y}$ if

$$L_t = E\Big[\frac{\mathrm{d}Q}{\mathrm{d}P}\,\Big|\,\mathcal{F}_t\Big]$$

for some equivalent martingale measure Q.

The main general result of [26] is that if U satisfies (4.2) then optimal elements $\hat{X}(x), \hat{Y}(y)$ exist for (4.1),(4.3) and the functions u, v are continuously differentiable. Taking $y = u'(x)$ we have

$$\hat{X}_T(x) = I\big(y\hat{Y}_t(y)\big)\,, \quad E\big[\hat{Y}_T(x)\hat{X}_T(x)\big] = x\,. \tag{4.4}$$

It may not however be the case that $\hat{Y}$ is the density of a martingale measure.

If there is cumulative 'random endowment' C_t then formally the dual problem is modified to

$$v_C(y) = \inf_{Y\in\mathcal{Y}} E\Big[V(y\,Y_T) + y\int_0^T Y_t\,\mathrm{d}C_t\Big] \tag{4.5}$$

though the exact duality relationship is complicated in this case. See Cuoco [5], Cvitanić, Schachermayer and Wang [6] and Hugonnier and Kramkov [20].

4.2 Example: Optimal investment with randomly terminating income

In the problem of Section 3.4 income at rate a is guaranteed for all time, so that the capital value of future income is a/r at any time. Thus investor's 'solvency constraint' is $X_t \geq -a/r$ for all t, a.s. The investor is able to borrow against future income because there is no doubt that this income will be received. The same idea applies much more generally. Suppose the investor has cumulative random income c_t which is hedgeable; i.e., can be replicated by trading in the market. Then the value of the income stream at time 0 is

$$p = E_Q\int_0^\infty e^{-rt}\,\mathrm{d}C_t$$

where Q is *any* equivalent martingale measure[2]. We assume $p < \infty$. We can then borrow p and form a replicating portfolio Z_t with initial value $Z_0 = -p$, so that

$$\int_0^t e^{r(t-s)}\,\mathrm{d}C_s + Z_t + E_Q\Big[\int_t^\infty e^{-r(s-t)}\,\mathrm{d}C_s\,\Big|\,\mathcal{F}_t\Big] = 0 \text{ a.s.}$$

Thus the mark-to-market value of the portfolio plus income is always zero, and the investor can use his enhanced initial capital $x+p$ for investment, achieving a maximum expected utility $v_0(x+p)$ where $v_0(x)$ is the maximum utility starting with capital x and no income. This argument applies equally well to the general incomplete-market finite-time horizon models of Section 4.1 above. When the income stream is not hedgeable, the investor will not be able to borrow against it in the same way, and the problem becomes more interesting (and realistic). The papers [5], [6], [20] mentioned above give existence results but few hints as to what optimal strategies are like. To get some insight, Davis and Vellekoop [10] consider Merton's problem, as described above, with constant income at rate a, but supposing that this income terminates at a random time τ, exponentially distributed

[2] More precisely, Q is equivalent to P on the σ-field $\mathcal{F}_T$ for any $T > 0$

with parameter η and independent of (w_t). The market is now incomplete since the random cancellation is an unhedgeable risk.

For this problem, the portfolio value should be thought of as a jump-diffusion process $\mathcal{X}_t = (X_t, Y_t)$ on the state space $R^+ \times \{0,1\}$ where X_t is the cash value of the portfolio as before and $Y_t = \mathbf{1}_{(t<\tau)}$. The process $\mathcal{X}_t$ starts at $(x,1)$ and jumps from $(X_t, 1)$ to $(X_t, 0)$ at $t = \tau$. Writing the value function of dynamic programming as $v(x,i) = v_i(x)$, $x \in R^+$, $i = 0, 1$, we see that $v_0(x)$ is the value function as derived previously with no income, while v_1 satisfies the Bellman equation

$$v_1' r\big(x + \tfrac{a}{r}\big) - \tfrac{1}{2}\big(\tfrac{\mu - r}{\sigma}\big)^2 \frac{(v_1')^2}{v_1''} + U\big(I(v_1')\big) - I(v_1')v_1' + \eta v_0 - (\delta + \eta)v_1 = 0\,. \quad (4.6)$$

There is no simple closed form solution to (4.6) for, say, logarithmic or power utility.

Turning to duality, the densities of martingale measures Q_λ take the form

$$\frac{\mathrm{d}Q_\lambda}{\mathrm{d}P}\bigg|_{\mathcal{F}_t} = \exp\Big\{\int_0^t \tfrac{r-\nu}{\sigma}\,\mathrm{d}w - \tfrac{1}{2}\int_0^t \big(\tfrac{r-\nu}{\sigma}\big)^2\,\mathrm{d}t$$
$$- \int_0^t (\lambda_s - \eta)(1 - N_s)\,\mathrm{d}s + \int_0^t \log\big(\tfrac{\lambda_s}{\eta}\big)\,\mathrm{d}N_s\Big\}\,.$$

Here $N_t = \mathbf{1}_{(t \geq \tau)} = 1 - Y_t$. Under measure Q_λ, N_t has hazard rate λ_t, replacing the hazard rate η under measure P. We find that the dual optimal hazard rate $\hat{\lambda}(\,\cdot\,)$ is the solution to the following *deterministic optimal control problem*: minimise

$$J(\lambda) = \int_0^\infty e^{-\alpha t}\big(g(\lambda(t)) + ay\,x(t)\big)\,\mathrm{d}t \quad (4.7)$$

over pairs $\big(\lambda(\,\cdot\,), x(\,\cdot\,)\big)$ satisfying

$$\frac{\mathrm{d}}{\mathrm{d}t}x(t) = -x(t)\big(r - \alpha + \lambda(t)\big)\,, \quad x(0) = 1\,, \quad (4.8)$$

where $\alpha = \delta + \eta$ and

$$g(\lambda) = \tfrac{\eta}{\delta}\Big(\frac{\lambda - \eta}{\eta} - \log\tfrac{\lambda}{\eta}\Big)\,.$$

(Note that $g(\lambda)$ has a global minimum of 0, achieved at $\lambda = \eta$.) Analysis of the opimal control problem (4.7), (4.8) proceeds by application of the Pontryagin maximum principle and is surprisingly delicate: the Hamiltonian and control functions have to lie on a certain manifold to secure stable solutions. Having realised this, the problem can be solved and the value function v_1 computed. (Naive attempts to solve the Bellman equation (4.6) numerically are doomed to failure.) Figure 1 shows the solution for some specific parameter values with logarithmic utility. It is notable that $v_1(0) > -\infty$ but $v_1'(0) = \infty$.

4.3 A general pricing formula

In incomplete markets most contingent claims involve unhedgeable risks. Their 'value' must be related to their use in constructing portfolios that are attractive from

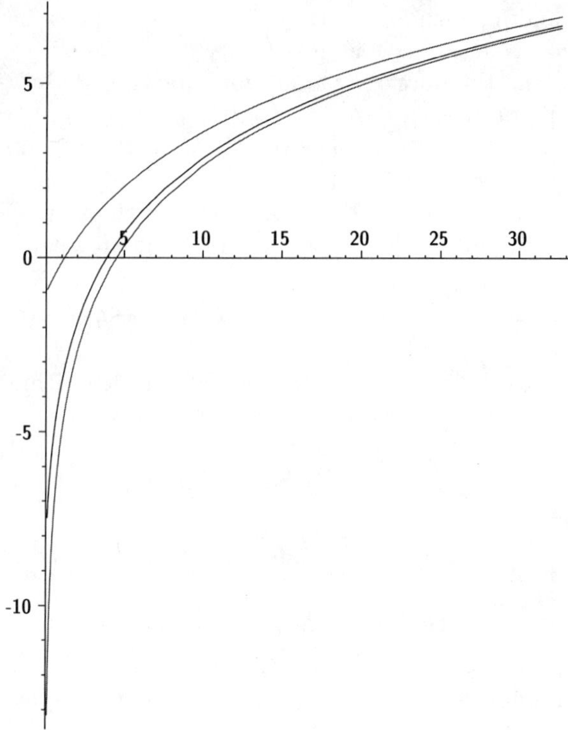

Figure 1. *Value function for randomly terminating income problem, with* $\mu = 0.05, \sigma = 0.3, \eta = 0.05, \delta = 0.15, r = 0.03, a = 0.05$. *Also shown are the value functions for constant income without termination (upper curve) and with no income (lower curve)*

an investment point of view. Considerations of this sort go back to the 'principle of equi-marginal utility' formulated by Jevons [23] in the 19th century and have been extensively developed in the economics literature; see Foldes [17] for an exposition. To formalise these ideas in the present framework, suppose an investor's objective is to maximise expected utility. As in Section 4 denote

$$u(x) = \sup_{X \in \mathcal{X}(x)} E\big[U(X_T)\big] .$$

In an option contract, we exchange a sure payment p at time zero for the random exercise value H of the option at time T. An 'indifference price' for the option is the number $\hat{p}$ such that

$$u(x) = u_H(x - \hat{p})$$

where

$$u_H(x) = \sup_{X \in \mathcal{X}(x)} E\big[U(X_T + H)\big] . \tag{4.9}$$

Thus indifference pricing is a special case of the 'random endowment' problem (4.5). The optimal 'hedge' is the maximising strategy for the problem (4.9). Note that $\hat{p}$ is a nonlinear pricing function: $2\hat{p}$ is not the indifference price for $2H$. To get a linear pricing rule, consider buying ϵ units at price p per unit and define

$$v(x,\epsilon,p) = \sup_{X\in\mathcal{X}(x-\epsilon p)} E\big[U(X_T+\epsilon H)\big]\,. \tag{4.10}$$

$\check{p}$ is the *marginal utility price* if

$$\frac{\partial}{\partial\epsilon}v(x,\epsilon,\check{p})\Big|_{\epsilon=0} = 0\,.$$

Formally differentiating (4.10) we obtain

$$\frac{\partial}{\partial\epsilon}v(x,\epsilon,\check{p})\Big|_{\epsilon=0} = -p\,u'(x) + E\big[U'(\hat{X}_T)H\big]\,,$$

where $\hat{X}_T$ is the optimal terminal wealth, giving the following formula for the marginal utility price:

$$\check{p} = E\Big[\frac{U'(\hat{X}_T)}{u'(x)}H\Big]\,. \tag{4.11}$$

Now recall the characterisation of the optimal investment portfolio $\hat{X}_T$ given at (4.4) above, namely $\hat{X}_T = I\big(y\hat{Y}_T\big)$ where $\hat{Y}_T$ is the solution of the dual minimisation problem. Since $I=(U')^{-1}$ and $y=u'(x)$

$$\hat{Y}_T = \frac{1}{u'(x)}\,U'\big(\hat{X}_T\big)\,. \tag{4.12}$$

Now suppose that $\hat{Y}_t$ happens to be a martingale (rather than just a supermartingale). Then the measure $\hat{Q}$ defined by

$$\frac{\mathrm{d}\hat{Q}}{\mathrm{d}P} = \hat{Y}_T \tag{4.13}$$

is an equivalent martingale measure, and the price $\check{p}$ of (4.11) is expressed as[3]

$$\check{p} = E_{\hat{Q}}[H]\,. \tag{4.14}$$

As was mentioned earlier, all arbitrage-free valuations lie in the interval $\big(\inf_{Q\in\mathcal{M}} E_Q[H]\,,\sup_{Q\in\mathcal{M}} E_Q[H]\big)$, so we see that the marginal utility price picks a specific value from this interval, justified on economic grounds and depending on the investor's preferences.

If the optimal dual supermartingale $\hat{Y}_t$ is *not* a martingale, then the marginal utility price is not uniquely defined; see Kramkov and Schachermayer [26].

[3]Recall we have taken $S_0(t)\equiv 1$, so no 'discount factor' appears.

5 Parameter Uncertainty

Recall the Black–Scholes price model

$$\mathrm{d}S_t = \mu S_t \,\mathrm{d}t + \sigma S_t \,\mathrm{d}w_t\,.$$

The Black–Scholes value at time $t < T$ for an option with exercise value $h(S_T)$ is a function $C_h(S, r, \sigma, T-t)$ where S is the price at time t and r is the riskless interest rate. It is a key point that C_h does not depend on the growth rate μ, and this is fortunate because μ is essentially impossible to estimate. Indeed, if σ is known, the minimum variance unbiased estimate of μ is

$$\hat{\mu}_t = \tfrac{1}{t}\log\frac{S_t}{S_0} + \tfrac{1}{2}\sigma^2 = \mu + \frac{\sigma w_t}{t}\,,$$

with variance $\mathrm{var}(\hat{\mu}_t) = \sigma^2/t$. A typical volatility might be $\sigma = 20\%$, so to achieve 95% confidence that $|\mu - \hat{\mu}| < 1\%$ we need $1.96\sigma/\sqrt{t} < 0.01$, i.e. $t > 1521$ years. For shorter periods, the estimation error is enormous; for example, at $t = 3$ years the standard deviation is 11.5%. Estimating the volatility σ is easier, not surprisingly since in contiuous time $\sigma^2\varepsilon$ is exactly observable as the quadratic variation of the sample path $\log S_t$ over any time interval of length ε. See Rogers [32] for properties of estimates based on sampled data.

A further source of estimation error is that it is surely not credible to assume that the parameters are constant over lengthy periods such as 20 or 30 years. Financial data is thus subject to a kind of "uncertainty principle". Parameters can never be estimated to arbitrarily high precision: if the sampling interval is short then estimation error variance is high, while if the samplying interval is long then there will be bias errors due to non-stationarity. Figure 2 shows sample estimates based on simulated data from the log-normal model with $\mu = 5\%, \sigma = 20\%$ using an increasing number of years of data back from the present, while similar estimates for the growth rate of the FTSE100 index, based on a log-normal model with the sample volatility, are shown in Figure 4. The index itself is shown in Figure 3. One can see how useless these estimates are, even in the artificial case with simulated data. The lesson is that parameter uncertainty must be explicitly allowed for in the model.

Optimimisation under parameter uncertainty uses one of two approaches:

- Stochastic control with random parameters;
- Stochastic programming using a small number of random scenarios.

We very briefly describe these approaches in the next subsections.

5.1 Stochastic control with random parameters

Here asset prices are modelled as

$$\mathrm{d}S_t = \mu(S_t, Z_t)\, S_t \,\mathrm{d}t + \sigma(S_t, Z_t)\, S_t \,\mathrm{d}w_t\,.$$

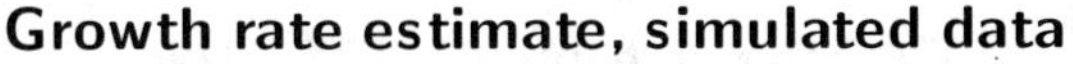

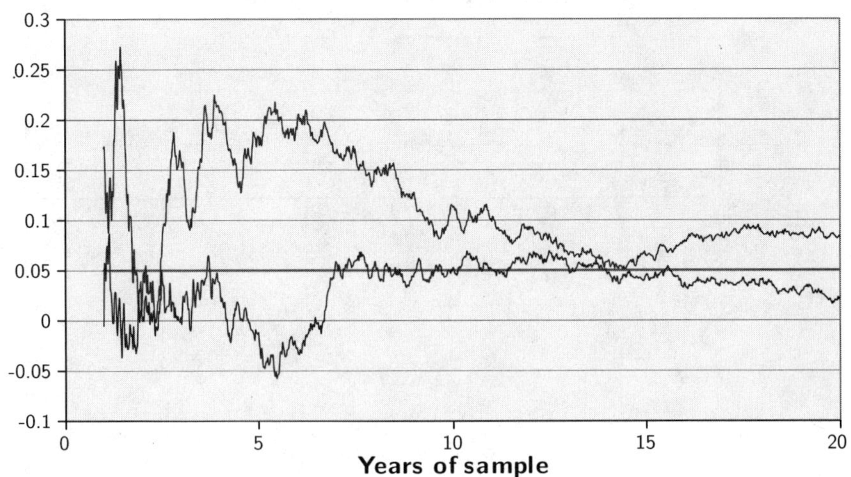

Figure 2. *Sample estimates of* μ*, simulated data*

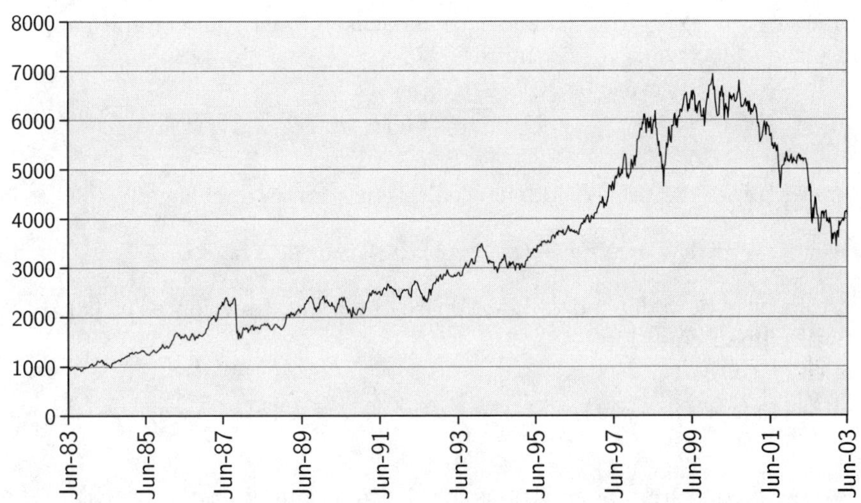

Figure 3. *FTSE100 Index 1983-2003*

w_t is vector Brownian motion and Z_t is a 'factor process' representing economic conditions. It could be another diffusion or, more tractably, a finite-state process $Z_t \in \{1, 2, \ldots, \kappa\}$ with transition rates

$$P\big[Z(t + \mathrm{d}t) = j \mid \mathcal{F}_t\big]\mathbf{1}_{(Z(t)=i)} = a_{ij}(S_t, Z_t)\,\mathrm{d}t + \mathrm{o}(\partial t)\,, \quad j \neq i$$

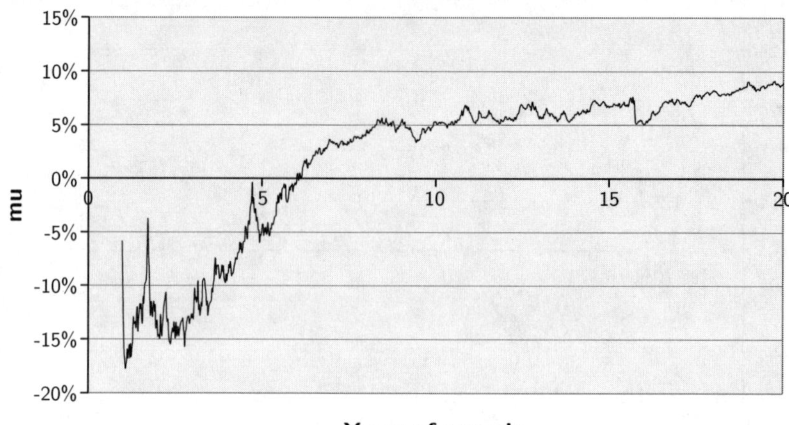

Figure 4. *Estimating μ for the FTSE100 Index*

We define $a_{ii} = -\sum_{j \neq i} a_{ij}$. Take the scalar case (1 risky asset) and suppose that as above the objective is to maximise

$$E \int_0^\infty e^{-\delta t} U(c_t) \, \mathrm{d}t \, .$$

If the factor process is observed, the wealth equation is unchanged:

$$\mathrm{d}X_t = X_t\big(r + (\mu - r)\pi\big) \, \mathrm{d}t - c_t \, \mathrm{d}t + X_t \pi_t \, \sigma \, \mathrm{d}w_t$$

except that μ, σ now depend on Z_t. Write the value function as $v_z(x) = v(x, z)$; then the Bellman equation is

$$\max_{u,c} \Big\{ \frac{\partial v_i}{\partial x} \big(x\big(r + (\mu - r)\pi\big) - c\big) + \tfrac{1}{2} u^2 \sigma^2 x^2 \frac{\partial^2 v_i}{\partial x^2} + U(c) + \sum_j a_{ij} v_j - \delta v_i \Big\} = 0 \, .$$

This is a coupled system of equations for the functions $v_1(x), \ldots, v_\kappa(x)$.

5.2 Stochastic programming

Stochastic programming is a discrete-time approach using a 'scenario tree' with a small number T of time periods. This is probably the best available method for analysing long-term asset and liability management for large insurance and pensions companies; see Mulvey and Ziemba [30] for a comprehensive treatment. A *scenario* is any path from left to right through the tree; in the example shown in Figure 5 there are $N = 6$ scenarios. In realistic applications, for example the pension planning model described in [36], a typical number is $N = 10000$.

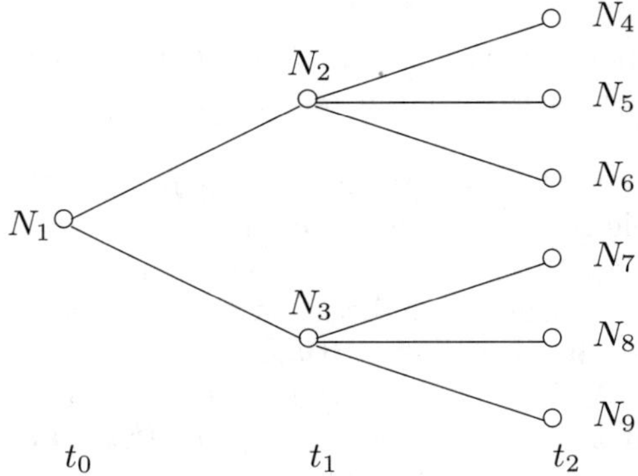

Figure 5. *Scenario tree for stochastic programming problem*

Asset classes available for investment are labelled $a \in \{1, \ldots, M\}$. For each scenario s, returns

$$\{R^s_{a,i}\,,\ a \in \{1, \ldots, M\},\ i \in \{1, \ldots, T\}\}$$

are specified (by sampling and quantising historical return distributions), so that \$1 invested in asset class a at time t_{i-1} becomes $\$R^s_{a,i}$ at time t_i. This scenario occurs with probability p_s. Investment decisions are made at each node at times $t_i < t_T$. Tree structure enforces 'non-anticipativity': if scenarios s, s' coincide up to time t_i then the investment decisions for s and s' at t_i are the same. The objective is to maximise

$$EU(X) = \sum_s \left[p_s\, U\big(X(s)\big)\right]$$

where $X(s)$ is the wealth at the terminal node of scenario s and U is a concave utility function.

Various penalties and constraints can be included, for example investment constraints (no shortselling etc.),penalties for not meeting growth targets, penalties for failing to pay liabilities, or transaction costs.

This problem can be formulated and solved as a *linear programming problem.* (The sub-problem of evaluating piecewise-linear approximations to the concave utility function U is also a linear program.) The art of this approach is to include a sufficiently rich set of scenarios to cover a wide range of favourable and adverse economic conditions. Then solutions are 'robust' in that they guard against adverse conditions. It turns out that choice of the probabilities p_s associated with the scenarios is less important.

6 Concluding Remarks

In a word, the message of this paper is as follows. "Traditional" mathematical finance, centred around the Black–Scholes model and its extensions, is a complete-market theory based on pricing by replication. In this setting only volatilities are important, and in applications most of the modelling effort is devoted to deriving appropriate volatility structure for particular products such as fixed-income derivatives.

In recent years there has been a huge increase in the trading of products involving unhedgeable risks. Examples include credit derivatives, counter party default risk, long-term equity risk and contracts on non-standard underlying variables such as the weather or insurance indices. In all these cases perfect hedging is not possible and therefore the investor's attitude to risk is an essential ingredient of the problem.

Our contention is that unhedgeable contracts should be thought of as investment opportunities, and this approach leads both to pricing methods based on utility indifference and 'hedging' in the form of optimal investment strategies to maximise utility and/or minimise the probability of failing to honour liabilities. The chief technical tool is then stochastic control theory, either in continuous time or using discrete-time scenario-based methods. Stylised problems, such as the annuity payment problem of section 3.5 or the terminating income problem of section 4.2 are useful in giving guidance as to the form of optimal strategies for specific purposes, and in the author's opinion the repertoire of solved problems of this sort is, at the present time, insufficiently large.

If one seeks to build a model for a specific application then an extra difficulty appears, namely parameter uncertainty. As we have pointed out, there is no way of estimating relevant parameters with any degree of precision. How to operate in this world of gross uncertainty is still far from settled. The stochastic programming method described in section 5.2 provides a systematic approach for long-term problems, but for shorter term problems such as hedging credit risk some fresh thinking is required to provide solutions that are both implementable and soundly based. This is the challenge for the future.

Acknowledgments

The author thanks Michel Vellekoop for helpful comments and suggestions

Bibliography

[1] D. Becherer, *The numéraire portfolio for unbounded semimartingales*, Finance & Stochastics **5** (2001) pp. 327–341.

[2] F. Black and M. Scholes, *The pricing of options and corporate liabilities*, J. Political Econ. **81** (1973) pp. 637–654.

[3] S. Browne, *Survival and growth with a liability: optimal portfolio strategies in continuous time*, Mathematics of Operations Research **22** (1997) pp. 468–493.

[4] J. C. Campbell and L. Viceira, *Strategic Asset Allocation*, Oxford University Press, 2001.

[5] D. Cuoco, *Optimal consumption and equilibrium prices with portfolio constraints and stochastic income*, J. Economic Theory, **72** (1997) pp. 33–73.

[6] J. Cvitanić, W. Schachermayer and H. Wang, *Utility maximisation in incomplete markets with random endowment*, Finance & Stochastics, **5** (2001) pp. 259–272.

[7] M. H. A. Davis, *Option Pricing in Incomplete Markets*, in *Mathematics of Derivative Securities*, eds M. A. H. Dempster and S. R. Pliska, Cambridge University Press 1997.

[8] M. H. A. Davis, *Optimal hedging with basis risk*, preprint, Imperial College, 2000.

[9] M. H. A. Davis and A. R. Norman, *Portfolio selection with transaction costs*, Mathematics of Operations Research, **15** (1990) pp. 676–713.

[10] M. H. A. Davis and M. H. Vellekoop, *An optimal investment problem with randomly terminating income*, preprint, Imperial College 2003.

[11] M. H. A. Davis, *Complete-market models of stochastic volatility*, Proc. Royal Society London (A), to appear 2004.

[12] F. Delbaen and W. Schachermayer, *A general version of the fundamental theorem of asset pricing*, Math. Ann., **300** (1994) pp. 463–520.

[13] R. J. Elliott and P. E. Kopp, *Mathematics of financial markets*, Springer-Verlag, Heidelberg 1999.

[14] T. S. Ferguson, *Mathematical Statistics*, Academic Press, New York 1967.

[15] W. H. Fleming and R. Rishel, *Deterministic and stochastic optimal control*, Springer-Verlag, New York 1975.

[16] L. Foldes, *Existence and uniqueness of an optimum in the infinite-horizon portfolio-cum-saving model with semimartingale investments*, Stochastics, **41** (1992) pp. 241–267.

[17] L. Foldes, *Valuation and martingale properties of shadow prices: an exposition*, Economic Dynamics and Control, **24** (2000) pp. 1641–1701.

[18] H. Föllmer and P. Leukert, *Efficient hedging: cost versus shortfall risk*, Finance & Stochastics, **2** (2000) pp. 117–146.

[19] V. Henderson, *Valuation of nontraded assets using utility maximization*, Mathematical Finance, **12** (2002) pp. 351–373.

[20] J. Hugonnier and D. Kramkov, *Optimal investment with random endownments in incomplete markets*, preprint HEC Montréal, 2002.

[21] J. Hugonnier, D. Kramkov and W. Schachermayer, *On the utility-based pricing of contingent claims in incomplete markets*, preprint HEC Montréal., 2002.

[22] J. Hull, *Options, Futures and Other Derivatives*, 5th ed., Prentice Hall, Upper Saddle River NJ, 2001.

[23] W. S. Jevons, *The Theory of Political Economy*, 1871, reprinted Pelican Books, London, 1970.

[24] I. Karatzas and S. E. Shreve, *Mathematical Finance*, Springer-Verlag, New York, 1998.

[25] D. O. Kramkov, *Optional decomposition of supermartingales and hedging contingent claims in incomplete security markets*, Probability Theory and Related Fields, **105** (1996) pp. 459–479.

[26] D. O. Kramkov and W. Schachermayer, *The asymptotic elasticity of utility functions and optimal investment in incomplete markets*, Annals of Applied Probability, **9** (1999) pp. 904–950.

[27] H. Markowitz, *Portfolio selection*, Journal of Finance, **7** (1952) pp. 77–91.

[28] R. Merton *Optimum consumption and portfolio rules in a continuous-time model*, J. Econ. Theory, **3** (1971) pp. 373–413.

[29] R. Merton *Continuous-time finance*, Blackwell, Oxford 1992.

[30] M. Musiela and T. Zariphopoulou *Indifference prices and related measures*, preprint, University of Texas, Austin, 2001.

[31] H. Nagai and S. Peng, *Risk-sensitive dynamic portfolio optimization with artial information on infinite time horizon*, Ann. Appl. Prob., **12** (2002) pp. 173–195.

[32] L. C. G. Rogers, *relaxed investor and parameter uncertainty*, Finance & Stochastics, **5** (2001) pp. 131–154.

[33] L. C. G. Rogers and D. Williams, *Diffusions, Markov Processes and Martingales, Volume 2: Ito Calculus*, 2nd ed., Cambridge University Press, 2000.

[34] P. Schönbucher, *Credit Derivatives Pricing Models: Models, Pricing and Implementation*, John Wiley, Chichester 2003.

[35] W. T. Ziemba and J. M. Mulvey (eds) *Worldwide asset and liability modelling*, Cambridge University Press, 1998.

[36] W. T. Ziemba, A. Geyer, W. Herold and K. Krontriner *The Innovest Austrian pension fund financial planning model*, preprint, University of British Columbia, 2001.

Invited speaker Franco Brezzi (Italy). photography by *Happy Medium Photo Co.*

James W Demmel is professor of computer science and mathematics at the UC Berkeley, and Chief Scientist of CITRIS, the Center for Information Technology Research in the Interest of Society. His research is in numerical linear algebra and tools for large scale simulation. He produces widely used numerical software such as LAPACK, collaborates with applications scientists in diverse areas in the use of high performance computing, and studies the complexity of fundamental numerical problems.

Demmel received his PhD in 1983, and was a faculty member at the Courant Institute until moving to Berkeley in 1990, where he holds the Richard Carl Dehmel Distinguished Prize in 1988 and jointly in 1991, and the J H Wilkinson Prize in Numerical Analysis in 1993. He was elected an ACM Fellow in 1998 and elected to the National Academy of Engineering in 1999.

Chapter 5

Accurate and Efficient Algorithms for Floating Point Computation*

James Demmel[†]
Plamen Koev[‡]

Abstract: Our goal is to find accurate and efficient algorithms, when they exist, for evaluating rational expressions containing floating point numbers, and for computing matrix factorisations (like LU, the singular value decomposition (SVD) and eigenvalue decompositions) of matrices with rational expressions as entries. More precisely, *accuracy* means the relative error in the output must be less than one (no matter how tiny the output is), and *efficiency* means that the algorithm runs in polynomial time. Our goal is challenging because our accuracy demand is much stricter than usual.

The classes of floating point expressions or matrices that we can accurately and efficiently evaluate or factor depend strongly on our model of arithmetic:

1. In the "Traditional Model" (TM), the floating point result of an operation like $a + b$ is $fl(a + b) = (a + b)(1 + \delta)$, where $|\delta|$ must be tiny.
2. In the "Long Exponent Model" (LEM) each floating point number $x = f \cdot 2^e$ is represented by the pair of integers (f, e), and there is no bound on the sizes of the exponents e in the input data. The LEM permits accurate and efficient computation of strictly larger classes of expressions or matrices than the TM.
3. In the "Short Exponent Model" (SEM) each floating point number $x = f \cdot 2^e$ is also represented by (f, e), but the input exponent sizes are bounded in terms of the sizes of the input fractions f. We believe the SEM permits accurate and efficient computation of strictly more expressions or matrices than the LEM.

[1]Funding for this paper was furnished by NSF grant ACI–9813362 and DOE grant DE–FG03–94ER25219, The information presented here does not necessarily reflect the position of the Government and no official endorsement should be inferred.

[†]Mathematics Department and Computer Science Division, University of California, Berkeley
[‡]Department of Mathematics, Massachusetts Institute of Technology, USA

These classes will be described by factorisability properties of the rational expressions, or of the minors of the rational matrices. For each such class, we identify new algorithms that attain our goals of accuracy and efficiency. These algorithms are often exponentially faster than prior algorithms, which would simply use a conventional algorithm with sufficiently high precision.

For example, we can factorise Cauchy matrices, Vandermonde matrices, many kinds of totally positive matrices, and suitably discretised differential and integral operators in all three models much more accurately and efficiently than before. But we provably cannot add three numbers accurately in the TM, even though it is easy in the other models.

Contents

1 Introduction

We will survey recent progress and describe open problems in the area of accurate floating point computation, in particular for matrix computations. A very short bibliography would include [18, 12, 15, 31, 24, 4, 17, 23, 5].

We consider the evaluation of multivariate rational functions $r(x)$ of floating point numbers $x = (x_1, x_2, \dots)$, and matrix computations on rational matrices $A(x)$, where each entry $A_{ij}(x)$ is such a rational function. Matrix computations will include computing determinants (and other minors), linear equation solving, performing Gaussian Elimination (GE) with various kinds of pivoting, computing the singular value decomposition (SVD), and computing eigenvalues, among others. Our goals are *accuracy* (computing each solution component with tiny relative error) and *efficiency* (the algorithm should run in time bounded by a polynomial function of the input size).

We consider three models of arithmetic, defined in the abstract, and for each one we try to classify rational expressions and matrices as to whether they can be evaluated or factored accurately and efficiently (we will say "compute(d) accurately and efficiently," or "CAE" for short).

In the Traditional "$1 + \delta$" Model (TM), we have $fl(a \otimes b) = (a \otimes b)(1 + \delta)$, $\otimes \in \{+, -, \times, \div\}$ and $|\delta| \leq \epsilon$, where $\epsilon \ll 1$ is called *machine precision.* It is the conventional model for floating point error analysis, and means that every floating point result is computed with a relative error δ bounded in magnitude by ϵ. The values of δ may be arbitrary real (or complex) numbers satisfying $|\delta| < \epsilon$, so that any algorithm proven to CAE in the TM must work for arbitrary real (or complex) number inputs and arbitrary real (or complex) $|\delta| \leq \epsilon$. The size of the input in the TM is the number of floating point words needed to describe it, independent of ϵ.

The Long Exponent (LEM) and Short Exponent (SEM) models, which are implementable on a Turing machine [34], make errors that may be described by the TM, but their inputs and δ's are much more constrained. Also, we compute the size of the input in the LEM and SEM by counting the number of bits, so that higher precision and wider range take more bits.

It will turn out that problems we can provably CAE in the TM are a strict subset of those we can CAE in the LEM, which in turn we conjecture are a strict subset of those we can CAE in the SEM. In all three models we will describe the classes of rational expressions and rational matrices in terms of the factorisation properties of the expressions, or of the minors of the matrices.

The reader may wonder why we insist on accurately computing tiny quantities with small relative error, since in many cases the inputs are themselves uncertain, so that one could suspect that the inherent uncertainty in the input could make even the signs of tiny outputs uncertain. Firstly, in a number of practical problems (e.g. computing vibrational frequencies or energy levels) it is the smallest eigenvalues that are of physical interest. Secondly, it will turn out that in the TM, the class we can CAE appears to be identical to the class where all the outputs are in fact accurately determined by the inputs, in the sense that small relative changes in the inputs cause small relative changes in the outputs. In other words, it is worthwhile computing the output with some relative accuracy, because that is the accuracy to which it is determined by the input. We discuss this conjecture in Section 6 below.

There are many ways to formulate the search for efficient and accurate algorithms [11, 8, 41, 7, 29, 40, 38]. Our approach differs in several ways. In contrast to either conventional floating point error analysis [29] or the model in [11], we ask that even the tiniest results have correct leading digits, and that zero be exact. In [11] the model of arithmetic allows a tiny absolute error in each operation, whereas in TM we allow a tiny relative error. Unlike [11] our LEM and SEM are conventional Turing machine models, with numbers represented as bit strings, and so we can take into account the cost of arithmetic on very large and very small numbers (i.e. those with many exponent bits). For these reasons we believe our models are closer to computational practice than the model in [11]. In contrast to [38], we (mostly) consider the input as given exactly, rather than as a sequence of ever better approximations. Finally, many of our algorithms could easily be modified to explicitly compute guaranteed interval bounds on the output [40].

2 Factorisability and Minors

We show here how to reduce the question of accurate and efficient matrix computations to accurate and efficient rational expression evaluation. The connection to LU factorisation and similar operations is elementary, but the SVD requires an algorithm from [18]. (Here an accurate SVD means that the singular values are known with guaranteed relative error less than one, but the error in a singular vector u_i

may be inversely proportional to the *relative gap* $\mathrm{relgap}(i) \equiv \min_{j \neq i} |\sigma_i - \sigma_j|/\sigma_i$ between its corresponding singular value σ_i and the other singular values σ_j.)

Proposition 1. *Being able to CAE the absolute value of the determinant* $|\det(A(x))|$ *is* necessary *to be able to CAE the following matrix computations on* $A(x)$*: LU factorisation (with or without pivoting), QR factorisation, all the eigenvalues* λ_i *of* $A(x)$*, and all the singular values of* $A(x)$*. Conversely, being able to CAE* all *the minors of* $(A(x))$ *is* sufficient *to be able to CAE the following matrix computations on* $A(x)$*:* A^{-1}*, LU factorisation (with or without pivoting), and the SVD of* $A(x)$*. This holds in any model of arithmetic.*

Proof. First consider necessity. $|\det(A(x))|$ may be written as the product of diagonal entries of the matrices L, U and R in these factorisations, or as the product of eigenvalues or singular values. If these entries or values can be CAE, then so can their product in a straightforward way.

Now consider sufficiency. The statement about A^{-1} is just Cramer's rule, which only needs $n^2 + 1$ different minors. The statement about LU factorisation depends on the fact that each nontrivial entry of L and U is a quotient of minors. The SVD is more difficult [18], and depends on the following two step algorithm: (1) Compute a *rank revealing* decomposition $A = X \cdot D \cdot Y$ where X and Y are "well-conditioned" (far from singular in the sense that $\|X\| \cdot \|X^{-1}\|$ is not too large); and (2) use a bisection-like algorithm to compute the SVD from XDY. □

Computing $\det(A(x))$ accurately is obviously necessary for LU factorisation and eigenvalues, but not necessarily for QR factorisation or the SVD. The sufficiency proof can be extended to other matrix computations like the QR decomposition and pseudoinverse by considering minors of matrices like $\begin{bmatrix} I & A \\ A^T & 0 \end{bmatrix}$. Furthermore, if we can CAE the minors of $C(x) \cdot A(x) \cdot B(x)$, and $C(x)$ and $B(x)$ are well-conditioned, then we can still CAE a number of matrix factorisations, like the SVD.

The SVD can be applied to get the eigendecomposition of symmetric matrices [25], but efficiently computing accurate eigenvalues of nonsymmetric matrices seems to require more conditions. We have recently shown that computing a small set of minors accurately is sufficient for computing accurate singular values *and* eigenvalues of totally positive matrices, and that if these minors can be computed efficiently then so can their singular values and eigenvalues [33]. We discuss this further in Section 5 below.

3 CAE in the Traditional Model

We begin by giving examples of expressions and matrix computations that we can CAE in the TM, and then discuss what we cannot do. The results will depend on details of the axioms we adopt, but for now we consider the minimal set of operations described in the abstract.

As long as we only do *admissible operations*, namely multiplication, division, addition of like-signed quantities, and addition/subtraction of (exact!) input data ($x \pm y$), then the worst case relative error only grows very slowly, roughly proportionally to the number of operations. It is when we subtract two like-signed approximate quantities *and* significant cancellation occurs, that the relative error can become large. So we may ask which problems we can CAE just using only admissible operations; i.e., which rational expressions factor in such a way that only admissible operations are needed to evaluate them, and which matrices have all minors with the same property.

Here are some examples [15] where we assume that the inputs are arbitrary real or complex numbers.

(1) The determinant of a Cauchy matrix $C_{ij} = 1/(x_i+y_j)$ is CAE using the classical expression $\prod_{i<j}(x_j - x_i)(y_j - y_i)/\prod_{i,j}(x_i + y_j)$, as is every minor. In fact, changing one line of the classical GE routine will compute each entry of the LU decomposition accurately in about the same time as the original inaccurate version.

(2) We can CAE all minors of sparse matrices; i.e., those with certain entries fixed at 0 and the rest independent indeterminates x_{ij}, if and only if the undirected bipartite graph presenting the sparsity structure of the matrix is *acyclic*; a one-line change to GE again renders it accurate. An important special case are bidiagonal matrices, which arise in the conventional SVD algorithm.

(3) The eigenvalue problem for the second centered difference approximation to a Sturm–Liouville ODE or elliptic PDE on a rectangular grid (with arbitrary rectilinear boundaries) can be written as the SVD of an "unassembled" problem $G = D_1 U D_2$ where D_1 and D_2 are diagonal (depending on "masses" and "stiffnesses") and U is *totally unimodular*; i.e., all its minors are ± 1 or 0. Again, a simple change to GE renders it accurate.

In contrast, one can show that it is impossible in the TM to add three numbers $x+y+z$ accurately in constant time; the proof involves showing that for *any* algorithm the rounding errors δ and inputs x, y, z can be chosen to have an arbitrarily large relative error [18]. This depends on the δ's being permitted to be arbitrary real numbers in our model. This impossibility is in stark contrast to the bit model, although computing accurate floating sums as efficiently as possible is more subtle [16].

Vandermonde matrices $V_{ij} = x_i^{j-1}$ are more subtle. Since the product of a Vandermonde matrix and the Discrete Fourier Transform (DFT) is Cauchy, and we can compute the SVD of a Cauchy, we can compute the SVD of a Vandermonde [13]. This fits in our TM model because the roots of unity in the DFT need only be known approximately, and so may be computed in the TM model. In contrast, one can use the result in the last paragraph to show that the inverse of a Vandermonde cannot be computed accurately. Similarly, polynomial Vandermonde matrices with $V_{ij} = P_i(x_j)$, P_i a (normalised) orthogonal polynomial, also permit accurate SVDs [19] but probably not inverses.

These results, with others to be discussed later, are summarised in Tables 1 and 2.

4 Adding Nonnegativity to the Traditional Model

If we further restrict the domain of (some) inputs to be nonnegative, then many more accurate and efficient computations are possible, $x+y+z$ as a trivial example. A more interesting example is the set of weakly diagonally dominant M-matrices, which arise as discretisations of PDEs; they must be represented as off-diagonal entries and the row sums, [31, 20].

More interesting is the class of *totally positive (TP) matrices*, all of whose minors are positive. Numerous structure theorems show how to represent such matrices as products of much simpler TP matrices, [15, Sec. 9] [1, 30]. Accurate formulas for the (nonnegative) minors of these simpler matrices combined with the Cauchy–Binet theorem yield accurate formulas for the minors of the original TP matrix, but typically at an exponential cost.

The most important structure theorem states that all TP matrices have a *bidiagonal or Neville factorisation*, an LDU factorisation where L (U) is the product of unit lower (upper) bidiagonal nonnegative matrices, and D is diagonal and positive. There are n^2 free positive parameters defining these matrices that parameterise the set of all TP matrices. If these positive parameters, which are products and quotients of minors of the matrix, can be computed accurately and efficiently, then many subsequent linear algebra operations (like the SVD, solving $Ax = b$, inversion, and the eigenvalue problem) can also be solved accurately and efficiently [33]. This all applies to nonsingular totally nonnegative, oscillatory, and sign-regular matrices (inverses of TP matrices) [26].

In fact, the set of totally positive matrices for which it is possible to compute the SVD and eigenvalues (as well as inverses, etc.) accurately and efficiently is closed under a large set of operations: Given any two TP matrices A and B with Neville factorisations, the Neville factorisation of $A \cdot B$, the scaled inverse $SA^{-1}S$ (where $S = \mathrm{diag}(\mathrm{sign}(\pm 1))$) and Schur complements of A can be CAE, so that the SVDs and eigenvalues of these new matrices can be CAE [32].

A great many TP matrices have efficient and accurate algorithms for computing their Neville factorisations. For example, the Björck–Pereyra method [6] was invented for Vandermonde systems, and has been extended to Cauchy [9], Cauchy–Vandermonde [36], and generalised Vandermonde matrices [21].

But for TP generalised Vandermonde matrices $G_{ij} = x_i^{\mu_j}$, where the μ_j form an increasing nonnegative sequence of integers, the problem is much more subtle. $\det(G)$ is known to be the product of $\prod_{i<j}(x_j - x_i)$ and a *Schur function* [35] $s_\lambda(x_i)$, where the sequence $\lambda = (\lambda_j) = (\mu_{n+1-j} - (n-j))$ is called a *partition*. Schur functions are polynomials with nonnegative integer coefficients, so since their arguments x_i are nonnegative, they can certainly be computed accurately. However, straightforward evaluation would have an exponential cost $\mathrm{O}(n^{|\lambda|})$, $|\lambda| = \sum_j \lambda_j$. But by exploiting combinatorial identities satisfied by Schur functions along with techniques of divide-and-conquer and memoisation, the cost of evaluating the determinant (or complete Neville factorisation) can be reduced to polynomial time $n^2\Lambda$ where $\Lambda \le 2(\lambda+1) \cdot (\#\lambda_i > 0) \cdot \prod_{j>1}(\lambda_j + 1)^2$. the λ_i are counted as part of the size of the input in this case [31, 22, 21].

Combined with the results on computing the SVD and eigenvalues in [33, 32], this positively settles an open question in [31, 14]: we can now accurately and efficiently compute the eigenvalues and singular values of TP generalised Vandermonde matrices. Questions remain, however, about the accuracy of the associated vectors.

In [14] we made a conjecture (Conjecture 1) in an attempt to characterise the set of homogeneous polynomials that can be evaluated accurately in the TM. Briefly, we conjectured that $f(x_1, ..., x_n)$ could be evaluated accurately (on a reasonable subset of the unit sphere) if and only if each of its factors were of the form x_i, $x_i - x_j$, $x_i + x_j$ or bounded away from 0 on the domain of evaluation. These properties are indeed sufficient, but not necessary, as the simple example $x_1^2 + (x_2 + x_3)^2$ shows. Indeed, any formulation of a condition for efficient evaluation must depend on more than the variety determined by f, since $f(x_1, x_2, x_3, x_4) = x_1^4 + x_1^2 \cdot (x_2 + x_3)^2$ and $g(x_1, x_2, x_3, x_4) = x_1^4 + x_1^2 \cdot (x_2 + x_3 + x_4)^2$ determine the same real variety $\{x_1 = 0\}$, yet it seems that only f can be evaluated accurately for all inputs, since any accurate algorithm for g would require the accurate evaluation of $(x_2 + x_3 + x_4)^2$ when x_1 is tiny enough, apparently contradicting the result in [18] cited in Section 4. Necessary conditions remain elusive.

5 Summary of Costs in the Traditional Model

Tables 1 and 2 summarise our state of knowledge of the cost of accurate matrix computations in the TM. There is one row for the most important classes of matrices discussed so far, as well as others. Furthermore, some totally positive (TP) matrix classes are considered as well, since this can completely change the cost of accurate computation. Details of matrix classes are explained in footnotes.

There is one column in either Table 1 or Table 2 for each matrix computation we consider. "Any minor" means any single arbitrary minor of the matrix A. GENP, GEPP and GECP stand for Gaussian elimination with no pivoting, partial pivoting, and complete pivoting, resp. SVD and EVD mean the singular value and eigenvalue decompositions, respectively; we only consider the problem of computing all singular value or eigenvalues accurately, the more difficult question of vectors is omitted (see the open questions in Section 8). NF refers to Neville (bidiagonal) factorisation, which is primarily of interest for TP matrices. Other details are explained in footnotes.

The table entry n^2 means $O(n^2)$, for example. The table entry "exp" means exponential time, based on formulas for evaluating arbitrary Schur polynomials [31, 35] in the case of TP (generalised) Vandermonde matrices, and based on the Cauchy–Binet formula for minors of products of matrices in the case of "any TP". The table entry "No" means impossible, because it would require the computation of an expression including $x+y+z$, which we know to be impossible. It is interesting that sometimes the SVD is possible in $O(n^3)$ time, but that inversion is impossible (Vandermondes).

Our results have an asterisk (*) in front of the citation. (Sometimes we cite one of our own papers if it is a convenient place to find a concise description of a classical result; there is no asterisk in this case).

Table 1. *State of Knowledge, Part 1 — Cost of Accurate Computation in TM*

Type of Matrix	det A	A^{-1}	Any minor	GENP	GEPP	GECP
Cauchy	n^2 [27]	n^2 [27]	n^2 [27]	n^2 [12]	n^3 [12]	n^3 * [12]
TP Cauchy	n^2 [27]	n^2 [27]	n^2 [27]	n^3 [12]	n^3 [12]	n^3 *[12]
Vandermonde	n^2 [6]	**No** *[18]	**No** *[18]	**No** *[18]	**No** *[18]	**No** *[18]
TP Vandermonde^7)	n^2 [6]	n^3 *[28, 31]	exp *[31]	n^2 *[31]	n^2 *[31]	exp *[31]
Confluent Vandermonde	n^2 [28]	**No** *[18]	No *[18]	No *[18]	No *[18]	No *[18]
TP Confluent Vandermonde	n^2 [28]	n^3 [28]	?	n^3 *[31]	?	?
Three-term Vandermonde Orth. Poly.	n^2 [28]	?	?	?	?	?
3-term Vand. Orth. Poly. + other cond.^5)	n^2 [28]	n^3 [28]	?	?	?	?
Generalized Vandermonde	No *[18]	No *[18]	No *[18]	No *[18]	No *[18]	No *[18]
TP generalized Vandermonde^4)	Λn^2 *[22, 21]	Λn^3 *[22, 21]	**exp** *[22, 21]	Λn^2 *[22, 21]	Λn^2 *[22, 21]	**exp** *[22, 21]
Any TP^6)	n [33]	n^3 *[33]	exp *[33]	n^3 *[33]	exp *[33]	exp *[33]
Wkly diag.dom. M-matrices	n^3 [37]	n^3 [3, 2]	No *[18]	n^3 *[20]	n^3 *[20]	n^3 *[20]

1) Each entry is meant in a big-O sense; i.e. n^2 means $O(n^2)$.

2) Below each entry is a citation to where the result (old or new) appears.

3) Our results have an asterisk (*) in front of the citation.

4) If $G_{ij} = x_i^{j-1+\lambda_{n-j}}$ then $\Lambda \leq 2(\lambda_1+1)(\lambda_2+1)^2 \ldots (\lambda_p+1)^2 p$.

5) Also need $0 < x_1 < x_2 < \ldots < x_n$, $d_i > 0, c_i \geq 0$ for all i, and $b_i - x_k \leq 0$ for all $i+k \leq n-1$.

6) Any TP matrix, specified by the entries of its bidiagonal decomposition, or by combining such matrices by taking products, Schur complements and scaled inverses as described in Section 4.

7) The contribution to computing the inverse of a TP Vandermonde in [31] improves the complexity of the result in [28] from $2n^3$ to $\frac{5}{12}n^3$.

Table 2. *State of Knowledge, Part 2 — Cost of Accurate Computation in TM*

Type of Matrix	SVD	EVD	NF	$Ax=b$ F[7)]	$Ax=b$ B[7)]
Cauchy	n^3 *[12]	?	n^2 [9]	n^2 [9]	?
TP Cauchy	n^3 *[12]	n^3 *[33]	n^2 [9]	n^2 [9]	n^2 [9]
Vandermonde	n^3 *[12, 19]	?	n^2 [6]	No *[18]	?
TP Vandermonde	n^3 *[12, 19]	n^3 *[33]	n^2 [6]	n^2 [28]	n^2 [28]
Confluent Vandermonde	?	?	n^2 [28]	No *[18]	?
TP Confluent Vandermonde	n^3 *[33]	n^3 *[33]	n^2 [28]	n^2 [28]	n^2 [28]
Three-term Vandermonde Orth. Poly.	n^3 *[19]	?	?	?	?
3-term Vand. Orth. Poly. + other cond.[5)]	n^3 *[19]	?	?	n^2 [28]	?
Generalized Vandermonde	No *[18]	?	No *[18]	No *[18]	?
TP generalized Vandermonde[4)]	Λn^3 *[33]	Λn^3 *[33]	Λn^2 *[22, 21]	Λn^2 *[22, 21]	Λn^2 *[22, 21]
Any TP[6)]	n^3 *[33]	n^3 *[33]	0	n^2 [28]	n^2 [28]
Wkly diag.dom. M-matrices	n^3 *[20]	?	?	?	?

1) Each entry is meant in a big-O sense; i.e. n^2 means $O(n^2)$.

2) Below each entry is a citation to where the result (old or new) appears.

3) Our results have an asterisk (*) in front of the citation.

4) If $G_{ij} = x_i^{j-1+\lambda_{n-j}}$ then $\Lambda \leq 2(\lambda_1 + 1)(\lambda_2 + 1)^2 \ldots (\lambda_p + 1)^2 p$.

5) Also need $0 < x_1 < x_2 < \ldots < x_n$, $d_i > 0, c_i \geq 0$ for all i, and $b_i - x_k \leq 0$ for all $i + k \leq n - 1$.

6) Any TP matrix, specified by the entries of its bidiagonal decomposition, or by combining such matrices by taking products, Schur complements and scaled inverses as described in Section 5.

7) "$Ax = b$: F" means x satisfies a *Forward* error bound $|x - \hat{x}| \leq O(\epsilon)|A^{-1}||b|$ (so $|x - \hat{x}| \leq O(\epsilon)|x|$, when $\text{sign}(x_i) = (\pm 1)^i$ has alternating sign pattern and A is TP. "$Ax = b$: B" means x has *Backward* error $|A - \hat{A}| \leq O(\epsilon)|A|$.

6 Extending the Traditional Model

So far we have considered the simplest version of the TM, where:

(1) we have only the input data, and no additional constants available, (not even integers, let alone arbitrary rationals or reals);

(2) the input data is given exactly (as opposed to within a factor of $1+\delta$); and

(3) there is no way to "round" a real number to an integer, and so convert the problem to the LEM or SEM models.

We note that in [11], (1) integers are available, (2) the input is rounded, and (3) there is no way to "round" to an integer. Changes to these model assumptions will affect the classes of problems we can solve. For example, if we (quite reasonably) were to permit exact integers as input, then we could CAE expressions like $x-1$, and otherwise presumably not. If we went further and permitted exact rational numbers, then we could also CAE $9x^2-1=9(x-\frac{1}{3})(x+\frac{1}{3})$. Allowing algebraic numbers would make $x^2-2=(x-\sqrt{2})(x+\sqrt{2})$ CAE.

If inputs were not given exactly, but rather first multiplied by a factor $1+\delta$, then we could no longer always accurately compute $x \pm y$ where x and y are inputs, eliminating Cauchy matrices and most others. But the problems we can solve with exact inputs in the TM still have an attractive property with inexact inputs: Small relative changes in the inputs *often* cause only a small relative change in the outputs, independent of their magnitudes. The output relative errors may be larger than the input relative error by a factor called a *relative condition number* κ_{rel}, which is at most a polynomial function of $\max(1/\text{rel\_gap}(x_i, \pm x_j))$. Here $\text{rel\_gap}(x_i, \pm x_j) = |x_i \mp x_j|/(|x_i|+|x_j|)$ is the *relative gap* between inputs x_i and $\pm x_j$, and the maximum is taken over all factors $x_i \mp x_j$ dividing the rational expression being computed. So if all the inputs differ in several of their leading digits, all the leading digits of the outputs are determined accurately. We note that κ_{rel} can be large, depending on the expression being evaluated, but it can only be unbounded when a relative gap goes to zero.

If a problem has this attractive property, we say that it possesses a relative perturbation theory. In practical situations, where only a few leading digits of the inputs x_i are known, this property justifies the use of algorithms that try to compute the output as accurately as we do.

In [14] we stated a conjecture (Conjecture 2) similar to Conjecture 1, that a relative perturbation theory for the evaluation of f existed if and only if f factored in a certain simple way. As with Conjecture 1, these conditions are in fact sufficient but not necessary for a relative perturbation theory to exist.

7 CAE in the Long and Short Exponent Models

Now we consider standard Turing machines, where input floating point numbers $x = f \cdot 2^e$ are stored as the pair of integers (f, e), so the size of x is $\text{size}(x) = \#\text{bits}(f) + \#\text{bits(e)}$. We distinguish two cases, the Long Exponent Model (LEM) where f and e may each be arbitrary integers, and the Short Exponent Model (SEM), where the length of e is bounded depending on the length of f. In the simplest case, when $e = 0$ (or lies in a fixed range) then the SEM is equivalent to taking integer inputs, where the complexity of problems is well understood. This is more generally the case if #bits(e) grows no faster than a polynomial function of #bits(f).

In particular it is possible to CAE the determinant of an integer (or SEM) matrix each of whose entries is an independent floating point number [10]. This is not possible as far as we know in the LEM (even for tridiagonal matrices with independent floating point entries) which accounts for a large complexity gap between the two models.

We start by illustrating some differences between the LEM and SEM, and then describe the class of problems that we can CAE in the LEM.

First, the number of bits in an expression with LEM inputs can be exponentially larger than the number of bits in the same expression when evaluated with SEM inputs. For example, the number of nonzero bits in $\prod_{i=1}^{n}(1 + 2^{e_i})$ grows like $O(n)$ if the e_i are bounded, but like 2^n if the e_i are arbitrary. In fact, LEM arithmetic can encode symbolic algebra, because if e_1 and e_2 have no overlapping bits, then we can recover e_1 and e_2 from their product $2^{e_1} \cdot 2^{e_2} = 2^{e_1+e_2}$, letting us effectively represent the indeterminates x_1 and x_2 by 2^{e_1} and 2^{e_2}, resp.

Second, the error of many conventional matrix algorithms is typically proportional to the condition number $\kappa(A) = \|A\| \cdot \|A^{-1}\|$. This means that a conventional algorithm run with $O(\log \kappa(A))$ extra bits of precision will compute an accurate answer. It turns out that if $A(x)$ has rational entries in the SEM model, then $\log \kappa(A)$ is at most a polynomial function of the input size, so conventional algorithms run in high precision will CAE the answer. However $\log \kappa(A)$ for LEM matrices can be exponentially larger, so this approach does not work. The simplest example is $\log \kappa(\mathrm{diag}(1, 2^e)) = e = 2^{\#\mathrm{bits}(e)}$. On the other hand $\log \log \kappa(A(x))$ is a lower bound on the complexity of any algorithm, because this is a lower bound on the number of exponent bits in the answer. One can show that $\log \log \kappa(A(x))$ grows at most polynomially large in the size of the input [18].

Finally, we consider the problem of computing an arbitrary bit in the simple expression $p = \prod_{i=1}^{n}(1 + x_i)$. When the x_i are in the SEM, then p can be computed exactly in polynomial time. However when the x_i are in the LEM, then one can prove that computing an arbitrary bit of p is as hard as computing the permanent, a well-known combinatorially difficult problem. Here is another apparently simple problem not known to even be in NP: testing singularity of a floating point matrix. In the SEM, we can CAE the determinant. But in the LEM, the obvious choice of a "witness" for singularity, a null vector, can have exponentially many bits in it, even if the matrix is just tridiagonal. We conjecture that deciding singularity of an LEM matrix is NP-hard.

So how do we compute efficiently in the LEM? The idea is to use *sparse arithmetic*, or to represent only the nonzero bits in the number. (A long string of 1s can be represented as the difference of two powers of 2 and similarly compressed). In contrast, in the SEM one uses *dense arithmetic*, storing all fraction bits of a number. For example, in sparse arithmetic $2^e + 1$ takes $O(\log e)$ bits to store in sparse arithmetic, but e bits in dense arithmetic. This idea is exploited in practical floating point computation, where extra precise numbers are stored as arrays of conventional floating point numbers, with possibly widely different exponents [39].

Now we describe the class of rational functions that we can CAE in the LEM. We say the rational function $r(x)$ is in *factored form* if $r(x) = \sum_{i=1}^{n} p_i(x_1, ..., x_k)^{e_i}$, where each e_i is an integer, and $p_i(x_1, ..., x_k)$ is written as an explicit sum of nonzero monomials. We say size(r) is the number of bits needed to represent it in factored form. Then by

(1) computing each monomial in each p_i exactly;

(2) computing the leading bits of their sum p_i using sparse arithmetic (the cost is basically sorting the bits [16]); and

(3) computing the leading bits of the product of the $p_i^{e_i}$ by conventional rounded multiplication or division, one can evaluate $r(x)$ accurately in time a polynomial in size(r) and size(x).

In other words, the class of rational expression that we can CAE are those that we can express in factored form in polynomial space.

Now we consider matrix computations. It follows from the last paragraph that if each minor $r(x)$ of $A(x)$ can be written in factored form of a size polynomial in the size of $A(x)$, then we can CAE all the matrix computations that depend on minors. So the question is which matrix classes $A(x)$ have all their minors (or just the ones needed for a particular matrix factorisation) expressible in a factored form no more than polynomially larger than the size of $A(x)$. The obvious way to write a minor like $\det\big(A(x)\big)$, with the Laplace expansion of $n!$ terms, is clearly exponentially larger than $A(x)$, so it is only specially structured $A(x)$ that will work.

All the matrices that we could CAE in the TM are also possible in the LEM. The most obvious classes of $A(x)$ that we can CAE in the LEM that were impossible in the TM are gotten by replacing all the indeterminates in the TM examples by arbitrary rational expressions of polynomial size. For example, the entries of an M-matrix can be polynomial-sized rational expressions in other quantities. Another class are Green's matrices (inverses of tridiagonals), which can be thought of as discretised integral operators, with entries written as $A_{ij} = x_i \cdot y_j$.

The obvious question is whether A each of whose entries is an independent number in the LEM falls in this class. We conjecture that it does not, as mentioned before.

8 Conclusions and Open Problems

Our goal has been to identify rational expressions (or matrices) that we can evaluate accurately (or on which we can perform accurate matrix computations), in polynomial time. Accurately means that we want to get a relative error less than 1, and polynomial time means in a time bounded by a polynomial function of the input size.

We have defined three reasonable models of arithmetic, the Traditional Model (TM), the Long Exponent Model (LEM) and the Short Exponent Model (SEM), and tried to identify the classes of problems that can or cannot be computed accurately and efficiently for each model. The TM can be used as a model to do proofs that also hold in the implementable LEM and SEM, but since it ignores the structure of floating point numbers as stored in the computer, it is strictly weaker than either the LEM or SEM. In other words, there are problems (like adding $x + y + z$) that are provably impossible in the TM but straightforward in the other two models.

We also believe that the LEM is strictly weaker than the SEM, in the sense that there appear to be computations (like computing the determinant of a general, or even tridiagonal, matrix) that are possible to do accurately in polynomial time in the SEM but not in the LEM. In the SEM, essentially all problems that can be written down in polynomial space can be solved accurately in polynomial time. For the LEM, only expressions that can be written in *factored form* in polynomial space can be computed efficiently in polynomial time.

A number of open problems and conjectures were mentioned in the paper, including the question marks in Table 1. We mention three additional ones here.

1. It is known that a small relative change $\eta \ll 1$ in any entry of a bidiagonal matrix can only perturb a singular vector by approximately $\eta/\text{relgap}(i)$ (see the beginning of Section 2). If this same result were true for the product of two nonnegative bidiagonals, namely that a small relative change η in any entry could only perturb a singular vector of the product by $\eta/\text{relgap}(i)$, then we could conclude that the singular vectors of totally positive matrices in Table 1 could be computed as accurately as expected, namely with accuracy proportional to machine precision divided by $\text{relgap}(i)$.

2. What can be said about the nonsymmetric eigenvalue problem, other than for totally positive matrices, all of which have positive and simple eigenvalues? In other words, what matrix properties, perhaps related to minors, guarantee that all eigenvalues of a nonsymmetric matrix can be computed accurately?

3. How do the complexity classes change when we consider randomness, either in the problems considered, or the algorithms, and tolerate a small probability of large relative error?

Acknowledgements

The authors acknowledge Benjamin Diament, Zlatko Drmač, Stan Eisenstat, Ming Gu, William Kahan, Yozo Hida, Ivan Slapničar and Kresimir Veselič for their collaboration over many years in developing this material.

Bibliography

[1] T. Ando. Totally positive matrices. *Lin. Alg. Appl.*, 90, pp. 165–219, 1987.

[2] S. A. Attahiru, X. Junggong, and Q. Ye. Accurate computation of the smallest eigenvalue of a diagonally dominant M-matrix. preprint, 2002.

[3] S. A. Attahiru, X. Junggong, and Q. Ye. Entrywise perturbation theory for diagonally dominant M-matrices with applications. *Numer. Math.*, 90, pp. 401–414, 2002.

[4] J. Barlow and J. Demmel. Computing accurate eigensystems of scaled diagonally dominant matrices. *SIAM J. Num. Anal.*, 27(3), pp. 762–791, June 1990.

[5] J. Barlow, B. Parlett, and K. Veselic, editors. *Proceedings of the International Workshop on Accurate Solution of Eigenvalue Problems*, volume 309 of *Linear Algebra and its Applications*. Elsevier, 2001.

[6] Å. Björck and V. Pereyra. Solution of Vandermonde systems of equations. *Math. Comp.*, 24(112), pp. 893–903, 1970.

[7] L. Blum, F. Cucker, M. Shub, and S. Smale. *Complexity and Real Computation*. Springer, 1997.

[8] L. Blum, M. Shub, and S. Smale. On a theory of computation and complexity over the real numbers: NP-completeness, recursive functions and universal machines. *AMS Bulletin (New Series)*, 21(1), pp. 1–46, July 1989.

[9] T. Boros, T. Kailath, and V. Olshevsky. The fast Björck-Pereyra-type algorithm for parallel solution of Cauchy linear equations. *Lin. Alg. Appl.*, 302/303, pp. 265–293, 1999.

[10] K. Clarkson. Safe and effective determinant evaluation. In *33rd Annual Symp. on Foundations of Comp. Sci.*, pages 387–395, 1992.

[11] F. Cucker and S. Smale. Complexity estimates depending on condition and roundoff error. *J. ACM*, 46(1), pp. 113–184, Jan 1999.

[12] J. Demmel. Accurate SVDs of structured matrices. *SIAM J. Mat. Anal. Appl.*, 21(2), pp. 562–580, 1999.

[13] J. Demmel. Accurate SVDs of structured matrices. *SIAM J. Mat. Anal. Appl.*, 21(2), pp. 562–580, 1999.

[14] J. Demmel. The complexity of accurate floating point computation. In *Proceedings of the 2002 International Congress of Mathematicians*, Beijing, 2002.

[15] J. Demmel, M. Gu, S. Eisenstat, I. Slapničar, K. Veselić, and Z. Drmač. Computing the singular value decomposition with high relative accuracy. *Lin. Alg. Appl.*, 299(1–3), pp. 21–80, 1999.

[16] J. Demmel and Y. Hida. Accurate and efficient floating point summation. to appear in SIAM. J. Sci. Comp,; `http://www.cs.berkeley.edu/~demmel/AccurateSummation.pdf`, 2002.

[17] J. Demmel and W. Kahan. Accurate singular values of bidiagonal matrices. *SIAM J. Sci. Stat. Comput.*, 11(5), pp. 873–912, September 1990.

[18] J. Demmel and P. Koev. Necessary and sufficient conditions for accurate and efficient rational function evaluation and factorizations of rational matrices. In V. Olshevsky, editor, *Special Issue on Structured Matrices in Mathematics, Computer Science and Engineering*, volume 281 of *Contemporary Mathematics*, pages 117–145. AMS, 2001.

[19] J. Demmel and P. Koev. Accurate SVDs of polynomial Vandermonde matrices involving orthonormal polynomials. to appear in Lin. Alg. Appl.; `http://www-math.mit.edu/~plamen/files/chebvand.ps`, 2002.

[20] J. Demmel and P. Koev. Accurate SVDs of weakly diagonally dominant M-matrices. submitted to Num. Math.; `http://www-math.mit.edu/~plamen/files/mmat.ps`, 2002.

[21] J. Demmel and P. Koev. The accurate and efficient solution of a totally positive generalized Vandermonde linear system. submitted to SIAM J. Mat. Anal. Appl., `http://www-math.mit.edu/~plamen/`, 2003.

[22] J. Demmel and P. Koev. Efficient and accurate evaluation of Schur and Jack polynomials. in preparation, `http://www-math.mit.edu/~plamen/files/p-scuhr.pdf`, 2002.

[23] J. Demmel and K. Veselić. Jacobi's method is more accurate than QR. *SIAM J. Mat. Anal. Appl.*, 13(4), pp. 1204–1246, 1992.

[24] I. S. Dhillon. *A New $O(n^2)$ Algorithm for the Symmetric Tridiagonal Eigenvalue/Eigenvector Problem.* PhD thesis, University of California, Berkeley, California, May 1997.

[25] F. M. Dopico, J. M. Molera, and J. Moro. An orthogonal high relative accuracy algorithm for the symmetric eigenproblem. to appear in SIAM J. Mat. Anal. Appl., `http://www.uc3m.es/uc3m/dpto/MATEM/molera/indice.html`, 2003.

[26] F. Gantmacher and M. Krein. *Oszillationsmatrizen, Oszillationskerne, und kleine Schwingungen mechanischer Systeme.* Akademie-Verlag, Berlin, 1960.

[27] I. Gohberg, T. Kailath, and V. Olshevsky. Fast Gaussian elimination with partial pivoting for matrices with displacement structure. *Math. Comp.*, 64(212), pp. 1557–1576, 1995.

[28] N. J. Higham. Stability analysis of algorithms for solving confluent Vandermonde-like systems. *SIAM J. Mat. Anal. Appl.*, 11(1), pp. 23–41, 1990.

[29] N. J. Higham. *Accuracy and Stability of Numerical Algorithms.* SIAM, Philadelphia, PA, 1996.

[30] S. Karlin. *Total Positivity.* Stanford University Press, 1968.

[31] P. Koev. *Accurate and efficient computations with structured matrices.* PhD thesis, University of California, Berkeley, California, May 2002.

[32] P. Koev. Accurate computations with totally nonnegative matrices. in preparation, `http://math.mit.edu/~plamen/`, 2003.

[33] P. Koev. Accurate eigenvalues and SVDs of totally positive matrices. in preparation, `http://www-math.mit.edu/~plamen/files/tpeig.pdf`, 2003.

[34] H. Lewis and C. Papadimitriou. *Elements of the theory of computation.* Prentice Hall, 2nd edition edition, 1997.

[35] I. G. MacDonald. *Symmetric functions and Hall polynomials.* Oxford University Press, 2nd edition, 1995.

[36] J. J. Martínez and J. M. Peña. Fast algorithms for Björck–Pereyra type for solving Cauchy–Vandermonde linear systems. *Appl. Num. Math.*, 26(3), pp. 343–352, 1998.

[37] C. O'Cinneide. Relative-error for the LU decomposition via the GTH algorithm. *Numer. Math.*, 73, pp. 507–519, 1996.

[38] M. Pour-El and J. Richards. *Computability in Analysis and Physics.* Springer-Verlag, 1989.

[39] D. Priest. Algorithms for arbitrary precision floating point arithmetic. In P. Kornerup and D. Matula, editors, *Proceedings of the 10th Symposium on Computer Arithmetic*, pages 132–145, Grenoble, France, June 26-28 1991. IEEE Computer Society Press.

[40] Reliable Computing (a journal). Kluwer. `http://www.cs.utep.edu/interval-comp/rcjournal.html`.

[41] S. Smale. Some remarks on the foundations of numerical analysis. *SIAM Review*, 32(2), pp. 211–220, June 1990.

Invited speaker Jennifer Chayes (Microsoft, USA). photography by *Happy Medium Photo Co.*

Peter Deuflhard is Professor of Scientific Computing in the Department of Mathematics and Computer Science at the Freie Universität Berlin. He is also President of the Konrad-Zuse-Zentrum für Informationstechnik Berlin, or Zuse Institut Berlin (ZIB) in short. His present research activities cover modelling, simulation, identification and optimisation in application areas like Chemical Engineering, Medicine and Biotechnology.

After graduating with a Diploma Degree in Physics from the Technical University of Munich (TUM), Deuflhard became a Scientific Assistant in Mathematics at the Cologne University, completing his Dissertation in 1972. During 1973–78 he was a Senior Scientific Assistant in Mathematics at Munich again. After receiving his Habilitation in 1977, he took the chair in Numerical Analysis, as a Full Professor at the University of Heidelberg. In 1986 he moved to FU Berlin and founded the ZIB as a scientific research institute, serving first as Vice-President, and since 1987 as President. At the same time he is Full Professor at the Freie Universität, holding the chair of Scientific Computing. In 1994 Peter was awarded the Gerhard Damköhler Medal for fundamental contributions to chemical engineering, and was awarded Docteur *honoris causa* from Université de Genève in 2000. He was elected as a Member of the Berlin-Brandenburgische Akademie der Wissenschaften in 2001.

Chapter 6

Molecular Conformation Dynamics and Computational Drug Design*

Peter Deuflhard[†‡]
Christof Schütte[‡]

Abstract: We survey recent progress in the mathematical modelling and simulation of essential molecular dynamics. Particular emphasis is put on computational drug design wherein time-scales of milliseconds up to minutes play the dominant role. Classical long-term molecular dynamics computations, however, would run into ill-conditioned initial-value problems already after time-spans of only psec $= 10^{-12}$ sec. Therefore, in order to obtain results for times of pharmaceutical interest, a combined deterministic–stochastic model is needed.

The concept advocated in this paper is the direct identification of *metastable conformations* together with their life times and their transition patterns. It can be interpreted as a *transfer-operator* approach corresponding to some underlying hybrid Monte Carlo process, wherein short-term trajectories enter. The spatial discretization of this operator is a hard problem of its own. In order to avoid the 'curse of dimension', the construction of appropriate spatial boxes requires careful consideration. Once this operator has been discretised a stochastic matrix arises. This matrix is then treated by *Perron cluster analysis*, a recently developed cluster analysis method involving the numerical solution of an eigenproblem for a cluster of eigenvalues called the Perron cluster. As a biomolecular example we present a rather recent SARS protease inhibitor.

---

[1]supported by DFG Research Center "Mathematics for Key Technologies" in Berlin
[†]Zuse Institute Berlin (ZIB)
[‡]Free University of Berlin, Department of Mathematics and Computer Science

Contents

Introduction

In recent years, *prion* diseases, like the mad cow disease, but also *viral* diseases such as HIV or SARS, have attracted much public and political interest. Whenever any new such disease shows up, there is a highly competitive race for new drugs against them. This race typically starts with computers.

For quite a while, algorithms from discrete mathematics or computer science have already played a publicly visible role; for example, in the decoding of the human genome. These approaches primarily aim at the geometry of the molecules under consideration; i.e., on the secondary or tertiary structure. A real understanding of *biological function*, however, requires detailed knowledge about the biomolecular *dynamics*.

In dynamics, the situation is characterised by the fact that real times of pharmaceutical interest are in the region of milliseconds up to minutes, whereas simulation times are presently in the region of nsec $= 10^{-9}$ sec with timesteps of less than $5\,\text{fsec} = 5 \times 10^{-15}$ sec. The established 'molecular dynamics' approach (usually just called MD) realises numerical integration of the *Hamiltonian dynamics* of the molecular systems — often limited by the available computer power. This kind of approach, however, has an even stricter mathematical limitation: the Hamiltonian trajectories to be computed are known to be asymptotically chaotic. Consequently, traditional long-term trajectory simulations may, at best, give information about time averages, which, under some ergodic hypothesis, are equivalent to statistical ensemble averages.

As a result of this insight, any investigation of the dynamics of molecular systems for time-scales of interest in drug design will require a rather different mathematical approach. In the past few years, the present authors and their joint research group have created such a different approach based on concepts of nonlinear dynamics; for early papers see, e.g., [9, 40, 39, 12]. This approach, now called *conformation dynamics*, has already been surveyed in articles like [7, 41]. The present paper updates the state of the art in this fast-moving research topic.

1 Transfer Operators and Metastable Conformations

Hamiltonian dynamics. We assume that the dynamics of the molecular system under consideration is characterised by a *separable* Hamilton function

$$H(q,p) = \tfrac{1}{2}\, p^{\mathsf{T}} M^{-1} p + V(q) \, ,$$

where the first term, the kinetic energy, only depends on the generalised momenta variables p, while the second term, the potential energy, only depends on the position variables q. From given H, the Hamiltonian differential equations for N atoms are defined as

$$q_i' = \frac{\partial H}{\partial p_i} \, , \quad p_i' = -\frac{\partial H}{\partial q_i} \, , \qquad i = 1, \ldots, N \, . \tag{1.1}$$

Of course, the quality of any molecular dynamics calculation is strongly dependent on the quality of the available potential data (we mostly use MMFF [27]). Details of the numerical integration of these ODEs are omitted here; they can be found, e.g., in the textbooks [37, 26] or, especially for the aspects of importance here, in Sections 1.2 and 4.3.4 of the textbook [8].

The unique solution of this initial-value problem can be written in terms of the flow Φ^t as

$$x(t) = \big(q(t)\,, p(t)\big) = \Phi^t x_0 \, .$$

The sensitivity of the solution, i.e. the solution perturbation $\delta x(t)$ versus the initial perturbation δx_0, is characterised by the *condition number* κ. Following [8, Sect. 3.1.2], this quantity is defined (in first-order perturbation analysis) as

$$\|\delta x(t)\| \mathrel{\dot{\le}} \kappa(t)\, \|\delta x_0\| \, , \quad \kappa(t) = \left\| \frac{\partial \Phi^t}{\partial x_0} \right\| .$$

As already discovered by H. Poincaré, Hamiltonian systems can be *chaotic.* In Numerical Analysis, we want to know the critical finite time, after which some kind of chaoticity occurs. In almost all molecular dynamics problems the condition number seems to grow exponentially such that almost all information concerning the initial state is lost after critical times t_{crit} no longer than a few psec. That is why the traditional MD with numerical long-term integration can only be interpreted as computing ensemble averages via time averages in the sense of the ergodic theorem — which need not hold in all cases.

On this basis, we are led to the following conclusion:

> *Instead of the* point concept *of classical mechanics based on deterministic trajectories, which is only able to model short-term dynamics, we need to derive some* set concept *including stochastic elements to model long-term dynamics.*

Smoluchowski or Langevin dynamics. In the literature, several stochastic dynamical systems are discussed as alternative models for certain aspects of molecular motion in a heat bath. The most prominent of these are the Langevin or Smoluchowski dynamics. For medium to large molecular systems these models are believed to describe the effective dynamical behavior well enough. The Smoluchowski system models the dynamics in the position space only. It defines a *reversible* Markov process by means of the stochastic differential equation

$$\gamma \dot{q} = -\nabla_q V(q) + \sigma \dot{W}_t . \tag{1.2}$$

Here $\gamma > 0$ denotes some friction constant and $F_{\text{ext}} = \sigma \dot{W}_t$ is the external forcing term given by a $3N$-dimensional Brownian motion W_t. The external stochastic force is assumed to model the influence of the heat bath surrounding the molecular system. The stochastic differential equation (1.2) defines a continuous-time Markov process Q_t on the state space Ω with invariant probability measure [36]

$$\mathcal{Q}(\,\mathrm{d}q) \propto \exp\bigl(-\beta V(q)\bigr)\,\mathrm{d}q\ .$$

There is a long history of using it as a simple toolkit for investigation of dynamical behavior in complicated energy landscapes [4]. We will herein use it for the same purpose; i.e., we will concentrate on the stochastic reformulation of Hamiltonian motion (see next section) but use Smoluchowski dynamics for simplified illustration and comparison.

Biomolecular conformations and metastable sets. Today, the effective dynamics of many biomolecules is understood to be governed by statistically-rare transitions between so-called *conformations* of the biomolecule (cf. [49]). In a conformation, the large-scale geometric structure of the molecule is understood to be conserved, whereas on smaller scales the system may well rotate, oscillate or fluctuate. Furthermore, transitions between conformations are rare events or, in other words, a typical trajectory of a molecular system stays for long periods of time within the conformation, while exits are long-term events. Hence, the term conformation includes both *geometric* and *dynamical* aspects. From the geometrical point of view, conformations are understood to represent all molecules with the same large-scale geometric structure and may thus be identified with a subset of the state space. From the dynamical point of view, a conformation typically persists for long periods of time (compared to the fastest molecular motions) such that the associated subset of the state space is *metastable* and the resulting *macroscopic dynamical behavior* can be described as a flipping process between the metastable subsets. Consequently, it is of utmost interest to decompose the state space of the molecular motion into certain metastable sets, evaluate the transition probabilities between them and perhaps learn about the corresponding transition pathways.

The standard biophysical explanation for the existence of conformations is as follows: The *free energy landscape* of a molecular system, say a protein or peptide, decomposes into particularly deep wells each containing huge numbers of local minima. These wells are separated from each other by relatively large barriers—as measured on the scale of the thermal energy ($\sim T$: temperature)—and represent different metastable conformations. The hierarchy of barrier heights induces a hierarchy of conformations [16, 20, 19]. The corresponding hierarchy of time-scales observed for conformational transitions seems to confirm the biophysical explanation for the existence of conformations [35]. However, this concept does *not* (at least not directly) refer to dynamical aspects but describes conformation transitions in terms of a thermodynamic quantity, the *free energy*. In Section 2 below we will show that the Smoluchowski dynamics is an ideal setting to discuss similarities and differences between this thermodynamic concept and the dynamical concepts to be presented herein.

1.1 Transfer operators

The just-mentioned set concept can be realised by virtue of some stochastic transfer operator (or Markov operator), which is discussed here in necessary detail.

Perron–Frobenius operator. Starting point for the new approach was the pioneering work of M. Dellnitz and co-workers [6] on the numerical approximation of invariant measures $\bar{\mu}$ and their corresponding invariant sets $\bar{B}$ via the (unitary) Perron–Frobenius operator $\mathbf{U}$. In terms of this operator, $\bar{\mu}$ and $\bar{B}$ are characterised by the eigenvalue problem

$$\mathbf{U}\bar{\mu} = \bar{\mu}\,, \quad \Phi^{-t}(\bar{B}) \subset \bar{B}\,, \quad \forall t \geq 0\,, \tag{1.3}$$

for the Perron eigenvalue $\lambda = 1$. Moreover, eigenvalues $\lambda \neq 1$ close to the Perron eigenvalue seemed to have an interpretation in terms of *almost-invariant sets* of the dynamical system.

The success of that approach was intimately linked to dynamical systems that asymptotically collapse to some dynamics on a low-dimensional manifold. This is definitely not the case in Hamiltonian dynamics, so that a generalisation to molecular dynamics is all but trivial. A first attempt in this direction has been published in [9]. However, the subdivision technique applied there caused some *curse of dimension* that restricted the applicability of the method to a domain far away from realistic molecules.

Self-adjoint transfer operator. In [40, 39] a new stochastic operator $\mathbf{T}$ has been constructed, which embeds $\mathbf{U}$ into a canonical distribution

$$f_0(q,p) = \frac{1}{Z}\exp\left[-\beta\left(\tfrac{1}{2}p^{\mathsf{T}}M^{-1}p + V(q)\right)\right],$$

with Z as normalisation factor and β proportional to the inverse temperature. For separable Hamiltonian H this distribution may be factorised according to

$$f_0 = \mathcal{P}\mathcal{Q}\,, \quad Z = Z_p Z_q\,, \qquad \int \mathcal{P}(p)\,\mathrm{d}p = \int \mathcal{Q}(q)\,\mathrm{d}q = 1\,, \tag{1.4}$$

where

$$\mathcal{P}(p) = \frac{1}{Z_p}\exp\left(-\tfrac{1}{2}\beta p^{\mathsf{T}}M^{-1}p\right), \quad \mathcal{Q}(q) = \frac{1}{Z_q}\exp\left(-\beta V(q)\right).$$

At this point recall that metastable conformations are understood to be objects in *position space* $q \in \Omega$ rather than in the whole phase space Γ. Let $A, B \subset \Omega$ be subsets in position space and define cylinders

$$\Gamma(A) = \{(q,p) : q \in A\} .$$

The required transfer operator may then be constructed integrating the Perron–Frobenius operator $\mathbf{U}$ over the cylinders $\Gamma(\cdot)$ — thus achieving an operator $\mathbf{T}^\tau$ that acts on functions in position space:

$$\mathbf{T}^\tau u(q) = \int_{\mathbb{R}^d} u\big(\Pi_q \Phi^{-\tau}(q,\xi)\big)\, \mathcal{P}(\xi)\, \mathrm{d}\xi , \tag{1.5}$$

where Π denotes the projection $\Pi(q,p) = q$ onto the position space. In the sequel we will often omit the superindex τ, if the time-scale τ that has been chosen is clear and does not change.

As has been shown in [39], $\mathbf{T}^\tau$ can be interpreted as the transfer operator associated with the Markov chain

$$q_{k+1} = \Pi\, \Phi^\tau(q_k, p_k) , \qquad p_k : \ \mathcal{P}\text{-distributed} . \tag{1.6}$$

often called *Hamiltonian system with randomised momenta.* For a schematic representation, see Fig. 1. This Markov chain combines a short-term deterministic model characterised by the flow Φ^τ with a statistical model characterised by the $\mathcal{P}$-distribution, the momentum part of the canonical distribution, which is just a Gaussian distribution due to the quadratic kinetic energy; see (1.4). For a discussion of the physical meaning of this stochastic model of the dynamics visit [42].

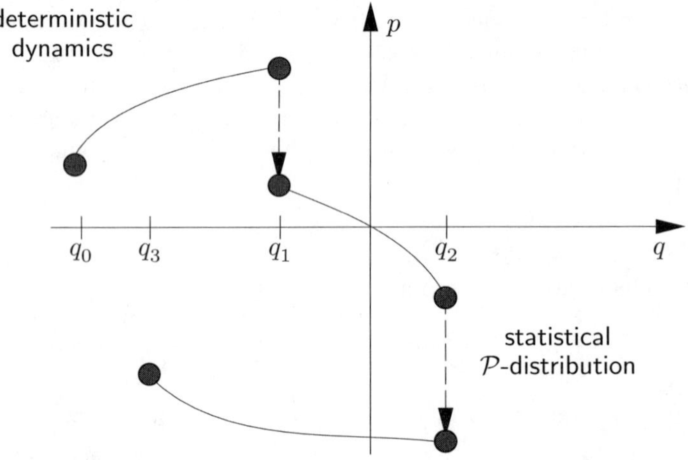

Figure 1. *Markov chain (1.6).*

The operator $\mathbf{T}^\tau$ is defined over the weighted spaces

$$L^r_{\mathcal{Q}}(\Omega) = \big\{u : \Omega \to \mathcal{C}, \ \int_\Omega \big|u(q)\big|^r \mathcal{Q}\ \mathrm{d}q < \infty\big\}, \quad r = 1, 2 .$$

The Hilbert space $L^2_{\mathcal{Q}}(\Omega)$ is naturally associated with the weighted inner product

$$\langle u, v\rangle_{\mathcal{Q}} = \int\limits_{\Omega} u(q)v(q)\mathcal{Q}(q)\,\mathrm{d}q\,. \tag{1.7}$$

Among the properties of $\mathbf{T}^\tau$ for all τ we mention (from [39]):

- $\mathbf{T}^\tau$ is a Markov operator on $L^1_{\mathcal{Q}}(\Omega)$.
- $\mathbf{T}^\tau$ is *self-adjoint* in $L^2_{\mathcal{Q}}(\Omega)$.

Hence, its spectrum satisfies $\sigma(\mathbf{T}^\tau) \subset [-1, 1]$. Moreover, under certain quite general conditions the existence of metastable sets is deeply related to a cluster of eigenvalues close to the Perron eigenvalue $\lambda = 1$, called the *Perron cluster*, which is well-separated from the remaining (continuous) part of the spectrum (see Theorem 1 for details). Discretisation of this operator (to be studied in Section 4 below) generates a stochastic sparse matrix T, which inherits the self-adjointness of the operator as symmetry with respect to a discrete analog of the weighted inner product $\langle\cdot,\cdot\rangle_{\mathcal{Q}}$.

With these preparations, we are ready to express all relevant information about the dynamical system. Let $\chi(A)$ denote the characteristic function of A, a set function that is 1 inside A and 0 outside. Then we obtain:

- The probability for the dynamical system to *be* within A is

$$\pi(A) = \int\limits_{\Gamma(A)} f_0(p,q)\,\mathrm{d}q\,\mathrm{d}p = \int\limits_{A} \mathcal{Q}(q)\,\mathrm{d}q = \langle\chi_A\,,\chi_A\rangle_{\mathcal{Q}}\,. \tag{1.8}$$

- The conditional probability for the system, once it is in A, to *move* from A to B during time τ can be defined by virtue of $\mathbf{T}^\tau$ as

$$w(A,B,\tau) = \frac{\langle\chi_A\,,\mathbf{T}^\tau\chi_B\rangle_{\mathcal{Q}}}{\langle\chi_A\,,\chi_A\rangle_{\mathcal{Q}}}\,. \tag{1.9}$$

- The probability for the system, once it is in A, to *stay* in A during time τ (more exactly: to be found in A at time $t=\tau$ after being in A at time $t=0$) comes out as

$$w(A,A,\tau) = \frac{\langle\chi_A\,,\mathbf{T}^\tau\chi_A\rangle_{\mathcal{Q}}}{\langle\chi_A\,,\chi_A\rangle_{\mathcal{Q}}}\,. \tag{1.10}$$

Given open sets A and B, we could compute these probabilities by means of long-term iteration of the Markov chain (1.6) associated with $\mathbf{T}^\tau$. Any realisation would yield a sequence of positions $\{q_k\}$ that can be proved to be distributed according to $\mathcal{Q}$ asymptotically [39]. The relative frequency of transitions from A to B in this sequence asymptotically approximates $w(A,B,\tau)$ (see Section 4 for algorithmic consequences and difficulties). In addition we get a *sequence of τ-sub-trajectories* of the Hamiltonian system under consideration. If this sequence is long enough, it will explore the state space entirely, and contain all necessary information about the dynamical features of the system.

Transfer operator for Smoluchowski dynamics. The transfer operator describes the evolution of probability densities under the dynamics in question. For the Smoluchowski system (1.2) the evolution of probability densities f (w.r.t. the Lebesgue measure) is governed by the Fokker–Planck equation

$$\partial_t f = \Big(\frac{\sigma^2}{2\gamma^2} \Delta_q + \frac{1}{\gamma} (\nabla_q V(q) \cdot \nabla_q \; + D^2 V(q)) \Big) u \, .$$

Upon introducing the probability distribution $v = u/\mathcal{Q}$, this evolution equation reads

$$\partial_t v \; = \; \mathbf{A}_{\mathrm{Smo}} \, v \; = \; \Big(\frac{\sigma^2}{2\gamma^2} \Delta_q - \frac{1}{\gamma} (\nabla_q V(q) \cdot \nabla_q) \Big) v \, .$$

Thus, the associated transfer operators $\mathbf{T}^t_{\mathrm{Smo}}$ form a semigroup. For twice continuously differentiable $u \in L^r_{\mathcal{Q}}(\Omega)$ with $1 \le r < \infty$, this semigroup admits $\mathbf{A}_{\mathrm{Smo}}$ as a strong generator such that in this case

$$\mathbf{T}^t_{\mathrm{Smo}} = \exp\big(t \mathbf{A}_{\mathrm{Smo}}\big) \, .$$

For details on $\mathbf{A}_{\mathrm{Smo}}$ see the theory of Fokker–Planck equations and Kolmogoroff forward and backward equations [36, 43, 28].

Hence the Smoluchowski case gives us the opportunity to study the relation between dominant eigenvectors of the transfer operator and metastable sets by means of partial differential operators. The fact that the Smoluchowski system has at least some relation to the Hamiltonian case is reflected in the following relation between the transfer operator $\mathbf{T}^\tau$ of the Hamiltonian system with randomised momenta and the Smoluchowski generator:

$$\mathbf{T}^\tau = \mathrm{Id} + \tau^2 \mathbf{A}_{\mathrm{Smo}} + \mathrm{O}(\tau^4) \, .$$

For $u \in L^2_{\mathcal{Q}}(\Omega)$ the reversibility of the Smoluchowski dynamics implies that $\mathbf{A}_{\mathrm{Smo}}$ is self-adjoint.

1.2 Dominant Spectra and Metastability

There are several recent articles on the relation between metastability and dominant eigenmodes of the transfer operator associated with the considered dynamical system [42, 30, 28, 6, 40]. Within these approaches, metastability is a *set-wise* notion and conceptually defined in the following way: some dynamical system is said to exhibit metastability or to have a *metastable decomposition*, if its state space can be decomposed into a finite (hopefully small) number of disjoint sets such that the *probability of exit* from each of these sets is extremely small [42, 6]. There are basically two different concepts of probability of exit:

(a) the probability of exit from a set is defined via an ensemble of systems and measures the fraction of systems that exit from the set during some fixed time interval [42, 40];

(b) in case of a stochastic process the probability of exit is measured from the distribution of exit times from the set. That is, the probability of exit is the smaller the larger is the expected exit time [3]; or, equivalently, the slower is the decay of the distribution of exit times [30].

However, both concepts (a) and (b) are related to the dominant eigenvectors of the transfer operator. Accordingly, the basic insight of the transfer-operator approach to metastability can be formulated as follows [42]:

Identification of metastable decompositions. *Metastable decompositions can be detected via the discrete eigenvalues of the transfer operator $\mathbf{T}^\tau$ close to its maximal eigenvalue $\lambda = 1$; they can be identified by exploiting structural properties of the corresponding eigenfunctions. In doing so, the number of sets in the metastable decomposition is equal to the number of eigenvalues close to 1, including $\lambda = 1$ and counting multiplicity.*

We will later learn about the identification algorithm constructed on this basis. Furthermore, we will present illustrative examples in Section 2. In the final paragraphs of this section, however, we will present one of several mathematical statements supporting this idea. To this end, recall the formula for the probability to remain within some set A during time-span τ:

$$w(A, A, \tau) = \frac{\langle \chi_A \, , \mathbf{T}^\tau \chi_A \rangle_Q}{\langle \chi_A \, , \chi_A \rangle_Q} \, .$$

The metastability of a set A may be measured by $w(A, A, \tau)$.

Definition: Metastability of a decomposition. For an arbitrary decomposition $\mathcal{D} = \{A_1, \ldots, A_m\}$ of the state space into m disjoint sets A_k, we define

$$w_m(\tau) = \sum_{i=1}^{m} w(A_i, A_i, \tau) \tag{1.11}$$

as the corresponding metastability.

The following crucial result is due to [31]; a specialised version for two subsets has been published by Huisinga in his thesis [28].

Theorem 1. *Let $T^\tau : L^2_Q(\Omega) \to L^2_Q(\Omega)$ denote a reversible transfer operator whose essential spectral radius is strictly less than 1 and for which the eigenvalue $\lambda = 1$ is simple. Then $\mathbf{T}^\tau$ is self-adjoint and the spectrum has the form*

$$\sigma(\mathbf{T}^\tau) \subset [a, b] \cup \{\lambda_m\} \cup \ldots \cup \{\lambda_2\} \cup \{1\} \, ,$$

with $-1 < a \le b < \lambda_m \le \ldots \le \lambda_1 = 1$ and isolated, not necessarily simple, eigenvalues of finite multiplicity that are counted according to multiplicity. Denote by $v_m, \ldots, v_1$ the corresponding eigenfunctions, normalised to $\|v_k\|_{L^2_Q(\Omega)} = 1$. Let Q

be the orthogonal projection of $L^2_Q(\Omega)$ *onto* $\mathrm{span}\{\chi_{A_1},\ldots,\chi_{A_m}\}$. *Then the following bounds hold:*

$$1+\kappa_2\lambda_2+\ldots+\kappa_m\lambda_m+c \;\le\; w_m(\tau) \;\le\; 1+\lambda_2+\ldots+\lambda_m\,,$$

where $\kappa_j = \left\|Qv_j\right\|^2_{L^2_Q(\Omega)} \le 1\,, j=1,\ldots,m\,,$ *and* $c = a\,(1-\kappa_2)\ldots(1-\kappa_m)$.

This theorem obviously holds for the transfer operator of the Hamiltonian system with randomised momenta as well as for the one related to Smoluchowski dynamics. Whenever the dominant eigenfunctions $v_2,\ldots,v_m$ are almost constant on the metastable subsets $A_1,\ldots,A_m$ — which then implies that $\kappa_j \approx 1$ and thus $c \approx 0$ — then the above lower and upper bounds are close. Moreover, Huisinga et al. [31] have even shown that both bounds are sharp and asymptotically exact. The idea to exploit locally almost-constant levels of the dominant eigenmodes lies at the heart of the algorithm to be presented below in detail in Section 3.

2 A Complete Picture in a Simplified Setting

In principle, the transfer-operator approach as presented so far allows us to identify an almost optimal metastable decomposition of the state space. In terms of the biochemical background this gives us the main conformations of the molecular system under consideration. However, this solves "only" one of the four most important biophysical problems. One may want to:

(a) identify the *dominant conformations*;

(b) characterise the *geometric and dynamical flexibility* of the molecule within one of its main conformations;

(c) estimate the *transition probabilities* between conformations or the *exit rates* from a single one; and

(d) characterise the *transition regions* and *pathways* on which the transitions between conformations will occur most probably.

The characterisation of the internal flexibility and (roughly) of the transition regions are automatic by-products of the algorithmic realisation of the transfer-operator approach via sequences of sub-trajectories exploring state space (see Section 4 below). Furthermore this algorithmic realisation will permit the direct computation of transition probabilities w.r.t. some prescribed time-span τ, by means of formula (1.9).

In order to illustrate the relation between the different concepts (transfer-operator approach, exit rates, transition pathways) we now work out details at a rather simple test case from *Smoluchoswki dynamics*.

Test system (2D). We consider the two-dimensional system given by the potential

$$\begin{aligned} V(x,y) = {}& 3\exp\big(-x^2-(y-\tfrac{1}{3})^2\big) \;-\; 3\exp\big(-x^2-(y-\tfrac{5}{3})^2\big) \\ & -5\exp\big(-(x-1)^2-y^2\big) \;-\; 5\exp\big(-(x+1)^2-y^2\big)\,. \end{aligned}$$

The potential is illustrated in Fig. 2. We observe that there are two equally important deep wells with minima at $(x, y) = (1, 0)$ and $(x, y) = (-1, 0)$, and a not so important one around $(x, y) = (0, \frac{5}{3})$. The barrier between the two dominant wells is substantially higher than the barrier between each of the dominant ones and the less important wells. The inverse temperature $\beta = 2\gamma/\sigma^2$ is set to $\beta = 3$ (with $\gamma = 1$) such that crossing the barriers in the potential certainly will be a rare event.

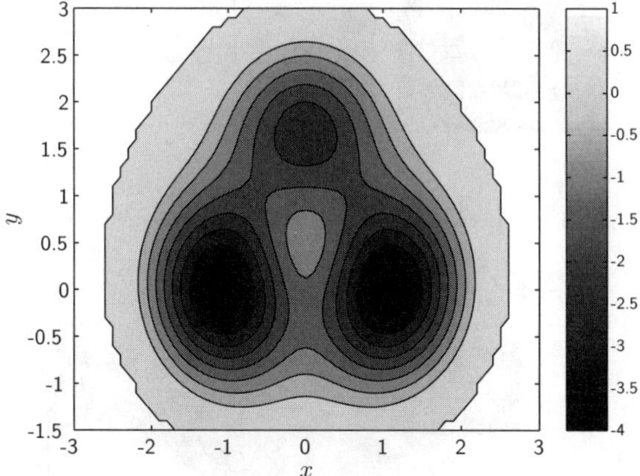

Figure 2. Smoluchowski test problem: *Contour lines of potential V. Regions between the contour lines are shaded according to average value of potential.*

Metastable decomposition. The eigenvalue problem of the generator $\mathbf{A}_{\text{Smo}}$ of the transfer operator $\mathbf{T}^t_{\text{Smo}}$ can be solved numerically by means of finite-element eigenvalue solvers for elliptic problems. This leads to the following numerical results for the dominant eigenvalues of $\mathbf{A}_{\text{Smo}}$ in $L^2_Q(\Omega)$:

λ_1	λ_2	λ_3	λ_4	...
0.000	−0.002	−0.144	−2.330	...

The eigenfunction associated with $\lambda_1 = 0$ obviously is the constant function. The eigenfunctions associated with λ_2 and λ_3 are shown in Fig. 3.

From Fig. 3 we observe: (a) the three metastable sets given by the three wells in the potential show up as regions of almost constancy of the three eigenfunctions; (b) the less important well (being coded into the third eigenfunction with a significantly larger eigenvalue) obviously exhibits less metastabilty than the other two.

In Section 3 below, we will present an algorithm for the identification of metastable decompositions as shown in Fig. 4.

Exit times and exit rates. The exit rate from a set A is defined as the decay rate of the exponential decay of the distribution of exit times from the set [30]. If A is

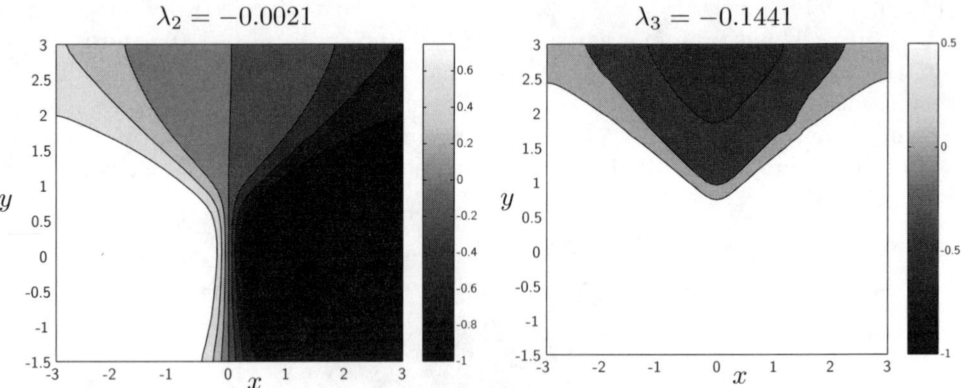

Figure 3. Smoluchowski test problem: *second and third eigenfunction. Illustration via contour lines as explained in Fig. 2.*

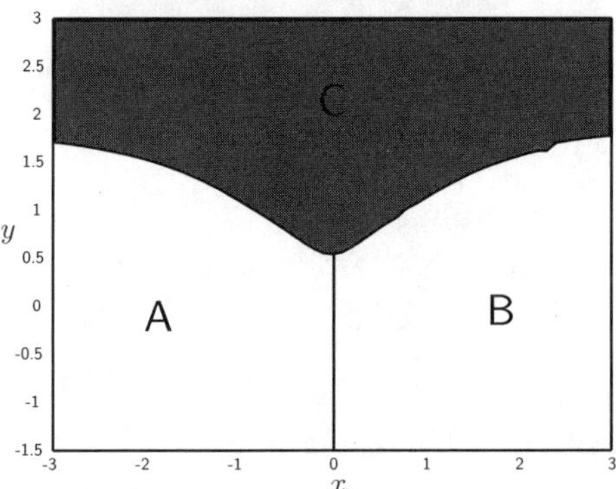

Figure 4. *State-space decomposition into sets A, B, C by identification algorithm as presented in Section 3.*

the open set defined by one of the strictly positive or negative components of an eigenfunction of $\mathbf{A}_{\text{Smo}}$, then the exit rate can be shown to equal the modulus of the eigenvalue associated with this eigenfunction [30]. If, for example, A is the left of the main wells of our example, the exit rate for $\beta = 3$ is given by $|\lambda_2| = 0.002$; i.e., most exit times will be in the hundreds of units. If β is asymptotically large, the expected exit time $\bar{\tau}$ is known to scale like

$$\bar{\tau} = C \exp(\beta \Delta V),$$

where ΔV denotes the smallest energy barrier via which the exit is possible. However, the preconstant C increases with the "narrowness" of the saddle-point region through which the exit occurs. Situations like in our example are of utmost interest: there are two such regions, one that is a little bit wider but whose energy barrier is a little higher than that of the other one (which is more narrow). If β is not asymptotically large exits will occur in both regions. For real-life applications this is the crucial problem of all algorithms designed to identify transition regions, pathways, or states: one always has to ask whether all important regions have been explored.

Transition pathways. Transition state theory tells us that the transition pathways between two (disjoint) wells W_1 and W_2 can be computed from some reaction coordinate $\xi : X \to \mathbb{R}$ where X may denote the important portion of the state space in which the wells $W_1 \subset X$ and $W_2 \subset X$ are dominant. In the case of our test system, W_1, and W_2 should be left and right main wells of the potential-energy landscape; i.e., the core sets of the metastable sets A and B of Fig. 4. In general, ξ is given by the following boundary-value problem [44]:

$$\begin{aligned} &\mathbf{A}_{\mathrm{Smo}}\,\xi = 0\,, \qquad \text{in } X \setminus \{W_1 \cup W_2\}\,, \\ &\xi|_{\partial W_1} = 0\,, \qquad \xi|_{\partial W_2} = 1\,, \qquad \partial_n \xi|_{\partial X} = 0\,. \end{aligned} \tag{2.1}$$

Under certain circumstances the solution is closely related to the dominant eigenvalues. For example, let μ denote the probability measure generated by the invariant density $\exp(-\beta V)$, and suppose that almost all weight is concentrated in W_1 and W_2; i.e., $\mu(W_1 \cup W_2) \approx 1$. Moreover, let there be only one negative eigenvalue λ_2 of $\mathbf{A}_{\mathrm{Smo}}$ very close to $\lambda_1 = 0$. Then we have approximately

$$\xi \approx \mu(W_1)\chi_X + \sqrt{\mu(W_1)\mu(W_2)}\, u_2\,,$$

where u_2 denotes the eigenfunction associated with λ_2.

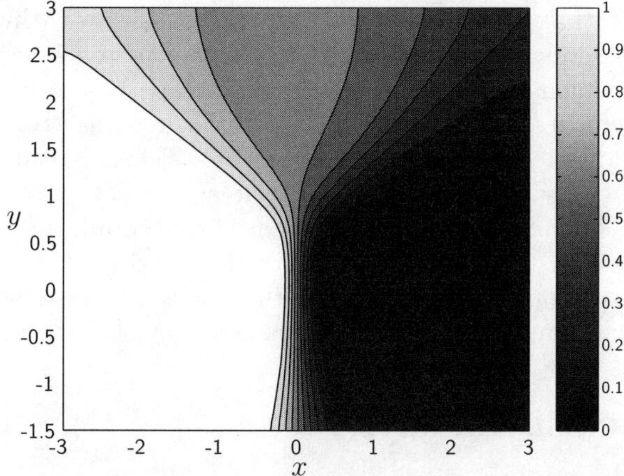

Figure 5. *Reaction coordinate ξ for the system under consideration with W_1 being the left and W_2 the right main well. Illustration via contour lines in the same sense as explained in Fig. 2.*

This is obviously the case in our test system, as we can see in Fig. 5 the solution of (2.1) for the case where W_1 and W_2 are identical to the left and right main wells, respectively. The figure exhibits the *level-set* $\xi = \xi_0$ for given ξ_0 of ξ. Transition-state theory tells us that the transition pathways intersect the level-sets of ξ perpendicularly [44].

From all possible transition paths only those are of importance that intersect the level-sets where the restricted invariant distributions

$$\nu\big|_{\xi_0} = \frac{1}{Z(\xi_0)} \exp(-\beta V)\big|_{\xi=\xi_0}, \quad Z(\xi_0) = \int \delta(\xi - \xi_0) \exp\bigl(-\beta V(x,y)\bigr)\, \mathrm{d}x\, \mathrm{d}y$$

are large enough.

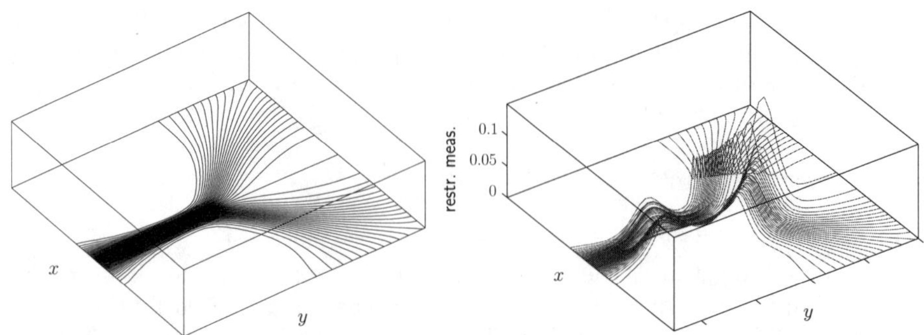

Figure 6. *Level-sets of ξ (left) and restricted invariant distribution $\nu|_{\xi_0}$ (right) on the level-sets $\xi = \xi_0$ for some $\xi_0 = 0.1, \ldots, 0.9$ for the system under consideration.*

Fig. 6 exhibits some of these restricted invariant distributions together with the level-sets of the reaction coordinate ξ for a transition from the left to the right main well of our test system. We observed that there are at least two different transition regions, that contain different optional transition pathways. In situations like this the usual concept of *free-energy landscapes* is not general enough; e.g., it is not clear over which variables the energy landscape has to be averaged in order to compute a useful free energy for both transition regions. However, it should be obvious that the identification of transition regions is closely related to the dominant eigenmodes of the transfer operator, and that a complete picture of the effective dynamical effects of the systems has to be based on the information coded in these dominant eigenmodes. A very promising direction of work that combines aspects of the "global" transfer-operator approach with a "localised", so-called *string method* [15], for the direct computation of transition pathways is presented in [44].

3 Perron cluster analysis

Suppose we have already discretised the above transfer operator $\mathbf{T}$ — a topic postponed to the subsequent Section 4, since it requires techniques to be presented first. Then, in order to identify m almost-invariant sets corresponding to m metastable chemical conformations, we need only deal with a stochastic (generalised symmetric) matrix T of dimension N. This is a problem of *cluster analysis* where, in addition, m is unknown in advance.

Comparable to (1.3), we start from the eigenvalue problem

$$\pi^{\mathsf{T}} T = \pi^{\mathsf{T}}, \quad Te = e, \qquad \pi^{\mathsf{T}} e = 1, \tag{3.1}$$

where the left eigenvector $\pi^{\mathsf{T}} = (\pi_1, \ldots, \pi_N)$ represents the discrete invariant measure and the right eigenvector $e^{\mathsf{T}} = (1, \ldots, 1)$ the characteristic function of the discrete invariant set; each corresponding to the Perron eigenvalue $\lambda_1 = 1$. The basic approach to be described requires an analysis of the *Perron cluster* of eigenvalues

$$\lambda_1 = 1\,, \quad \lambda_2 \approx 1\,, \quad \ldots\,, \quad \lambda_m \approx 1\,,$$

and their corresponding eigenvectors $V_m = [v_1, \ldots, v_m]$ with $v_1 = e$. For given $u, v \in \mathbb{R}^N$ we will use the special inner product and norm

$$\langle u\,, v\rangle_\pi = \sum_{l=1}^{N} u_l \pi_l v_l = u^{\mathsf{T}} D^2 v\,, \quad \|v\|_\pi = \langle v\,, v\rangle_\pi^{\frac{1}{2}}\,, \tag{3.2}$$

where $D = \operatorname{diag}\big(\sqrt{\pi_1}, \ldots \sqrt{\pi_N}\big)$ is a diagonal scaling matrix. Obviously, (3.2) is the discrete analog of the continuous inner product (1.7). Any reversible stochastic matrix T is symmetric under this inner product.

Algorithm PCCA. The first algorithm to tackle this problem has been the PCCA method (abbreviated from: **P**erron **C**luster **C**luster **A**nalysis), as worked out in detail in [12]; for a rather elementary introduction see also Section 5.5 of the latest edition of the undergraduate textbook [11]. We will also sketch the more robust variant called PCCA+, originally suggested by M. Weber [46, 47] and further improved by P. Deuflhard and M. Weber [13].

Uncoupled Markov chains. Let $\mathcal{S} = \{1, 2, \ldots, N\}$ denote the total index-set decomposed as

$$\mathcal{S} = \mathcal{S}_1 \oplus \cdots \oplus \mathcal{S}_m$$

into m disjoint index subsets, which represent m uncoupled Markov chains, each of which is running "infinitely long" within the corresponding subset. Then the total transition matrix T is strictly block-diagonal with block submatrices $\{T_1, \ldots, T_m\}$; see e.g., [33]. Each of these submatrices is stochastic and gives rise to a single Perron eigenvalue $\lambda(T_i) = 1$, $i = 1, \ldots, m$. Let the submatrices be primitive. Then, due to the Perron–Frobenius theorem, each block T_i possesses a unique right eigenvector $e_{\mathcal{S}_i} = (1, \ldots, 1)^{\mathsf{T}}$ of length $\dim(T_i)$ having unit entries over the index subset $\mathcal{S}_i$. Therefore, in terms of the total transition matrix T, the eigenvalue $\lambda = 1$ has multiplicity m and the corresponding eigenspace is spanned by the vectors

$$\chi_i = (0, \ldots, 0, e_{\mathcal{S}_i}^{\mathsf{T}}, 0, \ldots, 0)^{\mathsf{T}}\,, \qquad i = 1, \ldots, m\,.$$

Our notation deliberately emphasises that these eigenvectors can be interpreted as *characteristic functions* of the invariant index subsets (see Fig. 7, left).

In general, any Perron eigenbasis $V_m = \{v_1, \ldots, v_m\}$ can be written as a linear combination of the characteristic functions $\chi = [\chi_1, \ldots, \chi_m]$ such that

$$\chi = V_m \mathcal{A}\,, \quad V_m = \chi \mathcal{A}^{-1}, \tag{3.3}$$

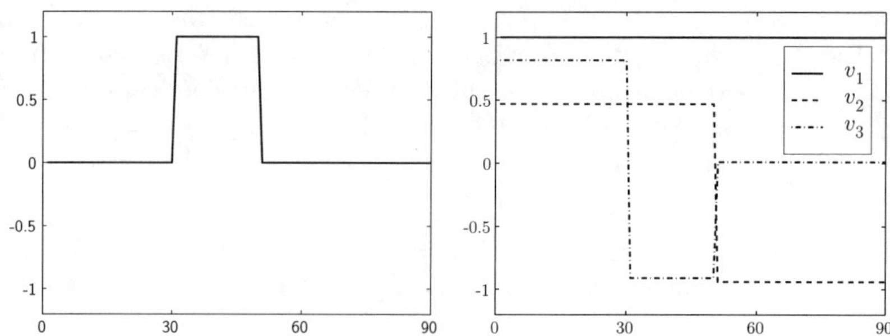

Figure 7. *Uncoupled Markov chain over* $m = 3$ *disjoint index subsets. The state space* $\mathcal{S} = \{s_1, \ldots, s_{90}\}$ *divides into the index subsets* $\mathcal{S}_1 = \{s_1, \ldots, s_{29}\}$, $\mathcal{S}_2 = \{s_{30}, \ldots, s_{49}\}$, *and* $\mathcal{S}_3 = \{s_{50}, \ldots, s_{90}\}$. Left: *Characteristic function* $\chi_2 = e^T \mathcal{S}_2$.
Right: *Perron eigenbasis* $V_3 = \{v_1, v_2, v_3\}$ *corresponding to 3-fold Perron eigenvalue* $\lambda = 1$.

wherein the (m, m)-matrix $\mathcal{A} = (\alpha_{ij})$ is nonsingular (due to $\dim\ker(\mathcal{A}) = 0$) so that $\mathcal{A}^{-1} = (a_{ij})$ exists. In PCCA, each subset $\mathcal{S}_i$ for $i = 1, \ldots, m$ is identified by some component-wise sign structure of the eigenvectors V_m, using the three values $\{+, 0, -\}$ for the sign function; compare with [12].

Uncoupled Markov chains. Suppose now we have m nearly uncoupled Markov chains, each of which is staying "for a long time" in one of the conformations i. For the transition probabilities (1.9) and (1.10) this means that

$$w(i, i, \tau) = 1 - \mathrm{O}(\epsilon), \qquad w(i, j, \tau) = \mathrm{O}(\epsilon), \quad i \neq j, \tag{3.4}$$

in terms of some perturbation parameter ϵ not further specified here. In this case the transition matrix T is (after some unknown permutation) block-diagonally *dominant*. As a perturbation of the m-fold Perron root in the uncoupled case $\epsilon = 0$, the *Perron cluster*

$$\lambda_1 = 1, \quad \lambda_i = 1 - \mathrm{O}(\epsilon), \quad i = 2, \ldots, m$$

arises. In the PCCA approach, the cluster identification is done exploiting the fact that, for $\epsilon = 0$, each cluster is clearly associated with the set of signs of the components of the eigenvectors $v_1, \ldots, v_m$, where $v_1 = e$ is set. Clearly, the signs of the components are preserved as long as the perturbation ϵ is 'small enough'; for 'too small' entries in an eigenvector v_i, $i > 1$, however, we will have to define some 'dirty zero' as a perturbation of the exact sign function value 0; compare v_3 over the index subset $\mathcal{S}_3$ in Fig. 7, right. Therefore, in PCCA, the least squares requirement

$$\left\|\chi - V_m \mathcal{A}\right\|_\pi = \min \tag{3.5}$$

is imposed and solved iteratively by successive reduction of the 'dirty zero' parameter. In this way, some discontinuity enters into the algorithm, which leads to some lack of robustness of the PCCA approach. As a consequence, we were eventually led to abandon this approach, but not the concept of Perron cluster analysis as a whole.

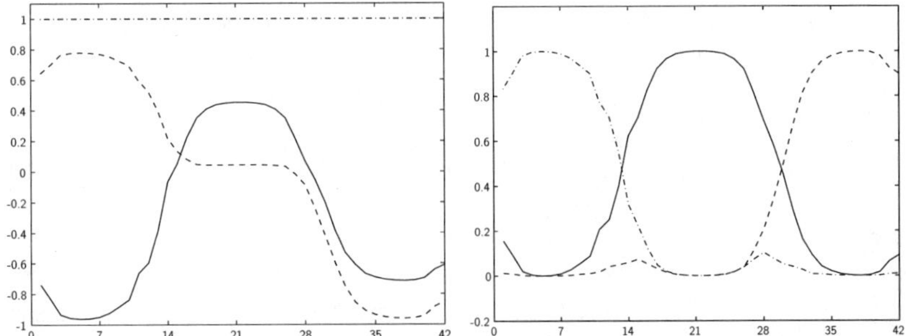

Figure 8. *Perron cluster* $\lambda = 1, 0.99, 0.98$ *in butane molecule.* Left: *Eigenbasis* v_1, v_2, v_3. Right: *Soft characteristic functions.*

Algorithm PCCA+. In this approach, the linear least-squares problem (3.5) is replaced by the relation

$$\widetilde{\chi} = V_m \widetilde{\mathcal{A}} \,, \tag{3.6}$$

wherein 'almost characteristic functions' $\widetilde{\chi}_i(\epsilon)$ enter instead of the 'exact' characteristic function $\widetilde{\chi}_i(0) = \chi_i$; in Fig. 8, right. Such quantities are represented schematically for a simple toy molecule. Almost characteristic functions are assumed to satisfy the *positivity* property:

$$\widetilde{\chi}_i(l) \geq 0 \,, \quad i = 1, \ldots, m \,, \quad l = 1, 2, \ldots, N \tag{3.7}$$

and the *partition of unity* property:

$$\sum_{i=1}^{m} \widetilde{\chi}_i(l) = 1 \,, \quad l = 1, 2, \ldots, N \,. \tag{3.8}$$

This approach has been interpreted as replacing the metastable sets by 'fuzzy sets', a notion that is all but clear. For an intuitive understanding we may naturally interpret them via the modified probabilities

$$(\pi_1 \widetilde{\chi}_i(1), \ldots, \pi_N \widetilde{\chi}_i(N)) \,, \tag{3.9}$$

which are anyway the basis for the visualization of conformation i (typically by volume rendering or isosurface representation). Physically speaking, the almost characteristic functions may be interpreted as "mixed states" generated by perturbation of "pure states" $\chi_1, \ldots, \chi_m$.

The Perron cluster analysis supplies m *metastable chemical conformations* via the m almost-characteristic functions $\widetilde{\chi}_1, \ldots, \widetilde{\chi}_m$ and, on this basis, yields the following information:

- the probabilities $\widetilde{\pi}_i$ for the system to *be* within state i as

  $$\widetilde{\pi}_i = \pi^{\mathsf{T}} \widetilde{\chi}_i = \langle \widetilde{\chi}_i \,, e \rangle_\pi \,, \tag{3.10}$$

  which is a variation of (1.8);

- the probabilities $w_{ii} = w(i,i,\tau)$ for the system, once it is in state i, to *stay* within i during time τ

$$w_{ii} = \frac{\langle \widetilde{\chi}_i \,, T\widetilde{\chi}_i \rangle_\pi}{\langle \widetilde{\chi}_i \,, e \rangle_\pi} = \frac{\langle \widetilde{\chi}_i \,, T\widetilde{\chi}_i \rangle_\pi}{\widetilde{\pi}_i}\,, \tag{3.11}$$

which is a variation of (1.10); and

- the probabilities $w_{ij} = w(i,j,\tau)\ , i \neq j$, for the system, once it is in state i, to *move* to state j,

$$w_{ij} = \frac{\langle \widetilde{\chi}_i \,, T\widetilde{\chi}_j \rangle_\pi}{\langle \widetilde{\chi}_i \,, e \rangle_\pi} = \frac{\langle \widetilde{\chi}_i \,, T\widetilde{\chi}_j \rangle_\pi}{\widetilde{\pi}_i}\,, \tag{3.12}$$

which is a variation of (1.9).

The actual computation of $\widetilde{\chi}$ is performed such that the corresponding *metastability*

$$w_m(\tau) = \sum_{i=1}^{m} w_{ii}\,, \quad \text{where all} \quad w_{ii} > 0.5\,,$$

is *maximized*; the link to Theorem 1 is obvious, even though the definition of metastability there differs slightly from the one here. More details can be found in the original paper [13].

As for the parameter ϵ used above without specification, we quote the definition

$$\epsilon = \max_{i=1,\ldots,m} \big(1 - w_{ii}\big) = 1 - \min_{i=1,\ldots,m} w_{ii}\,, \tag{3.13}$$

from [12] or, as an alternative, the more recent definition from [13]

$$\epsilon = 1 - \lambda_2\,. \tag{3.14}$$

Summarising, we may state the following:

> *Given a sufficiently accurate approximation matrix T of the transfer operator $\mathbf{T}$, the Perron cluster analysis supplies the number, the lifetimes, and the decay pattern of the metastable chemical conformations.*

4 Approximation of Stochastic Operator

The whole Perron cluster analysis as described in Section 3 will only work if the stochastic operator $\mathbf{T}$ can be approximated appropriately, which is the topic of this section. As has been shown in [39], $\mathbf{T}$ can be interpreted as a *transition operator* associated with the Markov chain (1.6).

Hybrid Monte Carlo method (HMC). First we want to briefly describe the mixed deterministic–stochastic process that directly mimics the Markov chain shown in Fig. 1. For details see references [14, 40].

In order to approximate the Hamiltonian flow Φ^τ in the definition of the transfer operator, we will have to discretise the Hamiltonian equations of motion (1.1). Suppose that this discretisation with time-step $h = \tau/k$ yields the discrete flow Ψ^h such that $\Phi^\tau x_0$ is approximated by

$$x_{j+1} = \Psi^h x_j \,, \quad j = 0, \ldots, k-1 \,.$$

All explicit discretisations with certain long-term stability properties, e.g., symplectic ones, do *not* exactly conserve the energy. Therefore, the chain

$$q_{k+1} = \pi \, (\Psi^h)^N (q_k \,, p_k) \,, \qquad p_k \,:\, \mathcal{P} - \text{distributed} \,,$$

will in general *not* sample the distribution $\mathcal{Q}$ of interest. In order to correct this, one has to use the Metropolis acceptance procedure. This yields the HMC chain, which leads to a chain of the same structure as the one shown in Fig. 1, has the correct invariant measure, and still contains good approximations of sub-trajectories of the Hamiltonian system.

Monte Carlo approximation of transition probabilities. Given a discretisation of the position space Ω, in terms of boxes $\{B_1, \ldots, B_N\}$ and a realisation $\{q_1, \ldots, q_M\}$ of the HMC chain, the elements T_{ij} of the transition matrix T can be computed by virtue of

$$T_{ij} = \frac{\#\{q_{k+1} \in B_j \wedge q_k \in B_i\}}{\#\{q_k \in B_i\}} \qquad i, j = 1, \ldots N \,.$$

By means of this we obtain an approximation $T^{(M)}$ with an error like

$$\left|T - T^{(M)}\right| \leq \frac{\gamma}{\sqrt{M}} \,,$$

where this estimate has to be understood in the sense of the central-limit theorem for Markov chains (under special conditions there are much sharper convergence results [34]). As in all Monte-Carlo-type processes, however, *trapping* within local minima will occur unless we take special precautions. In fact, the above constant γ exceeds any bound, if the spectral gap at the Perron root approaches 0. However, as we want to analyze Perron clusters, this is just the case treated here. Below we will present a temperature-embedding technique especially designed to deal with this situation.

Spatial box discretisations. The number N of spatial boxes is also the dimension of the arising transition matrix T. Of course, we must assure that N remains of moderate size even for larger molecular systems. From chemical insight into the problem, different conformations occur corresponding to the double- or triple-well structure in the torsion-angle potentials; see Fig. 9. Let s be the number of minima in the torsion potential ($s = 2$ or $s = 3$) and n the number of torsion angles ($n \approx 7$ per nucleotide), then our first-applied subdivision technique from [9] would have led to a number

$$N \approx s^n$$

of boxes. For the small RNA segment with 70 atoms and three genetic letters (ACC) given in [7], we have $n = 37$. This would have led to $N > 10^{11}$; which is, of course, intolerable! This combinatorial explosion is the well-known "curse of dimension".

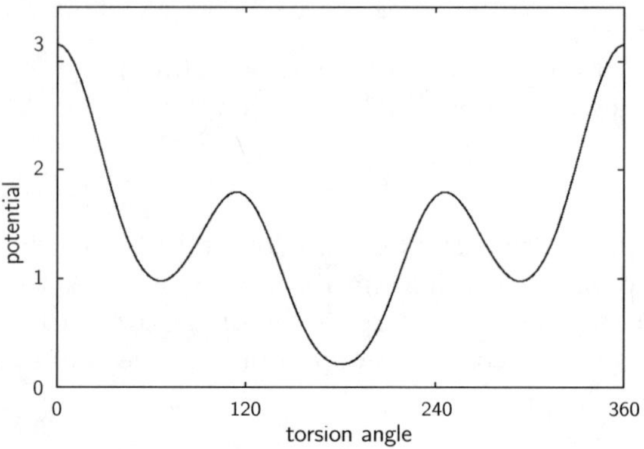

Figure 9. *Molecular torsion potential with triple well ($s = 3$).*

In order to overcome this undesirable effect, we have experimented with several heuristics. First, we modified the method suggested by Amadei et al. [1] to circular coordinates [29, 40]; this method identifies "essential degrees of freedom" by principal component analysis (PCA) of dynamical fluctuations. This technique turned out to lack robustness already for quite small molecules. Next, we tried self-organising maps (SOM) due to Kohonen [32] in combination with our PCCA: the speed-up of the combined cluster algorithm has been reported in [24]; an advanced multilevel version called self-organising *box* maps (SOBM) has been developed in detail by Galliat et al. [21, 22, 23]. Our present favorite box discretisation technique is a combination of the two heuristics to be described next.

Successive PCCA of dihedrals. This kind of box discretisation heuristics is due to Cordes *et al.*. [5]. It starts from the chemical insight that dihedrals (or torsion angles) are useful indicators for conformational changes. The principle of the algorithm is as follows:

> On the basis of a precomputed HMC series, we afford to construct rather fine discretisations for each of the dihedrals *separately*. This defines separate "dihedral transition matrices" T for each dihedral decomposition, which are analyzed in terms of PCCA+. Among these matrices, the one with eigenvalue λ_2 closest to $\lambda_1 = 1$ is selected and subdivided according to the PCCA+ strategy.

Upon applying this idea recursively to the remaining dihedral subspaces, a rather useful "coarse grid" is constructed, which is then taken as the box discretisation for the final transition matrix to be analyzed as a whole. In Fig. 10, a few steps of this recursive scheme are schematically presented in a two-dimensional dihedral plane.

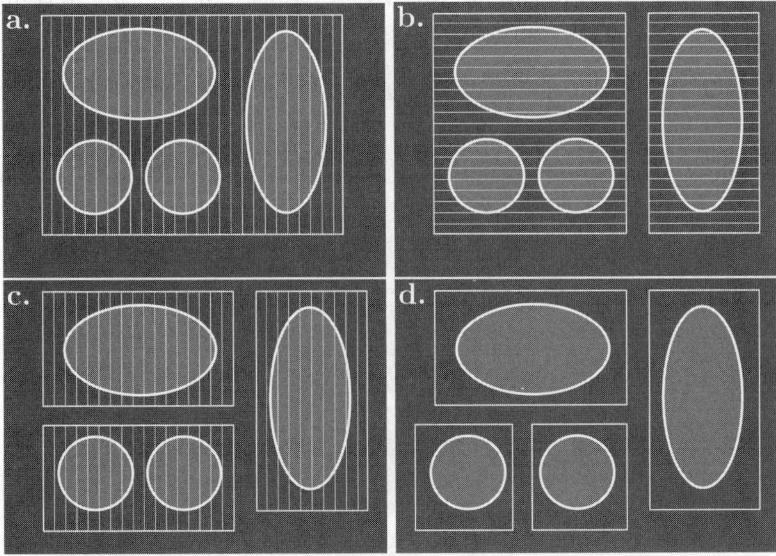

Figure 10. Algorithmic scheme for successive PCCA of dihedrals: *Four metastable regions are drawn as ellipses in a 2-dimensional dihedral space. Thin lines show the successive fine discretisations of each dihedral. Figures a. to d. illustrate the alternation between fine discretisation and coarse grid construction. The final coarse grid (Fig. d) consists of four spatial boxes.*

This rather simple strategy is surprisingly robust and works well even for rather complex molecules. It will clearly fail whenever there is a coupling between torsion angles that have successively been selected. At present, such an occurrence is detected and corrected at some later stage of the UCMC strategy to be described next. In order to make the techique more robust and self-consistent, we are planning to combine it with our former neural network strategy (SOM, see above) to avoid such a situation already at the level where it could occur. As an alternative, we also work on a combination of the present technique with a newly developed Hidden Markov Model approach to the identification of highly dimensional coupling by means of low-dimensional projections [45].

Uncoupling-Coupling Monte Carlo method (UCMC). This technique has been developed by A. Fischer et al. [18, 17]. From an abstract point of view, the algorithmic scheme is a Monte Carlo extension of aggregation/disaggregation techniques suggested in 1989 by C. D. Meyer [33]; there, however, the stationary distribution had been the object of interest, which in our context is given as input.

As the starting point for an algorithmic realisation of the transfer-operator approach we need a sample of the state space distributed according to the canonical distribution $\mathcal{Q}^* \propto \exp(-\beta_* V)$ at inverse temperature β_*. Yet, a direct sampling of the state space via the associated HMC Markov chain (1.6) will result in *slow mixing* and, hence, poor convergence caused by the presence of metastabilities—which we actually want to compute.

In order to address this problem, an iterative scheme of alternating uncoupling and coupling is applied, which realises the steps

(a) embedding $\mathcal{Q}^*$ in a series of canonical distributions of increasing temperatures, which decreases metastability;

(b) hierarchical decomposition of state space into metastable sets and restart of restricted Markov chains therein, applying a type of annealing strategy; and

(c) coupling the samples from restricted Markov chains for proper reweighing of the samples at $\mathcal{Q}^*$.

The sampling starts with one HMC Markov chain at the highest temperature level searching the whole state space. Step (b) already includes transfer-operator techniques for the identification of metastable sets, but within the annealing strategy the state space is decomposed as soon as some metastability emerges. By construction, all restricted HMC Markov chains exhibit *rapid mixing*, which speeds up the computation and, at the same time, increases robustness of the overall algorithm. In coupling step (c) we set up a coupling matrix by computing quotients of normalising constants between samples at neighboring temperatures with an overlapping domain in the hierarchy. Coupling factors connecting samples from different domains are then given by the entries of the stationary distribution of the coupling matrix. The situation is illustrated in Fig. 11.

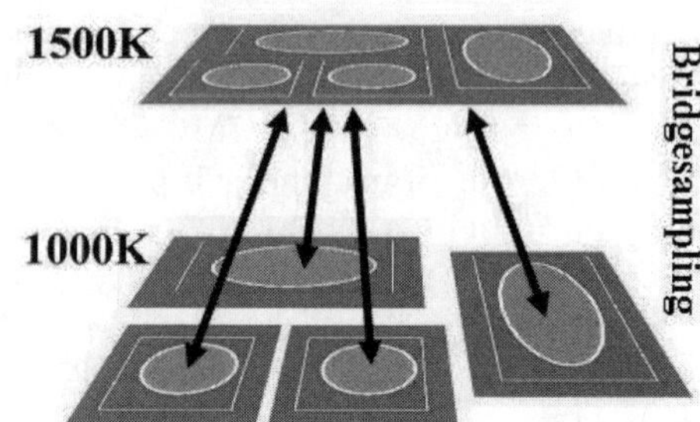

Figure 11. Hierarchical simulation protocol for UCMC: *After decomposition, the metastable subsets of the conformational space are sampled independently at a lower temperature level. Two temperature levels are connected via bridge samplings.*

As a result of the UCMC technique, we obtain a weighted sample, which is distributed according to $\mathcal{Q}^*$. Technical details of this quite complicated process can be found in [18, 17].

5 Example: SARS Protease Inhibitor

The here-described bunch of new mathematical methods for the identification of metastable conformations has been published in a series of papers by the research group of the authors, among which the surveys [7, 41] also contain numerical results for interesting biomolecules; e.g., the green tea molecule *epigallocatechine*, a suspected anti-cancer drug, and an HIV protease inhibitor.

In the present paper we restrict our attention to SARS (abbreviation for **S**evere **A**cute **R**espiratory **S**yndrome). The corresponding corona virus responsible for the sudden occurrence of the epidemics arose early this year, unknown until then. It is only since 30 May 2003, that the 3D structure of one of its enzymes, a protease, has been available on the internet [48]; this molecule takes part in the viral metabolism by cutting larger proteins into smaller peptide strands. The underlying biochemical experiments have been published by the research group of Hilgenfeld [2]. In Fig. 12, we show the result obtained from a homology model on top of an X-ray analysis of a similar molecule, which seemed to reveal some active site of the SARS protease; the associated molecule in the active site has been observed to fit into the molecular pocket, but is not expected to be a drug against SARS. Instead the search race continues at high speed.

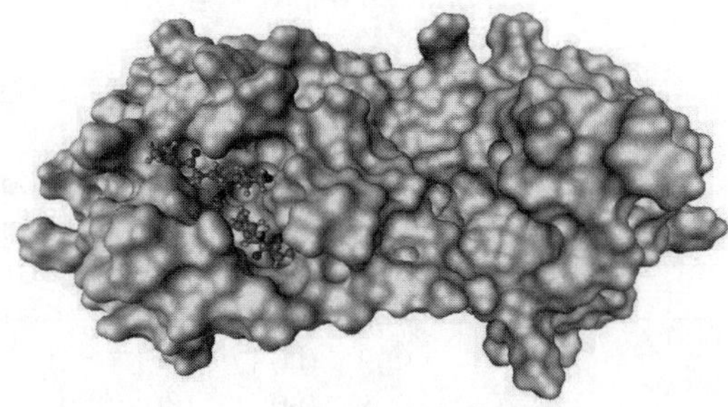

Figure 12. SARS protease: *active site as suspected from X-ray analysis*

Starting from the internet data, we investigated a molecule, the inhibitor AG7088, with our mathematical tools for conformation dynamics. Upon applying the UCMC technique for box discretisation at the temperatures 1500K, 1000K, 600K, and 300K, we obtained the results arranged in Table 1. On each of the subsets we ran the fast-mixing Markov chains based on HMC. The Perron clusters obtained from PCCA+ in connection with the box-discretisation technique of successive PCCA are included. As can be seen, we detected $m = 3$ metastable conformations at 1500K, which divide into 4, 6, and 5 conformations separately at 1000K. At 600K, only the metastable conformation with highest thermodynamical weight has been selected, which then decomposes into 5 subsets at room temperature or human body temperature 300K, respectively.

Of course, these data are mainly of interest for the drug designer. That is why, in

Table 1. SARS protease inhibitor: *hierarchical temperature sequence and coarse grid spectra, as obtained from UCMC and successive PCCA+. At 600K only the metastable conformation with highest thermodynamical weight has been selected, which then decomposes into 5 subsets at human body temperature 300K.*

T[K]	coarse spectrum	coupling matrix
1500	1.000 0.984 0.975 0.861	0.982 0.003 0.015 0.003 0.976 0.021 0.001 0.002 0.997
1000	1.000 0.994 0.987 0.971 0.955	0.992 0.001 0.005 0.002 0.000 0.966 0.024 0.010 0.001 0.014 0.982 0.003 0.001 0.006 0.003 0.990
1000	1.000 0.999 0.997 0.990 0.985 0.982 0.971	0.987 0.000 0.009 0.000 0.004 0.000 0.000 0.997 0.001 0.000 0.000 0.002 0.001 0.000 0.984 0.002 0.008 0.005 0.000 0.000 0.001 0.970 0.000 0.029 0.000 0.001 0.001 0.000 0.985 0.013 0.000 0.000 0.000 0.002 0.003 0.995
1000	1.000 0.995 0.992 0.990 0.988 0.982	0.978 0.002 0.016 0.004 0.000 0.002 0.976 0.000 0.014 0.008 0.003 0.000 0.987 0.009 0.001 0.000 0.002 0.005 0.986 0.007 0.000 0.001 0.001 0.017 0.981
600	1.000 0.998 0.994 0.988 0.987 0.979	0.959 0.032 0.001 0.000 0.008 0.008 0.981 0.003 0.002 0.006 0.001 0.012 0.980 0.002 0.005 0.000 0.001 0.000 0.966 0.033 0.000 0.001 0.000 0.006 0.993

Fig. 13, we additionally present an image of the molecule in the frame of conformation dynamics: there we combine a volume-rendering visualisation of the (discrete) invariant measure at 1500K together with one snapshot of the molecule in ball-and-stick representation. More insight into these conformations can be gained from the isosurfaces for the conformations as given in Fig. 14 for the dominant one, and in Fig. 15 for the subdominant one.

Remark. The authors are aware of the fact that in *prion* diseases (such as *scrapie* or the mad-cow disease) rather rare conformations with high probability to *stay* within may nevertheless well play a decisive role, as has been pointed out by Griffith [25] already in 1967.

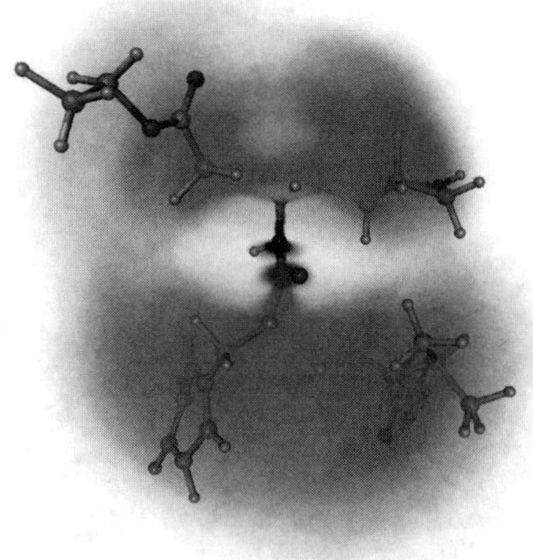

Figure 13. SARS protease inhibitor: *volume rendering representation of invariant measure at temperature $T = 1500K$. Insertion of ball-and-stick representation of two dominant conformations.*

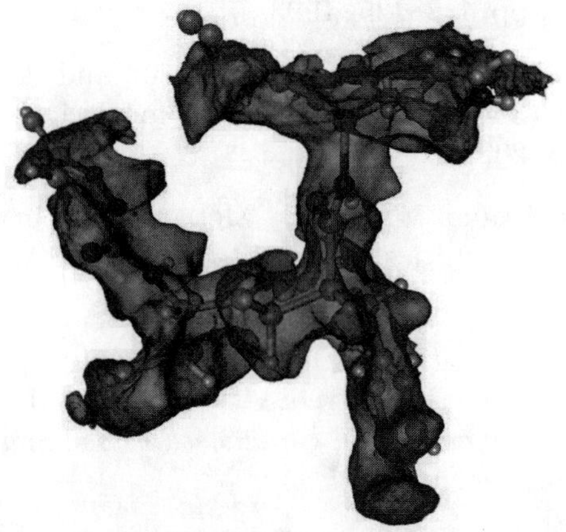

Figure 14. SARS protease inhibitor: *isosurface representation of dominant conformation (probability* $\sim 56.5\,\%$ *to* be *within).*

Acknowledgements. The authors want to thank all of their coworkers for their collaboration in this fascinating field, in particular Frank Cordes, Alexander Fischer, Wilhelm Huisinga, and Marcus Weber for invaluable groundwork to this article.

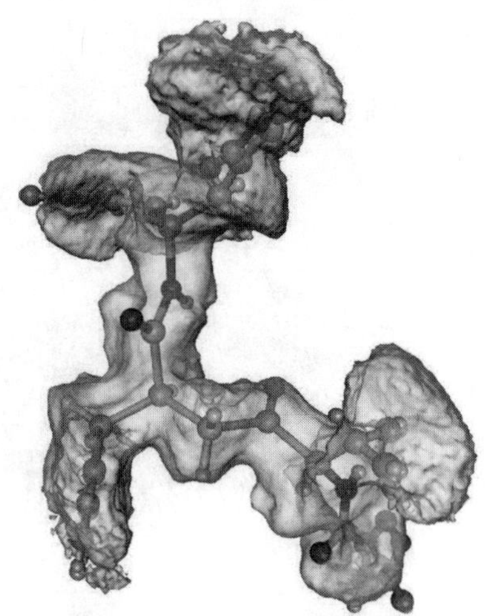

Figure 15. SARS protease inhibitor: *isosurface representation of subdominant conformation (probability* $\sim$ 35.6 % *to* be *within).*

Bibliography

[1] A. Amadei, A. B. M. Linssen, and H. J. C. Berendsen. Essential dynamics of proteins. *Proteins*, 17, pp. 412–425, 1993.

[2] K. Anand, J. Ziebuhr, P. Wadhani, J. R. Mesters, and R. Hilgenfeld. Corona virus main proteinase (3clpro) structure: Basis for design of anti-SARS drugs. *Science*, 300, pp. 1763, 2003.

[3] A. Bovier, V. Gayrard, and M. Klein. Metastability in reversible diffusion processes II: Precise asymptotics for small eigenvalues. WIAS preprint, Sept. 2002.

[4] D. Chandler. Finding transition pathways: throwing ropes over rough montain passes, in the dark. In B.J. Berne, G. Ciccotti, and D.F. Coker, editors, *Classical and Quantum Dynamics in Condensed Phase Simulations*, pp. 51–66. Singapure: World Scientific, 1998.

[5] F. Cordes, M. Weber, and J. Schmidt-Ehrenberg. Metastable Conformations via successive Perron Cluster Cluster Analysis of dihedrals. Technical Report ZIB 02–40, Zuse Institute Berlin, 2002.

[6] M. Dellnitz and O. Junge. On the approximation of complicated dynamical behavior. *SIAM J. Num. Anal.*, 36(2), pp. 491–515, 1999.

[7] P. Deuflhard. From molecular dynamics to conformational dynamics in drug design. In M. Kirkilionis, S. Krömker, R. Rannacher, and F. Tomi, editors, *Trends in Nonlinear Analysis*, pp. 269–287. Springer, 2003.

[8] P. Deuflhard and F. Bornemann. *Scientific Computing with Ordinary Differential Equations*, volume 42 of *Texts in Applied Mathematics*. Springer, New York, 2002.

[9] P. Deuflhard, M. Dellnitz, O. Junge, and Ch. Schütte. Computation of essential molecular dynamics by subdivision techniques. In *[10]*, pp. 98–115, 1999.

[10] P. Deuflhard, J. Hermans, B. Leimkuhler, A. E. Mark, S. Reich, and R. D. Skeel, editors. *Computational Molecular Dynamics: Challenges, Methods, Ideas*, volume 4 of *Lecture Notes in Computational Science and Engineering*. Springer-Verlag, 1999.

[11] P. Deuflhard and A. Hohmann. *Numerical Analysis in Modern Scientific Computing: An Introduction*, volume 43 of *Texts in Applied Mathematics*. Springer, New York, 2003.

[12] P. Deuflhard, W. Huisinga, A. Fischer, and Ch. Schütte. Identification of almost-invariant aggregates in reversible nearly uncoupled Markov chains. *Lin. Alg. Appl.*, 315, pp. 39–59, 2000.

[13] P. Deuflhard and M. Weber. Robust Perron Cluster Analysis in Conformation Dynamics. Technical Report ZIB 03–19, Zuse Institute Berlin, 2003.

[14] S. Duane, A. D. Kennedy, B. J. Pendleton, and D. Roweth. Hybrid Monte Carlo. *Phys. Lett. B*, 195(2), pp. 216–222, 1987.

[15] W. E, W. Ran, and E. Vanden-Eijnden. Probing multiscale energy landscapes using the string method. *Phys. Rev. Lett.*, to appear, 2002.

[16] R. Elber and M. Karplus. Multiple conformational states of proteins: A molecular dynamics analysis of Myoglobin. *Science*, 235, pp. 318–321, 1987.

[17] A. Fischer. *An Uncoupling–Coupling Method for Markov chain Monte Carlo simulations with an application to biomolecules.* PhD thesis, Free University Berlin, 2003.

[18] A. Fischer, Ch. Schütte, P. Deuflhard, and F. Cordes. Hierarchial uncoupling–coupling of metastable conformations. In *[38]*, pp. 235–259, 2002.

[19] H. Frauenfelder and B. H. McMahon. Energy landscape and fluctuations in proteins. *Ann. Phys. (Leipzig)*, 9(9–10), pp. 655–667, 2000.

[20] H. Frauenfelder, P. J. Steinbach, and R. D. Young. Conformational relaxation in proteins. *Chem. Soc.*, 29A (145–150), 1989.

[21] T. Galliat. *Adaptive Multilevel Cluster Analysis by Self-Organizing Box Maps.* PhD thesis, FU Berlin, March 2002.

[22] T. Galliat and P. Deuflhard. Adaptive hierarchical cluster analysis by self-organizing box maps. Konrad–Zuse–Zentrum, Berlin. Report SC–00–13, 2000.

[23] T. Galliat, P. Deuflhard, R. Roitzsch, and F. Cordes. Automatic identification of metastable conformations via self-organized neural networks. In *[38]*, pp. 260–284, 2002.

[24] T. Galliat, W. Huisinga, and P. Deuflhard. Self-organizing maps combined with eigenmode analysis for automated cluster identification. In H. Bothe and R. Rojas, editors, *Neural Computation*, pp. 227–232. ICSC Academic Press, 2000.

[25] J. Griffith. Self-replication and scrapie. *Nature*, 215, pp. 1043–1044, 1967.

[26] E. Hairer, Ch. Lubich, and G. Wanner. *Geometric Numerical Integration. Structure-Preserving Algorithms for Ordinary Differential Equations*, volume 31 of *Springer Series in Computational Mathematics*. Springer, Berlin, Heidelberg, New York, 2002.

[27] T.A. Halgren. Merck molecular force field. *J. Comp. Chem.*, 17(I–V), pp. 490–641, 1996.

[28] W. Huisinga. *Metastability of Markovian systems: A transfer operator based approach in application to molecular dynamics.* PhD thesis, Free University Berlin, 2001.

[29] W. Huisinga, Ch. Best, R. Roitzsch, Ch. Schütte, and F. Cordes. From simulation data to conformational ensembles: Structure and dynamic based methods. *J. Comp. Chem.*, 20(16), pp. 1760–1774, 1999.

[30] W. Huisinga, S. Meyn, and Ch. Schütte. Phase transitions & metastability in Markovian and molecular systems. accepted in Ann. Appl. Probab., 2002.

[31] W. Huisinga and B. Schmidt. Metastability and Dominant Eigenvalues of Transfer Operators, in preparation, 2002.

[32] T. Kohonen. *Self-Organizing Maps.* Springer, Berlin, 2nd edition, 1997.

[33] C. D. Meyer. Stochastic complementation, uncoupling Markov chains, and the theory of nearly reducible systems. *SIAM Rev.*, 31, pp. 240–272, 1989.

[34] S. P. Meyn and R. L. Tweedie. *Markov Chains and Stochastic Stability.* Springer, Berlin, 1993.

[35] G. U. Nienhaus, J. R. Mourant, and H. Frauenfelder. Spectroscopic evidence for conformational relaxation in Myoglobin. *PNAS*, 89, pp. 2902–2906, 1992.

[36] H. Risken. *The Fokker–Planck Equation.* Springer, New York, 2nd edition, 1996.

[37] J. M. Sanz-Serna and M. P. Calvo. *Numerical Hamiltonian Problems.* Chapman and Hall, London, UK, 1994.

[38] T. Schlick and H. H. Gan, editors. *Computational Methods for Macromolecules: Challenges and Applications — Proc. of the 3rd Intern. Workshop on Algorithms for Macromolecular Modelling*, Berlin, Heidelberg, New York, 2000. Springer.

[39] Ch. Schütte. *Conformational Dynamics: Modelling, Theory, Algorithm, and Application to Biomolecules.* Habilitation Thesis, Fachbereich Mathematik und Informatik, Freie Universität Berlin, 1999.

[40] Ch. Schütte, A. Fischer, W. Huisinga, and P. Deuflhard. A direct approach to conformational dynamics based on hybrid Monte Carlo. *J. Comput. Phys., Special Issue on Computational Biophysics*, 151, pp. 146–168, 1999.

[41] Ch. Schütte and W. Huisinga. Biomolecular conformations as metastable sets of Markov chains. In R. S. Sreenivas and D. L. Jones, editors, *Proceedings of the Thirty-Eighth Annual Allerton Conference on Communication, Control, and Computing, Monticello, Illinois*, pp. 1106–1115. University of Illinois at Urbana–Champaign, 2000.

[42] Ch. Schütte and W. Huisinga. Biomolecular conformations can be identified as metastable sets of molecular dynamics. In P. G. Ciarlet and C. Le Bris, editors, *Handbook of Numerical Analysis*, volume X. Special Volume Computational Chemistry; pp. 699–744. North–Holland, 2002.

[43] Ch. Schütte, W. Huisinga, and P. Deuflhard. Transfer operator approach to conformational dynamics in biomolecular systems. In B. Fiedler, editor, *Ergodic Theory, Analysis, and Efficient Simulation of Dynamical Systems*, pp. 191–223. Springer, 2001.

[44] E. Vanden-Eijnden. Metastability and effective dynamics in ergodic systems, 2003.

[45] S. Waldhausen, A. Fischer, and Ch. Schütte. A Hidden Markov Model for the Identification of Biomolecular Conformations from Incomplete Observations. Technical report, Freie Universität Berlin, 2003. Submitted to J. Comp. Chem.; available via `http://www.math.fu-berlin.de/~biocomp/`.

[46] M. Weber. Improved Perron Cluster Analysis. Technical Report ZIB 03–04, Zuse Institute Berlin, 2003.

[47] M. Weber and T. Galliat. Characterization of transition states in conformational dynamics using Fuzzy sets. Technical Report 02–12, Konrad–Zuse–Zentrum (ZIB), Berlin, March 2002.

[48] A. Wiley and Gh. Deslongchamps. Homology model of SARS–CoV Mpro protease, 2003.

[49] H. X. Zhou, S. T. Wlodec, and J. A. McCammon. Conformation gating as a mechanism for enzyme specificity. *Proc. Nat. Acad. Sci. USA*, 95, pp. 9280–9283, 1998.

Yoshikazu Giga is a Professor of Mathematics at the Hokkaido University and one of the leading experts in the mathematical analysis for nonlinear parabolic partial differential equations. In his 23-year research career, his research interests have included analysis on the Navier–Stokes equations, blow-up phenomena for semilinear heat equations, surface evolution equations, and viscosity solutions.

Giga was born in Tokyo, Japan, and studied at the University of Tokyo before taking his PhD there. In his early career he stayed at the Courant Institute as a postdoctoral researcher, as well as at other institutions, before he moved to the mathematics department of the Hokkaido University in 1986. He was an Andrejewski lecturer of the Humbolt University and a Rudolph-Lipschitz lecturer of the University of Bonn in 1997. He was awarded the Nissan Science Prize for his analysis of singularities in nonlinear parabolic equations in 1997. He was awarded the Autumn Prize 1999 by the Mathematical Society of Japan for his contribution on surface evolution equations. He has served as an editor of several scientific journals, most notably of *Mathematische Annalen*, and of *SIAM Journal on Mathematical Analysis.* He has been a member of the executive committee of the Mathematical Society of Japan.

Chapter 7

Singular Diffusivity — Facets, Shocks and more

Yoshikazu Giga*

Abstract: There is a class of nonlinear evolution equations with singular diffusivity, so that diffusion effect is nonlocal. A simplest one-dimensional example is a diffusion equation of the form $u_t = \delta(u_x)u_{xx}$ for $u = u(x,t)$, where δ denotes Dirac's delta function. This review paper is intended to provide an overview of analytic aspects of such equations, as well as various applications. Equations with singular diffusivity are applied to describe several phenomena in the applied sciences, and to provide several devices in technology, especially image processing. A typical example is a gradient flow of the total variation of a function, which arises in image processing, as well as in material science to describe the motion of grain boundaries. In the theory of crystal growth the motion of a crystal surface is often described by an anisotropic curvature flow equation with a driving force term. At low temperature the equation includes a singular diffusivity, since the interfacial energy is not smooth. Another example is a crystalline algorithm to calculate curvature flow equations in the plane numerically, which is formally written as an equation with singular diffusivity.

Because of singular diffusivity, the notion of solution is not *a priori* clear, even for the above one-dimensional example. It turns out that there are two systematic approaches. One is variational, and applies to divergence type equations. However, there are many equations like curvature flow equations which are not exactly of divergence type. Fortunately, our approach based on comparision principles turns out to be succesful in several interesting problems. It also asserts that a solution can be considered as a limit of solution of an approximate equation. Since the equation has a strong diffusivity at a particular slope of a solution, a flat portion with this slope is formed. In crystal growth ploblems this flat portion is called a *facet*. The discontinuity of a solution (called a *shock*) for a scalar conservation law is also considered as a result of singular diffusivity in the vertical direction.

*Partly supported by the Grant-in-Aid for Scientific Research, No. 14204011, No. 15634008, Japan Society for the Promotion of Science.

1 Introduction

This is a review paper on nonlinear evolution equations with singular diffusivity, in which the diffusion effect is nonlocal. Recently, such equations become popular in material sciences and image analysis, although its mathematical analysis is far from being satisfactory to cover all important problems. Our goal is to provide an overview of mathematical analysis of these types of problem, especially

(i) unique solvability of the initial value problem;

(ii) stability under approximation of both equations and data

for various typical problems.

To see the difficulty of the problem we start with the simplest one-dimensional example, having the form

$$u_t = \delta(u_x)u_{xx}, \quad x \in \mathbb{R}, \ t > 0 \tag{1.1}$$

for $u{=}u(x,t)$, where δ denotes Dirac's delta function and $u_t{=}\partial u/\partial t$, $u_x{=}\partial u/\partial x$, $u_{xx}{=}\partial^2 u/\partial x^2$. The multiplier $\delta(u_x)$ is not well-defined as a distribution, even if u is smooth. So even the notion of solution is unclear, even in a naïve sense. Fortunately, this example can be regarded as the gradient flow of (the half of) the total variation $\int |u_x|\,\mathrm{d}x$. So we are able to apply the nonlinear semigroup theory [10, 5] for maximal monotone operators (initiated by Kōmura [56]) to conclude unique global solvability under suitable boundary conditions [49, 53, 31]. This example also has stability under approximation of the functionals [11, 71].

An explicit form of a solution is often computable. For example, if the initial data is $\cos x$, then a flat portion (often called a *facet*) instanteneously forms at the maximum and minimum points of $\cos x$. In fact, as in [53] we have

$$u(x,t) = \min\bigl(\cos x\,, h(t)\bigr)\,, \quad |x| \le \tfrac{1}{2}\pi\,,$$
$$\frac{\mathrm{d}h}{\mathrm{d}t} = -L(t)^{-1}\,, \quad h(0) = 1\,,$$

where $L(t)$ is the length of a facet (i.e. the length of the interval on which $h(t) < \cos x$), until h becomes zero. (By an elemetary observation this gives an explicit formula of u [22, 53]). It is not difficult to see that $h(t)$ becomes zero within finite time. After that time, $u(x,t) \equiv 0$.

We learn from this example that the speed on the facet is determined by its length, so the speed is nonlocal. It is formally obtained in the following way. Let $\bigl(a(t), b(t)\bigr)$ be the interval corresponding to the facet. Assume that u_t is constant on $\bigl(a(t), b(t)\bigr)$. (This assumption is called facet-stay-as-facet hypothesis.) Since our equation (1.1) can be written as

$$u_t = \tfrac{1}{2}(\operatorname{sgn} u_x)_x\,, \tag{1.2}$$

integrating over $\bigl(a(t)-\varepsilon, b(t)+\varepsilon\bigr)$ and sending $\varepsilon(>0)$ to zero yields

$$u_t\bigl(b(t)-a(t)\bigr) = \tfrac{1}{2}\bigl(\operatorname{sgn}(-0) - \operatorname{sgn}(+0)\bigr) = -1$$

if u is concave near $\big(a(t), b(t)\big)$. This yields the desired speed.

One may write the equation (1.2) in the form

$$u_t = \tfrac{1}{2}\xi_x\,, \quad \xi = \operatorname{sgn} u_x\,.$$

However, $\operatorname{sgn} p$ should be regarded multivalued at $p = 0$ with values $[-1, 1]$ as a maximal monotone function. Thus, instead of $\xi = \operatorname{sgn} u_x$ one has to write

$$\xi(x,t) \in \operatorname{sgn} u_x(x,t)\,.$$

At the first glance there should be a huge ambiguity. However, honest regularity of u implies the uniqueness of a solution, thanks to the theory of nonlinear semigroups. Although the equation itself may have an ambiguity, the solution knows how to evolve.

Unfortunately, many important equations are not of the form of a gradient flow. A typical example is a level-set equation of an anisotropic curvature flow, which is of the form

$$\psi_t + |\nabla\psi| \operatorname{div} \xi\Big(\frac{-\nabla\psi}{|\nabla\psi|}\Big) = 0\,, \quad x \in \mathbb{R}^n\,, \quad t > 0 \tag{1.3}$$

for $\psi = \psi(x,t)$, where $\nabla\psi = \big(\partial\psi/\partial x_1, \ldots, \partial\psi/\partial x_n\big)$ and $n \geq 2$. Here ξ is the gradient of the *interfacial energy density* $\gamma : \mathbb{R}^n \to [0, \infty)$ which is convex and positively homogeneous of degree one. If $\gamma(p) = |p|$ so that γ is isotropic, then (1.3) becomes the level-set equation

$$\psi_t = |\nabla\psi| \operatorname{div}\Big(\frac{\nabla\psi}{|\nabla\psi|}\Big) \tag{1.4}$$

of the mean curvature flow equation $V = H$; i.e., each level-set of ψ moves by its mean curvature H, at least formally. Here V denotes the normal velocity in the direction of the normal $\mathbf{n} = -\nabla\psi/|\nabla\psi|$. In general, (1.3) is the level-set equation of an anisotropic curvature-flow equation:

$$V = \Lambda_\gamma(\mathbf{n}) \quad \text{with} \quad \Lambda_\gamma(\mathbf{n}) := -\operatorname{div}_{\Gamma_t} \xi(\mathbf{n}) \quad \text{on} \quad \Gamma_t \tag{1.5}$$

for an evolving hypersurface $\{\Gamma_t\}$. The quantity $\Lambda_\gamma(\mathbf{n})$ is called a *weighted mean curvature.* The level-set equation is important to solve equations like (1.5) globally in time beyond singularities, both analytically [15, 20] and numerically [60]; see also books [64, 36, 59] and review articles [32, 33]. A classical result based on the theory of viscosity solutions implies unique global solvability for any continuous (periodic) initial data for (1.3), provided that γ is smooth enough, say C^2 [15]. However, in crystal growth problems γ may not be C^1 [29]. A typical example is that

$$\operatorname{Frank} \gamma = \big\{p \in \mathbb{R}^n \;\big|\; \gamma(p) \leq 1\big\} \tag{1.6}$$

is a polyhedra. In this case γ is called a *crystalline energy* and (1.5) is called a *crystalline curvature flow* (equation). This problem has a feature which is similar to (1.1). The diffusion effect is very strong in the direction corresponding to vertices of Frank γ. Indeed, if $n = 2$ and $\gamma(p) = \frac{1}{2}(|p_1| + |p_2|)$ for $p = (p_1, p_2)$, then (1.5) becomes (1.1) when Γ_t is represented as the graph of a function $u(x_1, t)$ of one variable x_1.

Crystalline flow problems are first proposed by [4] and [68] independently for planar curve evolutions; i.e., $n = 2$. They restricted a class of solution into special polygonal evolution and reduced the problem to a system of ordinary differential equations (ODEs) under the *facet-stay-as-facet* hypothesis. Their solution is called a crystalline flow (for (1.5)) [26]. However, it was not clear at that time that this facet-stay-as-facet hypothesis was a natural one. This problem has been solved affirmatively in [22, 25, 27] by extending the notion of viscosity solutions [17] so that it applies to the nonlocal problem. In particular, the facet-stay-as-facet hypothesis is actually obtained as the limit of a smoother problem. Moreover, the level-set method for (1.3) with singular γ has been established in [26, 27]. The key observation is that the evolution is order-preserving, despite the fact that the diffusion effect is nonlocal. In section 2 we highlight these results more precisely.

For a higher-dimensional problem [9] the facet-stay-as-facet hypothesis contradicts the comparison principle, so such a hypothesis is no longer natural [73]. It is even not clear, when $n \geq 3$, what *is* the reasonable notion so that (1.3) is solvable globally-in-time for crystalline γ. (Of course, if one admits the facet-stay-as-facet hypothesis, the problem is solvable [38]). Several notions of solution have been proposed [8, 41]. However, it is not known that the initial-value problem is solvable, even locally in time.

Instead of (1.5) one has to consider an equation with driving force, such as

$$V = \Lambda_\gamma(\mathbf{n}) + C(x),$$

where C is a given function. Such an equation is important in studying the Stefan-type problem with Gibbs–Thomson and kinetic supercooling effects. If $C(x)$ is not a constant, so that the problem is spatially inhomogeneous, again the facet-stay-as-facet hypothesis may contradict the comparison principle [23]. An expected, speed is computable [24]. However, local existence of a solution is also not known for this problem. A Stefan-type problem with crystalline γ has been studied [42, 43]. Local existence of solution is obtained under the facet-stay-as-facet hypothesis for a cylindrical crystal [42]. It is expected that the facet-stay-as-facet hypothesis should be true for a small crystal, but work is still in progress.

Besides these rather classical fields, the singular diffusivity is important to understand shocks for scalar first-order equations. Let us start with an example

$$\varphi_t + y\,\varphi_x = M\left|\nabla_{x,y}\,\varphi\right| \frac{\partial}{\partial y}\left(\frac{\varphi_y}{|\varphi_y|}\right), \quad x \in \mathbb{R}, \quad y \in \mathbb{R}, \quad t > 0 \tag{1.7}$$

for $\varphi = \varphi(x, y, t)$, where $M > 0$ is a constant. Without M-term this is the level-set equation for the graph of a solution of the Burgers equation:

$$u_t + uu_x = 0\,. \tag{1.8}$$

However, the level-set of φ may overturn, and it does not represent the graph of an entropy solution after it develops jump discontinuities called *shocks*. It turns out that for sufficiently large M the level-set of φ of (1.6) represents the graph of an entropy solution. This is analytically proved in [28] and numerically confirmed in [70]. The right-hand side of (1.7) plays a role if $\varphi_y = 0$, otherwise it does not play any role. So this term represents singular vertical diffusion to prevent overturning. The method is not limited to conservation laws but equations with nonconservative type, for which the level-set is expected to represent a proper viscosity solution introduced by [34]. Since the problem is not spatially homogeneous, there might still be a chance that the level-set may overturn, so the facet-stay-as-facet hypothesis is not expected to hold for small M. The notion of solution for (1.7) has not yet been fully established. In section 3 we highlight a level-set method to track the graph of solutions with shocks by vertical singular diffusivity.

We now return to the gradient flow of total variation, whose one-dimensional version is (1.1). To remove the result of noise from images it has been proposed in [62] to use the gradient flow of total variation of a grey-level function, under the constraint $\int |u - u_0|^2 \, dx = \text{const.}$, where u_0 is a given image. There are several models related to this problem. For example, the total variation flow with values in a sphere is important for removing noise from direction fields of color grey-level mappings, $u = (u_1, u_2, u_3)$, while maintaining overall strength [67]. The explicit form is

$$u_t = \operatorname{div}\Big(\frac{\nabla u}{|\nabla u|}\Big) + |\nabla u|\, u\,. \tag{1.9}$$

These problems do not have order-preserving structure, so nonlinear semigroup theory does not apply as the nonlinear operator is not necessarily maximal monotone. For a background of problems related to image analysis see [63]. A similar model for a direction field is also important to model the evolution of multigrain [55, 72], where u represents the orientation of the grain. In section 5 we review several analytic results for a gradient flow of total variation.

A fourth-order parabolic equation with singular diffusivity is proposed for example in [66] to model an evolution by surface diffusion with facet. Its analysis was started by Y. Kashima [51]. For more references see the papers cited in [51].

2 Anisotropic curvature flow with singular interfacial energy

A curvature flow equation is now popularly used to describe the motion of phase-boundaries, such as the motion of grain boundaries in materials science (see e.g.[47]). It is also used in image processing to remove noises from images (see [63, 12]). Anisotropic effects are often important in the crystal-growth problem [54]. In image processing it is also important to use an equation with anisotropy depending on local feature of the images (e.g. [61]).

Here is a general form of an anisotropic curvature-flow equation for evolving curves $\{\Gamma_t\}$ in $\mathbb{R}^2$, when it is spatially homogeneous and time-independent:

$$V = g\big(\mathbf{n}, \Lambda_\gamma(\mathbf{n})\big) \quad \text{on} \quad \Gamma_t\,, \tag{2.1}$$

where g is nondecreasing in the second variable so that the problem in parabolic.

Recently, for non-differentiable γ including crystalline energy, we have been able to establish a level-set method. First let us write down the level-set equation for (2.1). It is of the form:

$$\varphi_t - |\nabla\varphi|\, g\Big(\frac{-\nabla\varphi}{|\nabla\varphi|}, \Lambda_\gamma\big(\frac{-\nabla\varphi}{|\nabla\varphi|}\big)\Big) = 0\,. \tag{2.2}$$

Let $\mathcal{I}$ denote the set of all convex interfacial energy-densities $\gamma(\geq 0)$ such that the boundary curve of Frank γ is a closed piecewise C^2 curve which has at most finitely-many singularities. Let $\mathcal{G}_0$ denote the set of all continuous functions on $S^1 \times \mathbb{R}$ such that $\lambda \mapsto g(p,\lambda)$ is non-decreasing and

$$\|g\|_1 = \sup\Big\{\frac{|g(p,\lambda)|}{|\lambda|+1}\,,\ \ p \in S^1\,,\ \ x \in \mathbb{R}\Big\} < \infty\,.$$

We state a periodic version of the main results of [27]. Let $\mathbf{T}^n$ denote a flat torus defined by $\mathbf{T}^n = \prod_{i=1}^{n} \big(\mathbb{R}/\omega_i\mathbb{Z}\big)$ with $\omega_i > 0$.

Theorem 2.1. *Assume that $\gamma \in \mathcal{I}$ and $g \in \mathcal{G}_0$. There is an explicit notion of viscosity-like solution (consistent with usual viscosity solutions) such that the following properties are valid:*

Global Unique Solvability. *For $\varphi_0 \in C\big(\mathbf{T}^2\big)$ there exists a unique solution $\varphi \in C\big(\mathbf{T}^2 \times [0,\infty)\big)$ of (2.2) with $\varphi|_{t=0} = \varphi_0$.*

Convergence Property. *Assume that $\gamma_\varepsilon \in \mathcal{I}$ and $g_\varepsilon \in \mathcal{G}_0$ for $\varepsilon \in (0,1)$. Assume that $\gamma_\varepsilon \to \gamma$, $g_\varepsilon \to g$ locally uniformly as $\varepsilon \to 0$. Assume that $\sup_{0<\varepsilon<1} \|g_\varepsilon\|_1 < \infty$. Let φ_ε be the solution of*

$$\varphi_t - |\nabla\varphi|\, g_\varepsilon\Big(\frac{-\nabla\varphi}{|\nabla\varphi|}, \Lambda_{\gamma_\varepsilon}\big(\frac{-\nabla\varphi}{|\nabla\varphi|}\big)\Big) = 0$$

with $\varphi|_{t=0} = \varphi_{0\varepsilon} \in C\big(\mathbf{T}^2\big)$. If $\varphi_{0\varepsilon} \to \varphi_0$ in $C\big(\mathbf{T}^2\big)$, then $\varphi_\varepsilon \to \varphi$ locally uniformly in $\mathbf{T}^2 \times [0,\infty)$ as $\varepsilon \to 0$.

As expected the (super) level-set $\{\varphi \geq c\}$ is uniquely determined from the initial level-set $\{\varphi_0 \geq c\}$ and the comparison principle still holds. However, one should notice that the proof is not easy since the problem is of a nonlocal nature. We refer to [27] for the proof.

We shall discuss a few applications of these powerful results. For this purpose, following [26] we recall the notion of crystalline flow [4, 68] of (2.1) when γ is crystalline; i.e., Frank γ is a convex m-polygon. Let q_i $(i = 1,2,...,m)$ be its vertices, and let $\mathcal{N} \subset S^1 = \big\{p \in \mathbb{R}^2 \;\big|\; |p| = 1\big\}$ be the set of all unit vectors of the form $q_i/|q_i|$ $(i = 1,2,...,m)$. A simple polygonal curve S in $\mathbb{R}^2$ is called an *admissible crystal* if all outward normal (orientation) vectors belong to $\mathcal{N}$ and the orientations of adjacent segments (facet) point to vertices adjacent in Frank γ. A family of polygons $\{S_t\}_{t\in J}$ is an *admissible evolving crystal* if S_t moves at least C^1 in time, where J is a time interval. This last requirement implicitly assumes that the numbers of facets and the orientation at each facet are independent of time. In other words, S_t is of the form $S_t = \cup_{j=1}^r S_j(t)$ and $S_j(t)$ is a maximal nontrivial closed segment (facet) of S_t and the orientation $\mathbf{n}_j$ of $S_j(t)$ is independent of time. To fix the idea we number facets clockwise.

Let $\{S_t\}_{t\in J}$ be an admissible evolving crystal with $J = [0, T)$. We say that $\{S_t\}_{t\in J}$ is a *γ-regular flow* of (2.1) if

$$V_j = g\Big(\mathbf{n}_j \;;\; \chi_j \frac{\Delta(\mathbf{n}_j)}{L_j(t)}\Big) \quad \text{on} \quad S_j(t) \tag{2.3}$$

for $j = 1, 2, ..., r$, where V_j denotes the normal velocity of $S_j(t)$. The quantity $\chi_j \Delta(\mathbf{n}_j)/L_j(t)$ is a nonlocal weighted curvature $\Lambda_\gamma(\mathbf{n}_j)$, where $L_j(t)$ denotes the length of $S_j(t)$ and $\Delta(\mathbf{m}_i) = \tilde{\gamma}'(\theta_i + 0) - \tilde{\gamma}'(\theta_i - 0)$, $\mathbf{m}_i = (\cos\theta_i, \sin\theta_i) \in \mathcal{N}$ with $\tilde{\gamma}(\theta) = \gamma(\cos\theta, \sin\theta)$. The quantity χ_j is a transition number. It takes $+1$ (resp. -1) if S_t is concave (resp. convex) in the direction of $\mathbf{n}_j$ near S_j; we use the convention that $\chi_j = -1$ for all $j = 1, ..., r$ if S_t is a convex polygon. Otherwise we set $\chi_j = 0$. It is well-known that $\Delta(\mathbf{m}_i)$ is the length of a facet of the Wulff shape

$$W_\gamma = \{x \in \mathbb{R}^2 \mid x \cdot p \le \gamma(p) \quad \text{for all } p \in \mathbb{R}^2\}$$

with outward normal $\mathbf{m}_i \in \mathcal{N}$. By this convention the weighted curvature Λ_γ of a Wulff shape is -1, independent of the facet. For smooth γ the curvature Λ_γ of the corresponding Wulff shape is always -1, so it is a substitute for a circle when γ is isotropic. Thus the definition of nonlocal curvature is quite natural, to the extent that one admits the facet-stay-as-facet hypothesis as derived from (1.1).

Note that L_j always fulfills a transport equation

$$\frac{\mathrm{d}L_j}{\mathrm{d}t}(t) = (\cot\sigma_{j+1} + \cot\sigma_j)V_j - (\sin\sigma_j)^{-1}V_{j-1} - (\sin\sigma_{j+1})^{-1}V_{j+1} \tag{2.4}$$

as observed in [4, 68]. Here $\sigma_j = \theta_j - \theta_{j-1}$ for $\mathbf{n}_j = (\cos\theta_j, \sin\theta_j)$ and V_j denotes the normal velocity of $S_j(t)$; the index j is considered modulo r. Combining (2.3) and (2.4) we get an r-system of ordinary differential equations for the L_js. A local existence theorem of ODEs yields local existence of γ-regular flow.

Proposition 2.2. *Assume that $\lambda \mapsto g(\mathbf{m}_i, \lambda)(\mathbf{n}_i \in \mathcal{N})$ is locally Lipschitz continuous on $\mathbb{R} \setminus \{0\}$. Let S_0 be an admissible crystal. Then there is a constant $T > 0$ and a unique γ-regular flow $\{S_t\}_{t\in J}$ of (2.1) with initial data S_0 where $J = [0, T)$.*

During evolution, some facets may disappear at the maximal existence time T of γ-regular flow. As discussed in [26], for a wide class of equations (2.3) (or (2.1)) S_t becomes still admissible at T, so one can construct γ-regular flow starting from S_T. A *crystalline flow* $\{S_t\}$ of (2.1) is obtained by repeating this procedure. For simplicity we consider (1.5) with symmetric crystalline γ; i.e., $\gamma(p) = \gamma(-p)$. For this equation a unique crystalline flow exists until it shrinks to a point without self-intersection if initially S_0 is an admissible crystal. Moreover, this crystalline flow is consistent with a level-set flow defined by $\{\varphi = c\}$ for φ solving (1.3) [26].

We now mention two typical applications of our convergence theorem (Theorem 2.1). Let $\mathcal{I}_s$ denote the set of $\gamma \in \mathcal{I}$ such that Frank γ has a smooth boundary with nonzero curvature and that γ is symmetric; i.e. $\gamma(p) = \gamma(-p)$. For $\gamma \in \mathcal{I}_s$ it is well-known [16] that (1.5) admits a smooth solution $\{\Gamma_t\}$ starting from a simple, closed, smooth curve Γ_0 until it shrinks to a point. As is well-known, this $\{\Gamma_t\}$ is also a level-set flow of (1.3); in particular, no fattening occurs [37, 36]. It is easy to see that for a symmetric crystalline γ there is a family $\{\gamma_\varepsilon\} \subset \mathcal{I}_s$ such that $\gamma_\varepsilon \to \gamma$ locally uniformly as $\varepsilon \to 0$. It is also easy to see that for $\gamma \in \mathcal{I}_s$ there is a family $\{\gamma_\varepsilon\}$ of crystalline energies such that $\gamma_\varepsilon \to \gamma$. Theorem 2.1 (with help of [27, Corollary 8.3]) provides various approximation results; see e.g. [26]. We give a simple example.

Theorem 2.3. ***(Convergence of crystalline algorithm)*** *For $\gamma \in \mathcal{I}_s$ let γ_ε be a symmetric crystalline energy such that $\gamma_\varepsilon \to \gamma$ locally uniformly as $\varepsilon \to 0$. Let $\{S_t^\varepsilon\}_{t\in J}$ be the crystalline flow of* (1.5) *with γ replaced by γ_ε starting from S_0^ε. For a smooth, closed, simple curve Γ_0 let $\{\Gamma_t\}_{t\in J}$ be the solution of* (1.5) *with $\Gamma_t|_{t=0} = \Gamma_0$. If $d_{\mathrm{H}}(S_0^\varepsilon, \Gamma_0) \to 0$ as $\varepsilon \to 0$, then $\sup_{0\le t\le T'} d_{\mathrm{H}}(\Gamma_t, S_t^\varepsilon) \to 0$ for $T' < T$, where $J = [0, T)$. Here d_{H} denotes the Hausdorff distance and T is the time when Γ_t shrinks to a point.*

Theorem 2.4. ***(Approximation by a smooth problem)*** *For a symmetric crystalline energy let $\gamma_\varepsilon \in \mathcal{I}_s$ satisfy $\gamma_\varepsilon \to \gamma$ as $\varepsilon \to 0$ locally uniformly. Let $\{\Gamma_t^\varepsilon\}_{t\in J}$ be the solution of* (1.5) *with γ replaced by γ_ε starting from a smooth, closed, simple curve Γ_0^ε. Let $\{S_t\}_{t\in J}$ be the crystalline flow of* (1.5) *starting from an admissible crystal S_0. If $d_{\mathrm{H}}(\Gamma_0^\varepsilon, S_0) \to 0$ as $\varepsilon \to 0$, then $\sup_{0\le t\le T'} d_{\mathrm{H}}(\Gamma_t^\varepsilon, S_t) \to 0$ for $T' < T$, where $J = [0, T)$. Here T is the time when S_t shrinks to a point.*

There has been some work on convergence of the crystalline algorithm [21, 46, 25] for graph-like curves. For isotropic energy, i.e. for a curve-shortening equation, the convergence is shown for a convex curve [45] and a general curve [50]. Theorem 2.3 is a generalisation of these results. For a more general statement and more references, see [26] and the references cited therein.

Theorem 2.4 justified the crystalline flow as a limit of a smoother problem. There are several previous works (including [21, 25]) concerning graph-like curves. This result is 'essentially' known from [27, Corollary 8.3] and consistency results; however, it seems that it has not been stated explicitly in the literature.

Our level-set flow for a crystalline energy provides a solution starting from a non-admissible polygon. It turns out that it evolves as a polygon and is still computable. We do not further investigate this problem here, leaving it to [30], and its application to image analysis to [48].

Finally, we mention that a singular interfacial energy really arises when one considers a crystal at low temperature [29]. There it is shown that an evaporation dynamics (with a facet) proposed by [66] is approximated by a smoother problem, as an application of a convergence result in [25].

3 Shocks and vertical diffusion

We consider the initial-value problem for a nonlinear first-order equation of the form

$$u_t + H(u, \nabla u) = 0\,, \quad x \in \mathbb{R}^n\,, \quad t > 0 \tag{3.1}$$

$$u(x,0) = u_0(x)\,, \tag{3.2}$$

where $u = u(x,t)$ is a real-valued function and $H = H(r,p)$ is continuous. Let us derive a level-set equation for the graph of u. Let $\varphi = \varphi(x,y,t)$ be a function such that $\varphi\big(x, u(x,t), t\big) = 0$ for $x \in \mathbb{R}^n, t > 0$. To fix the idea we assume that $\varphi_y \leq 0$. Then $u_t = -\varphi_t/\varphi_y$, $\partial u/\partial x_i = -(\partial\varphi/\partial x_i)/\varphi_y$ so (3.1) yields

$$\varphi_t - \varphi_y\, H\Big(y\,, \frac{-\nabla_x \varphi}{\varphi_y}\Big) = 0 \tag{3.3}$$

on the zero level-set of φ. We consider (3.2) in $\mathbb{R}^n \times \mathbb{R} \times (0,T)$ rather than on the zero level-set of φ. To solve the initial-value problem (3.1)–(3.2) we rather solve φ with initial data φ_0 such that the zero level-set of φ_0 agrees with the graph of u_0. One advantage of (3.3) over (3.1) is that (3.3) does not contain the unknown φ, while (3.1) depends explicitly on the unknown u. This idea goes back to Jacobi [13] to solve (3.1)–(3.2) by the method of characteristics. Such a level-set formulation is studied by Evans [19] for the heat equation. A related level-set approach is proposed by Osher [58] for a stationary Hamilton–Jacobi equation. The author with M.-H. Sato [44], applies the above level-set method to solve (3.1) globally-in-time in a viscosity sense when u_0 is not necessarily continuous, assuming that $r \mapsto H(r,p)$ is nondecreasing (under some linear growth condition in p of $H(r,p)$). Since the solution is not expected to be continuous, the uniqueness of a solution is not generally valid. It turns out that this problem is related to 'fattening' in the level-set method for a curvature flow equation [6]. The level-set method [44] provides the largest and the smallest solutions of (3.1)–(3.2).

If the monotonicity condition on $H(r,p)$ w.r.t. r is removed, the situation is quite different. A typical example is the conservation law

$$u_t + \operatorname{div} \mathbf{F}(u) = 0$$

with $\mathbf{F} = (F_1, ..., F_n)$, which includes the Burgers equation as a special example. The solution of the initial value problem (3.1) may develop discontinuity (called *shock*) in finite time, even if the initial data is smooth. (Under the monotonicity condition on H a solution stays continuous if initial data is (uniformly) continuous.) For conservation laws there is a unique way to extend a solution globally-in-time after it develop singularities. This special weak solution is called an *entropy solution* [18, 57]. If one tracks the zero level-set of (3.3) it may 'overturn' and it cannot be viewed as the graph of single-valued function after the solution u develops shocks. In what way should one track the graph of an entropy solution?

In [34, 35] we propose to consider

$$\varphi_t - \varphi_y H\Big(y\,, \frac{-\nabla_x \varphi}{\varphi_y}\Big) = M|\nabla_{x,y}\varphi|\,\frac{\partial}{\partial y}\Big(\frac{\varphi_y}{|\varphi_y|}\Big) \tag{3.4}$$

with some $M > 0$ instead of (3.3). This is again the equation with singular diffusivity, and the notion of solution is unclear. Each level-set $\{\Gamma_t\}$ of φ is moved by

$$V = -n_{n+1}\, H\Big(y\,, \frac{-(n_1, .., n_n)}{n_{n+1}}\Big) - M\ \mathrm{div}_{\Gamma_t}\, \xi(\mathbf{n}) \quad \text{on } \Gamma_t\,, \tag{3.5}$$

where $\mathbf{n} = (n_1, ..., n_{n+1})$ is the upward normal of Γ_t and $\xi = \nabla\gamma$ with $\gamma(p) = |p_{n+1}|$ for $p = (p_1, ..., p_{n+1})$. The equation (3.4) is the level-set equation of the anisotropic curvature-flow equation (3.5) with singular interfacial energy. However, the theory in Section 2 does not apply even for $n = 1$, since (3.5) is spatially inhomogeneous.

Instead of developing a general theory for (3.4) or (1.7) we rather study its approximation. We consider a simple example (1.8) with initial data

$$u(x,0) = u_0(x) = \begin{cases} a & \text{for } x > 0 \\ a+d & \text{for } x < 0\,, \end{cases}$$

for $a \in \mathbb{R}$ and $d > 0$. Then it is well-known that the entropy solution of (1.8) is of the form

$$u_E(x,t) = \begin{cases} a & \text{for } x > ct \\ a+d & \text{for } x < ct\,, \end{cases} \tag{3.6}$$

with $c = \big(f(a+d) - f(a)\big)/d$, where $f(r) = \frac{1}{2}r^2$. Consider

$$\varphi_t + y\varphi_x = M|\nabla_{x,y}\,\varphi|\ \mathrm{div}\, \xi_\varepsilon\Big(\frac{-\nabla\varphi}{|\nabla\varphi|}\Big) \tag{3.7}$$

with $\xi_\varepsilon = \nabla\gamma_\varepsilon$ such that γ_ε approximates γ. Initial data φ_0 of φ is taken so that it is uniformly continuous with

$$\big\{(x,y) \mid \varphi_0(x,y) \le 0\big\} = \big\{(x,y) \mid y \le u_0(x)\big\} \tag{3.8}$$

and that $y \mapsto \varphi_0(x,y)$ is non-increasing. If γ_ε is convex and C^2, it is known that (3.7) admits a unique continuous solution φ_ε [36]. The next result indicates the role of the vertical-diffusion term.

Let φ_ε be the solution (3.7) satistying (3.8). For technical reasons we assume that $M < \frac{1}{8}d^2$.

Theorem 3.1. *([28].) There is a sequence of convex, C^2 functions $\{\gamma_\varepsilon\}$ converging to $\gamma(p) = |p_{n+1}|$ locally uniformly such that*

$$E_\varepsilon = \big\{(x,y,t) \in \mathbb{R}^2 \times [0,\alpha) \bigm| \varphi_\varepsilon(x,y,t) \le 0\big\}$$

converges to

$$\big\{(x,y,t) \in \mathbb{R}^2 \times [\partial,\infty) \bigm| y \le u_E(x,t)\big\}$$

in the Hausdorff distance, as $\varepsilon \to 0$, if and only if $M \ge \frac{1}{16}d^2$. If $M < \frac{1}{16}d^2$, then the limit E_ε cannot be viewed as the graph of a single-valued function.

The class of approximation such that the convergence is valid is given explicitly in [28]. For technical reasons we assume that γ_ε is positively homogeneous of degree one and $M < \frac{1}{8}d^2$, but these assumptions are expected to be removed after further development of the theory. The threshold value $\frac{1}{16}d^2$ is consistent with one obtained in the theory of nonlinear semigroups [35, 28]. It does not depend on our method of approximation. The limit of E_ε for $M < \frac{1}{16}d^2$ is also explicitly described in [28].

When the equation (3.1) is not a conservation law, it is not clear *a priori* what is a reasonable notion of weak solution after u develops shocks. A typical example is

$$u_t - u\bigl(1 + |\nabla u|^2\bigr)^{\frac{1}{2}} = 0 \tag{3.9}$$

It controls the speed V of the graph Γ_t of $u(\,\cdot\,, t)$ as $V = y$ in xy-space. Evidently, u may develop a jump discontinuity within finite time for some smooth initial data. Classical theory for viscosity solutions does not apply to such a problem. In [34] we introduced a notion of a *proper viscosity solution* which is a special viscosity solution [17] with control over the speed of the shock. It turns out this is a suitable notion for solving the initial-value problem globally in time, and it is consistent with the notion of an entropy solution when the equation is a conservation law. A proper viscosity solution is obtained by a vanishing viscosity method such as an entropy solution [34].

The equation (3.4) is useful to track the evolution of the graph of a proper viscosity solution of (3.1). Although this is still work in progress, it is expected that a set like E_ε converges to the subgraph set of a (maximal) proper viscosity solution for sufficiently large M, at least for bounded solutions. Thus it is reasonable to define the notion of a solution of the level-set equation of the form

$$\varphi_t - \varphi_y H\Bigl(y\,, \frac{-\nabla_x \varphi}{\varphi_y}\Bigr) = \infty \bigl|\nabla_{x,y}\varphi\bigr| \frac{\partial}{\partial y}\Bigl(\frac{\varphi_y}{|\varphi_y|}\Bigr). \tag{3.10}$$

For simplicity we consider a periodic function in x.

Definition 3.2. *Assume that $\varphi = \varphi(x, y, t)$ is non-increasing in y and φ is upper-semicontinuous in $Q = \mathbf{T}^n \times \mathbb{R} \times (0, T)$. We say that φ is a viscosity subsolution of* (3.10) *if the height function*

$$u_c^\sharp(x, t) = \sup\{y \mid \varphi(x, y, t) \ge c\}$$

of each super level-set $\{\varphi \ge c\}$ is a proper viscosity subsolution [34] of (3.1). *The definition of supersolution for a lower-semicontinuous function is obtained by replacing* sup *by* inf, *$\{\varphi \ge c\}$ by $\{\varphi \le c\}$, and subsolution by supersolution.*

Since proper viscosity solutions enjoy a weak comparison principle [34, Theorem 4.1], it is not difficult prove a comparison principle for (3.10).

Theorem 3.3. *Assume that H is continuous and that H satisfies*

$$\bigl|H(r, p) - H(r', p)\bigr| \le C|r - r'|\,\bigl(|p| + 1\bigr)$$

for all $r, r' \in \mathbb{R}$ satisfying $|r|, |r'| \le K$ and for all $p \in \mathbb{R}^n$ with C depending only on K. Let φ_1 and $-\varphi_2$ be upper-semicontinuous in $\bar{Q}$. Assume that $\lambda H(r, p/\lambda)$

converges locally uniformly in $(r,p) \in \mathbb{R}^n \times \mathbb{R}$ *as* $\lambda \to 0$. *Assume that* $y \mapsto \varphi_i(x,y,t)$ $(i=1,2)$ *is non-increasing. Assume that* φ_1 *and* φ_2 *are, respectively, a sub- and supersolution of* (3.10) *in* Q. *Then* $\varphi_1 \le \varphi_2$ *in* Q *if* $\varphi_1 \le \varphi_2$ *at* $t=0$ *(provided that each super and sub level-set of* φ_1 *and* φ_2 *are bounded in the* y *or* $-\,y$ *direction.)*

Proof. Suppose that the conclusion were to be false. Then there is $c \in \mathbb{R}$ such that

$$\{\varphi_1 \ge c\} \cap \{\varphi_2 \le c\}$$

contains an interior point in Q, where we abbreviate $\{(x,y,t) \in Q \mid \varphi_1(x,y,t) \ge c\}$ by $\{\varphi_1 \ge c\}$ etc., as before. Thus there is $c' < c$ close to c such that

$$\{\varphi_1 \ge c\} \cap \{\varphi_2 \le c'\} \ne \emptyset\,. \tag{3.11}$$

Since $\varphi_1 \le \varphi_2$ at $t=0$, then

$$\{\varphi_1|_{t=0} \ge c\} \cap \{\varphi_2|_{t=0} \le c'\} = \emptyset \quad \text{in } \mathbf{T}^n \times \mathbb{R}\,. \tag{3.12}$$

By definition

$$\begin{aligned} u^\sharp(x,t) &= \sup\{y \mid \varphi_1(x,y,t) \ge c\} \\ v_\sharp(x,t) &= \inf\{y \mid \varphi_2(x,y,t) \le c'\} \end{aligned}$$

are, respectively, a proper viscosity sub and supersolution of (3.1).

Both functions are bounded. By (3.12) $u^\sharp < v_\sharp$ at $t=0$. We now apply the weak comparison principle [34, Theorem 4.1] to conclude that $u^\sharp < v_\sharp$ for all $t \in [0,T)$. However, this contradicts (3.11). So we conclude that $\varphi_1 \le \varphi_2$ in Q. (A similar idea is found in [36, Chapter 5] to discuss the relation of comparison principles of solutions of a level-set equation and of their slices.) □

If each level-set of φ is bounded in the y direction, and the difference of upper bound and lower bound is bounded independent of levels, it is expected that for a sufficiently large M equation (3.4) and (3.10) are the same for such φs. This is despite the fact that the definition of a solution for (3.4) is not yet established. The equation (3.4) is also convenient to calculate a proper viscosity solution numerically, as developed in [70, 69].

4 Gradient flow of total variation

The gradient flow of total variation is of the form

$$u_t = \operatorname{div}\Big(\frac{\nabla u}{|\nabla u|}\Big) \tag{4.1}$$

for $u = u(x,t)$, $x \in \Omega \subset \mathbb{R}^n$ with a bounded domain Ω. As mentioned for (1.1) under the Dirichlet (or Neumann) boundary condition for this problem classical nonlinear semigroup theory yields a global unique solution [53, 31] for the initial-boundary value problem for (4.1). In [49] the global unique solvability is established for equations including (4.1) with time dependent Dirichlet boundary data. It is also shown in [49] that the solution tends to a minimiser of total variation as time tends infinity when the boudary data is time independent. Note that (4.1) is different from the level-set mean-curvature flow equation (1.4) [15, 20] where the singularity at $\nabla\varphi = 0$ is weaker than (4.1) so that equation (1.4) is still a local equation. As well as L^2 theory, L^1-theory has been established in [3, 1, 2]. They conclude that the solution semigroup is an L^1-contraction as well as being an L^2 and L^∞-contraction. However, a detailed behaviour of a facet is not well-studied for $n \geq 2$, except for in [7]. What is known is that the facet-stay-as-facet hypothesis is no longer valid. In other words u_t may not be a constant on a facet of slope zero [8, 9] when the space dimension $n \geq 2$.

For applications to both image processing and multigrain motion [55, 72] it is important to consider a spatially inhomogeneous equation of the form

$$b(x)\, u_t = \operatorname{div}\Big(a(x)\, \frac{\nabla u}{|\nabla u|}\Big) \tag{4.2}$$

with $a(x) > 0$, $b(x) > 0$. In [53, 31] the one-dimensional version of (4.2) with Dirchlet condition is studied. It provides a necessary and sufficient condition that a facet (called a *plateau* in [53, 31]) may break. Note that this problem is still in the realm of nonlinear semigroup theory for maximal monotone operators in L^2. Thus the notion of solution is *a priori* defined. When a is piecewise linear with $b \equiv 1$, the whole evolution for piecewise-constant initial data, with jumps included in the singular set of a, is easily computed by solving ODEs; here, the facet-stay-facet-hypothesis is still valid [31]. In [31] the evolution of piecewise-constant functions is studied in detail. In the meantime the Allen–Cahn type equation with top-order term $\operatorname{div}(\nabla u/|\nabla u|)$, and with obstacle-type double-well potential, has been studied in detail. For example, classification of a stationary solution and its stability is discussed in [52, 65] for $n = 1$. Recently, a higher-dimensional problem has also been studied by M. Kimura and K. Shirakawa.

There are several analytic results reflecting constraint $\int |u - u_0|^2\, dx = \text{const}$. For the equation

$$u_t = \operatorname{div}\Big(a(x)\, \frac{\nabla u}{|\nabla u|}\Big) - b(u - u_0) \quad (b > 0)$$

the global unique solvability has been established with Neumann data; see [14]. However, the gradient flow of total variation, with the constraint $\int |u - u_0|^2 = \text{const}$, seems not to have been studied from the point of view of mathematical analysis. For a value-constraint problem like (1.9) little is known. In [40] the initial-value problem for (1.9) with the Dirichlet problem is studied when $n = 1$ and the constraint is $u_1^2 + u_2^2 = 1$ (with $u_3 \equiv 0$). For piecewise-constant initial data it is shown that the solution tends to some stationary solution within finite time [40]. It is expected that a solution exists globally-in-time for general initial data, after suitable interpretation of the equation (1.9). Recently, local solvability has been proved in [39], when initial data is smooth and its total variation is small under periodic boundary conditions.

Bibliography

[1] F. Andreu, C. Ballester, V. Caselles and J. M. Mázon, *Minimizing total variation flow*, Diff. Int. Eq., **14** (2001), pp. 321–360.

[2] F. Andreu, C. Ballester, V. Caselles and J. M. Mázon *The Dirichlet problem for the total variation flow*, J. Funct. Anal., **180** (2001), pp. 347–403.

[3] F. Andreu-Vaillo, V. Caselles and J. M. Mázon, *Existence and uniqueness of a solution for a parabolic quasilinear problem for linear growth functionals with L^1 data*, Math. Ann., **322** (2002), pp. 139–206.

[4] S. B. Angenent and M. E. Gurtin, *Multiphase thermomechanics with interfacial structure 2. Evolution of an isothermal interface*, Arch. Rational Mech. Anal., **108** (1989), pp. 323–391.

[5] V. Barbu, *Nonlinear Semigroups and Differential Equations in Banach spaces*, Noordnoff, Grøningen, 1976.

[6] G. Barles, H. M. Soner and P. E. Souganidis, *Front propagation and phase field theory*, SIAM J. Control Optim., **31** (1993), pp. 439–469.

[7] G. Bellettini, V. Caselles and M. Novaga, *The total variation flow in $\mathbb{R}^N$*, J. Differential Equations, to appear.

[8] G. Bellettini and M. Novaga, *Approximation and comparison for non-smooth anisotropic motion by mean curvature in $\mathbb{R}^N$*, Math. Mod. Methods. Appl. Sci., **10** (2000), pp. 1–10.

[9] G. Bellettini, M. Novaga and M. Paolini, *Facet-breaking for three dimensional crystal evolving by mean curvature*, Interfaces and Free Boundaries, **1** (1999), pp. 39–55.

[10] H. Brezis, *Operateurs Maximaux Monotone*, North-Holland, Amsterdam, 1973.

[11] H. Brezis and A. Pazy, *Convergence and approximation of semigroups of nonlinear operators in Banach spaces*, J. Funct. Anal., **9** (1972), pp. 63–74.

[12] F. Cao, *Geometric Curve Evolution and Image Processing*, Lecture Notes in Math., Springer 1805, 2003.

[13] C. Carathéodory, *Calculus of Variations and Partial Differential Equations of the First Order*, Part I, §49, Holden–Day, San Francisco, London, Amsterdam, 1965.

[14] Y. Chen and T. Wunderli, *Adaptive total variation for image restoration in BV space*, J. Math. Anal. Appl., **272** (2002), pp. 117–137.

[15] Y.-G. Chen, Y. Giga and S. Goto, *Uniqueness and existence of viscosity solutions of generalized mean curvature flow cquations*, J. Diff. Geom., **33** (1991), pp. 749–786.

[16] K.-S. Chou and X.-P. Zhu, *The Curve Shortening Problem*, Chapman and Hall / CRC, 2001.

[17] M. G. Crandall, H. Ishii and P.-L. Lions, *User's guide to viscosity solutions of second order partial differential equation*, Bull. Amer. Math. Soc., **27** (1992), pp. 1–67.

[18] C. Dafermos, *Hyperbolic Conservation Laws in Continuum Physics*, Grundlehren der Mathematischen Wissenschaften 325, Springer, Berlin 2000.

[19] L. C. Evans, *A geometric interpretation of the heat equation with multivalued initial data*, SIAM J. Math. Anal., **27** (1996), pp. 932–958.

[20] L. C. Evans and J. Spruck, *Motion of level sets by mean curvature I*, J. Diff. Geom., **33** (1991), pp. 635–681.

[21] T. Fukui and Y. Giga, *Motion of a graph by nonsmooth weighted curvature*, in World Congress of Nonlinear Analysis '92, V. Lakshmikantham, ed., de Gruyter, Berlin, vol. I (1996), pp.47–56.

[22] M.-H. Giga and Y. Giga, *Evolving graphs by singular weighted curvature*, Arch. Rational Mech. Anal., **141** (1998), pp. 117–198.

[23] M.-H. Giga and Y. Giga, *Remark on convergence of evolving graphs by nonlocal curvature*, in Progress in Partial Differential Equations, vol.1, H. Amann et al. eds., Pitman Reseach Notes in Math. Ser., **383** (1998), pp. 99–116.

[24] M.-H. Giga and Y. Giga, *A subdifferential interpretation of crystalline motion under nonuniform driving force*, in Proc. of the International Conference in Dynamical Systems and Differential Equations, Springfield Missouri, (1996), "Dynamical Systems and Differential Equations", W.-X. Chen and S.-C. Hu eds., Southwest Missouri Univ. vol.**1** (1998), pp. 276–287.

[25] M.-H. Giga and Y. Giga, *Stability for evolving graphs by nonlocal weighted curvature*, Commum. in Partial Differential Equations., **24** (1999), pp. 109–184.

[26] M.-H. Giga and Y. Giga, *Crystalline and level set flow — Convergence of a crystalline algorithm for a general anisotropic curvature flow in the plane*, in Free Boundary Problems: Theory and Applications, (N. Kenmochi ed.), Gakuto International Series, Math. Sciences and Appl., vol **13** (2000), pp. 64–79.

[27] M.-H. Giga and Y. Giga, *Generalized motion by nonlocal curvature in the plane*, Arch. Rational. Mech. Anal., **159** (2001), pp. 295–333.

[28] M.-H. Giga and Y. Giga, *Minimal vertical singular diffusion preventing overturning for the Burges equation*, Contem. Math., to appear.

[29] M.-H. Giga and Y. Giga, *A PDE approach for motion of phase-boundaries by a singular interfacial energy*, Hokkaido Univ, Preprint Series in Math., ♯584 (2003).

[30] M.-H. Giga, Y. Giga and H. Hontani, *Selfsimilar expanding solutions in a sector for a crystalline flow*, in preparation.

[31] M.-H. Giga, Y. Giga and R. Kobayashi, *Very singular diffusion equations*, Proc. of Taniguchi Conf. on Math., Advanced Studies in Pure Mathematics, **31** (2001), pp. 93–125.

[32] Y. Giga, *A level set method for surface evolution equations*, Sūgaku **47** (1995), pp. 321–340; English translation, Sugaku Exposition, **10** (1997), pp. 217–241.

[33] Y. Giga, *Anisotropic curvature effect in interface dynamics*, Sūgaku **52** (2000), pp. 113–127; English translation, to appear.

[34] Y. Giga, *Viscosity solutions with shocks*, Comm. Pure Appl. Math., **55** (2002), pp. 431–480.

[35] Y. Giga, *Shocks and very strong vertical diffusion*, Proc. of international conference on partial differential equations in celebration of the seventy fifth birthday of Professor Louis Nirenberg, Taiwan, 2001.

[36] Y. Giga, *Surfaces Evolution Equations — a level set method*, Hokkaido University Technical Report Series in Mathematics, ♯71, Sapporo, 2002, also Lipschitz Lecture Notes, **44**, University of Bonn, 2002.

[37] Y. Giga and S. Goto, *Geometric evolution of phase-boundaries*, in On the evolution of phase-boundaries, M. Gurtin and G. McFadden eds., IMA vol. Math. Appl., Springer **43** (1992), pp. 51–65.

[38] Y. Giga, M. E. Gurtin and J. Matias, *On the dynamics of crystalline motions*, Japan J. Indust. Appl. Math., **15** (1998), pp. 7–50.

[39] Y. Giga, Y. Kashima and N. Yamazaki, *Local solvability of a constrained gradient system of total variation*, preprint.

[40] Y. Giga and R. Kobayashi, *On constrained equations with singular diffusivity*, Hokkaido Univ. Preprint Series in Math., ♯588 (2003).

[41] Y. Giga, M. Paolini and P. Rybka, *On the motion by singular interfacial energy*, Japan J. Indust. Appl. Math., **18** (2001), pp. 231–248.

[42] Y. Giga and P. Rybka, *Quasi-static evolution of 3-D crystal grown from supersaturated vapor*, Diff. Int. Eq., **15** (2002), pp. 1–15.

[43] Y. Giga and P. Rybka, *Berg's effect*, Adv. Math. Sci. Appl., to appear.

[44] Y. Giga and M.-H. Sato, *A level set approach to semicontinuous viscosity solutions for Cauchy problems*, Comm. Partial Differential Equations, **26** (2001), pp. 813–839.

[45] P. M. Girão, *Convergence of a crystalline algorithm for the motion of a simple closed convex curve by weighted curvature*, SIAM J. Numer. Anal., **32** (1995), pp. 886–899.

[46] P. M. Girão and R. V. Kohn, *Convergence of a crystalline algorithm for the heat equation in one dimension and for the motion of a graph by weighted curvature,* Numer. Math., **67** (1994), pp. 41–70.

[47] M. E. Gurtin, *Thermomechanics of Evolving Phase Boundaries in the Plane*, Oxford, Clarendon Press, 1993.

[48] H. Hontani, Mi-Ho. Giga, Y. Giga and K. Deguchi, *A computation of a crystalline flow starting from non-admissible polygon using expanding selfsimilar solutions*, preprint.

[49] R. Hardt and X. Zhou, *An evolution problem for linear growth functionals*, Comm. Partial Differential Equations, **19** (1994), pp. 1879–1907.

[50] K. Ishii and H. M. Soner, *Regularity and convergence of crystalline motion*, SIAM J. Math. Anal., **30** (1999), pp. 19–37.

[51] Y. Kashima *A subdifferential formulation of fourth order singular diffusion equations*, Adv. Math. Sci Appl., to appear.

[52] N. Kenmochi and K. Shirakawa, *A variational inequality for total variation functional with constraint*, Nonlinear Anal., **46** (2001), pp. 435–455.

[53] R. Kobayashi and Y. Giga, *Equations with singular diffusivity*, J. Stat. Phys., **95** (1999), pp. 1187–1220.

[54] R. Kobayashi and Y. Giga, *On anisotropy and curvature effects for growing crystals*, Japan J. Indust. Appl. Math., **18** (2001), pp. 207–230.

[55] R. Kobayashi, J. A. Warren and W. C. Carter, *A Continuum Model of Grain Boundaries*, Physica D, **140** (2000), pp. 141–150.

[56] Y. Kōmura, *Nonlinear semi-groups in Hilbert space*, J. Math. Soc. Japan., **19** (1967), pp. 493–507.

[57] P. D. Lax, *Hyperbolic Systems of Conservation Laws and Mathematical Theory of Shock Waves*, SIAM, Philadelphia, Pa., 1973.

[58] S. Osher, *A level set formulation for the solution of the Dirichlet problem for Hamilton–Jacobi equations*, SIAM J. Math. Anal., **24** (1993), pp. 1145–1152.

[59] S. Osher and R. Fedkiw, *Level Set Methods and Dynamic Implicit Surfaces*, Applied Math. Ser. 153, Springer, 2003.

[60] S. Osher and J. A. Sethian, *Front propagation with curvature dependent speed: Algorithm based on Hamilton–Jacobi formulations*, J. Comput. Phys., **79** (1988), pp. 12–49.

[61] T. Preusser and M. Rumpf, *A level set method for anisotropic geometric diffusion in 3D image processing*, SIAM J. Appl. Math., **62** (2002), pp. 1772–1793.

[62] L. I. Rudin, S. Osher and E. Fatemi, *Nonlinear total variation based noise removal algorithms*, Physica D, **60** (1992), pp. 259–268.

[63] G. Sapiro, *Geometric Partial Differential Equations and Image Analysis*, Cambridge University Press, United Kingdom, 2001.

[64] J. A. Sethian, *Level Set Methods, Evolving Interfaces in Geometry, Fluid Mechanics, Computer Vision, and Materials Science*, Cambridge Univ. Press, 1996.

[65] K. Shirakawa, *Parabolic variational inequality associated with the total variation functional*, Nonlinear Anal., **47** (2001), pp. 3195–3206.

[66] H. Spohn, *Surface dynamics below the roughening transition*, J. Phys. I France, **3** (1993), pp. 69–81.

[67] B. Tang, G. Sapiro and V. Caselles, *Color image enhancement via chromaticity diffusion*, IEEE Trans. on Image Processing, **10** (2001), pp. 701–707.

[68] J. Taylor, *Constructions and conjectures in crystalline nondifferential geometry*, in Differential Geometry, B. Lawson and K. Tanenblat, eds., Proceedings of the Conference on Differential Geometry, Rio de Janeiro, Pitman Monograph Surveys Pure Appl. Math., **52** (1991), pp. 321–336.

[69] Y.-H. R. Tsai and Y. Giga, *A numerical study of anisotropic crystal growth with bunching under very singular vertical diffusion*, preprint.

[70] Y.-H. R. Tsai, Y. Giga and S. Osher, *A level set approach for computing discontinuous solutions of a class of Hamilton–Jacobi equations*, Math. Comp., **72** (2003), pp. 159–181.

[71] J. Watanabe, *Approximation of nonlinear problems of a certain type, in Numerical Analysis of Evolution Equations*, H. Fujita and M. Yamaguti, eds., Lecture Notes in Num. Appl., 1, Kinokuniya, Tokyo, (1979), pp. 147–163.

[72] J. A. Warren, R. Kobayashi and W. C. Carter, *Modeling grain boundaries using a Phase-field technique*, J. Cryst. Growth, **211** (2000), pp. 18–20.

[73] J. Yunger, *Facet stepping and motion by crystalline curvature*, PhD Thesis, Rutgers University, 1998.

At the Icebreaker reception: Mark Davis (invited speaker, UK), Diana Wolfe (media consultant) and Cathy Sage (media consultant). photography by *Happy Medium Photo Co.*

Philippe Toint is a full professor at the University of Namur. His interests include the practical and theoretical aspects of nonlinear optimisation, and, more specifically, of large-scale problems. He pioneered the use of sparse linear algebra in quasi-Newton methods and contributed to the study of decomposition techniques, including the notion of partially separable functions.

Toint is director of his university's Transportation Research Group, whose interests range from urban traffic modelling, both static and dynamic, to behavioural studies of a household's mobility. He participated in various projects of the European Union on the impact of technology on transport and conducted the last Belgian National Travel Survey. He is currently working on activity-based transportation analysis techniques and on their application at various levels within the European Union. Until September 2003 Professor Toint was also head of the University of Namur Computing Services, an activity that allowed him to strengthen his interests in the many applications of computer technology in research and society.

Toint obtained his PhD in 1978 from the University of Namur in Belgium after having worked on large-scale nonlinear optimisation with Professor M J D Powell at the University of Cambridge. Together with Andrew Conn and Nicholas Gould, he produced the LANCELOT package, a Fortran program for the solution of large-scale optimisation problems, which was awarded the 1994 Beale–Orchard–Hays Prize of the International Mathematical Programming Society. With the same collaborators, he recently published a complete reference book on trust-region methods, a recognised class of optimisation methods. His current research in this field includes sequential quadratic programming methods and derivative free algorithms, with applications in design of electrical networks and progressive optical lenses, amongst others. He is currently vice-chair of the SIAM Activity Group on Optimization.

Chapter 8

How Mature is Nonlinear Optimization?

*Nick Gould**
Philippe L. Toint†

Abstract: Numerical methods for solving nonlinear optimization problems have been developed for over 50 years. Has this field reached maturity? What are the current research frontiers and ongoing challenges? These are questions that this paper attempts to clarify, if not fully answer. The discussion does not explore the technical intricacies of nonlinear optimization techniques, but instead focusses on concepts and research practice. The field's vibrant nature is illustrated by a number of applications, such as adaptive lens design for spectacles, the piloting of injection of dangerous drugs to patients, the identification of parameters in biochemical models of neurons, food sterilisation or animation techniques for video games.

Contents

*Rutherford Appleton Laboratory, England
†University of Namur, Belgium

1 Introduction

Numerical optimization, that is the corpus of methods and techniques for the solution of mathematically posed problems where one wishes to optimise one "objective" subject to a number of "constraints", has a rich history and continues to be an active research field. The purpose of the present paper is to consider an important subfield, nonlinear optimization, and to propose some thoughts about its level of maturity, both from the research and applications points of view.

Nonlinear optimization is concerned with the solution of continuous problems expressed in the form

$$\min_x f(x) \quad \text{subject to} \quad \begin{cases} c_{\mathcal{E}}(x) = 0, \\ c_{\mathcal{I}}(x) \geq 0, \end{cases} \tag{1.1}$$

where $f : \mathbb{B}^n \to \mathbb{B}$, $c_{\mathcal{E}} : \mathbb{B}^n \to \mathbb{B}^e$ and $c_{\mathcal{I}} : \mathbb{B}^n \to \mathbb{B}^i$ are smooth. This description is somewhat simplified, as, for instance, the level of smoothness of the involved functions may vary or the constraints may involve convex sets, but it is adequate for the purpose of our discussion. The formulation (1.1) also hides a number of interesting special cases and distinctions, the most important being that between convex and nonconvex problems. We will discuss these issues in due course. At this point, we simply note that, except in a few isolated cases, all methods for solving (1.1) are iterative in nature, in the sense that they produces a potentially infinite sequence of iterates that (hopefully) converges to a desired solution.

2 A field no longer in infancy

That nonlinear optimization is no longer a new or young field of scientific activity barely needs discussion. Its rich history includes, for instance, the famous paper by Cauchy (1847), in the middle of the 19th century, but is also associated with other luminaries such as Euler, Gauss and Lagrange, for instance. It really became a field of its own immediately after World War II, along with the birth of the new field of "operations research". The development of the nonlinear least-squares fitting techniques by Levenberg (1944) may be considered as seminal from this point of view. The field had (and still has) special connections with linear algebra (the Cauchy paper being a good example, as are other important contributions such as the Conjugate-Gradient method by Hestenes and Stiefel, 1952 or the exact solution of the trust-region subproblem by Moré and Sorensen, 1983, and many others). Since our purpose is elsewhere, we will thus simply note that the field has a respectable history, a clear sign of being no longer in infancy.

At what we think is a more fundamental level, the methodological focus has also evolved from immediate (and vital) needs, such as effectively solving small problems, to more "long-term" questions. For instance, the early concern of designing methods that are asymptotically fast (that is converge quickly when started in a, possibly very small, neighbourhood of the solution) has progressively shifted to that of methods that are robustly *globally convergent*, in the sense that they are guaranteed to converge to a solution irrespective of the chosen starting point.

The field has become more conscious of itself, as new journals published a growing number of specialised contributions. Although some of the high quality publications of the early days such as *Mathematical Programming* continue to play an important role, it is telling that the long association of nonlinear optimization research with the *SIAM Journal on Numerical Analysis* has been mostly replaced by the highly successful *SIAM Journal on Optimization*. But, as also happens with teenagers, this growing self awareness went along with a clearer and more urgent realisation of the dependence on the rest of the world. In particular, the strong and fruitful interaction of nonlinear optimization with a number of scientific domains where its techniques are applied has become even more crucial. The links between good numerical optimization methods and good software have also emerged as an important research topic. Other new journals, such as, for instance, *Optimization Methods and Software* and *Computational Optimization and Applications*, testify to this evolution, along with the continuing success of older sources like the *Journal of Optimization Theory and Applications* and the *Transactions of the ACM on Mathematical Software*.

Finally, the focus of the problems being solved has evolved from "toy problems", typically involving a very small number (typically less than 10) of variables and/or constraints, to larger and often more realistic instances, that currently feature possibly hundreds of thousands or even millions of variables and constraints. This is not to say that all small problems are uninteresting or easy, but the increasing size of the problems that can realistically be solved is, in our view, indicative of the field's evolution.

3 Some signs of maturity

We next review some elements that we believe testify of the maturity of nonlinear optimization.

3.1 An adequate theoretical understanding

We first look at the state of the theoretical understanding of the problems and numerical procedures to solve them. It is to us very noticeable that the role of theory itself has evolved to occupy a place which we believe is well balanced with practice. In what concerns the problems themselves, the gap between necessary and sufficient optimality conditions has now been shown to be tiny in general, and non-existent for problems such as quadratic programming. When applied to the theory of numerical algorithms, this balance manisfests itself in two complementary developments.

We first note that most of today's best practical algorithms are backed with a suitable convergence theory. This trend is not new, since it started with the convergence studies of variable-metric and quasi-Newton algorithms for unconstrained optimization in the 1970s (see, for instance Powell (1970, 1976) and with the analysis of penalty methods for constrained problems (see Fiacco and McCormick, 1968), soon followed by augmented Lagrangian (see Powell, 1969, Rockafellar, 1974 and Tapia, 1977) and sequential quadratic programming (SQP) (see Han, 1977, and Powell,

1978) methods. Irrespective of what these methods actually are, it is enough to say that they were (and for some, still are) at the leading edge of numerical nonlinear optimization at the time where they were studied. Thus researchers in the field have come to agree that providing convergence theory for successful algorithms is a very important part of making them even more robust and reliable. We continue today to hold the view that such a theory is a *necessary*[1], *while by no means sufficient, condition* for a successful algorithm. Remarkably, today's best algorithms and packages (see next paragraph) are also supported by an adequate convergence theory. In a number of cases, this theory provides results on the crucial issue of global convergence to critical points, but also on the ultimate speed at which this convergence occurs. We also note that the two traditionally distinct (or, even, competing) algorithmic paradigms, known as linesearch and trust-region methods, may today be viewed in a unifying framework (see Section 10.3 of Conn, Gould and Toint, 2000*a*), which we also consider as a sign of maturity.

The second important such sign, as far as theory is concerned, is the development of an improved theory for the simpler but very important subclass of convex problems. For a long time, and although the best available algorithms were often more efficient for such problems, their supporting theory was typically unable to provide stronger or finer results for the convex case. The theory of self-scaling functions pioneered by Nesterov and Nemirovsky (1993) has changed this state of affairs considerably for the better. It indeed gives a much better insight on the *global speed of convergence*, that is even from the early iterations, when the iterates may still be far away from the (in that case, unique) solution. This has allowed the development of very efficient methods that are specific to convex problems. Again we see as a sign of maturity that the arguedly most important distinction between nonlinear problems (conxevity vs. non-convexity) is now reflected in our theoretical understanding. Equally important is that theoretical and practical improvements for the convex case are distilling into the non-convex world, a prime example being the global adoption of primal-dual rather than primal models in interior point algorithms for constrained optimization.

3.2 Improved software testing

A second important element in our analysis is the clear improvement in the quality of software testing, itself resulting in better software reliability. At first sight, software testing and comparision may seem a rather mundane and unchallenging part of the algorithmic development process, but fortunately this view has now been widely replaced with the realisation of its crucial nature.

Testing nonlinear optimization software rests on two important and complementary topics: test problems and comparison methododogy. Both of these have matured considerably over the past ten years.

When nonlinear optimization was young, and most problems treated were small-scale "toys", exchanging the formulation of test cases was easy, as one could

[1]Honesty forces us to acknowledge a few remarkable exceptions to this rule, like the BFGS variable-metric algorithm for nonconvex unconstrained minimisation (Broyden, 1970, Fletcher, 1970, Goldfarb, 1970 and Shanno, 1970) or the MINOS algorithm (Murtagh and Saunders, 1978).

write their analytic description "on the back of an envelope", or publish them in a paper (see Hock and Schittkowski, 1981 and Moré, Garbow and Hillstrom, 1981 for influential publications of that kind). When larger problems became the norm, the likelihood of introducing coding errors or slight variations in test problems grew and made software comparison very awkward. An electronically transferable format for test problems was therefore desirable. The first such widely used format originated as a by-product of the development of the nonlinear programming package LANCELOT (see Conn, Gould and Toint, 1992) in the early 1990s. This format, ambitiously (and, with hindsight, perhaps rather arrogantly) called the Standard Input Format, or SIF, was designed as a direct extension to nonlinear problems of the highly successful MPS format for linear programs. As such, it missed several features of more advanced modelling languages (such as sets), but had and continues to have the advantages of merely existing and of coming with free decoding programs. A complete testing environment, the Constrained and Unconstrained Testing Environment, or CUTE, was made available (without cost) to the research community by Bongartz, Conn, Gould and Toint (1995), with a large collection of test problems already coded in SIF, interfacing tools between this format and a number of existing packages and an extensive set of tools to facilitate testing of codes still at the development stage. These combined adavantages have probably contributed to outweight SIF's limitations and the use of the CUTE test problems and environment quickly became ubiquitous. For having talked with package developers, we believe that CUTE has increased the level of testing of software packages significantly, helping to track down coding bugs and providing a better assessment of code reliability. It is important to note that the CUTE test problem collection has continued to grow to include other sets (see Moré, 1989, Averick and Moré, 1992, Bondarenko, Bortz and Moré, 1999, Maros and Meszaros, 1999) and a number of problems arising directly from applications. It currently contains over a thousand problems of varying size, structure and difficulty. The CUTE environment has recently been superseded by a substantially improved avatar, named CUTEr (see Gould, Orban and Toint, 2003*b*).

Another positive development along this line is the growing success of the more complete modelling languages AMPL (see Fourer, Gay and Kernighan, 2003) and GAMS (see Brooke, Kendrick and Meeraus, 1988). There is no doubt that their modelling power considerably exceed that of SIF, but their generalisation remains, in our view, somewhat hampered by their non-trivial cost.

The second pilar of nonlinear optimization software testing is the methodology used for comparing algorithms. This has long been a matter of debate, as providing combined measures of both reliability (the capacity of a package to effectively solve a problem) and efficiency (its speed in obtaining the solution) has always been difficult. The initial attempts by the Mathematical Programming Committe on Algorithms (COAL) did not result in any consensus in the community, and reporting of numerical experience with new algorithms has been *ad hoc* for a long time. It is only recently that Dolan and Moré (2001) have proposed the concept of a *performance profile*, which seems to have gained increasing acceptance as a suitable way to compare reliability and efficiency of different algorithms. Suppose that a given algorithm i from a set $\mathcal{A}$ reports a statistic $s_{ij} \geq 0$ when run on example j

from a problem test set $\mathcal{T}$, and that the smaller this statistic the better the variant is considered. Let

$$k(s, s^*, \sigma) = \begin{cases} 1 & \text{if } s \leq \sigma s^*, \\ 0 & \text{otherwise.} \end{cases}$$

Then, the performance profile of algorithm i is the function

$$p_i(\sigma) = \frac{\sum_{j \in \mathcal{T}} k(s_{i,j}, s_j^*, \sigma)}{|\mathcal{T}|} \qquad (\sigma \geq 1),$$

where $s_j^* = \min_{i \in \mathcal{A}} s_{ij}$. Thus $p_i(1)$ gives the fraction of the number of examples for which algorithm i was the most effective (according to statistics s_{ij}), $p_i(2)$ gives the fraction of the number for which algorithm i is within a factor of 2 of the best, and $\lim_{\sigma \to \infty} p_i(\sigma)$ gives the fraction of the examples for which the algorithm succeeded. Thus the performance profile gives comparative information on both efficiency and reliability. We believe that such profiles provide a very effective means of comparing the relative merits of different algorithms. This is important when designing new algorithms or improved variants, and clearly helps in establishing a more balanced (shall we say mature?) relative appraisal of today's nonlinear optimization packages, like KNITRO (Byrd, Hribar and Nocedal, 2000*b*), LOQO (Vanderbei and Shanno, 1999), SNOPT (Gill, Murray and Saunders, 2002), IPOPT (Wächter, 2002), filterSQP (Fletcher and Leyffer, 1998) or the GALAHAD library (Gould, Orban and Toint, 2003*c*). This appraisal is further clarified by independent comparative algorithms benchmarking, as H. Mittelman's initiative (see `http://plato.asu.edu/bench.html`).

3.3 A world of applications

While giving all the above considerations their proper place in the argument, the most obvious sign of maturity of nonlinear optimization remains the vast range of its applications to various branches of scientific research. Reviewing them, even briefly, is totally impossible here. A limited list of references to applications (of trust-region methods only) is available in Section 1.3 of Conn et al. (2000*a*). It is enough to mention here these applications cover fields as diverse as applied mathematics, physics, chemistry, biology, geology, engineering, computer science, medicine, economics, finance, sociology, transportation, ...; but this enumeration is far from being exhaustive.

In what follows, we briefly outline five applications that we find interesting. We do not expect the reader to follow every detail of these problems (as we do not supply it), but their description or mathematical formulation is intended to illustrate the diversity of applications being considered, as well as the level of complexity that can be tackled with today's techniques. The interested reader is also invited to consult Averick and Moré (1992), Bondarenko et al. (1999) or R. Vanderbei's fascinating Web site `http://www.princeton.edu/~rvdb`.

Progressive adaptive lens design

Our first application is the use of nonlinear optimization for the design of "progressive adaptive lenses" (PAL). In its simplest form, the PAL problem is to design the surface of a lens whose optical power must be smooth and is specified in different parts of the lens (low for far vision in the middle and high for near vision in the bottom part, see Figure 1), while at the same time minimising astigmatism. Different formulations of the problem are possible (constrained or unconstrained), but they are all strongly nonlinear and nonconvex. Indeed, if the equation of the lens surface is given as the smooth function $z(x,y)$, then the optical power at (x,y) is given by

$$p(x,y) = \tfrac{1}{2}N(x,y)^3 \left[\left(1 + \left[\frac{\partial z}{\partial x}(x,y)\right]^2\right) \frac{\partial^2 z}{\partial y^2}(x,y) + \left(1 + \left[\frac{\partial z}{\partial y}(x,y)\right]^2\right) \frac{\partial^2 z}{\partial x^2}(x,y) - 2\frac{\partial z}{\partial x}(x,y)\,\frac{\partial z}{\partial y}(x,y)\,\frac{\partial^2 z}{\partial x \partial y}(x,y)\right],$$

where $N(x,y)$ is the z component of the vector normal to the surface; that is

$$N(x,y) = \frac{1}{\sqrt{1 + \left[\frac{\partial z}{\partial x}(x,y)\right]^2 + \left[\frac{\partial z}{\partial y}(x,y)\right]^2}}.$$

The surface astigmatism at (x,y) is then given by

$$a(x,y) = -2\sqrt{p(x,y) - N(x,y)^4 \left(\frac{\partial z}{\partial x}(x,y)\,\frac{\partial z}{\partial y}(x,y) - \left[\frac{\partial^2 z}{\partial x \partial y}(x,y)\right]^2\right)},$$

which is even more nonlinear than the optical power.

Controlled drug injection

Discretised optimal control problem also constitute a growing source of applications for nonlinear optimization. Problems that involve constraints on the state variables (as opposed to constraints on the control variables only) are of special interest.

The controlled drug-injection problem, whose full description can be found in of Maurer and Wiegand (1992), is a control problem based on the kinetic model of Aarons and Rowland for drug displacement, which simulates the interaction of the two drugs (warfarin and phenylnutazone) in a patient bloodstream. The state variable are the concentrations of unbound warfarin and phenylbutazone. The problem is to control the rate of injection of the pain-killing phenylbutazone so that both drugs reach a specified steady-state in minimum time and the concentration of warfarin does not rise above a given toxicity level. This last constraint therefore applies to the state variables of the problem, making the use of nonlinear programming techniques attractive. The differential equation describing the evolution of the drug concentrations in the bloodstream is discretised using a simple trapezoidal rule. The intrinsic nonlinearities of the model are non-convex.

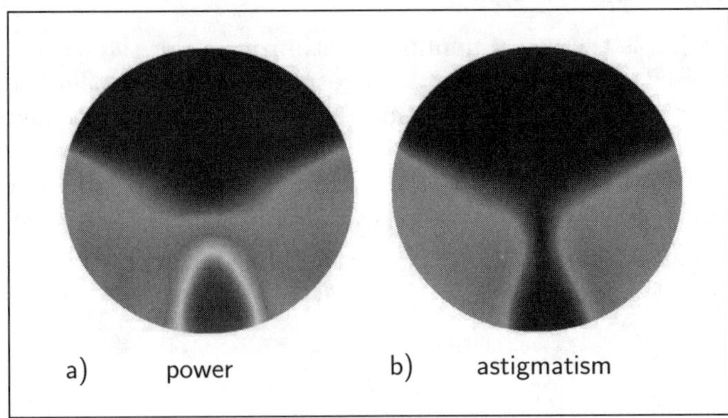

Figure 1. *Optical power and astigmatism in a typical PAL design, with a smooth transition from low values in blue to high values in red (source: Loos et al., 1997).*

Food sterilisation

Another interesting discretised control problem is that of piloting the process of food sterilisation in industrial autoclaves, as described in Kleis and Sachs (2000), where a full discussion of the problem and its solution can be found.. The idea is that the food to be sterilised is placed in closed autoclaves (see Figure 2) where it is heated (typically by hot water or steam).

Figure 2. *An autoclave for food sterilisation*

The question is then to optimise this heating in order to minimise the loss of vitamins but subject to the constraint that a certain fraction of undesired micro-organisms are killed and that every part of the food must reach a minimum temperature and not exceed a maximal one. The destruction of micro-organims and other nutrients of interest is described by

$$\frac{\partial C}{\partial t}(x,t) = -K\big[\theta(x,t)\big]C(x,t)\,,$$

where $C(x,t)$ is the concentration of living micro-organisms or nutrients and $\theta(x,t)$ is the absolute temperature, at point x and time t. We also have that the function K depends on the temperature via the Arrhenius equation, that is

$$K[\theta] = K_1 e^{-K_2\left(\frac{1}{\theta}-\frac{1}{\theta_r}\right)}\,,$$

where K_1, K_2 and θ_r are suitable constants. The evolution of temperature in the food container within the autoclave is described by a nonlinear heat equation of the form

$$\rho c(\theta)\frac{\partial\theta}{\partial t} = \nabla\cdot\big[k(\theta)\nabla\theta\big]\,,$$

with suitable boundary conditions. Due to symmetry of the autoclaves, this 3D-problem can be reduced to 2D. The heat equation is discretised using finite elements for the spatial variables and the backward Euler method for time. This problem is also mentioned in Sachs (2003), where the reader will find an interesting discussion of PDE constrained optimization. There is an increasing awareness in the PDE community of the power of optimization, and an ongoing project to foster further links in this direction (see `http://plato.asu.edu/pdecon.html`).

Biological parameters estimation

We next consider a biological parameter identification problem discussed in Toint and Willms (2003). The problem is to identify parameters in a model of the voltage across a neuron membrane in the presence of a single passive current and a single voltage-activated current with Hodgkin–Huxley channel gating (see Figure 3). That is, the activation of p independent gates and total inactivation divided into n_h groups of partial activations with identical steady-state characteristics but different kinetic properties to give multi-exponential decay characteristics.

The ODEs for the voltage $v(t)$, the activation $m(t)$ and the partial inactivations $h_i(t)$ are

$$\begin{aligned}
C\frac{\mathrm{d}v}{\mathrm{d}t}(t) &= -g_a m(t)^p h(t)\big(v(t)-E_a\big) - g_\ell\big(v(t)-E_\ell\big) + I(t)\,,\\
\frac{\mathrm{d}m}{\mathrm{d}t} &= \alpha_m\big[v(t)\big]\big(1-m(t)\big) - \beta_m\big[v(t)\big]m(t)\,,\\
\frac{\mathrm{d}h_i}{\mathrm{d}t} &= \alpha_{h_i}\big[v(t)\big]\big(1-m(t)\big) - \beta_{h_i}\big[v(t)\big]m(t)\,, \quad (i=1,\dots,n_h)\,,
\end{aligned}$$

where C is the membrane capacitance, g_a is the (time independent) active conductance, g_ℓ is the (time independent) passive conductance, E_a is the (time independent) active current reversal potential, E_ℓ is the (time independent) passive current reversal potential, $I(t)$ is the injected current, and where the total inactivation $h(t)$ is the sum of the different partial inactivations

$$h(t) = \sum_{i=1}^{n_h} f_i h_i(t) + f_{n_h+1}$$

for all t, and where the inactivation fractions f_i satisfy

$$0 \leq f_i \leq 1 \quad (i = 1, \dots, n_h) \text{ and } \sum_{i=1}^{n_h+1} f_i = 1 .$$

The functions $\alpha_*(v)$ and $\beta_*(v)$ are Boltzmann functions of the form

$$\alpha_*[v] = \frac{1}{\tau_{\alpha,*}(1 - e^{(v-u_{\alpha,*})/\sigma_{\alpha,*}})}$$

and

$$\beta_*[v] = \frac{1}{\tau_{\beta,*}(1 - e^{(v-u_{\beta,*})/\sigma_{\beta,*}})}$$

with $*$ being m or h_i $(i = 1, \dots, n_h)$. Additionally, the parameters of the Boltzmann functions have to satisfy, for $i = 1, \dots, n_h$,

$$\tau_{\alpha,h_i} = \epsilon_i \tau_{\alpha,h} , \qquad u_{\alpha,h_i} = u_{\alpha,h} , \qquad \sigma_{\alpha,h_i} = \sigma_{\alpha,h}$$

$$\tau_{\beta,h_i} = \epsilon_i \tau_{\beta,h} , \qquad u_{\beta,h_i} = u_{\beta,h} , \qquad \sigma_{\beta,h_i} = \sigma_{\beta,h}$$

where the scaling factors ϵ_i are constrained by

$$1 = \epsilon_1 < \epsilon_2 < \dots < \epsilon_{n_h} .$$

The ODE's are discretised using a 5 steps Backward Differentiation Formula with constant time stepping. The objective function is to minimise the least-squares distance between the voltages satisfying those equations and observed voltage values for a number of experiments (or sweeps). The experimental data is for a potassium A current in a pyloric dilator cell of the stomatogastric ganglion of the Pacific spiny lobster (see Figure 4). As can be seen from the equations, the problem is non-convex.

In its current formulation, the problem uses four experimental sweeps and involves around 16,000 variables and about the same number of constraints, only one of which is linear.

Mechanics and video games

Finally, we would like to mention here an application in a fairly different area: that of video animation and video-games. In an intereting paper, Anitescu and Potra (1996) have formulated the problem of representing the motion of multiple rigid objects in space, including their interaction (friction) when they hit each other. The formulation used is that of a time-dependent linear complementarity problem.

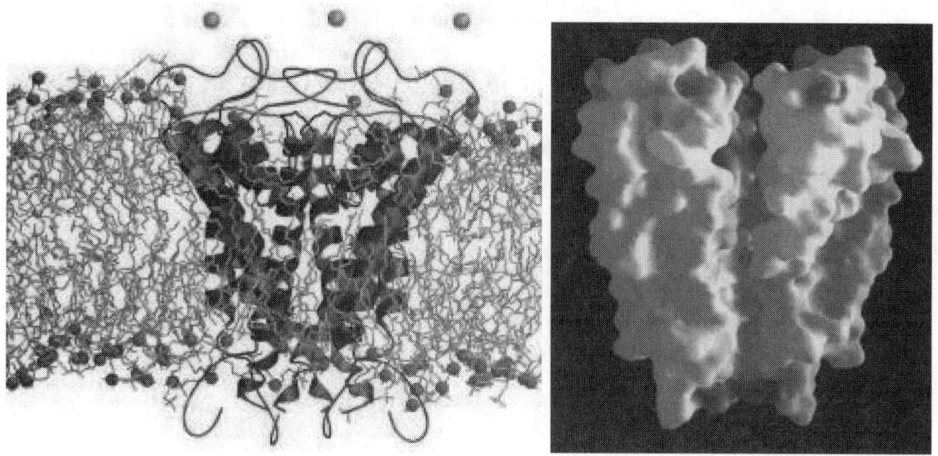

Figure 3. *The ribbon structure of the K^+ channel molecule showing its insertion the membrane (the blue ions on top are at the exterior of the cell) and a solid rendering of this molecule (source: right picture from Sansom, 2001, left picture from Doyle et al., 1998)*

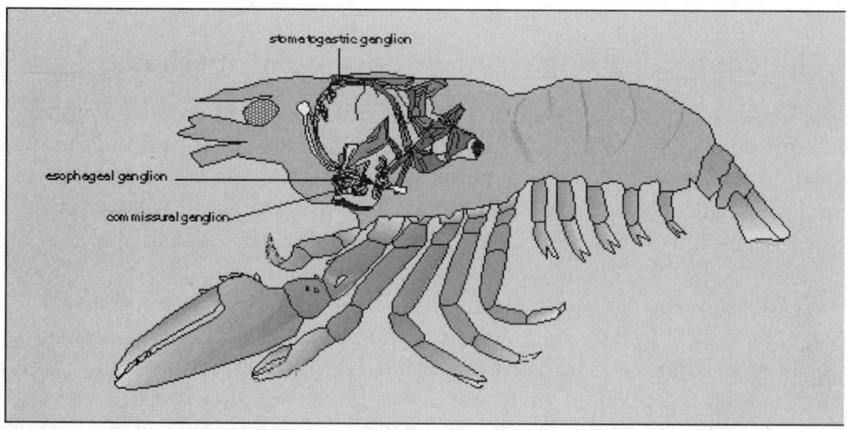

Figure 4. *The position of the stomatogastric ganglion within the Pacific spiny lobster (source: Simmers et al., 1995)*

While this problem is at the boundary of linear and nonlinear problems (it is solved by a variant of Lemke's algorithm), it is nevertheless of interest to us because it can be seen as the problem of finding a feasible solution, at each time t, of the nonlinear set of inequalities

$$\nabla_q \Phi\big[q(t)\big] v(t) \geq 0\,, \quad \Phi\big(q(t)\big) \geq 0$$

where $q(t)$ is vector of states (positions) of the multi-body system at time t, $v(t) = \frac{dq}{dt}(t)$ is the vector of velocities, and the second inequality expresses the contact constraints (the fact that the problem bodies do not interpenetrate) for some smooth function Φ. This formulation is not only elegant, but is also amenable to practical implementation. It is in fact, in an implementation by MathEngine, at the heart of video-games such as the Vivid Image Actor, and provides a very realistic real-time simulation of shocks between rigid objects. The hidden presence of nonlinear problems in environments as ubiquitous as video-games also testify of its interest and reinforce our argument.

4 Is senility lurking?

Cynical observers may thus accept the maturity of nonlinear optimization as a discipline. They might also wonder if it already shows dangerous signs of aging and obsolescence; these signs typically include a more self-centered discourse or the repetition of older ideas instead of the creation of new ones. Although we acknowledge that self-centered contributions do exist[2], we hope that the variety of applications we have exposed in the previous section is convincing enough to dismiss the case of a narrower interaction with the world at large. We therefore focus, in what follows, on indicating that new directions and ideas continue to sustain the field's creativity.

4.1 The continuing impact of interior point methods

The first active current of research was initiated by the revival of interior point methods in linear and semi-definite programming. This generated a number of new contributions that attempted to adapt these ideas initially to nonlinear convex problems, and subsequently to nonconvex ones. The main difficulty in adapting to the latter is that the first-order optimality conditions, for minimisation, which are necessary and sufficient for linear and convex problems, are insufficient for nonconvex ones. Indeed, they can be satisfied at saddle points or even at maximisers.

We believe it is fair to say that the numerous contributions[3] on this topic are far from having exhausted the question or solved all practical problems. Outstanding issues include the efficient handling of nonlinear equality constraints, the effect of constraint scaling, suitable preconditioning techniques and extrapolation along the (possibly bizarre) central path for nonconvex problems. Moreover, the relative merits of interior point methods compared to more traditional SQP approaches are still a matter of lively research and debate (for a recent non-technical discussion of this topic, see Gould, 2003).

[2]There are, in our view, too many papers presenting convergence proofs for algorithms that have never been and will probably never be properly implemented, or even tried on simple examples.

[3]See, for instance, Bonnans and Bouhtou (1995), Lasdon, Plummer and Yu (1995), Coleman and Li (1996*a*, 1996*b*), Bonnans and Pola (1997), Simantiraki and Shanno (1997), Forsgren and Gill (1998), Dennis, Heinkenschloss and Vicente (1998), Gay, Overton and Wright (1998), Vanderbei and Shanno (1999), Byrd et al. (2000*b*), Conn, Gould, Orban and Toint (2000*b*), Byrd, Gilbert and Nocedal (2000*a*), Chapter 13 of Conn et al. (2000*a*), or Gould, Orban, Sartenaer and Toint (2001), amongst many others.

4.2 The revival of derivative free optimization

Algorithms for nonlinear programming that do not make use of derivative information have also come back in the foreground of research, after a long eclipse. Very popular in the infancy of the field, with classics like the simplex method of Nelder and Mead (1965), interest in these methods has been revived by significant recent progress in two different directions: interpolations methods and pattern search methods.

The first class of methods attempts to build a (typically quadratic) model of the function to be minimised, using multivariate interpolation techniques. The resulting algorithms (see Powell, 1994, 2000, 2002, or Conn, Scheinberg and Toint, 1997, 1998) are typically very efficient, and exploitation of problem structure is currently being successfully experimented (Colson and Toint, 2001, 2002, 2003).

The second class of derivative free methods use a prespecified or adaptive "pattern" to sample the variable space and compute minimisers. These methods are also the subject of much ongoing research (see Dennis and Torczon, 1991, Torczon, 1997, Coope and Price, 2000 and 2001, or Audet and Dennis, 2003). Extension of these techniques to large-scale problems is also being investigated (see Price and Toint, 2003).

Much remains to be done in this challenging sector, including better algorithms to handle larger problems with constraints.

4.3 Filter methods

We could not conclude this section of the new exciting ideas in nonlinear programming without briefly covering the filter methodology introduced by Fletcher and Leyffer (2002). This technique aims at promoting global convergence to minimisers of constrained problems without the need for a penalty function. Instead, the new concept of a "filter" is introduced which allows a step to be accepted if it reduces either the objective function or the constraint violation function. This simple yet powerful idea may be, in our view, the most significant progress in the past five years, and has already generated, in a very short time, a flurry of related research, both on algorithmic aspects (Ulbrich, Ulbrich and Vicente, 2000, Chin and Fletcher, 2001, Fletcher and Leyffer, 2003, Gonzaga, Karas and Vanti, 2002, Gould and Toint, 2002, Gould, Leyffer and Toint, 2003*a*) and on its theoretical underpinnings (Wächter and Biegler, 2001, Fletcher, Leyffer and Toint, 2002*b*, Fletcher, Gould, Leyffer, Toint and Wächter, 2002*a*) and inspired the organisation of conferences and workshops devoted to this topic.

To illustrate its power, and at the same time that of the performance profiles of Dolan and Moré, in Figure 5 we present a CPU time comparison of a classical trust-region method and FILTRANE, a multidimensional filter method (Gould and Toint, 2003), on a large set of nonlinear feasibilty problems from the CUTEr collection.

We see in this figure that the classical pure trust-region algorithm (one of the very best options before the filter idea) is slightly less reliabile than FILTRANE, and that the latter code is best (or tied best) on around 88% of the problems, a very significant advantage when compared to approximately 66% of the problems where

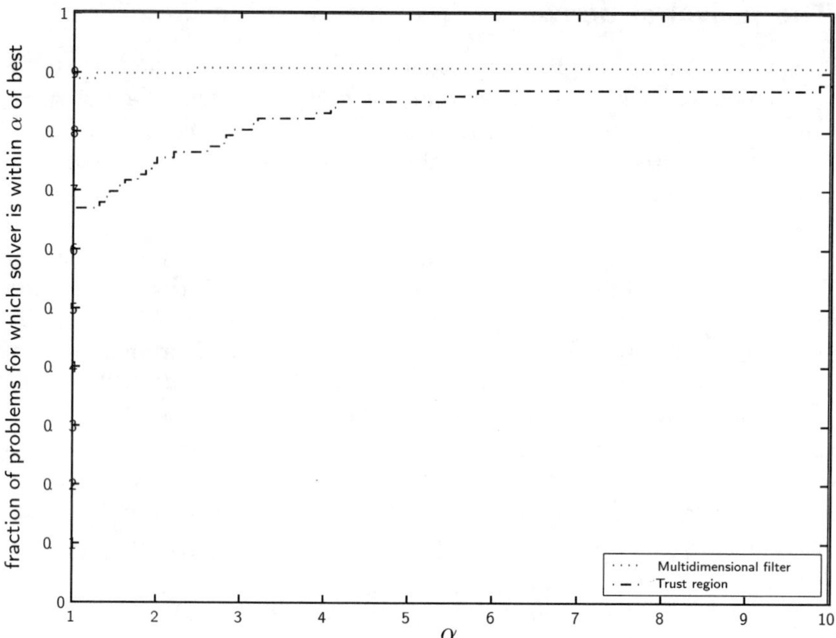

Figure 5. *CPU time performance profile for multidimensional filter algorithm vs. classical trust-region algorithm on a set of 106 nonlinear feasibility problems*

the while the trust-region method is best. Furthermore, FILTRANE is within a factor 2 of the best on approximately 89% and within a factor 5 for approximately 91% of the problems, again an excellent performance. This kind of numerical results is really encouraging and stimulating, and one may therefore expect even more research activity in the domain of the filter methods. If it were only for that, it would already be enough to indicate the continuing vitality of nonlinear optimization.

5 Conclusion: the future's challenges

We have presented some arguments to vindicate our view that nonlinear optimization is a mature but not yet senile domain of research. Of course, these arguments are biased by our own experience and work, but we believe they are shared by a number of actors in the field. The last issue of the SIAG/OPT Views-and-News[4] provides additional elements that concur with ours, and also points to other domains where nonlinear optimization is developing strongly, like problems with equilibrium constraints, DAE-constrained problems, or even more challengingly, nonlinear optimization with discrete variables.

[4] "Large-Scale Nonconvex Optimization", volume 14(1), April 2003, guest editors: S. Leyffer and J. Nocedal.

What are the future's challenges? Besides the continuing improvement of methods and software, we feel that the successful specialisation of nonlinear optimization to problem subclasses (like discretised optimal control problem or DAE constrained identification problems) constitutes a fruitful evolution and will in due course become important. The quest for methods that can solve problems that today are intractable because of their size, nonlinearity or because they involve too many discrete variables, is not yet anywhere near its end — a very invigorating perspective.

Acknowledgements

The second author is indebted to a number of colleagues who have helped supplying some of the material in this talk. In particular, thanks are due to Katia Demaseure, Mevlut Dinc at Vivid Image, Helmut Maurer, Ian Rycroft and Dilip Sequeira at MathEngine, Ekkehard Sachs, Andy Wathen and Allan Willms.

Bibliography

M. Anitescu and F. Potra. Formulating dynamic multi-rigid-body contact problems with friction as solvable linear complementarity problems. Reports on Computational Mathematics 93, Department of Mathematics, University of Iowa, Iowa City, USA, 1996.

C. Audet and J. E. Dennis. Analysis of generalized pattern searches. *SIAM Journal on Optimization*, **13**(3), pp. 889–903, 2003.

B. M. Averick and J. J. Moré. The Minpack-2 test problem collection. Technical Report ANL/MCS-P153-0694, Mathematics and Computer Science, Argonne National Laboratory, Argonne, Illinois, USA, 1992.

A. S. Bondarenko, D. M. Bortz, and J. J. Moré. COPS: Large-scale nonlinearly constrained optimization problems. Technical Report ANL/MCS-TM-237, Mathematics and Computer Science, Argonne National Laboratory, Argonne, Illinois, USA, 1999.

I. Bongartz, A. R. Conn, N. I. M. Gould, and Ph. L. Toint. CUTE: Constrained and Unconstrained Testing Environment. *Transactions of the ACM on Mathematical Software*, **21**(1), pp. 123–160, 1995.

J. F. Bonnans and M. Bouhtou. The trust region affine interior-point algorithm for convex and nonconvex quadratic-programming. *RAIRO-Recherche Opérationnelle—Operations Research*, **29**(2), pp. 195–217, 1995.

J. F. Bonnans and C. Pola. A trust region interior point algorithm for linearly constrained optimization. *SIAM Journal on Optimization*, **7**(3), pp. 717–731, 1997.

A. Brooke, D. Kendrick, and A. Meeraus. *GAMS: a User's Guide.* The Scientific Press, Redwood City, USA., 1988.

C. G. Broyden. The convergence of a class of double-rank minimization algorithms. *Journal of the Institute of Mathematics and its Applications*, **6**, pp. 76–90, 1970.

R. H. Byrd, J. Ch. Gilbert, and J. Nocedal. A trust region method based on interior point techniques for nonlinear programming. *Mathematical Programming, Series A*, **89**(1), pp. 149–186, 2000*a*.

R. H. Byrd, M. E. Hribar, and J. Nocedal. An interior point algorithm for large scale nonlinear programming. *SIAM Journal on Optimization*, **9**(4), pp. 877–900, 2000*b*.

A. Cauchy. Méthode générale pour la résolution des systèmes d'équations simultannées. *Comptes Rendus de l'Académie des Sciences*, pp. 536–538, 1847.

C. M. Chin and R. Fletcher. Convergence properties of SLP-filter algorithms that takes eqp steps. Numerical Analysis Report NA/199, Department of Mathematics, University of Dundee, Dundee, Scotland, 2001.

T. F. Coleman and Y. Li. An interior trust region approach for nonlinear minimization subject to bounds. *SIAM Journal on Optimization*, **6**(2), pp. 418–445, 1996*a*.

T. F. Coleman and Y. Li. A reflective Newton method for minimizing a quadratic function subject to bounds on some of the variables. *SIAM Journal on Optimization*, **6**(4), pp. 1040–1058, 1996*b*.

B. Colson and Ph. L. Toint. Exploiting band structure in unconstrained optimization without derivatives. *Optimization and Engineering*, **2**, pp. 349–412, 2001.

B. Colson and Ph. L. Toint. A derivative-free algorithm for sparse unconstrained optimization problems. *in* A. H. Siddiqi and M. Kočvara, eds, 'Trends in Industrial and Applied Mathematics', pp. 131–149, Dordrecht, The Netherlands, 2002. Kluwer Academic Publishers.

B. Colson and Ph. L. Toint. Exploiting problem structure in derivative-free algorithms for unconstrained optimization. Technical Report (in preparation), Department of Mathematics, University of Namur, Namur, Belgium, 2003.

A. R. Conn, N. I. M. Gould, and Ph. L. Toint. LANCELOT*: a Fortran package for large-scale nonlinear optimization (Release A).* Number 17 *in* 'Springer Series in Computational Mathematics'. Springer Verlag, Heidelberg, Berlin, New York, 1992.

A. R. Conn, N. I. M. Gould, and Ph. L. Toint. *Trust-Region Methods.* Number 01 *in* 'MPS–SIAM Series on Optimization'. SIAM, Philadelphia, USA, 2000*a*.

A. R. Conn, N. I. M. Gould, D. Orban, and Ph. L. Toint. A primal-dual trust-region algorithm for minimizing a non-convex function subject to bound and linear equality constraints. *Mathematical Programming*, **87**(2), pp. 215–249, 2000*b*.

A. R. Conn, K. Scheinberg, and Ph. L. Toint. Recent progress in unconstrained nonlinear optimization without derivatives. *Mathematical Programming, Series B*, **79**(3), pp. 397–414, 1997.

A. R. Conn, K. Scheinberg, and Ph. L. Toint. A derivative free optimization algorithm in practice. Technical Report TR98/11, Department of Mathematics, University of Namur, Namur, Belgium, 1998.

I. D. Coope and C. J. Price. Frame-based methods for unconstrained optimization. *Journal of Optimization Theory and Applications*, **107**, pp. 261–274, 2000.

I. D. Coope and C. J. Price. On the convergence of grid-based methods for unconstrained optimization. *SIAM Journal on Optimization*, **11**, pp. 859–869, 2001.

J. E. Dennis and V. Torczon. Direct search methods on parallel machines. *SIAM Journal on Optimization*, **1**(4), pp. 448–474, 1991.

J. E. Dennis, M. Heinkenschloss, and L. N. Vicente. Trust-region interior-point SQP algorithms for a class of nonlinear programming problems. *SIAM Journal on Control and Optimization*, **36**(5), pp. 1750–1794, 1998.

E. D. Dolan and J. J. Moré. Benchmarking Optimization Software with Performance Profiles. *Mathematical Programming*, **91**(2), pp. 201–213, 2002.

D. A. Doyle, J. Morais Cabral, R. A. Pfuetzner, A. Kuo, J. M. Gulbis, S. L. Cohen, B. T. Chait, and R. MacKinnon. Molecular basis of K^+ conduction and selectivity. *Science*, **280**(5360), pp. 69–77, 1998.

A. V. Fiacco and G. P. McCormick. *Nonlinear Programming: Sequential Unconstrained Minimization Techniques*. J. Wiley and Sons, Chichester, England, 1968. Reprinted as *Classics in Applied Mathematics 4*, SIAM, 1990.

R. Fletcher. A new approach to variable metric algorithms. *Computer Journal*, **13**, pp. 317–322, 1970.

R. Fletcher and S. Leyffer. User manual for filterSQP. Numerical Analysis Report NA/181, Department of Mathematics, University of Dundee, Dundee, Scotland, 1998.

R. Fletcher and S. Leyffer. Nonlinear programming without a penalty function. *Mathematical Programming*, **91**(2), pp. 239–269, 2002.

R. Fletcher and S. Leyffer. Filter-type algorithms for solving systems of algebraic equations and inequalities. *in* G. Di Pillo and A. Murli, eds, 'High Performance Algorithms and Software in Nonlinear Optimization', pp. 259–278, Dordrecht, The Netherlands, 2003. Kluwer Academic Publishers.

R. Fletcher, N. I. M. Gould, S. Leyffer, Ph. L. Toint, and A. Wächter. Global convergence of trust-region SQP-filter algorithms for nonlinear programming. *SIAM Journal on Optimization*, **13**(3), pp. 635–659, 2002*a*.

R. Fletcher, S. Leyffer, and Ph. L. Toint. On the global convergence of a filter-SQP algorithm. *SIAM Journal on Optimization*, **13**(1), pp. 44–59, 2002*b*.

A. Forsgren and P. E. Gill. Primal-dual interior methods for nonconvex nonlinear programming. *SIAM Journal on Optimization*, **8**(4), pp. 1132–1152, 1998.

R. Fourer, D. M. Gay, and B. W. Kernighan. *AMPL: A modeling language for mathematical programming.* Brooks/Cole–Thompson Learning, Pacific Grove, California, USA, second edn, 2003.

D. M. Gay, M. L. Overton, and M. H. Wright. A primal-dual interior method for nonconvex nonlinear programming. *in* Y. Yuan, ed., 'Advances in Nonlinear Programming', pp. 31–56, Dordrecht, The Netherlands, 1998. Kluwer Academic Publishers.

P. E. Gill, W. Murray, and M. A. Saunders. SNOPT: An SQP algorithm for large-scale constrained optimization. *SIAM Journal on Optimization*, **12**(4), pp. 979–1006, 2002.

D. Goldfarb. A family of variable metric methods derived by variational means. *Mathematics of Computation*, **24**, pp. 23–26, 1970.

C. C. Gonzaga, E. Karas, and M. Vanti. A globally convergent filter method for nonlinear programming. Technical report, Department of Mathematics, Federal University of Santa Catarina, Florianopolis, Brasil, 2002.

N. I. M. Gould. Some reflections on the current state of active-set and interior point methods for constrained optimization. *SIAG/OPT Views-and-News*, **14**(1), pp. 2–7, 2003.

N. I. M. Gould and Ph. L. Toint. Global convergence of a non-monotone trust-region filter algorithm for nonlinear programming. Technical Report (in preparation), Department of Mathematics, University of Namur, Namur, Belgium, 2002.

N. I. M. Gould and Ph. L. Toint. FILTRANE, a Fortran 95 filter-trust-region package for solving systems of nonlinear equalities, nonlinear inequalities and nonlinear least-squares problems. Report 03/17, Department of Mathematics, University of Namur, Namur, Belgium, 2003.

N. I. M. Gould, S. Leyffer, and Ph. L. Toint. A multidimensional filter algorithm for nonlinear equations and nonlinear least-squares. Technical Report TR–2003–004, Rutherford Appleton Laboratory, Chilton, Oxfordshire, England, 2003*a*.

N. I. M. Gould, D. Orban, and Ph. L. Toint. CUTEr, a contrained and unconstrained testing environment, revisited. *Transactions of the ACM on Mathematical Software*, (to appear), 2003*b*.

N. I. M. Gould, D. Orban, and Ph. L. Toint. GALAHAD—a library of thread-safe Fortran 90 packages for large-scale nonlinear optimization. *Transactions of the ACM on Mathematical Software*, (to appear), 2003*c*.

N. I. M. Gould, D. Orban, A. Sartenaer, and Ph. L. Toint. On the local convergence of a primal-dual trust-region interior-point algorithm for constrained nonlinear programming. *SIAM Journal on Optimization*, **11**(4), pp. 974–1002, 2001.

S. P. Han. A globally convergent method for nonlinear programming. *Journal of Optimization Theory and Applications*, **15**, pp. 319–342, 1977.

M. R. Hestenes and E. Stiefel. Methods of conjugate gradients for solving linear systems. *Journal of the National Bureau of Standards*, **49**, pp. 409–436, 1952.

W. Hock and K. Schittkowski. *Test Examples for Nonlinear Programming Codes.* Springer Verlag, Heidelberg, Berlin, New York, 1981. Lectures Notes in Economics and Mathematical Systems 187.

D. Kleis and E. W. Sachs. Optimal control of the sterilization of prepackaged food. *SIAM Journal on Optimization*, **10**, pp. 1180–1195, 2000.

L. S. Lasdon, J. Plummer, and G. Yu. Primal-dual and primal interior point algorithms for general nonlinear programs. *ORSA Journal on Computing*, **7**(3), pp. 321–332, 1995.

K. Levenberg. A method for the solution of certain problems in least squares. *Quarterly Journal on Applied Mathematics*, **2**, pp. 164–168, 1944.

J. Loos, G. Greiner, and H.-P. Seidel. Computer aided spectacle lens design. Technical Report 5, Department of Computer Science, University of Erlangen, Erlangen, Germany, 1997.

I. Maros and C. Meszaros. A repository of convex quadratic programming problems. *Optimization Methods and Software*, **11–12**, pp. 671–681, 1999.

H. Maurer and M. Wiegand. Numerical solution of a drug displacement problem with bounded state variables. *Optimal Control Applications and Methods*, **13**, pp. 43–55, 1992.

J. J. Moré. A collection of nonlinear model problems. Technical Report ANL/MCS-P60-0289, Mathematics and Computer Science, Argonne National Laboratory, Argonne, Illinois, USA, 1989.

J. J. Moré and D. C. Sorensen. Computing a trust region step. *SIAM Journal on Scientific and Statistical Computing*, **4**(3), pp. 553–572, 1983.

J. J. Moré, B. S. Garbow, and K. E. Hillstrom. Testing unconstrained optimization software. *Transactions of the ACM on Mathematical Software*, **7**(1), pp. 17–41, 1981.

B. A. Murtagh and M. A. Saunders. Large-scale linearly constrained optimization. *Mathematical Programming*, **14**, pp. 41–72, 1978.

J. A. Nelder and R. Mead. A simplex method for function minimization. *Computer Journal*, **7**, pp. 308–313, 1965.

Y. Nesterov and A. Nemirovsky. *Self-concordant functions and polynomial-time methods in convex programming.* SIAM, Philadelphia, USA, 1993.

M. J. D. Powell. A method for nonlinear constraints in minimization problems. *in* R. Fletcher, ed., 'Optimization', pp. 283–298, London, 1969. Academic Press.

M. J. D. Powell. A new algorithm for unconstrained optimization. *in* J. B. Rosen, O. L. Mangasarian and K. Ritter, eds, 'Nonlinear Programming', pp. 31–65, London, 1970. Academic Press.

M. J. D. Powell. Some global convergence properties of a variable metric algorithm for minimization without exact line searches. *in* 'SIAM–AMS Proceedings 9', pp. 53–72, Philadelphia, USA, 1976. SIAM.

M. J. D. Powell. A fast algorithm for nonlinearly constrained optimization calculations. *in* G. A. Watson, ed., 'Numerical Analysis, Dundee 1977', number 630 *in* 'Lecture Notes in Mathematics', pp. 144–157, Heidelberg, Berlin, New York, 1978. Springer Verlag.

M. J. D. Powell. A direct search optimization method that models the objective by quadratic interpolation. Presentation at the 5th Stockholm Optimization Days, Stockholm, 1994.

M. J. D. Powell. UOBYQA: unconstrained optimization by quadratic interpolation. Technical Report NA14, Department of Applied Mathematics and Theoretical Physics, Cambridge University, Cambridge, England, 2000.

M. J. D. Powell. Least Frobenius norm updating of quadratic models that satisfy interpolation conditions. Technical Report NA02, Department of Applied Mathematics and Theoretical Physics, Cambridge University, Cambridge, England, 2002.

C. J. Price and Ph. L. Toint. Exploiting problem structure in pattern search methods for unconstrained optimization. Technical Report (in preparation), Department of Mathematics, University of Namur, Namur, Belgium, 2003.

R. T. Rockafellar. Augmented Lagrangian multiplier functions and duality in nonconvex programming. *SIAM Journal on Control and Optimization*, **12**(2), pp. 268–285, 1974.

E. W. Sachs. PDE constrained optimization. *SIAG/OPT News-and Views*, **14**(1), pp. 7–10, 2003.

M. S. P. Sansom. Laboratory journal 2001 web page. `http://biop.ox.ac.uk/www/lj2001/sansom/sansom.html`, Laboratory of Molecular Biophysics, University of Oxford, Oxford, UK, 2001.

D. F. Shanno. Conditioning of quasi-Newton methods for function minimization. *Mathematics of Computation*, **24**, pp. 647–657, 1970.

E. M. Simantiraki and D. F. Shanno. An infeasible-interior-point method for linear complementarity problems. *in* I. Duff and A. Watson, eds, 'The State of the Art in Numerical Analysis', pp. 339–362, Oxford, England, 1997. Oxford University Press.

J. Simmers, P. Meyrand, and M. Moulins. Dynamic networks of neurons. *American Scientist*, **83**, pp. 262–268, 1995.

R. A. Tapia. Diagonalized multiplier methods and quasi-Newton methods for constrained optimization. *Journal of Optimization Theory and Applications*, **22**, pp. 135–194, 1977.

Ph. L. Toint and A. Willms. Numerical estimation of the parameters in a model of the voltage across a neuron membrane in the presence of a single passive current and a single voltage-activated current with Hodgkin-Huxley channel gating. Technical Report (in preparation), Department of Mathematics and Statistics, University of Canterbury, Christchurch, New Zealand, 2003.

V. Torczon. On the convergence of pattern search algorithms. *SIAM Journal on Optimization*, **7**(1), pp. 1–25, 1997.

M. Ulbrich, S. Ulbrich, and L. Vicente. A globally convergent primal-dual interior point filter method for nonconvex nonlinear programming. Technical Report TR00–11, Department of Mathematics, University of Coimbra, Coimbra, Portugal, 2000.

R. J. Vanderbei and D. F. Shanno. An interior point algorithm for nonconvex nonlinear programming. *Computational Optimization and Applications*, **13**, pp. 231–252, 1999.

A. Wächter. *An Interior Point Algorithm for Large-Scale Nonlinear Optimization with Applications in Process Engineering.* PhD thesis, Department of Chemical Engineering, Carnegie Mellon University, Pittsburgh, USA, 2002.

A. Wächter and L. T. Biegler. Global and local convergence of line search filter methods for nonlinear programming. Technical Report CAPD B-01-09, Department of Chemical Engineering, Carnegie Mellon University, Pittsburgh, USA, 2001.

Alice Guionnet is a researcher at the Ecole Normale Supérieure de Lyon, France. In the last few years she has worked on particle systems in random interaction, particle approximations to nonlinear filtering and studied large random matrices theory, in particular, its link with free probability.

Guionnet was born in Paris where she entered the prestigious Ecole Normale Supérieure in 1989. She was soon interested in probability theory and studied, for her PhD thesis, some problems from statistical physics related to particle systems in random interaction. She was offered a position at Centre National de Recherche Scientifique for this work. After a few years at the University of Orsay, where she studied particle approximations to non-linear filtering, she moved to Ecole Normale Supérieure, Paris, then finally to Lyon in 2000. She was awarded the Oberwolfach Prize in 1998 and the Rollo Davidson Prize in 2003.

Chapter 9

Aging in Particle Systems

Alice Guionnet[*]

Abstract: The aging of physical systems out of equilibrium has recently attracted great interest in physics and mathematics. A system is said to age if the older it gets, the longer it takes to forget its past. Aging phenomena have been observed experimentaly in dilute media or granular matter, but the mathematical study is still restricted to very few models, since the underlying processes are generally non Markovian and nonlinear.

Experimentally, one begins with a medium at high temperature at time $t{=}0$, and freezes it to a temperature below the critical temperature T_c. One then measures an order parameter $q(t_w, t_w + t)$, where t_w is the time when the observation started (the age of the system) and $t_w + t$ the time when it finished. A system is said to age when $q\big(t_w, t_w + h(t_w)\big)$ converges to a nonzero constant as t_w goes to infinity, for some nontrivial increasing function h. Mathematically, $q(s,t)$ is often the covariance $E(X_tX_s) - E(X_t)\,E(X_s)$ of an observable X or the probability $P(X_t{=}X_s)$.

We describe some physical systems, especially disordered systems, for which aging is expected. Then, we will review the existing mathematical results, and the two main different phenomenologies causing aging, reporting results established within the five last years by G. Ben Arous, A. Bovier, J. Cerny, A. Dembo, A. Gayrard, R. Fontes, M. Isopi, T. Mountford, D. Stein, O. Zeitouni and myself.

Contents

[*]UMPA, Ecole Normale Supérieure de Lyon

1 Introduction

Statistical mechanics is devoted to the study of the thermodynamical properties of physical systems. Classical literature on this topic often concerns their static or equilibrium properties. However, most systems in nature are not in equilibrium (see [25] for a discussion on this subject) and such a study can at best be a good approximation to reality. Even more, equilibrium can be completely irrelevant for some systems which can only be observed out of equilibrium. One can distinguish at least two classes of such systems. The first describes systems which are naturally out of equilibrium because they are submitted to a gradient of temperature, of potential etc... The second concerns systems which relax to equilibrium so slowly that the equilibrium will never be reached during the experiment or the simulation. For instance, glasses, jelly, toothpaste are example of media which, even though they seem in our everyday life much alike solids in equilibrium, still evolve on very long time scales. These systems are called glasses; they appear when some parameter (such as temperature, pressure, etc.) is changed in such a way that their relaxation time to equilibrium diverges. Such systems are very diverse and we shall later be more specifically interested in spin glasses. A canonical example of spin glasses is a metal with dilute magnetic impurities, which were shown to exhibit a rather peculiar behaviour by De Nobel and Chantenier. Such a medium can be modelled by a system of particles in random interaction or with a random external field (the randomness coming from the randomness of the distribution of the impurities in a given sample). These models are called disordered and we shall detail them later in this survey. There are many other materials that exhibit a glass phase; let us quote some physics literature on the glass phase of supraconductors [30], granular materials [8, 9] etc.

One of the relevant properties which has been investigated recently for out of equilibrium dynamics is aging. A system is said to age if the older it gets, the longer it will take to forget its past. The age of the system is the time spent since the system reached its glass phase, which is often obtained by freezing it below the critical temperature. The experiment exhibiting aging is usually as follows. One considers a medium at time $t = 0$ at high temperature and freeze it at a temperature below the critical temperature T_c. One then measures a parameter $q(t_w, t_w+t)$ where t_w is the age of the system (i.e the time spent since the system was frozen in its glass phase) and $t + t_w$ the measurement time. The parameter $q(s, t)$ is often the covariance $E(X_t X_s) - E(X_t)E(X_s)$ of the observable X or the probability $P(X_t = X_s)$. Then, a system is said to age when $q(t_w, t_w + h(t_w))$ converges to a non zero constant as t_w goes to infinity, for some non trivial increasing function h. One usually observes the following. At large temperature, the system quickly equilibrates and the order parameter should rapidly become stationary; $q(s, t) \approx q(s - t)$ for t, s reasonably large. At lower temperature, one observes usually data as represented in figure 1;

the experimental covariances are not functions of $t - t_w$ only, but also depends on the age t_w of the system, and are therefore a more complicated function of t and t_w that one can investigate.

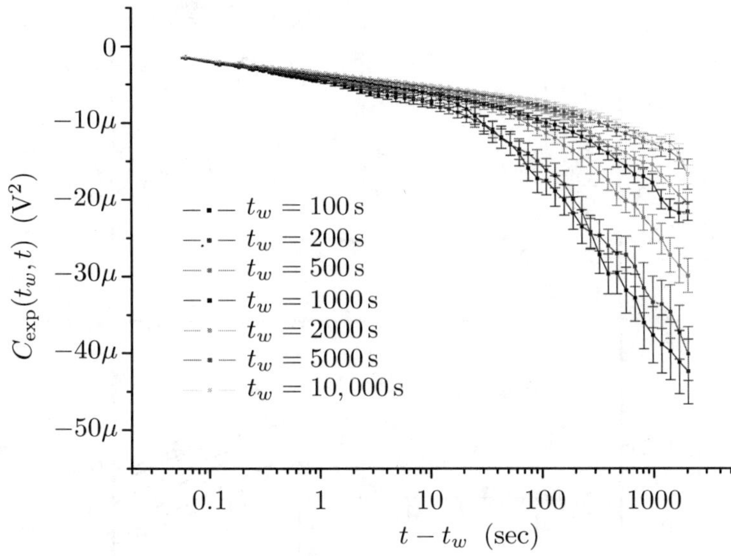

Figure 1. *Experimental covariances* $C_{\exp}(t_w, t) - C_0 = \langle X(t_w)\, X(t)\rangle\langle X(t_w)^2\rangle$ *in the insulating spin glass* $CdCr_{1.7}In_{0.3}S_4$*, measured by D. Herisson and M. Ocio [26, 27].*

For instance, it was observed in [27] that the covariance becomes approximately a function of $\zeta = (1-\mu)^{-1}(t^{1-\mu} - t_w^{1-\mu})$ with $\mu = 0.87$, as shown in figure 2.

Let us notice that the figures above are already taken in such a time scale that they do not show what happens for short times t. A more detailed study usually shows that at least two phenomena are going on; on a short scale, when $t - t_w$ goes to infinity while t_w stays small enough, the system reaches a state where q is approximately given by a constant q_{EA} (whose value is represented by the initial flat part in the covariance diagrams above) and stays in this state quite a long time so that the systems seems to be in equilibrium and the dynamics looks stationary. However, on a longer scale, the system will undergo dramatic changes which will drive the parameter q to zero. The existence of different time scales related to slower and quicker processes is also a description of aging.

The mathematical understanding of aging has been undertaken only very recently and is still very limited. For the time being, aging phenomenon could be analysed for very few disordered models. The two main phenomenologies that have been isolated as a source of aging can be illustrated by two toys models: the so-called Bouchaud's trap model and the spherical Sherrington–Kirkpatrick model. Since these two models were introduced to understand the dynamics for the Sherrington–

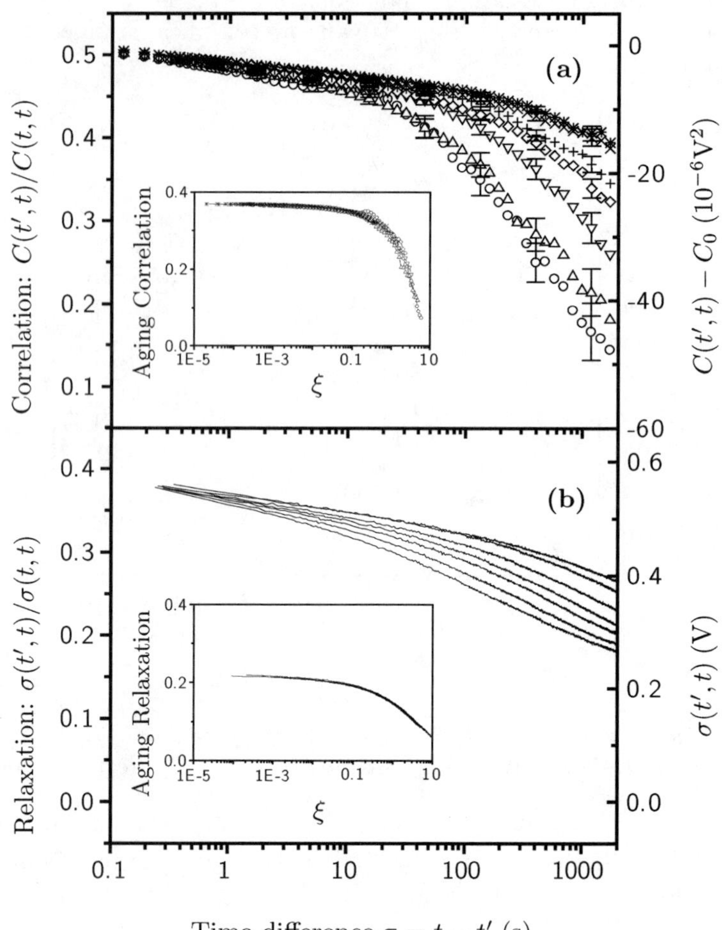

Figure 2. *Experimental covariance in the insulating spin glass* $CdCr_{1.7}In_{0.3}S_4$ *at low temperature [26, 27].*

Kirkpatrick model of spin glass, we shall first describe this model. It is given by the quadratic Hamiltonian

$$H_J(\mathbf{x}) = \sum_{1 \le i < j \le N} J_{ij}\, x_i\, x_j$$

where $\mathbf{x} = \big(x_i\,, 1 \le i \le N\big)$ represent the particles or spins, which belong to a set M. M can be either discrete, for instance $M = \{-1, +1\}$ in the Ising model, or continuous, for instance $M = \mathbb{R}$ or M is a compact Riemaniann manifold such as a sphere in $\mathbb{R}^d$. The J_{ij}s are centered independent random variables with variance N^{-1}, often assumed to be Gaussian for simplicity. If μ is a probability measure on M, a Gibbs (or equilibrium) measure for the Sherrington–Kirkpatrick model at temperature $T = \beta^{-1}$ is given by

$$\mu_N(\mathrm{d}\mathbf{x}) = \frac{1}{Z_N} e^{\beta H_J(\mathbf{x})} \prod_{i=1}^{N} \mathrm{d}\mu(x_i) \quad \text{with} \quad Z_N = \int e^{\beta H_J(\mathbf{x})} \prod_{i=1}^{N} \mathrm{d}\mu(x_i).$$

In the case $M = \mathbb{R}$, the associated Langevin dynamics (see section 2) were considered by Sompoliski and A. Zippelius (see [32, 29] and then by G. Ben Arous and myself [7, 21]). It is proved that the empirical measure $N^{-1}\sum_{i=1}^{N}\delta_{x^i_{[0,T]}}$ on path space converges as N goes to infinity for every time $T > 0$. Its limit is not Markovian (even though at finite N, its law is Markovian, it losts this property at the limit by self-averaging, the average of Markov laws being not necessarily Markovian) and given by a nonlinear equation. This limiting law is so complicated that the behaviour of its covariance could not be analysed so far, neither in the mathematics or in the physics literature. A similar work was achieved by M. Grunwald for Ising spins and standard Glauber dynamics [20]. However, it is expected that the Langevin dynamics for Sherrington–Kirkpatrick dynamics ages and actually with infinitely many time scales.

Since this already simple model of spin glass was already to difficult to analyse, toys models were introduced to try to understand why aging could appear. Their study allowed to point out two major situations to generate aging.

(a) The first is a flat energy space; the particle system has a single ground state characterised by the lowest possible energy E_0, but there are many other states with energy E_n which is very close to E_0, more precisely $E_n - E_0$ is of the order of N^{-1} if N is the number of particles. Then, the dynamics will be likely to visit all these states in a finite time (independent of N) before finding the ground state. This process will create a long time memory of the history and aging. Hence, aging is here caused by the flatness of the bottom of the most attracting valley in the energy landscape, and the consequent difficulty for the system to find its most favorable state within this valley. It will in fact find it typically in a time depending on the age of the system, time after which it will begin to forget its past.

    This phenomenon describes the spherical Sherrington–Kirkpatrick model, but should also describe the spherical p-spins model of Sherrington–Kirkpatrick. It is believed also that it should explain aging of the dynamics of the original Sherrington–Kirkpatrick model.

(b) The trap model; in this case, the evolution of the particle system is represented by a Markov process in a random energy landscape. The process will spend most time into deep valleys of lowest energy where it will be "trapped" and its evolution will be mostly driven by the seek of deeper valleys. The time spent in these valleys is random and aging will appear when the mean time spent in these valleys diverges.

    This model was originally introduced by Bouchaud to understand aging in the Random Energy Model (REM) introduced by Derrida as a simplification of the

Sherrington–Kirkpatrick model of spin glass. It was shown by G. Ben Arous, A. Bovier and V. Gayrard [2, 3] that this picture is indeed relevant. It also describes aging in the Sinaï model [17].

Note that in both cases, the main point is that the system has infinitely many favorable states which it can reach in finite time; this can be opposed to usual stationnary systems where ground states are separated by an energy barrier which diverges with the size of the system, forbidding the infinite system to visit several of them in a finite time.

2 Spherical model of spin glass

If $U : \mathbb{R} \rightarrow \mathbb{R}$ is some potential going to infinity fast enough at infinity, the Langevin dynamics at temperature $T = \beta^{-1}$ for the Sherrington–Kirkpatrick model are defined by the stochastic differential system

$$\mathrm{d}x_t^i = -\tfrac{1}{2}\beta\,\partial_{x^i} H_J^N(\underline{x}_t)\,\mathrm{d}t - U'(x_t^i) + \mathrm{d}B_t^i$$

with prescribed initial data. Here, $\big(B^i, 1 \le i \le N\big)$ are i.i.d Brownian motions.

One way to simplify considerably this system is to consider instead a smooth spherical constraint

$$\mathrm{d}x_t^i = -\tfrac{1}{2}\beta\,\partial_{x^i} H_J^N(\underline{x}_t)\,\mathrm{d}t - U'\Big(\frac{1}{N}\sum_{j=1}^{N}(x_t^j)^2\Big)x_t^i\,\mathrm{d}t + \mathrm{d}B_t^i \tag{2.1}$$

with a function U on $\mathbb{R}^+$ such that

$$\limsup_{x\rightarrow\infty}\frac{U(x)}{x} = +\infty$$

in order to insure the almost sure boundedness of the empirical covariance under the dynamics (2.1). A hard spherical constraint was considered in [14] where a similar study was undertaken.

The great simplification offered by the spherical model is that the empirical covariance

$$K_N(s,t) = \frac{1}{N}\sum_{i=1}^{N} x_s^i\, x_t^i\,, \tag{2.2}$$

satisfies, at the large N limit, an autonomous equation. Indeed, one computes

$$\begin{aligned} K_N(t,s) =& \frac{1}{N}\,\mathrm{tr}\Big(\big(e^{-\int_0^t \mathbf{V_N}(u)\mathrm{d}u}x_0 + \int_0^t e^{-\int_v^t \mathbf{V_N}(u)\mathrm{d}u}\,\mathrm{d}B_v\big) \\ &\times \big(e^{-\int_0^s \mathbf{V_N}(u)\mathrm{d}u}x_0 + \int_0^s e^{-\int_v^s \mathbf{V_N}(u)\mathrm{d}u}\,\mathrm{d}B_v\big)\Big) \end{aligned} \tag{2.3}$$

with $\mathbf{V_N}(u)$ the $N \times N$ matrix given by $\mathbf{V_N}(u) := U'\big(K_N(u,u)\big)\mathbf{I} - \beta\mathbf{J}$ if $\mathbf{J}$ is the symmetric matrix with entries $\{J_{ij}\,, 1 \le i \le j \le N\}$ above the diagonal. From this formula, it is easily seen from this formula that the long time behaviour of the covariance will be driven by the largest eigenvalues of the matrix $\mathbf{J}$. The eigenvalues $\lambda_1 \geq \lambda_2 \geq \cdots \geq \lambda_N$ of the Wigner matrix $\mathbf{J}$ are well known; λ_1 converges almost surely towards 2, but the difference of the next eigenvalues with λ_1 are of order N^{-1} so that

$$\lim_{N\to\infty} \frac{1}{N} \sum_{i=1}^{N} \delta_{\lambda_i} = \sigma \qquad \text{a.s.}$$

with σ the semi-circle law $\sigma(\mathrm{d}x) = C\sqrt{4-x^2}\,\mathrm{d}x$, which is absolutely continuous w.r.t Lebesgue measure, in particular in the neighborhood of 2. From this asymptotic, one deduces that if the $(x_0^i, 1 \leq i \leq N)$ are independent equidistributed variables with law μ_0 (which corresponds to an infinite temperature initial condition), K_N converges almost surely towards K, solution of the renewal equation

$$\begin{aligned} K(t,s) =& R_\beta(t) R_\beta(s)\, \mathcal{L}\big(\beta(t+s)\big) \int x^2\, \mathrm{d}\mu_0(x) \\ &+ \int_0^s \frac{R_\beta(t)\, R_\beta(s)}{R_\beta(v)^2} \mathcal{L}\big(\beta(t+s-2v)\big)\, \mathrm{d}v \end{aligned} \tag{2.4}$$

with, for $\theta > 0$ and $t \geq 0$,

$$\mathcal{L}(\theta) = \int e^{\theta\lambda}\, \mathrm{d}\sigma(\lambda) \qquad R_\beta(t) = e^{-\int_0^t U'(K(s,s))\mathrm{d}s}$$

One can analyze this equation when

$$U(x) = \tfrac{1}{2} c\, x^2$$

for some $c > 0$. We find that for the solution of (2.4), if we let

$$\beta_c = \frac{c}{2\lambda^*} \int \frac{1}{\lambda^* - \lambda}\, \mathrm{d}\sigma(\lambda)\,. \tag{2.5}$$

and assume that

$$\sigma(\lambda^* - \lambda \leq \theta) \underset{\theta \to 0}{\simeq} b_1 \theta^q \tag{2.6}$$

for some $q > 1$ and $b_1 > 0$ (and hence $\beta_c < \infty$), then, the unique solution K to (2.4) satisfies

1. For $\beta < \beta_c$, there exists $\delta_\beta > 0$ and $c_\beta \in \mathbb{R}^+$ so that for all $t, s \in \mathbb{R}^+$,

$$\big|K(t,s)\big| \leq c_\beta\, e^{-\delta_\beta |t-s|}\,. \tag{2.7}$$

2. For $\beta = \beta_c$, $q \neq 2$, $t \gg s \gg 1$, we have the polynomial decay

$$K(t,s) \sim \begin{cases} (t-s)^{1-q} & \text{for } t/s \text{ bounded} \\ \dfrac{s^{1-\frac{1}{2}\psi_q}}{t^{q-\frac{1}{2}\psi_q}} & \text{otherwise}\,, \end{cases} \tag{2.8}$$

where $\psi_q = \max(2-q, 0)$.

3. When $\beta > \beta_c$ we get that

$$K(t,s) \sim \left(\frac{s}{t}\right)^{\frac{1}{2}q}, \tag{2.9}$$

so $K(t,s) \to 0$ if and only if $t/s \to \infty$.

Note that in the case where σ is the semi-circle appearing in the asymptotics of the spectral measure of $\mathbf{J}$, $q = \frac{3}{2}$. Hence, we see that aging appears for $\beta > \beta_c$ when the particles are initially independent. When starting from the top eigenvector, this phenomenon disappears (the system stays in the basin of attraction of the top eigenvector); in fact, for any fixed $\beta > \beta_c$, regardless of the way in which $t-s$ and s approach infinity,

$$K(t,s) \approx \frac{1}{2\beta}\Big(\frac{2\beta^2}{c}\lambda^* - \int \frac{1}{\lambda^* - \lambda}\, d\sigma(\lambda)\Big) > 0$$

There is thus no aging regime for this initial condition, which underlines the fact that aging phenomenon is very dependent on the initial conditions.

Note here that two factors were crucial to prove aging; the flatness of the energy landscape near the ground state but also the fact that the interaction between the particles results with a nonlinear equation for the covariance (indeed, without this nonlinearity, it could be checked that the covariance would be asymptotically stationnary [22]). In fact, the randomness of the matrix $\mathbf{J}$ is not necessary, provided its eigenvalues distribution (which could be deterministic) is sufficiently flat next to the maximum eigenvalue.

On a technical point of view, it was crucial that the covariance C satisfies an autonomous equation. It was pointed out by L. Cugliandolo and J. Kurchan [15] that an autonomous system of equations could be obtained for the covariance and the so-called response function for p spherical models, leading to an analysis of aging phenomenon for these systems. In particular, they believe that in some cases, these models lead to more than two different time scales. I recently derived rigorously with G. Ben Arous and A. Dembo the same system of equations, but we have not yet achieved its long time analysis.

3 Bouchaud's trap model; an energy trap model

Bouchaud's random walk is a simple model of a random walk trapped by random wells. It was proposed as an approximation of the evolution of a more complex system in an energy lanscape with favorable valleys, located at sites given by a discrete set V, and with energies $\{E_x, x \in V\}$.

Let $G = (V, B)$ be a graph described by its set of vertices V and its bonds B. Two vertices are said to be neighbours if they are related by a bond.

Bouchaud's simplest random walk X is a Markov process who jumps from a site x to its neighbours $y : (x,y) \in B$ with a rate

$$w_{x,y} = e^{-\beta E_x}$$

and $w_{x,y}=0$ if (x,y) are not neighbours. The $\{E_x\,, x\in V\}$ are independent random variables with exponential law;

$$P(E_x > t) = e^{-t}\,, \qquad \forall t>0\,.$$

Let $P^{\mathbf{E}}$ denote the quenched law of the Markov chain X (i.e given a realisation of the energies $\mathbf{E}=\{E_x\,, x\in V\}$) and P its annealed law (i.e is the average over the randomness of the energies of the P^Es; $P=\langle P^E\rangle$

The natural order parameters to consider here are either the two times probability

$$R^E(t_w, t_w+t) = P^E\big(X(t_w)=X(t_w+t)\big)$$

or its annealed version

$$R(t_w, t_w+t) = P\big(X(t_w)=X(t_w+t)\big)\,,$$

or can be the probability that the process did not jump between time t_w and time t_w+t;

$$\Pi^E(t_w,t_w+t) = P^E\big(X(t_w)=X(t_w+s)\mid s<t\big)\,,$$
$$\Pi(t_w,t_w+t) = P\big(X(t_w)=X(t_w+s)\mid s<t\big)\,.$$

Aging for such a model was first studied in the mathematics literature by Fontes, Isopi and Newman [18] in the case where $V=\mathbb{Z}$. They proved that, when $\beta>1$,

$$\lim_{t_w\to\infty} R\big(t_w,(1+\theta)t_w\big) = f(\theta)$$

with a well-defined function f, showing an aging regime in the scale of the age of the system. On the other hand, it was shown (see [4]) that Π satisfies

$$\lim_{t_w\to\infty} \Pi\big(t_w, t_w+\theta\, t_w^{\gamma}\big) = q(\theta)$$

with a well defined function q and $\gamma=(1+\beta)^{-1}$.

Combining these two results shows that the process will be able to quit a deep trap in a time of order t_w^{γ} but will not find a deeper trap before a time of order t_w.

4 Sinaï model

Bouchaud's trap model on $\mathbb{Z}$ describes also the long time behaviour of Sinaï's random walk in random environment; it is described as follows. Let $\mathbf{p}=\big(p_i\,, i\in\mathbb{Z}\big)\in[0,1]^{\mathbb{Z}}$ be independent equidistributed variables with law μ. Sinai's Markov chain $X^{\mathbf{p}}$ is then given by

$$P\big(X^{\mathbf{p}}_{n+1}=i+1\mid X^{\mathbf{p}}_n=i\big) = 1-P\big(X^{\mathbf{p}}_{n+1}=i-1\mid X^{\mathbf{p}}_n=i\big)=p_i\,,$$
$$P\big(X^{\mathbf{p}}_0=0\big)=1\,.$$

Let $\rho_i := \dfrac{1-p_i}{p_i}$ and assume that $\mathbb{E}\big[\log\rho_0\big] = \int\log(x^{-1}-1)\,\mathrm{d}\mu(x)$ is well defined. It is well known that if $\mathbb{E}\big[\log\rho_0\big]\neq 0$, the Markov chain is transient and will go to infinity when time goes to infinity. When $\mathbb{E}[\log\rho_0] = 0$, Sinaï [31] proved that the Markov chain $X^{\mathbf{P}}$, correctly renormalised, converges almost surely towards the deepest valley designed by the random environment that it could visit. More precisely, if we let

$$W^n(t) = \frac{1}{\log n}\sum_{i=0}^{\lfloor(\log n)^2 t\rfloor} \log\rho_i \cdot (\operatorname{sign} t)\,,$$

W^n will converge towards a Brownian motion W on $\mathbb{R}$. Then, the random walk $X^{\mathbf{P}}$, once divided by $(\log n)^2$, will converge towards the nearest point to the origin which corresponds to a well of depth greater or equal to one designed by W as shown in figure 3.

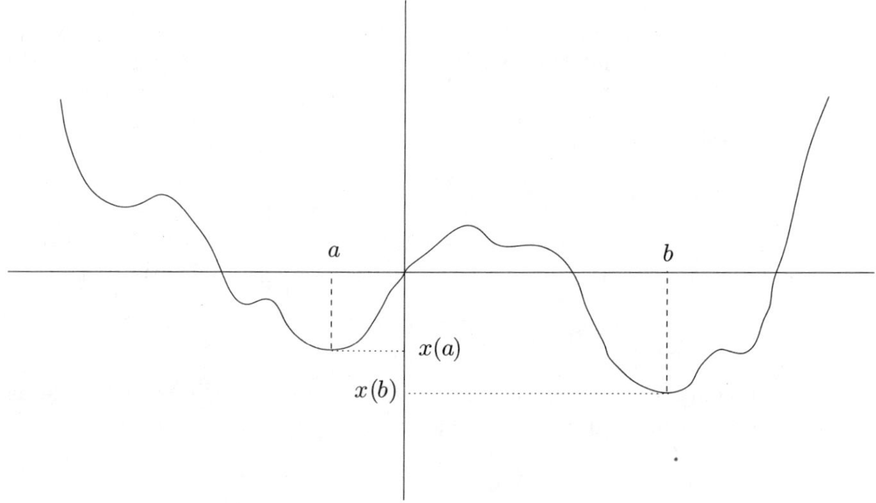

Figure 3. *If $x_a > 1$, then $(\log n)^{-2}X^{\mathbf{P}}(n)$ converges to a, if $x_a < 1$, $x_b > 1$, it converges towards b, etc.*

The aging phenomenon will then also occur since when time is going on, the random walk will have found better and better attractors and will therefore like to stay longer there; it was indeed shown (see [28, 19, 17]) that for any $h > 1$,

$$\lim_{\eta\to 0}\lim_{n\to\infty} P\Big(\frac{|X_{n^h} - X_n|}{(\log n)^2} < \eta\Big) = \frac{1}{h^2}\big[\tfrac{5}{3} - \tfrac{2}{3}e^{-(h-1)}\big] \tag{4.1}$$

5 Bouchaud's trap model on a complete graph

Generalisation of Bouchaud's model can be found in [4, 5], where jump rates depending not only on the site where the walk stands but also from the energy of the site where it wants to jump, higher dimension models are considered as well.

Bouchaud's random walk on a complete graph is also of interest since it is related with Derrida's random energy model. If G is the complete graph on M points, we denote $\Pi_M(t_w, t_w + t)$ the annealed probability that the walk stays in a given well during time t, then it was shown (see [10, 2, 3]) that Π_M converges as M goes to infinity. Moreover, its limit Π satisfies

$$\lim_{t_w \to \infty} \Pi(t_w, t_w + \theta t_w) = H(\theta)$$

with

$$H(\theta) = \left(\pi \operatorname{cosec}\left(\frac{\pi}{\beta}\right)\right)^{-1} \int_\theta^\infty (1+x)^{-1} x^{-\beta} \, dx \, .$$

6 Aging for the Random Energy Model

Let us finally describe the Random Energy Model (**REM**) introduced by Derrida; noticing that for any given $\mathbf{x}$, the Hamiltonian $H_J(\mathbf{x})$ for the Sherrington–Kirkpatrick model is a centered Gaussian variable and thinking that the $\mathbf{x}$ are Ising spins taking values $+1$ or -1, Derrida considered the Gibbs measure on $\{-1, +1\}^N$ given, for $\sigma \in \{-1, +1\}^N$, by

$$\mu_{\beta,N}(\sigma) = \frac{1}{Z_{\beta,N}} e^{\beta \sqrt{N} E_\sigma} \quad \text{with} \quad Z_{\beta,N} = \sum_{\sigma \in \{-1,+1\}^N} e^{\beta \sqrt{N} E_\sigma} .$$

Here, $\{E_\sigma, \sigma \in \{-1, +1\}^N\}$ are independent centered Gaussian variables with variance one, the independence hypothesis resulting with a great simplification with respect to the original Sherrington–Kirkpatrick model. A standard Glauber dynamic for this model is given by the transition kernel $p(\sigma, \eta)$ on $\{-1, +1\}^N$ which is null if σ and η differ at more than one site, given by $N^{-1} e^{-\beta \sqrt{N} E_\sigma^+}$ if σ and η only differ by a spin-flip, and $1 - e^{-\beta \sqrt{N} E_\sigma^+}$ if $\sigma = \eta$. Then, it was shown in [2, 3] that the motion of these dynamics when seen only on the deepest traps created by the energies $\{E_\sigma, \sigma \in \{-1, +1\}^N\}$ will be described by Bouchaud's random walk on a complete graph of large number of vertices. In fact, with a well-chosen threshold $u_N(E) \approx \sqrt{2N \log(2)} + \dfrac{E}{\sqrt{2N \log(2)}}$ and a natural scaling $c_{N,E} \approx e^{\beta \sqrt{N} u_N(E)}$, they proved that

$$\lim_{t_w \to \infty} \lim_{E \to -\infty} \lim_{N \to \infty} P\left(\left| \frac{\Pi_N(c_{N,E} t_w, c_{N,E}(t_w + t))}{H(t t_w^{-1})} - 1 \right| > \epsilon \right) = 0$$

for any $\epsilon > 0$.

Bibliography

[1] Aizenman M., Lebowitz J. L., Ruelle D.; Some rigorous results on the Sherrington–Kirkpatrick spin glass model. *Commun. Math. Phys.*, **112**, pp. 3–20 (1987).

[2] Ben Arous G., Bovier A., Gayrard V.; Glauber dynamics of the random energy model I. Metastable motion on the extreme states. *Comm. Math. Phys.*, **235**, pp. 379–425 (2003).

[3] Ben Arous G., Bovier A., Gayrard V.; Glauber dynamics of the random energy model II; Aging below the critical temperature. To appear in *Comm. Math. Phys.* (2003).

[4] Ben Arous G., Cerny J.; Bouchaud's model exhibits two aging regime in dimension one. Preprint (2002).

[5] Ben Arous G., Cerny J., Mountford T.; Aging for Bouchaud's model in dimension 2. Preprint (2002).

[6] Ben Arous G., Dembo A., Guionnet A.; Aging of spherical spin glasses. *Prob. Theo. Rel. Fields*, **120**, pp. 1–67 (2001).

[7] Ben Arous G., Guionnet A.; Symmetric Langevin spin glass dynamics. *Ann. Probab.*, **25**, pp. 1367–1422 (1997).

[8] Biljakovic K., Lasjaunias J. C., Monceau P.; Aging effects and non exponential energy relaxations in charge-density wave systems, *Phys. Rev. Lett*, **62**, p. 1512 (1989).

[9] Bouchaud J. P.; Granular media: some ideas from statistical physics; `http://www.arxiv.org/cond-mat/0211196`.

[10] Bouchaud J. P., Dean D.; Aging on Parisi tree, *Journal of Physics I*, **5**, p. 265 (1995).

[11] Bouchaud J. P., Cugliandolo L. F., Kurchan L. F., Mezard M.; Out of equilibrium dynamics in spin-glasses and other glassy systems. *Spin glass dynamics and Random Fields*, A. P. Young Editor (1997).

[12] Bovier A.; Picco P. (ed.); Mathematical aspects of spin glass and neural networks. *Birkhauser*, Prog. Probab. **41**, (1998).

[13] Cugliandolo L.; Dynamics of glassy systems; `http://xxx.lanl.gov/abs/cond-mat/0210312`.

[14] Cugliandolo L., Dean D. S.; Full dynamical solution for a spherical spin-glass model. *J. Phys. A*, **28**, p. 4213 (1995).

[15] Cugliandolo L., Kurchan J.; Analytical solution of the off-equilibrium dynamics of a long range spin-glass model, *Phys. Rev. Lett.*, **71**, p. 173 (1993).

[16] Cugliandolo L., Kurchan J.; On the out of equilibrium relaxation of the Sherrington–Kirkpatrick model, *J. Phys. A*, **27**, p. 5749 (1994).

[17] Dembo A., Zeitouni O., Guionnet A.; Aging properties of Sinai's random walk in random environment, *XXX preprint archive*, math.PR/0105215 (2001).

[18] Fontes L. R. G., Isopi M., Newman C.; Random walks with strongly inhomogeneous rates and singular diffusions: convergence, localization and aging in one dimension. *Ann. Prob.*, **30**, pp. 579–604 (2002).

[19] Golosov A. O.; On limiting distributions for a random walk in a critical one. *Comm. Moscow Math. Soc.*, **199**, (1985) pp. 199–200.

[20] Grunwald M.; Sanov results for Glauber spin-glass dynamics. *Prob. Th. Rel. Fields*, **106**, pp. 187–232 (1996).

[21] Guionnet A.; Annealed and quenched propagation of chaos for Langevin spin glass dynamics. *Prob. Th. Rel. Fields*, **109**, pp. 183–215 (1997).

[22] Guionnet A.; Non-Markovian limit diffusions and spin glasses *Fields Inst. Comm.*, **34**, pp. 65–74 (2002).

[23] Guionnet A., Zegarlinski B.; Decay to equilibrium in random spin systems on a lattice. *Commun. Math. Phys.*, **181**, No 3, pp. 703–732 (1996).

[24] Guionnet A., Zegarlinski B.; Decay to equilibrium in random spin systems on a lattice. *Journal of Stat.*, **86**, pp. 899–904 (1997).

[25] Ma S. K.; Statistical Mechanics. *World Scientific, Singapore* (1985).

[26] Herisson D.; Mesure de relations de fluctuation-dissipation dans un verre de spins. Thesis; `http://tel.ccsd.cnrs.fr/documents/archives0/00/00/20/55/`.

[27] Herisson D., Ocio M.; Fluctuation-dissipation ratio of a spin glass in the aging regime. *Phys. Rev. Lett.*, **88**, p. 2572 (2002).

[28] Le Doussal P., Monthus C. and Fisher D. S.; Random walkers in one-dimensional random environments: exact renormalization group analysis. *Phys. Rev. E*, **59**, (1999) pp. 4795–4840.

[29] Mezard M., Parisi G., Virasoro M.; Spin glass theory and beyond. *World Scientific Lecture Notes in Physics* (1987).

[30] Nattermann, Scheidi S.; Vortex glass phases in type-II superconductors, *Adv. Phys.*, **49**, p. 607 (2000).

[31] Sinai Ya. G.; The limiting behavior of a one-dimensional random walk in random environment, *Theor. Prob. and Appl.*, **27**, (1982) pp. 256–268.

[32] Sompolinsky H., Zippelius A.; Dynamic Theory of the Spin-Glass Phase, *Phys. Rev. Lett.* **47**, (1981) pp. 359–362.

Thomas Yizhao Hou is a Professor of Applied and Computational Mathematics at CalTech, and is one of the leading experts in numerical analysis for free boundary and multiscale problems. In his 18-year research career his research interests have centred around developing and analysing effective numerical methods for vortex dynamics, interfacial flows, and multiscale problems.

Hou was born in Canton, China, and studied at the South China Institute of Technology before taking his PhD from UCLA. Upon obtaining his PhD in 1987, he joined the Courant Institute as a postdoctoral researcher and then became a faculty member in 1989. He moved to the applied mathematics department at CalTech in 1993, and is currently a professor and the executive officer of applied mathematics. Professor Hou was awarded the J H Wilkinson Prize in Numerical Analysis and Scientific Computing in 2001, the François N Frenkiel Award from the American Physical Society in 1998, and the Feng Kang Prize in Scientific Computing in 1997. He was also an invited speaker at the International Congress of Mathematicians in Berlin in 1998, and a Sloan Foundation Research Fellowship holder from 1990 to 1992.

Chapter 10

Multiscale Modelling and Computation of Incompressible Flow*

Thomas Y. Hou[†]

Abstract: Many problems of fundamental and practical importance contain multiple-scale solutions. Composite materials, flow and transport in porous media, and turbulent flow are examples of this type. Direct numerical simulations of these multiscale problems are extremely difficult due to the range of length scales in the underlying physical problems. Here, we introduce a dynamic multiscale method for computing nonlinear partial differential equations with multiscale solutions.

The main idea is to construct semi-analytic multiscale solutions local in space and time, and use them to construct the coarse-grid approximation to the global multiscale solution. Such an approach overcomes the common difficulty associated with the memory effect and the non-unqiueness in deriving the global averaged equations for incompressible flows with multiscale solutions. It provides an effective multiscale numerical method for computing incompressible Euler and Navier–Stokes equations with multiscale solutions.

In a related effort, we introduce a new class of numerical methods to solve the stochastically-forced Navier–Stokes equations. We will demonstrate that our numerical method can be used to accurately compute high-order statistical quantites more efficiently than the traditional Monte-Carlo method.

Contents

[1]Research was in part supported by the National Science Foundation through a grant DMS–0073916 and an ITR grant ACI–0204932.

[†]Applied Mathematics, 217–50, Caltech, Pasadena, USA

1 Introduction

Many problems of fundamental and practical importance have multiple-scale solutions. Composite materials, wave propagation in random media, flow and transport through heterogeneous porous media, and turbulent flow are examples of this type. The direct numerical solution of multiple-scale problems is difficult due to the wide range of scales in the solution. It is almost impossible to resolve all the small-scale features by direct numerical simulations due to the limited capacity in computing power. On the other hand, from an engineering perspective, it is often sufficient to predict the macroscopic properties of the multiscale systems, such as the effective conductivity, elastic moduli, permeability, and eddy diffusivity. Therefore, it is desirable to develop a coarse-grid method that captures the small-scale effect on the large scales, but does not require resolving all the small-scale features.

In recent years, we have introduced a multiscale finite-element method (MsFEM) for solving partial differential equations with multiscale solutions [27, 28, 20, 9]. The central goal of this approach is to obtain the large-scale solutions accurately and efficiently without resolving the small-scale details. The main idea is to construct finite-element base functions which capture the small-scale information within each element. The small-scale information is then brought to the large scales through the coupling of the global stiffness matrix. Thus, the effect of small scales on the large scales is captured correctly. In our method, the base functions are constructed from the leading-order differential operator of the governing equation within each element. This leading-order operator is typically an elliptic operator with highly-oscillatory coefficients for composite materials, flow in porous media, wave propagation in random media, or convection-dominated transport with multiscale velocity field. As a consequence, the base functions are adapted to the local microstructure of the differential operator. In the case of two-scale periodic structures, we have proved that the multiscale method indeed converges to the correct solution, independent of the small scale in the homogenization limit [28, 20, 9].

One of the main difficulties in deriving effective multiscale methods is to derive accurate *local* microscopic boundary conditions that connect the small-scale solution from one coarse-grid block to the neighboring coarse-grid blocks. If one naïvely imposes a smooth boundary condition for multiscale bases at the boundary of a coarse-grid element, it will create a mismatch between the global multiscale solution and the approximate solution constructed by the multiscale numerical method. Using homogenization theory, we have identified a resonance error which manifests as the ratio between the physical small scale and the coarse-grid mesh size [27, 20]. Our analysis indicates that if we use inappropriate microscopic boundary conditions for the multiscale bases, it will generate a boundary layer in the boundary corrector of the multiscale base, which seems to be responsible for generating the resonance error. In the case when the coefficient has scale separation and periodic structure, we can solve the periodic-cell problem to construct the 'ideal' microscopic boundary condition which eliminates the artificial boundary layer in the boundary corrector. In this special case, we also obtain an analytic formulation for the multiscale bases. However, this approach can not be generalised to problems with many or a continuous spectrum of scales. On the other hand, our analysis indicates that interactions of small scales are strongly localised for elliptic or parabolic problems. Motivated by this observation, we propose an over-sampling technique which can effectively reduce the resonance error [27]. This over-sampling technique is quite general and can be applied to problems with many or a continuous spectrum of scales.

We have applied the multiscale finite-element method with the over-sampling technique to several applications, ranging from problems in composite materials, to wave propagation in random media, convection-dominated transport, and two-phase flow in heterogeneous porous media. The agreements between the coarse-grid multiscale finite-element calculations and the corresponding well-resolved calculations are striking. We remark that from a practical application view-point, it is important that multiscale computational methods can be applied to problems with an infinite number of scales that are not separable. In many applications, such as transport of flow through heterogeneous porous media, the multiscale coefficient (such as the permeability tensor) has a continuous spectrum of scales without scale separation or periodic structure. Therefore it is essential that we do not make explicit use of the assumption on scale separation and periodic structure in our multiscale finite-element method.

We remark that the idea of using base functions governed by the differential equations has been used in the finite-element community; see e.g., [3]. In particular, the multiscale finite-element method is similar in spirit to the residual-free bubble finite-element method [6, 44] and the variational multiscale method [33, 7]. There are also other multiscale methods that explore homogenization theory or separation of scales to derive effective coarse-grid methods; see e.g., [13, 37, 38, 24, 8, 10, 21].

While a lot of progress has been made in developing multiscale methods to solve elliptic or diffusion-dominated problems, there is only limited success in developing effective multiscale methods for convection-dominated transport in heterogeneous media [39, 17, 35, 48, 19, 31]. One of the common difficulties for this problem is the so-called nonlocal memory effect [46]. For nonlinear convection problems, it is also difficult to characterise how small scales propagate in time and what kind of small-scale structure is preserved by the flow dynamically.

Recently, together with Dr Danping Yang [30, 32], we have developed a systematic multiscale analysis for the 3-D incompressible Euler equations with a highly-oscillating initial-velocity field. The understanding of scale interactions for 3-D incompressible Euler and Navier–Stokes equations have been a major challenge. For high Reynolds number flows, the degrees of freedom are so high that it is almost impossible to resolve all small scales by direct numerical simulations. Deriving an effective equation for the large-scale solution is very useful in engineering applications. The nonlinear and nonlocal nature of the incompressible Euler or Navier–Stokes equations makes it difficult to construct a properly-posed multiscale solution.

The key idea in constructing our multiscale solution for the Euler equation is to reformulate the problem using a new phase variable to characterise the propagation of small scales. This phase variable is essentially the backward-flow map. The multiscale structure of the solution becomes apparent in terms of this phase variable. Our analysis is strongly motivated by the pioneering work of McLaughlin–Papanicolaou–Pironneau (MPP for short) [39]. The main difference is that MPP assumed that the small scales are convected by the mean flow, while we believe that the small scales are convected by the full velocity field. In fact, by using a Lagrangian description of the Euler equation, we can see that small scales are indeed propagated by the Lagrangian-flow map. By using a Lagrangian description, we can characterise the nonlinear convection of small scales exactly and turn a convection-dominated transport problem into an elliptic problem for the stream function. Thus, traditional homogenization results for elliptic problems can be used to obtain a multiscale expansion for the stream function. At the end, we derive a coupled multiscale system for the flow map and the stream function. In order for the homogenized system to be well-posed, we need to impose a solvability condition, which is to ensure that there is no secular growth term in the first-order correction of the flow map. The solvability condition can be interpreted as a projection or filtering to remove the resonant-velocity component. Such a resonant-velocity component prevents the flow from fully mixing and can lead to the development of the nonlocal memory effect [46].

For computational purposes, it is more convenient to derive a homogenized equation in the Eulerian formulation. By using the key observation from our multiscale analysis in the Lagrangian formulation, we derive a well-posed homogenized equation in the velocity–pressure formulation. In our multiscale analysis in the Eulerian frame, we use the phase variable to describe the propagation of small scales in the velocity field, but use the Eulerian variable to describe the large-scale solution. Since we treat the convection of small scales exactly by the phase variable, there is no convection term in the cell problem for the small-scale velocity field. As a consequence, we can solve for the cell problem with a relatively large time-step. Moreover, for fully mixed flow, we expect that the small-scale solution would reach a statistical equilibrium relatively quickly in time. In this case, we may need only to compute a small number of time-steps in the cell problem to evaluate the Reynolds stress term in the homogenized equation for the averaged velocity. Moreover, we may express the Reynolds stress term as the product of an eddy diffusivity and the deformation tensor of the averaged velocity field; see e.g., [45, 36, 12, 25]. For fully mixed homogeneous flow, the eddy diffusivity is almost constant in space. In this case, we need only to solve one representative cell problem and use the solution of this representative cell problem to evaluate the eddy diffusivity. This would give a self-consistent coarse-grid model that couples the evolution of the small and large scales dynamically.

The rest of the paper is organised as follows. In Section 2, we review the multiscale finite-element method and describe the issue of microscopic boundary conditions for the multiscale bases. We then introduce the over-sampling technique and discuss the convergence property of the method. In Section 3, we present several applications of the multiscale finite-element methods, including wave propagation in periodic and random media, convection-enhanced diffusion, and flow and transport in heterogeneous porous media. We also discuss how to use our multiscale method to upscale one-phase and two-phase flow. In Section 4, we describe some recent work in deriving nonlinear homogenization for the 3-D incompressible Euler equation.

2 Multiscale Finite-Element Method

In this section, we briefly review the multiscale finite-element method which was introduced in [27, 28] and has been applied to compute elliptic problems with highly-oscillating coefficients, wave propagation in multiscale media, convection-enhanced diffusion, and transport of flow in strongly-heterogeneous porous media. The multiscale finite-element method (MsFEM for short) is designed to effectively capture the large-scale behaviour of the solution without resolving all the small-scale features. The main idea of our multiple-scale finite-element method consists of construction of finite-element base functions which contain the small-scale information within each coarse-grid element. The small-scale information is then brought into the large scales through the coupling of the global stiffness matrix.

It should be noted that MsFEM is different from the traditional domain-decomposition method in an essential way, although the two methods appear to be similar. First of all, the design purposes are different. MsFEM is used as a method to obtain the correct discretisation of the large-scale problem on a relatively coarse grid, while the domain decomposition is an iterative method for solving the problem on a fine grid which resolves the small scales. One of the key features of MsFEM is that the construction of the base functions is a *local* operation within the coarse-grid elements. Thus, the construction of the base function in one element is *decoupled* from that in another element. In contrast, in domain-decomposition methods, the decomposed subdomains are still coupled together.

The decoupled construction of the multiple-scale bases provides some advantages in the computation. First, the construction can be carried out perfectly in parallel. In effect, we break a large-scale computation into many smaller and *independent* pieces. Thus, the method is automatically adapted to parallel computers. In addition, there is a great advantage in computer memory usage. Once the small-scale information within an element is gathered into the global stiffness matrix, the memory used for those base functions can be reused to construct bases of the next element. Thus, we can sequentially sample a large amount of fine-scale information from many elements with limited memory. Therefore, MsFEM is less constrained by the limit of computer memory than the direct methods. Another important feature of this approach is that small-scale solutions can be reconstructed *locally* from the coarse-grid computation by using the multiscale bases as interpolation bases. This feature is especially useful when we try to upscale two-phase flow in heterogeneous media. Moreover, by constructing the multiscale bases adaptively in space and time, we can recover the fine-scale detail using only a fraction of time that is required for a direct fine-grid simulation.

2.1 MsFEM for elliptic problems with oscillating coefficients

We will use the elliptic problem with highly-oscillating coefficients as an example to illustrate the main idea of MsFEM. We consider the following elliptic problem:

$$L_\epsilon u := -\nabla \cdot (a^\epsilon(\mathbf{x})\nabla u) = f \quad \text{in } \Omega\,, \qquad u = 0 \quad \text{on } \partial\Omega\,, \tag{2.1}$$

where $a^\epsilon(\mathbf{x}) = \left(a^\epsilon_{ij}(\mathbf{x})\right)$ is a positive definite matrix. This model equation represents a common difficulty shared by several physical problems. For flow in porous media, it is the pressure equation through Darcy's law. The coefficient a^ε represents the permeability tensor. For composite materials, it is the steady heat-conduction equation and the coefficient a^ε represents the thermal conductivity. For steady-transport problems with divergence-free velocity field, it is a symmetrised form of the governing equation. In this case, the coefficient a_ε is a combination of the transport velocity and viscosity tensor.

To simplify the presentation of the finite-element formulation, we assume the domain is a unit square $\Omega = (0,1) \times (0,1)$. The variational problem of (2.1) is to seek $u \in H^1_0(\Omega)$ such that

$$a(u,v) = f(v), \qquad \forall v \in H^1_0(\Omega), \tag{2.2}$$

where

$$a(u,v) = \int_\Omega a^\epsilon_{ij} \frac{\partial v}{\partial x_i}\frac{\partial u}{\partial x_j}\,\mathrm{d}\mathbf{x} \quad \text{and} \quad f(v) = \int_\Omega fv\,\mathrm{d}\mathbf{x}\,,$$

where we have used the Einstein summation notation. A finite-element method is obtained by restricting the weak formulation (2.2) to a finite-dimensional subspace of $H^1_0(\Omega)$. For $0 < h \le 1$, let $\mathcal{K}^h$ be a partition of Ω by a collection of rectangles K with diameter $\le h$, which is defined by an axi-parallel rectangular mesh. In each element $K \in \mathcal{K}^h$, we define a set of nodal basis $\{\phi^i_K, i = 1, \ldots, d\}$ with d being the number of nodes of the element. The subscript K will be neglected when bases in one element are considered. In our multiscale finite-element method, ϕ^i satisfies

$$L_\epsilon \phi^i = 0 \qquad \text{in } K \in \mathcal{K}^h\,. \tag{2.3}$$

Let $\mathbf{x}_j \in \overline{K}$ $(j = 1, \ldots, d)$ be the nodal points of K. As usual, we require $\phi^i(\mathbf{x}_j) = \delta_{ij}$. One needs to specify the boundary condition of ϕ^i to make (2.3) a well-posed problem (see below). For now, we assume that the base functions are continuous across the boundaries of the elements, so that

$$V^h = \operatorname{span}\{\phi^i_K : i = 1, \ldots, d; K \in \mathcal{K}^h\} \subset H^1_0(\Omega)\,.$$

In the following, we study the approximate solution of (2.2) in V^h, i.e., to find $u^h \in V^h$ such that

$$a(u^h, v) = f(v), \qquad \forall v \in V^h. \tag{2.4}$$

Note that this formulation of the multiscale method is not restricted to the rectangular elements. It can also be applied to triangular elements which are more flexible in modelling complicated geometries.

2.2 Microscopic boundary conditions for multiscale bases

The choice of boundary conditions in defining the multiscale bases will play a crucial role in approximating the multiscale solution. Intuitively, the boundary condition for the multiscale base function should reflect the multiscale oscillation of the solution u across the boundary of the coarse-grid element. To gain insight, we first consider the special case of periodic homogenization; i.e., when $a^\epsilon(\mathbf{x}) = a(\mathbf{x}, \mathbf{x}/\epsilon)$, with $a(\mathbf{x}, \mathbf{y})$ being periodic in $\mathbf{y}$. Using standard homogenization theory [4], we can perform a multiscale expansion for the base function, ϕ^ϵ, as follows ($\mathbf{y} = \mathbf{x}/\epsilon$):

$$\phi^\epsilon = \phi_0(\mathbf{x}) + \epsilon\phi_1(\mathbf{x}, \mathbf{y}) + \epsilon\theta^\epsilon(\mathbf{x}) + O(\epsilon^2),$$

where ϕ_0 is the effective solution, ϕ_1 is the first-order corrector. The boundary corrector θ^ϵ is chosen so that the boundary condition of ϕ^ϵ on ∂K is exactly satisfied.

By solving a periodic-cell problem for χ^j,

$$\nabla_{\mathbf{y}} \cdot a(\mathbf{x}, \mathbf{y}) \nabla_{\mathbf{y}} \chi^j = \frac{\partial}{\partial y_i} a_{ij}(\mathbf{x}, \mathbf{y}), \tag{2.5}$$

with zero mean, we can express the first-order corrector ϕ_1 as follows: $\phi_1(\mathbf{x}, \mathbf{y}) = -\chi^j \dfrac{\partial \phi_0}{\partial x_j}$. The boundary corrector, θ^ϵ, then satisfies

$$\nabla_{\mathbf{x}} \cdot a(\mathbf{x}, \tfrac{\mathbf{x}}{\epsilon}) \nabla_{\mathbf{x}} \theta^\epsilon = 0 \quad \text{in } K,$$

with boundary condition

$$\epsilon\theta^\epsilon|_{\partial K} = \left(\phi^\epsilon - (\phi_0 + \epsilon\phi_1(\mathbf{x}, \tfrac{\mathbf{x}}{\epsilon}))\right)\Big|_{\partial K}.$$

In general, the boundary condition of θ^ϵ has O(1) oscillations. This leads to formation of a boundary layer of thickness $O(\epsilon)$ near ∂K [4]. This boundary layer is completely numerical. It is due to the fact that we try to approximate the oscillatory solution of a global elliptic problem by linear superposition of a collection of local base functions whose boundary conditions are not conformed with the oscillatory solution across the edge of the local element.

Note that ϕ_0 corresponds to the base function of the homogenized equation, which does not contain any multiscale feature. Thus, we can approximate ϕ_0 by a simple linear finite-element base. If we impose a linear boundary condition for $\phi^\epsilon(\mathbf{x})$ over ∂K, i.e., $\phi^\epsilon|_{\partial K} = \phi_0|_{\partial K}$, then this will induce an oscillatory boundary condition for θ^ϵ:

$$\theta^\epsilon|_{\partial K} = -\phi_1(\mathbf{x}, \tfrac{\mathbf{x}}{\epsilon})|_{\partial K}.$$

As we mentioned earlier, this will introduce a numerical boundary layer to θ^ϵ, which will lead to the so-called resonance error (see discussion below) [27, 20]. To avoid this resonance error, we need to incorporate the multi-dimensional oscillatory information through the cell problem into our boundary condition for ϕ^ϵ; i.e., to set $\phi^\epsilon|_{\partial K} = (\phi_0 + \epsilon\phi_1(\mathbf{x}, \frac{\mathbf{x}}{\epsilon}))|_{\partial K}$. In this case, the boundary condition for $\theta^\epsilon|_{\partial K} = 0$. Therefore, we have $\theta^\epsilon \equiv 0$. In this case, we have an analytic expression for the multiscale base functions ϕ^ϵ as follows:

$$\phi^\epsilon = \phi_0(\mathbf{x}) + \epsilon\phi_1\big(\mathbf{x}, \tfrac{\mathbf{x}}{\epsilon}\big)\,, \tag{2.6}$$

with $\phi_1(\mathbf{x}, \mathbf{y}) = -\chi^j(\mathbf{x}, \mathbf{y})\dfrac{\partial\phi_0}{\partial x_j}$ and χ^j is the solution of the cell problem (2.5).

The above example is to illustrate the difficulty in designing the appropriate boundary condition for the base function. Of course, except for problems with periodic structure, we can not use this approach to compute the multiscale base functions in general. Later we introduce a more effective over-sampling technique to overcome the difficulty of designing the appropriate microscopic boundary conditions for the base functions.

2.3 Convergence Analysis

Convergence analysis has been carried out for the multiscale finite-element method in the case when the coefficient, $a^\epsilon(\mathbf{x})$, has a scale separation and periodic structure, although this assumption is not required by our method. What distinguishes our multiscale finite-element method from the traditional finite-element method is that MsFEM gives a convergence result uniform in ϵ as ϵ tends to zero. To obtain a sharp convergence rate, we need to use the multiscale solution structure given by the homogenization theory [4]. In particular, we rely on a sharp homogenization estimate which uses the boundary corrector [41]. In the case when the boundary conditions of the base functions are linear, we have proved the following convergence result in [28].

Theorem 2.1. *Let $a^\epsilon(\mathbf{x}) = a(\mathbf{x}/\epsilon)$ with $a(\mathbf{y})$ being periodic in $\mathbf{y}$ and smooth. Let $u \in H^2(\Omega)$ be the solution of (2.1) and u_h be the multiscale finite-element approximation obtained from the space spanned by the multiscale bases with linear boundary conditions. Then we have*

$$\|u - u_h\|_{H^1} \le C(h+\epsilon)\|f\|_{L^2} + C\big(\tfrac{\epsilon}{h}\big)^{\frac{1}{2}}\|u_0\|_{H^2}\,, \tag{2.7}$$

where $u_0 \in H^2(\Omega) \cap W^{1,\infty}(\Omega)$ is the solution of the homogenized equation.

We refer to [28] for the detail of the analysis. We remark that convergence analysis for elliptic problems with multiple scales and for problems with random coefficients has been obtained by Efendiev in his PhD dissertation [18]. Moreover, he proved that the above convergence theorem is still valid when the coefficient $a(\mathbf{y})$ is only piecewise smooth. We would like to point out that in the one-dimensional case the multiscale finite-element method can reproduce the exact solution at the coarse-grid nodal points without making any assumption of the scale separation and periodic structure of the coefficient [28].

2.4 The over-sampling technique

As we can see from the above theorem, MsFEM indeed gives the correct homogenized result as ϵ tends to zero. This is in contrast with the traditional finite-element method, which does not give the correct homogenized result as $\epsilon \to 0$. For the linear finite element method, the error would grow like $O(h^2/\epsilon^2)$. On the other hand, we also observe that when $h \sim \epsilon$, the multiscale method attains a large error in both H^1 and L^2 norms. This is what we call the *resonance* effect between the grid scale, h, and the small scale, ϵ, of the problem. As we indicated earlier, the boundary layer in the first-order corrector seems to be the main source of the resonance effect. By a judicious choice of boundary conditions for the base functions, we can eliminate the boundary layer in the first-order corrector. This would give a nice conservative difference structure in the discretisation, which in turn leads to *cancellation of resonance errors* and gives an improved rate of convergence.

Motivated by our convergence analysis, we propose an *over-sampling* technique to overcome the difficulty due to scale resonance [27]. The idea is quite simple and easy to implement. The main observation is that the boundary layer in the boundary correction θ^ϵ is *strongly localised*, with the width of order $O(\varepsilon)$. If we sample in a domain with size larger than $h+\varepsilon$ and use only the interior sampled information to construct the bases, we can reduce significantly the influence on the base functions of the boundary layer in the larger sample domain. As a consequence, we obtain an improved rate of convergence.

Specifically, let ψ^j be the base functions satisfying the homogeneous elliptic equation in the larger domain $S \supset K$. We then form the actual base ϕ^i by linear combination of ψ^j,

$$\phi^i = \sum_{j=1}^{d} c_{ij} \psi^j .$$

The coefficients c_{ij} are determined by the condition $\phi^i(\mathbf{x}_j) = \delta_{ij}$. The corresponding θ^ε for ϕ^i are now free of boundary layers. Our extensive numerical experiments have demonstrated that the over-sampling technique does improve the numerical error substantially in many applications.

Note that the over-sampling technique results in a *non-conforming* MsFEM method. In [20], we perform a careful estimate of the non-conforming errors in both the H^1-norm and the L^2-norm. The analysis shows that the non-conforming error is indeed small, consistent with our numerical results [27, 29]. Our analysis also reveals a cell resonance, which is the mismatch between the mesh size and the 'perfect' sample size. In case of a periodic structure, the 'perfect' sample size is the length of an integer multiple of the period. This cell resonance was first revealed by Santosa and Vogelius in [40]. When the sample size is an integer multiple of the period, then the cell-resonance error is identically zero [40, 20]. In the error expansion, this resonance effect appears as a *higher-order* correction. In numerical computations, we found that the cell-resonance error is generically small, and is almost negligible for random coefficients. Nonetheless, it is possible to completely eliminate this cell-resonance error by using a Petrov–Galerkin formulation [52]; i.e., to use the over-sampling technique to construct the base functions, but using piecewise-linear functions as test functions. This reduces the nonconforming error and eliminates the resonance error completely.

We remark that the over-sampling technique is different from the overlapping domain-decomposition method. The domain-decomposition method is an iterative method to solve the fine-grid solution globally, while MsFEM with over-sampling is a method to derive an accurate coarse-grid approximation by capturing the effect of small scales on large scales locally. On the other hand, in collaboration with Aarnes [1], we have shown that the multiscale finite-element method can be used to construct nearly optimal preconditioner for domain-decomposition methods applied to elliptic problems with highly-oscillating and high aspect-ratio coefficients. A multiscale finite-element method has also been used to upscale absolute permeability [51], where we analyse the source of upscaling error in some existing upscaling methods and demonstrate how the over-sampling technique can be used effectively to reduce the upscaling error.

2.5 Convergence and Accuracy

Except for special cases when the coefficient has periodic structure or is separable in space variables, in general we need to compute the multiscale bases numerically using a subgrid mesh. To assess the accuracy of our multiscale method, we compare MsFEM with a traditional linear finite-element method (FEM for short) using a subgrid mesh, $h_s = h/M$. The multiscale bases are computed using the same subgrid mesh. Note that MsFEM only captures the solution at the coarse grid h, while FEM tries to resolve the solution at the fine grid h_s. Our extensive numerical experiments demonstrate that the accuracy of MsFEM on the coarse grid h is comparable to that of the corresponding well-resolved FEM calculation at the same coarse grid. In some cases, MsFEM is even more accurate than FEM (see below and the next section).

First, we demonstrate the convergence in the case when the coefficient has scale separation and periodic structure. In Table 1 we present the result for

$$a\left(\tfrac{\mathbf{x}}{\varepsilon}\right) = \frac{2 + P\sin(2\pi x_1/\varepsilon)}{2 + P\cos(2\pi x_2/\varepsilon)} + \frac{2 + \sin(2\pi x_2/\varepsilon)}{2 + P\sin(2\pi x_1/\varepsilon)} \quad (P = 1.8)\,, \tag{2.8}$$

$$f(\mathbf{x}) = -1 \qquad \text{and} \qquad u|_{\partial\Omega} = 0\,. \tag{2.9}$$

The convergence of three different methods are compared for fixed $\varepsilon/h = 0.64$, where '-L' indicates that linear boundary condition is imposed on the multiscale base functions, 'os' indicates the use of over-sampling, and 'LFEM' stands for linear FEM.

We see clearly the scale resonance in the results of MsFEM-L and the (almost) first-order convergence (i.e., no resonance) in MsFEM-os-L. Evident also is the error of MsFEM-os-L being smaller than those of LFEM obtained on the fine grid. In [28, 27], more extensive convergence tests have been presented.

N	ε	MsFEM-L		MsFEM-os-L		LFEM	
		$\|E\|_{l^2}$	rate	$\|E\|_{l^2}$	rate	MN	$\|E\|_{l^2}$
16	0.04	3.54e-4		7.78e-5		256	1.34e-4
32	0.02	3.90e-4	-0.14	3.83e-5	1.02	512	1.34e-4
64	0.01	4.04e-4	-0.05	1.97e-5	0.96	1024	1.34e-4
128	0.005	4.10e-4	-0.02	1.03e-5	0.94	2048	1.34e-4

Table 1. *Convergence for periodic case.*

Next, we illustrate the convergence of the multiscale finite-element method when the coefficient is random and has no scale separation nor periodic structure. In Figure 1, we show the results for a log-normally distributed a^ε. In this case, the effect of scale resonance shows clearly for MsFEM-L; i.e., the error increases as h approaches ε. Here $\varepsilon \sim 0.004$ roughly equals the correlation length. Even the use of an oscillatory boundary conditions (MsFEM-O), which is obtained by solving a reduced 1-D problem along the edge of the element, does not help much in this case. On the other hand, MsFEM with over-sampling agrees very well with the well-resolved calculation. One may wonder why the errors do not decrease as the number of coarse-grid elements increases. This is because we use the same subgrid mesh size as the well-resolved grid size to construct the base functions for various coarse-grid sizes ($N = 32, 64, 128$, etc). If we use multiscale bases that are obtained analytically or computed with very high precision, then the errors indeed decay as the coarse-grid mesh decreases. The above calculations demonstrate that by using locally-constructed multiscale finite-element bases, we can recover the well-resolved calculation at the coarse grid with comparable accuracy. This is quite remarkable since the local bases are *not* the restrictions of the well-resolved solution of the global elliptic problem to the coarse-grid elements. Here the well-resolved calculation corresponds to a 2048 by 2048 linear finite-element calculation. The error is computed by extrapolating the 2048×2048 and the 4096×4096 linear finite-element solutions. The accuracy of the extrapolated solution is of the order of 10^{-6}.

3 Applications of MsFEM

In this section, we apply the multiscale finite-element method to a few applications. The applications we consider are wave propagation in heterogeneous media, convection-enhanced diffusion, and flow and transport in heterogeneous porous media.

3.1 Wave Propagation in Heterogeneous Media

The multiscale finite-element method can be easily extended to the time-dependent wave equation. Wave propagation in heterogeneous media is an important problem that has rich multiscale phenomena and a wide range of applications in geoscience

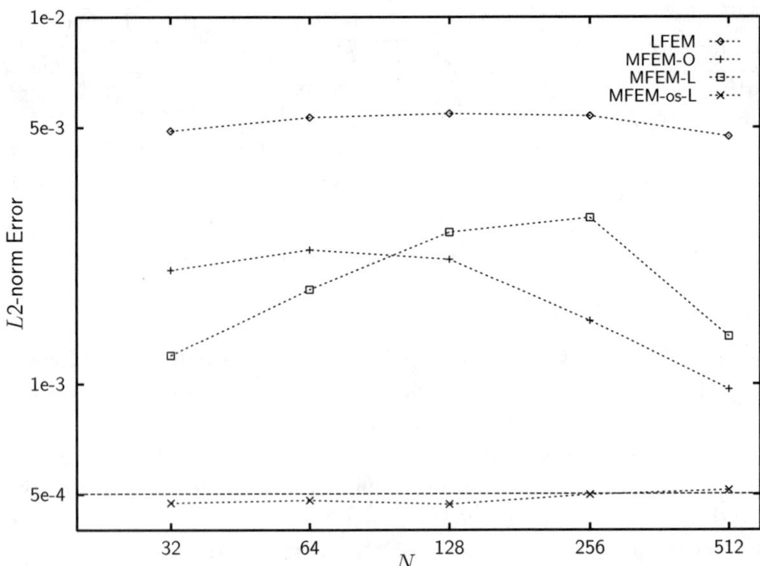

Figure 1. *The l^2-norm error of the solutions using various schemes for a log-normally distributed permeability field.*

and medical imaging.

Consider the wave equation in a heterogeneous media:

$$\frac{\partial^2 u}{\partial t^2} = \nabla \cdot a^\epsilon(\mathbf{x})\nabla u + F(\mathbf{x}, t) , \quad u|_{t=0} = f(\mathbf{x}), \quad (u_t)|_{t=0} = g(\mathbf{x}) .$$

For this wave equation, we can construct the multiscale finite-element bases in the same way as we did for the elliptic problem; i.e.,

$$\nabla \cdot a^\epsilon(\mathbf{x})\nabla \phi^\epsilon = 0 ,$$

with appropriate boundary conditions on the edge of each element K (e.g., using the over-sampling technique).

Using a similar coefficient as in Table 1 and with zero forcing, we have performed the following convergence study (the work described in this subsection was carried out by a former postdoc, Dr Yu Zhang). In this study, we choose $\epsilon = 0.005$, and the well-resolved calculation is obtained by using a 2048×2048 fine grid. We compare the multiscale finite element calculation with both the well-resolved solution, denoted as u, and the homogenized solution, denoted as u_0. We can see that the multiscale finite-element calculations converge to the well-resolved solution with a rate comparable to that for the homogenized solution.

We have also computed the wave equation with a random coefficient which has a continuous spectrum of scale and has a fractal dimension of 2.8 (see [27] for a description of this random medium). The initial condition is given as a symmetric Gaussian pulse with zero initial velocity. For deterministic homogeneous media, it is known that the solution will remain symmetric in time. But for random

N	$\|u_h - u\|$	$\|u_h - u_0\|$
32	0.03068	0.030370
64	0.01435	0.013994
128	0.0067795	0.00646126
256	0.0031569	0.00281998

Table 2. *Errors of multiscale finite-element calculations for the wave equation with periodic oscillating coefficients.*

homogeneous media, we found that the solution develops asymmetry in its wave-front dynamically, which is completely due to the randomness of the wave-speed coefficient.

The computational cost for the wave propagation using MsFEM is significantly reduced compared with the direct simulation using a fine grid. This is because the computational cost in computing the multiscale bases is only at time $t = 0$. Once we have generated the multiscale bases initially, we can compute the corresponding stiffness matrix for the coarse grid. The subsequent calculations are all done using a coarse spatial grid and a coarse time-step. In comparison, a fine grid with small time-step must be used for each fine-grid simulation. The saving can be quite significant.

3.2 Convection-Enhanced Diffusion

Another interesting application is the large-time behaviour of the convection–diffusion equation with a rapidly-oscillating velocity field and a slowly-varying initial condition:

$$\frac{\partial T}{\partial t} + \mathbf{u}_\delta(\mathbf{x}) \cdot \nabla T = \epsilon\, \Delta T\,, \quad T(\mathbf{x}\,,0) = T_0(\delta \mathbf{x})\,,$$

where the velocity field $\mathbf{u}_\delta$ is divergence-free and δ characterises the small scale in the velocity field. After rescaling the space and time variables, $\mathbf{x}' = \mathbf{x}/\delta$ and $t' = t/\delta^2$, we obtain the rescaled convection diffusion as follows (we still use $\mathbf{x}$ and t):

$$\frac{\partial T}{\partial t} + \mathbf{u}(\tfrac{\mathbf{x}}{\delta}) \cdot \nabla T = \epsilon\, \Delta T\,, \quad T(\mathbf{x}\,,0) = T_0(\mathbf{x})\,.$$

Under appropriate assumptions on $\mathbf{u}$ (see e.g., [23]), it can be shown that T^δ converges to an effective solution T^* as δ tends to zero for each $\epsilon > 0$ fixed:

$$\frac{\partial T^*}{\partial t} = \sigma_\epsilon\, \Delta T^*\,, \quad T^*(\mathbf{x}\,,0) = T_0(\mathbf{x})\,.$$

We call σ_ϵ the effective diffusivity. In general, we have $\sigma_\epsilon > \epsilon$. How σ_ϵ scales with ϵ, as ϵ tends to zero, is a problem of considerable interest. The answer depends on the geometry of the streamlines associated with the velocity field $\mathbf{u}$.

One of the well-known cases is the cellular flow in which the stream function is given by

$$H(\mathbf{x}) = \frac{1}{4\pi^2} \sin\left(2\pi \frac{x_1}{\delta}\right) \sin\left(2\pi \frac{x_2}{\delta}\right) ,$$

and the velocity field is given by $\mathbf{u} = (-H_{x_2}, H_{x_1})$. In this case, it has been shown analytically that $\sigma_\epsilon \sim C\sqrt{\epsilon}$ as $\epsilon \to 0$ (see e.g., [22]).

In the Ph.D. dissertation of Dr Peter Park [42], the multiscale finite-element method has been applied to compute the effective diffusivity as $\epsilon \to 0$. The multiscale bases are constructed by satisfying the following steady equation:

$$\mathbf{u}\left(\frac{\mathbf{x}}{\delta}\right) \cdot \nabla \phi^\epsilon = \epsilon \, \Delta \phi^\epsilon ,$$

with appropriate boundary conditions. By applying the Galerkin finite-element method with the above multiscale bases and discretising in time implicitly, we can obtain a finite-element discretisation similar to that considered before.

To compute the effective diffusivity, we use the following formula [22]

$$\sigma_\epsilon = \lim_{t \to \infty} \frac{1}{4t} \iint \left(|\mathbf{x}|^2 \, T(\mathbf{x}, t) \right) \, d\mathbf{x} ,$$

with the initial condition that $T_0 = \delta(\mathbf{x})$, the Dirac δ-function. With a relatively coarse grid (16×16) and a subgrid $h_s = 1/512$, we obtain the following results for the effective diffusivity σ_ϵ for different values of δ and ϵ in Table 3 [42].

ϵ	σ_ϵ	ratio	σ_ϵ	ratio	σ_ϵ	ratio
	$\delta = 1$		$\delta = 0.1$		$\delta = 0.05$	
0.04	0.03252		0.03265		0.03275	
0.02	0.02092	0.6432	0.02316	0.7093	0.02355	0.7191
0.01	0.01141	0.5454	0.01643	0.7094	0.01665	0.7070
0.005	0.00619	0.5425	0.01300	0.7912	0.01233	0.7405
0.0025	0.00344	0.5557	0.01046	0.8046	0.00872	0.7072

Table 3. *The diffusivity scaling for the cellular flow.*

As we can see from Table 3, when the velocity field is smooth, i.e. $\delta = 1$, the effective diffusivity is of the same order as the molecular diffusivity ϵ; i.e. $\sigma_\epsilon \sim C\epsilon$. But for a highly-oscillating velocity field with small δ ($\delta = 0.01$), we see that the effective diffusivity reveals the scaling $\sigma_\epsilon \sim C\sqrt{\epsilon}$ as ϵ tends to zero (note that $\frac{1}{\sqrt{2}} \approx 0.7071$). This verifies the theoretical result for the effective diffusivity for the cellular flow. Since the velocity field is steady, the multiscale bases are computed only at $t = 0$. The subsequent computation in time uses a coarse grid in space ($H = 1/16$) and a large time-step. This is much cheaper than performing the calculation using a fine space-grid which has to resolve δ and uses a small time-step.

In this particular example, we can solve the corresponding cell problem to construct the multiscale bases, which leads to additional computational saving. In general, the velocity field is generated by solving the Navier–Stokes equation, whose solution would not have scale separation nor periodic structure. For a certain random velocity, the convection–diffusion equation may exhibit anomalous diffusion behaviour (see e.g., [23]). The effective diffusivity may depend on some nonlocal geometric property of the velocity field. We have also investigated the anomalous diffusion induced by a nonlocal random velocity field by supplementing the local multiscale bases with a nonlocal base to capture the nonlocal interaction of small scales [42].

3.3 Flow and Transport in Porous Media

The flow and transport problems in porous media are considered in a hierarchical level of approximation. At the microscale, the solute transport is governed by the convection–diffusion equation in a homogeneous fluid. However, for porous media, it is very difficult to obtain full information about the pore structure. A certain averaging procedure has to be carried out, and the porous medium becomes a continuum with certain macroscopic properties, such as the porosity and permeability. Through the use of sophisticated geological and geostatistical modelling tools, engineers and geologists can now generate highly-detailed, three-dimensional representations of reservoir properties. Such models can be particularly important for reservoir management, as fine-scale details in formation properties, such as thin, high permeability layers or thin shale barriers, can dominate reservoir behaviour. The direct use of these highly-resolved models for reservoir simulation is not generally feasible because their fine level of detail (tens of millions grid blocks) places prohibitive demands on computational resources. Therefore, the ability to coarsen these highly-resolved geologic models to levels of detail appropriate for reservoir simulation (tens of thousands grid blocks), while maintaining the integrity of the model for the purposes of flow simulation (i.e., avoiding the loss of important details), is clearly needed.

We consider a heterogeneous system which represents two-phase immiscible flow. Our interest is in the effect of permeability heterogeneity on two-phase flow. Therefore, we neglect the effects of compressibility and capillary pressure, and consider porosity to be constant. This system can be described by writing Darcy's law for each phase (all quantities are dimensionless):

$$\mathbf{v}_j = \frac{k_{rj}(S)}{\mu_j}\mathcal{K}^\epsilon \nabla p\,, \tag{3.1}$$

where $\mathbf{v}_j$ are the Darcy's velocity for the phase j ($j = o, w$ oil, water), p is pressure, S is water saturation, K^ϵ is the permeability tensor, k_{rj} is the relative permeabilities of each phase and μ_j is the viscosity of the phase j. Darcy's law for each phase coupled with mass conservation can be manipulated to give the pressure and saturation

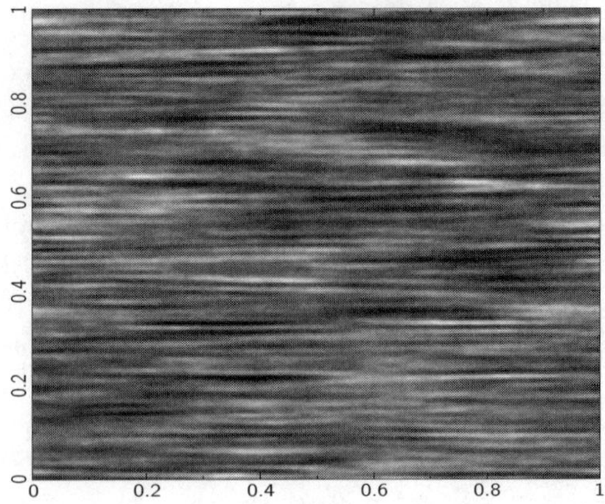

Figure 2. *A random porosity field with layered structure.*

equations:

$$\nabla \cdot \big(\lambda(S)\mathcal{K}^\epsilon \, \nabla p\big) = 0\,, \tag{3.2}$$

$$\frac{\partial S}{\partial t} + \mathbf{u}^\epsilon \cdot \nabla f(S) = 0\,, \tag{3.3}$$

which can be solved subject to some appropriate initial and boundary conditions. The parameters in the above equations are given by

$$\lambda = \frac{k_{rw}(S)}{\mu_w} + \frac{k_{ro}(S)}{\mu_o}\,, \tag{3.4}$$

$$f(S) = \frac{k_{rw}(S)/\mu_w}{k_{rw}(S)/\mu_w + k_{ro}(S)/\mu_o}\,, \tag{3.5}$$

$$\mathbf{u}^\epsilon = \mathbf{v}_w + \mathbf{v}_o = \, -\lambda(S)\, K^\epsilon \nabla p\,. \tag{3.6}$$

Fine-Scale Recovery

To solve transport problems in the subsurface formations, as in oil reservoir simulations, one needs to compute the velocity field from the elliptic equation for pressure; i.e., $\mathbf{u}^\epsilon = \, -\lambda(S)\, \mathcal{K}^\epsilon \, \nabla p$. In some applications involving isotropic media, the cell-averaged velocity is sufficient, as shown by some computations using the local upscaling methods (*cf.* [15]). However, for anisotropic media, especially layered ones (Figure 2), the velocity in some thin channels can be much higher than the cell average, and these channels often have dominant effects on the transport solutions. In this case, the information about fine-scale velocity becomes vitally important. Therefore, an important question for all upscaling methods is how to take those fast-flow channels into account.

For MsFEM, the fine-scale velocity can be easily recovered from the multiscale base functions, noting that they provide interpolations from the coarse h-grid to

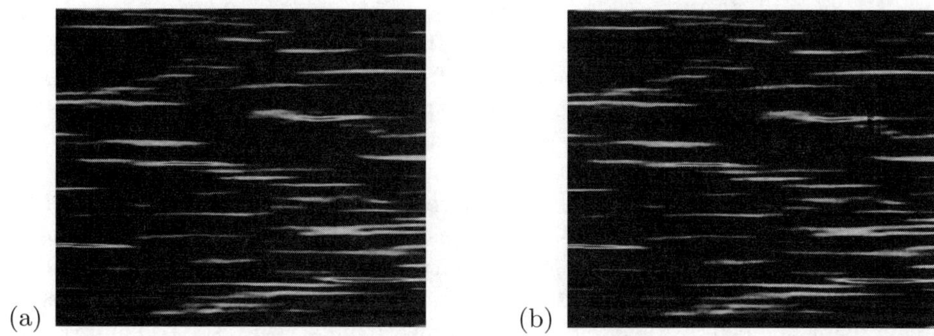

Figure 3. *(a): Fine-grid horizontal velocity field, N=1024. (b): Recovered horizontal velocity field from the coarse-grid calculation (N=64) using multiscale bases.*

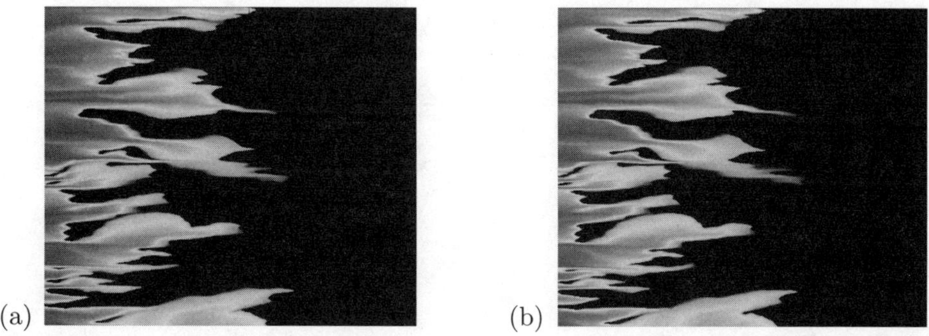

Figure 4. *(a): Fine-grid saturation at t=0.06, N=1024. (b): Saturation computed using the recovered velocity field from the coarse-grid calculation (N=64) using multiscale bases.*

the fine h_s-grid. To demonstrate the accuracy of the recovered velocity and effect of small-scale velocity on the transport problem, we show the fractional flow result of a 'tracer' test using the layered medium in Figure 2: a fluid with red colour originally saturating the medium is displaced by the same fluid with blue colour injected by flow in the medium at the left boundary, where the flow is created by a unit horizontal pressure drop. The transport equation is solved to compute the saturation of the red fluid (see [16] for more details).

To illustrate that we can recover the fine-grid velocity field from the coarse-grid pressure calculation, we plot the horizontal velocity fields obtained by two methods. In Figure 3a, we plot the horizontal velocity field obtained by using a fine-grid ($N = 1024$) calculation. In Figure 3b, we plot the same horizontal velocity field obtained by using the coarse-grid pressure calculation with $N = 64$ and using the multiscale finite-element bases to interpolate the fine-grid velocity field. We can see that the recovered velocity field captures very well the layer structure in the fine-grid velocity field. Further, we use the recovered fine-grid velocity field to compute the saturation in time. In Figure 4a, we plot the saturation at $t = 0.06$ obtained by the fine-grid calculation with $N = 1024$. Figure 4b shows the corresponding saturation obtained using the recovered velocity field from the coarse-grid calculation with $N = 64$. Most detailed fine-scale fingering structures in the well-resolved saturation are captured very well by the corresponding calculation using the recovered velocity field from the coarse-grid pressure calculation. The agreement is striking.

We also check the fractional-flow curves obtained by the two calculations. The fractional flow of the red fluid, defined as $F = \int S_{\text{red}}\, u_1^\epsilon \,\mathrm{d}y \,/ \int u_1^\epsilon \,\mathrm{d}y$ (S being the saturation, u_1^ϵ being the horizontal velocity component), at the right boundary is shown in Figure 5. The top pair of curves are the solutions of the transport problem using the cell-averaged velocity obtained from a well-resolved solution and from MsFEM; the bottom pair are solutions using well-resolved fine-scale velocity and the recovered fine-scale velocity from the MsFEM calculation. Two conclusions can be made from the comparisons. First, the cell-averaged velocity may lead to a large error in the solution of the transport equation. Second, both the recovered fine-scale velocity and the cell-averaged velocity obtained from MsFEM give faithful reproductions of respective direct numerical solutions.

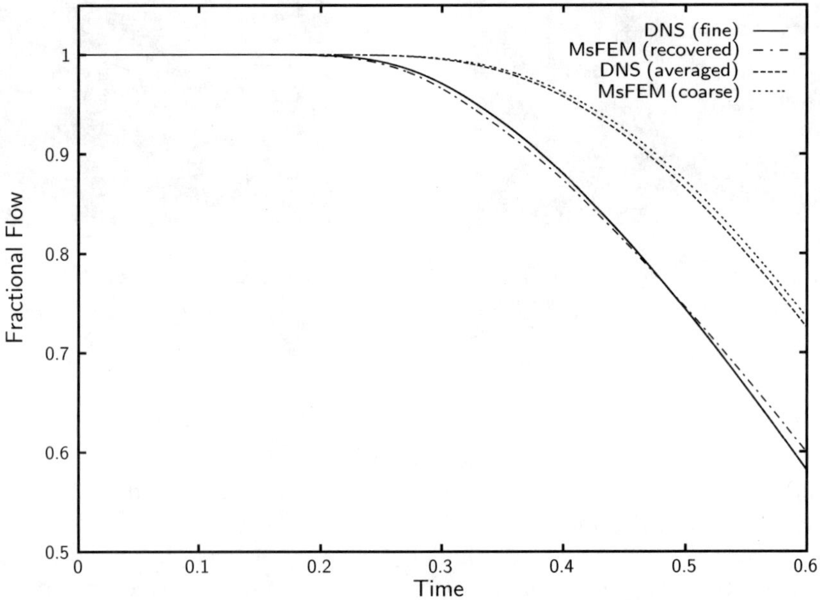

Figure 5. *Variation of fractional flow with time. DNS: well-resolved direct numerical solution using LFEM ($N = 512$). MsFEM: over-sampling is used ($N = 64$, $M = 8$).*

We remark that a finite-volume version of the multiscale finite-element method has been developed by the petroleum engineers from Chevron Petroleum Technology Co. [34]. They also found that by updating the multiscale bases adaptively in space and time, they can recover the fine-scale detail of the well-resolved calculation. The percentage of the multiscale bases that need to be updated is small (only a few percent of the total number of bases). In some sense, the multiscale finite-element method also offers an efficient approach to capture the fine-scale details using only a small fraction of the computational time required for a direct numerical simulation using a fine grid.

3.4 Scale-up of one-phase flows

The multiscale finite-element method has been used in conjunction with some moment-closure models to obtain an upscaled method for one-phase flows; see e.g., [18, 19, 9]. Note that the multiscale finite-element method presented above does not conserve mass exactly. For long time integration, it may lead to loss of mass. This is an undesirable feature of the method. In a recent work with Dr Zhiming Chen [9], we have designed and analysed a mixed multiscale finite-element method. We then applied this mixed method to study the scale-up of one-phase flows and found that mass is conserved very well, even for long time integration. Below we describe our results in some detail.

In its simplest form, neglecting the effect of gravity, compressibility, capillary pressure, and considering constant porosity and unit mobility, the governing equations for the flow transport in highly-heterogeneous porous media can be described by the following partial differential equations ([35],[47],[18])

$$\nabla \cdot \big(\mathcal{K}^\epsilon(\mathbf{x})\nabla p\big) = 0\,, \tag{3.7}$$

$$\frac{\partial S}{\partial t} + \mathbf{u}^\epsilon \cdot \nabla S = 0\,, \tag{3.8}$$

where p is the pressure, S is the water saturation, $\mathcal{K}^\epsilon(\mathbf{x}) = \big(\mathcal{K}^\epsilon_{ij}(\mathbf{x})\big)$ is the relative permeability tensor, and $\mathbf{u}^\epsilon = -\mathcal{K}(\mathbf{x})\nabla p$ is the Darcy velocity.

Now we describe how the (mixed) multiscale finite element can be combined with the existing upscaling technique for the saturation equation (3.8) to get a complete coarse-grid algorithm for the problem (3.7)–(3.8). The numerical upscaling of the saturation equation has been under intensive study in the literature [16, 19, 35, 26, 48, 49]. The problem is very challenging mathematically, without capillary pressure, because of the nonlocal memory effect [46]. On the other hand, if the capillary effect becomes important, the saturation equation becomes diffusion-dominated transport. In this case, an homogenization result has been obtained [5] and a numerical upscaling method can be designed [2].

In many oil-reservoir applications, the capillary effect is so small that it is neglected in practice. So we must face the nonlocal memory effect induced by the convection-dominated transport. Here we use the upscaling method proposed in [19] and [18] to design an overall coarse-grid model for the problem (3.7)–(3.8). The work of [19] for upscaling the saturation equation involves a moment-closure argument. The velocity and the saturation are separated into a local mean quantity and a small-scale perturbation with zero mean. For example, the Darcy velocity is expressed as $\mathbf{u}^\epsilon = \mathbf{u}_0 + \mathbf{u}'$ in (3.8), where $\mathbf{u}_0$ is the average of velocity $\mathbf{u}$ over each coarse element, $\mathbf{u}'$ is the deviation of the fine-scale velocity from its coarse-scale average. If one ignores the third-order terms containing the fluctuations of velocity and saturation, one can obtain an average equation for the saturation S as follows [19]:

$$\frac{\partial S}{\partial t} + \mathbf{u}_0 \cdot \nabla S = \frac{\partial}{\partial x_i}\Big(D_{ij}(\mathbf{x},t)\frac{\partial S}{\partial x_j}\Big), \tag{3.9}$$

where the diffusion coefficients $D_{ij}(\mathbf{x},t)$ are defined by

$$D_{ii}(\mathbf{x},t) = \big\langle\big|\mathbf{u}_i'(\mathbf{x})\big|\big\rangle L_i^0(\mathbf{x},t), \quad D_{ij}(\mathbf{x},t) = 0, \quad \text{for } i \neq j,$$

where $\langle|\mathbf{u}_i'(\mathbf{x})|\rangle$ stands for the average of $|\mathbf{u}_i'(\mathbf{x})|$ over each coarse element. The function $L_i^0(\mathbf{x},t)$ is the length of the coarse-grid streamline in the x_i direction which starts at time t at point $\mathbf{x}$; i.e.

$$L_i^0(\mathbf{x},t) = \int_0^t y_i(s)\,\mathrm{d}s,$$

where $\mathbf{y}(s)$ is the solution of the following system of ODEs

$$\frac{\mathrm{d}\mathbf{y}(s)}{\mathrm{d}s} = \mathbf{u}_0\big(\mathbf{y}(s)\big), \quad \mathbf{y}(t) = \mathbf{x}.$$

Note that the hyperbolic equation (3.8) is now replaced by a convection–diffusion equation. One should note that the induced diffusion term is history-dependent. In some sense, it captures the nonlocal history-dependent memory effect described by Tartar in the simple-shear flow problem [46]. The convection–diffusion equation (3.9) will be solved by the characteristic linear finite-element method [14, 43] in our simulation.

The mixed multiscale finite-element method can be readily combined with the above upscaling model for the saturation equation. The local fine-grid velocity $\mathbf{u}'$ can be reconstructed from the multiscale finite-element bases. We perform a coarse-grid computation of the above algorithm on the coarse 64×64 mesh. The fractional-flow curve using the above algorithm is depicted in Figure 6. It gives excellent agreement with the 'exact' fractional flow curve which is obtained using a fine 1024×1024 mesh. We also compare the contour plots of the saturation S obtained by the upscaled equation with those obtained by using a well-resolved calculation. We found that the upscaled saturation captures the overall large-scale structure of the fine-grid saturation, but the sharp oil/water interfaces are smeared out. This is due to the parabolic nature of the upscaled equation (3.9).

The main cost in the above algorithm lies in the computation of multiscale bases which can be done *a priori* and completely in parallel. This algorithm is particularly attractive when multiple simulations must be carried out due to the change of boundary and source distribution, as it is often the case in engineering applications. In such a situation, the cost of computing the multiscale base functions is just an overhead. Moreover, once these base functions are computed, they can be used for subsequent time integration of the saturation. Because the evolution equation is now solved on a coarse grid, a larger time-step can be used. This also offers additional computational saving. For many oil-recovery problems, due to the excessively large fine-grid data, upscaling is a necessary step before performing many simulations and realisations on the upscaled coarse-grid model. If one can coarsen the fine grid by a factor of 10 in each dimension, the computational saving of the coarse-grid model over the original fine model could be as large as a factor 10,000 (three space dimensions plus time).

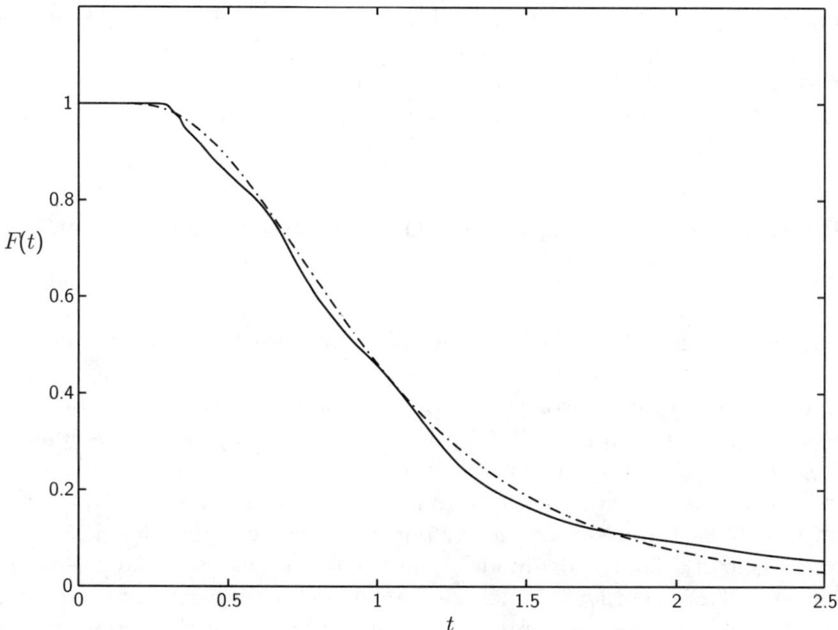

Figure 6. *The accuracy of the coarse-grid algorithm. Solid line is the well-resolved fractional-flow curve. The slash-dotted line is the fractional-flow curve using above coarse-grid algorithm.*

Upscaling the two-phase flow

Upscaling the two-phase flow is more difficult due to the dynamic coupling between the pressure and the saturation. One important observation is that the fluctuation in saturation is relatively small away from the oil/water interface. In this region, the multiscale bases are essentially the same as those generated by the corresponding one-phase flow (time-independent). In practice, we can design an adaptive strategy to update the multiscale bases in space and time. The percentage of multiscale bases that need to be updated is relatively small (a few percent of the total number of the bases). The upscaling of the saturation equation based on a moment-closure argument can be generalised to the two-phase flow as long as the fluctuation of the velocity field $\mathbf{u}'$ can be accurately recovered from the coarse-grid computation [35, 19]. As we discussed in the previous subsection, this is indeed one of the main advantages of the multiscale finite-element method.

It remains to develop a more systematic multiscale analysis to upscale the two-phase flow with heterogeneous media. While the upscaled saturation equation based on te moment-closure approximation is simple and easy to implement, it is hard to estimate its modelling error as the fluctuations in velocity or saturation are not small in practice. With Dr Danping Yang [31], we have recently developed a new multiscale analysis for the convection-dominated transport equation which allows us to upscale the saturation without making the moment-closure approximation. It is based on a delicate multiscale analysis of the transport equation. We are now trying to generalise the analysis to the two-phase flow [50]. The objective is to derive a systematic upscaling for the two-phase flow in heterogeneous porous media which do not have scale separation or periodic structure.

4 Nonlinear homogenization for 3-D Euler equations

The upscaling of the nonlinear transport equation in two-phase flows shares some of the common difficulties in deriving the effective equations for the incompressible Euler equations. With Dr Danping Yang [30], we have recently developed a systematic multiscale analysis for the incompressible 3-D Euler equations. The understanding of scale interactions for 3-D incompressible Euler and Navier–Stokes equations has been a major challenge. For flows with high Reynolds number, the degrees of freedom are so high that it is almost impossible to resolve all small scales by direct numerical simulations. Deriving an effective equation for the large-scale solution is very useful in engineering applications; see e.g., [45, 36, 12, 25]. In deriving a large-eddy simulation model, one usually needs to make certain closure assumptions. The accuracy of such closure models is hard to measure *a priori*. It varies from application to application. For many engineering applications, it is desirable to design a subgrid-based large-scale model in a systematic way so that we can measure and control the modelling error. The main difficulty in deriving effective equations systematically is the strong nonlinear interaction of small scales and the lack of scale separation. Our ultimate goal is to develop a new multiscale analysis that can effectively average infinitely-many scales without assuming scale separation. The challenge is to do this in such a way that we can account for the modelling error systematically when averaging out small scales.

We consider the 3-D incompressible Navier–Stokes equation

$$\mathbf{u}_t^\epsilon + (\mathbf{u}^\epsilon \cdot \nabla)\mathbf{u}^\epsilon + \nabla p^\epsilon = \nu\, \Delta \mathbf{u}^\epsilon\,, \tag{4.1}$$

$$\nabla \cdot \mathbf{u}^\epsilon = 0\,, \tag{4.2}$$

with multiscale initial data $\mathbf{u}^\epsilon(\mathbf{x},0) = \mathbf{u}_0^\epsilon(\mathbf{x})$. Here $\mathbf{u}^\epsilon(t,\mathbf{x})$ and $p^\epsilon(t,\mathbf{x})$ are velocity and pressure respectively, while ν is viscosity. For the time being, we consider only the infinite domain and assume that the solution decays to zero sufficiently fast at infinity. As a first step, we assume that the initial condition has scale separation and periodic structure. That is, we assume that $\mathbf{u}^\epsilon(\mathbf{x},0) = \mathbf{U}(\mathbf{x}) + \mathbf{W}(\mathbf{x},\mathbf{x}/\epsilon)$. Moreover we assume that $\mathbf{U}$ and $\mathbf{W}$ are smooth, $\mathbf{W}(\mathbf{x},\mathbf{y})$ is periodic in $\mathbf{y}$ and has zero mean. In order for the small-scale solution to survive the diffusion, the viscosity ν must be of the order $O(\epsilon^2)$. In this case, the diffusion term will not end the homogenized equation nor the cell problem. To simplify the derivation, we consider only the inviscid case with $\nu = 0$. The question of interest is how to derive a homogenized equation for the averaged velocity field as $\epsilon \to 0$.

The homogenization of the Euler equation with oscillating data was first studied by McLaughlin–Papanicolaou–Pironneau (MPP for short) in 1985 [39]. Since the Euler equation is nonlinear and nonlocal, it is not clear whether the solution will preserve the two-scale structure of the initial data dynamically. Intuitively, a nonlinear interaction may generate new scales dynamically. Therefore constructing an appropriate multiscale expansion for the nonlinear Euler equation is a difficult task. Motivated by some physical arguments, MPP made the assumption that the small-scale oscillation is convected by the mean flow. Based on this assumption, they made the following multiscale expansions for velocity and pressure:

$$\mathbf{u}^\epsilon(\mathbf{x},t) = \mathbf{u}(\mathbf{x},t) + \mathbf{w}\big(\mathbf{x},t,\frac{\theta(\mathbf{x},t)}{\epsilon},\frac{t}{\epsilon}\big) + \epsilon\,\mathbf{u}_1\big(\cdot,\cdot,\frac{\theta(\mathbf{x},t)}{\epsilon},\frac{t}{\epsilon}\big) + \cdots,$$

$$p^\epsilon(\mathbf{x},t) = p(\mathbf{x},t) + \pi\big(\mathbf{x},t,\frac{\theta(\mathbf{x},t)}{\epsilon},\frac{t}{\epsilon}\big) + \epsilon\,p_1\big(\cdot,\cdot,\frac{\theta(\mathbf{x},t)}{\epsilon},\frac{t}{\epsilon}\big) + \cdots,$$

where $\mathbf{w}(\mathbf{x},t,\mathbf{y},\tau)$, $\mathbf{u}_1(\mathbf{x},t,\mathbf{y},\tau)$, π, and p_1 are assumed to be periodic in both $\mathbf{y}$ and τ, and the phase θ is convected by the mean velocity field $\mathbf{u}$

$$\frac{\partial\theta}{\partial t} + \mathbf{u}\cdot\nabla_{\mathbf{x}}\,\theta = 0\,, \quad \theta(\mathbf{x},0) = \mathbf{x}\,. \tag{4.3}$$

By substituting these multiscale expansions into the Euler equation and equating coefficients of the same order, MPP obtained a periodic-cell problem for $\mathbf{w}(\mathbf{x},t,\mathbf{y},\tau)$, $\mathbf{u}_1(\mathbf{x},t,\mathbf{y},\tau)$, π, and p_1. On the other hand, it is not clear whether the resulting cell problem for $\mathbf{w}(\mathbf{x},t,\mathbf{y},\tau)$, etc. has a solution that is periodic in both $\mathbf{y}$ and τ. Even if such a solution exists, it may not be unique. Additional assumptions were imposed on the solution of the cell problem in order to derive a variant of the $k-\epsilon$ model. Instead of using the rate of energy dissipation ϵ, they used the mean helicity together with the mean kinetic energy k to derive a closure for the mean-velocity equation.

The idea of introducing a phase function in the multiscale expansion to describe the propagation of the small-scale velocity field is ingenious and insightful. If one naïvely looks for a multiscale expansion in the form of $\mathbf{w}(t,\mathbf{x},t/\epsilon,\mathbf{x}/\epsilon)$, then one would not be able to derive a well-posed cell problem for the $(\tau,\mathbf{y})$ variables (here $\tau = t/\epsilon, \mathbf{y} = \mathbf{x}/\epsilon$).

With Dr Danping Yang [30], we have recently revisited this problem. Our study shows that the phase variable θ^ϵ should be convected by the full oscillatory velocity field, $\mathbf{u}^\epsilon$:

$$\frac{\partial\theta^\epsilon}{\partial t} + \mathbf{u}^\epsilon\cdot\nabla_{\mathbf{x}}\,\theta^\epsilon = 0\,, \quad \theta^\epsilon(\mathbf{x},0) = \mathbf{x}\,. \tag{4.4}$$

This becomes obvious when we formulate the 2-D Euler equations in vorticity form

$$\omega(\mathbf{x},t) = \tilde{\omega}_0\big(\theta^\epsilon(\mathbf{x},t),\frac{\theta^\epsilon(\mathbf{x},t)}{\epsilon}\big)\,,$$

where $\tilde{\omega}_0$ is the initial vorticity. A similar conclusion can be drawn for the 3-D Euler equation, see below. On the other hand, it is not clear what is the multiscale structure of $\theta^\epsilon(\mathbf{x},t)$ since its structure is coupled to the multiscale structure of $\mathbf{u}^\epsilon$. By allowing θ^ϵ to develop multiscale structure, we embed multiscale structure within multiscale expansions. This approach is quite unconventional. To derive a well-posed cell problem is quite a challenge. If one naïvely expands $\theta^\epsilon(\mathbf{x},t)$ into

$$\theta^\epsilon(\mathbf{x},t) = \theta_0(\mathbf{x},t) + \epsilon\theta_1\big(\mathbf{x},t,\tfrac{\mathbf{x}}{\epsilon},\tfrac{t}{\epsilon}\big) + \cdots, \tag{4.5}$$

it will generate an infinite number of scales at $t > 0$. On the other hand, it is clear that if we substitute a multiscale expansion of θ^ϵ into the velocity field, the first-order correction term θ_1 would have O(1) contribution to the oscillatory velocity field, $\mathbf{w}$, which in turn will contribute to the averaged solution $\mathbf{u}$.

Lagrangian description of the Euler equations

The key idea in our multiscale analysis for the Euler equation is to reformulate the problem using θ^ϵ as a new variable. The multiscale structure of the solution becomes very apparent in terms of this θ^ϵ variable. This amounts to using a Lagrangian description of the Euler equations. Specifically, we introduce a change of variables from $\mathbf{x}$ to α: $\alpha = \theta^\epsilon(\mathbf{x},t)$. It is easy to see that the inverse of this map, denoted by $\mathbf{x} = \mathbf{X}^\epsilon(\alpha,t)$, is the Lagrangian flow map:

$$\frac{\partial \mathbf{X}^\epsilon(\alpha,t)}{\partial t} = \mathbf{u}^\epsilon\big(\mathbf{X}^\epsilon(\alpha,t),t\big), \quad \mathbf{X}^\epsilon(\alpha,0) = \alpha. \tag{4.6}$$

In terms of the α variable, the vorticity of the 3-D Euler equation has a simple expression (see e.g., page 32 of [11]):

$$\omega^\epsilon\big(\mathbf{X}^\epsilon(\alpha,t),t\big) = \frac{\partial \mathbf{X}^\epsilon}{\partial \alpha}(\alpha,t)\,\tilde{\omega}_0\big(\alpha,\tfrac{\alpha}{\epsilon}\big), \tag{4.7}$$

where $\omega^\epsilon = \nabla_x \times \mathbf{u}^\epsilon$ is vorticity, and $\tilde{\omega}_0$ is the initial vorticity. Velocity can be computed from the stream function ψ^ϵ; i.e., $\mathbf{u}^\epsilon = \nabla_x \times \psi^\epsilon$. Further, the stream function, ψ^ϵ, satisfies

$$-\Delta_{\mathbf{x}}\psi^\epsilon = \omega^\epsilon.$$

In terms of the α variable, we have

$$-\nabla_\alpha \cdot \mathcal{A}\mathcal{A}^\top \nabla_\alpha \psi^\epsilon = \frac{\partial \mathbf{X}^\epsilon}{\partial \alpha}(\alpha,t)\,\tilde{\omega}_0\big(\alpha,\tfrac{\alpha}{\epsilon}\big), \tag{4.8}$$

where $\mathcal{A} = \left(\frac{\partial \mathbf{X}^\epsilon}{\partial \alpha}\right)^{-1}$. Using $\left|\frac{\partial \mathbf{X}^\epsilon}{\partial \alpha}\right| = 1$, we can express $\mathcal{A}$ in terms of $\frac{\partial \mathbf{X}^\epsilon}{\partial \alpha}$. Now we can see clearly that the small-scale solution is propagated along the Lagrangian trajectory as a function of α/ϵ.

4.1 Multiscale analysis in the Lagrangian frame

The objective of our multiscale analysis is to obtain an averaged equation for the well-mixing long-time solution of the Euler equations. For this reason, we look for multiscale solutions of the form ($\mathbf{y} = \alpha/\epsilon$, $\tau = t/\epsilon$):

$$\psi^{\epsilon} = \psi^{(0)}(t,\alpha) + \epsilon\psi^{(1)}(t,\alpha,\tau,\mathbf{y}) + \mathrm{O}(\epsilon^2)\,,$$
$$\mathbf{X}^{\epsilon} = \mathbf{X}^{(0)}(t,\alpha) + \epsilon\mathbf{X}^{(1)}(t,\alpha,\tau,\mathbf{y}) + \mathrm{O}(\epsilon^2)\,,$$

where $\psi^{(1)}$ and $\mathbf{X}^{(1)}$ are periodic functions with respect to $\mathbf{y}$ and have zero mean.

By performing careful multiscale analysis, we can obtain the homogenized equations for both 2-D and 3-D Euler equations. For simplicity, we only state the main result for the 2-D Euler equation. In the 3-D Euler equation, there will be interaction between the small scales in the Jacobian of the Lagrangian flow map and the multiscale initial vorticity, which reflects the effect of vortex stretching. Such an effect is absent in the 2-D Euler equation. For the 2-D Euler equation, the homogenized equations for $\mathbf{X}^{(0)}$, $\mathbf{X}^{(1)}$, and $\psi^{(0)}$, $\psi^{(1)}$ are given as follows:

$$\partial_t\mathbf{X}^{(0)} - \nabla_\alpha\mathbf{X}^{(0)}\nabla_\alpha^{\perp}\psi^{(0)} = 0\,, \quad \mathbf{X}^{(0)}|_{\tau=0} = \alpha\,, \tag{4.9}$$
$$\partial_\tau\mathbf{X}^{(1)} - \nabla_{\mathbf{y}}\mathbf{X}^{(1)}\left(\nabla_\alpha^{\perp}\psi^{(0)} + \nabla_{\mathbf{y}}^{\perp}\psi^{(1)}\right) = \nabla_\alpha\mathbf{X}^{(0)}\nabla_{\mathbf{y}}^{\perp}\psi^{(1)}\,, \quad \mathbf{X}^{(1)}|_{\tau=0} = \mathbf{0}\,, \tag{4.10}$$
$$\nabla_{\mathbf{y}}\cdot\left(\mathcal{A}_0\mathcal{A}_0^{\top}\nabla_{\mathbf{y}}\psi^{(1)}\right) + \nabla_{\mathbf{y}}\cdot\left(\mathcal{A}_0\mathcal{A}_0^{\top}\nabla_\alpha\psi^{(0)}\right) = \omega_0\,, \tag{4.11}$$
$$\nabla_\alpha\cdot\left(\langle\mathcal{A}_0\mathcal{A}_0^{\top}\rangle\nabla_\alpha\psi^{(0)}\right) = \langle\omega_1\rangle - \nabla_\alpha\cdot\left(\langle\mathcal{A}_0\mathcal{A}_0^{\top}\nabla_{\mathbf{y}}\psi^{(1)}\rangle\right), \tag{4.12}$$

where $\nabla_\alpha^{\perp} = \left(-\partial_{\alpha_2}, \partial_{\alpha_1}\right)$, $\omega_0 = \mathrm{curl}_{\mathbf{y}}\,\mathbf{W}(\mathbf{x},\mathbf{y})$, $\omega_1(\mathbf{y}) = \mathrm{curl}_{\mathbf{x}}\,\mathbf{U}(\mathbf{x}) + \mathrm{curl}_{\mathbf{x}}\,\mathbf{W}(\mathbf{x},\mathbf{y})$, $\mathcal{A}_0 = \nabla_\alpha\mathbf{X}^{(0)} + \nabla_{\mathbf{y}}\mathbf{X}^{(1)}$ is the leading-order term in the expansion of the Jacobian matrix of $\mathcal{A}$, and $\langle f\rangle$ stands for averaging of f with respect to $\mathbf{y}$ over one period. It can be shown that $\left|\mathcal{A}_0\right| \equiv 1$, which implies the well-posedness of the elliptic equations for the stream functions ψ^0 and ψ^1. Note that equation (4.11) is a second-order elliptic equation for the first-order corrector $\psi^{(1)}$ as a function of $\mathbf{y}$ with periodic boundary condition. It plays a similar role as the cell problem for the standard elliptic homogenization. Since the vorticity is of order $1/\epsilon$, it enters the cell problem as a forcing term in the right-hand side of the elliptic cell problem. As in elliptic homogenization, we can factor out the slowly-varying component $\nabla_\alpha\psi^{(0)}$ from equation (4.11) for $\psi^{(1)}$. Then we can derive an effective equation for $\psi^{(0)}$ whose effective coefficient is defined through a cell problem that is independent of $\psi^{(0)}$.

The elliptic problems (4.11)–(4.12) for $\psi^{(0)}$ and $\psi^{(1)}$ are clearly solvable with appropriate boundary condition. However, to avoid the secular growth of the multiscale expansion, we need to ensure that $\epsilon\mathbf{X}^{(1)} \to 0$ as $\epsilon \to 0$. For this purpose, we need to derive a solvability condition for the $\mathbf{X}^{(1)}$ equation (4.10).

A solvability condition

As we mentioned above, we need to ensure that $\epsilon\mathbf{X}^{(1)}\left(\cdot,\cdot,\frac{\alpha}{\epsilon},\frac{t}{\epsilon}\right) \to 0$ as $\epsilon \to 0$. Let $\mathbf{w} = \nabla_{\mathbf{y}}^{\perp}\psi^{(1)}$ be the cell velocity, and $\mathbf{Y}(\mathbf{y},\tau)$ be the cell characteristic; i.e., for each α and t fixed, $\mathbf{Y}(\mathbf{y},\tau)$ satisfies

$$\frac{d\mathbf{Y}}{d\tau} = \mathbf{w}(\mathbf{Y},\tau)\,, \quad \mathbf{Y}(\mathbf{y},0) = \mathbf{y}\,.$$

We decompose the cell velocity field into two parts:

$$\mathbf{w}(t,\alpha,\tau,\mathbf{y}) = \mathbf{w}_1(t,\alpha,\tau,\mathbf{y}) + \mathbf{w}_2(t,\alpha,\tau,\mathbf{y}) .$$

To avoid the secular-growth term, we need to remove the non-mixable part of the cell velocity, $\mathbf{w}_2$, which corresponds to

$$\lim_{T\to\infty} \frac{1}{T}\int_0^T \mathbf{w}_2(t,\alpha,\tau,\mathbf{Y}(\mathbf{y},\tau))\, d\tau \to \eta(t,\alpha,\mathbf{y}) \neq 0 .$$

In other words, we will use the following projection method for $\mathbf{w}$:

$$\mathbf{w}(t,\alpha,\tau,\mathbf{y}) \quad \leftarrow \quad \mathbf{w}(t,\alpha,\tau,\mathbf{y}) - \lim_{T\to\infty} \frac{1}{T}\int_0^T \mathbf{w}(t,\alpha,\tau,\mathbf{Y}(\mathbf{y},\tau))\, d\tau .$$

For the Navier–Stokes equations, viscosity and random forcing play the role to eliminate the non-mixable component of the flow velocity. Eliminating this non-mixable component is essential for the flow to be fully mixed, and to reveal certain universality and scale-similarity. For inviscid Euler equations without external forcing, the projection method provides a systematic way to eliminate the non-mixable component. It can be also viewed as an acceleration method for the flow to be fully mixed.

4.2 Multiscale analysis in the Eulerian frame

For computational purposes, it is more convenient to derive a homogenized equation in the Eulerian formulation. As we see from the multiscale analysis in the Lagrangian frame, the key is to introduce a new phase variable, $\theta^\epsilon(t,\mathbf{x})$, to characterize the propagation of the small scales in the fluid flow. This phase variable is convected by the full flow velocity. The multiscale structure of the solution becomes apparent using this new variable. In deriving the multiscale analysis in the Eulerian frame, we use the new phase variable to describe the propagation of the small-scale component of the velocity field, but use the Eulerian variable to describe the large-scale averaged solution.

One advantage of this approach is that we can characterise the nonlinear convection of small scales exactly using this phase variable and turn a convection-dominated transport problem into an elliptic problem for the pressure. Thus, our multiscale finite-element methods for elliptic problems described earlier can be used to obtain a multiscale approximation for the small-scale pressure equation.

Guided by our multiscale analysis in the Lagrangian frame, we look for multiscale expansions of the velocity field and the pressure of the following form:

$$\mathbf{u}^\epsilon(t,\mathbf{x}) = \mathbf{u}(t,\mathbf{x}) + \mathbf{w}(t,\mathbf{x},\tau,\mathbf{y}) + \epsilon\mathbf{u}^{(1)}(t,\mathbf{x},\tau,\mathbf{y}) + \epsilon^2\mathbf{u}^{(2)}(t,\mathbf{x},\tau,\mathbf{y}) \cdots , \tag{4.13}$$

$$p^\epsilon(t,\mathbf{x}) = p(t,\mathbf{x}) + q(t,\mathbf{x},\tau,\mathbf{y}) + \epsilon p^{(1)}(t,\mathbf{x},\tau,\mathbf{y}) + \epsilon^2 p^{(2)}(t,\mathbf{x},\tau,\mathbf{y}) \cdots , \tag{4.14}$$

where $\tau = t/\epsilon$ and $\mathbf{y} = \theta^\epsilon(t,\mathbf{x})/\epsilon$. We assume that $\mathbf{w}$, and q have zero mean with respect to $\mathbf{y}$ and τ. Note that we do not assume that $\mathbf{w}$ and q are periodic in τ. The phase function θ^ϵ satisfies the following evolution equation:

$$\theta^\epsilon_t + \mathbf{u}^\epsilon \cdot \nabla\theta^\epsilon = \mathbf{0}, \quad \theta^\epsilon(0,\mathbf{x}) = \mathbf{x} . \tag{4.15}$$

Based on the multiscale analysis in the Lagrangian frame, we can show that θ^ϵ has the following multiscale expansion:

$$\theta^\epsilon = \theta(t, \mathbf{x}) + \epsilon\theta^{(1)}(t, \mathbf{x}, \tau, \frac{\theta^\epsilon}{\epsilon}) + \cdots . \tag{4.16}$$

It is important that the multiscale expansion for θ^ϵ is defined implicitly. If one tries to expand θ^ϵ as a function of $\mathbf{x}/\epsilon$ and t/ϵ as in (4.5), one will not be able to obtain a well-posed cell problem.

Expanding the Jacobian matrix, we get $\nabla_x \theta^\epsilon = \mathcal{B}^{(0)} + \epsilon\mathcal{B}^{(1)} + \cdots$. Substituting the expansion into the Euler equation and matching the terms of the same order, we obtain

$$\epsilon^{-1} : \partial_\tau \mathbf{w} + \mathcal{B}^{(0)^{\mathsf{T}}} \nabla_y q = \mathbf{0}, \tag{4.17}$$

$$\begin{aligned} \epsilon^{0} \;\; : \partial_\tau \mathbf{u}^{(1)} + \mathcal{B}^{(0)^{\mathsf{T}}} \nabla_y p^{(1)} = & - \big(\partial_t(\mathbf{u} + \mathbf{w}) + (\mathbf{u} + \mathbf{w}) \cdot \nabla_x(\mathbf{u} + \mathbf{w}) \\ & + \nabla_x (p + q) + \mathcal{B}^{(1)^{\mathsf{T}}} \nabla_y q\big) , \end{aligned} \tag{4.18}$$

where $\mathcal{B}^{(0)^{\mathsf{T}}}$ stands for the transpose of $\mathcal{B}^{(0)}$.

Further, we note that a necessary condition for the solvability of $\mathbf{u}^{(1)}$ is that the right-hand side of (4.18) has zero mean with respect to $\mathbf{y}$ and τ. Since $\mathbf{w}$ and q have zero mean in $\mathbf{y}$ and τ, we obtain the following homogenized equation for $\mathbf{u}$ by averaging the right-hand side of (4.18) with respect to $\mathbf{y}$ and τ,

$$\partial_t \mathbf{u} + \mathbf{u} \cdot \nabla_x \mathbf{u} + \big\langle (\mathbf{w} \cdot \nabla_x) \mathbf{w} \big\rangle^* + \big\langle \mathcal{B}^{(1)^{\mathsf{T}}} \nabla_y q \big\rangle^* = - \nabla_x p , \tag{4.19}$$

where $\langle f \rangle^*$ stands for average of f with respect to $\mathbf{y}$ and τ. Moreover, we can show by using the weak formulation of the Euler equation and the multiscale expansions of $\mathbf{u}^\epsilon$ and p^ϵ that $\big\langle \mathcal{B}^{(1)^{\mathsf{T}}} \nabla_y q \big\rangle^* = \big\langle \mathbf{w} \nabla_x \cdot \mathbf{w} \big\rangle^*$. Thus the effective equation for $\mathbf{u}$ becomes

$$\partial_t \mathbf{u} + \mathbf{u} \cdot \nabla_x \mathbf{u} + \nabla_x \cdot \langle \mathbf{w}\mathbf{w} \rangle^* = - \nabla_x p , \quad \mathbf{u}|_{t=0} = \mathbf{U}(\mathbf{x}) , \tag{4.20}$$

$$\nabla \cdot \mathbf{u} = 0 . \tag{4.21}$$

where $\mathbf{w}\mathbf{w}$ is a matrix with entry in the ith row and jth column equal to $w_i w_j$.

We remark that it is essential to combine the multiscale analysis in the strong form with the weak formulation of the Euler equation in order to derive a closed homogenized equation. The reason is because we allow multiscale structure in θ^ϵ within the multiscale expansion of the velocity field. If we consider only coefficient equations derived by substituting the multiscale expansions into the Euler equation and matching each order of ϵ, we will lose certain important conservative properties of the original Euler equation. We will not be able to derive a closed set of homogenized equations. In some sense, the weak formulation provides us with additional information that complements the multiscale analysis using asymptotic expansions.

The equation for the small-scale velocity field $\mathbf{w}$ is given by (4.17):

$$\partial_\tau \mathbf{w} + \mathcal{B}^{(0)^{\mathsf{T}}} \nabla_y q = 0 , \quad \tau > 0; \tag{4.22}$$

$$(\mathcal{B}^{(0)^{\mathsf{T}}} \nabla_y) \cdot \mathbf{w} = 0 , \quad \mathbf{w}|_{\tau=0} = \mathbf{W}(\mathbf{x}, \mathbf{y}) . \tag{4.23}$$

Moreover, it can be shown that $\mathcal{B}^{(0)^{\mathrm{T}}}\nabla_y q$ has zero mean in $\mathbf{y}$ and τ respectively. This guarantees the existence of a periodic solution for $\mathbf{w}$. Moreover, $\mathbf{w}$ will be bounded for large τ since $\mathcal{B}^{(0)^{\mathrm{T}}}\nabla_y q$ has zero mean in τ.

Further, we can derive the evolution equations for θ and $\theta^{(1)}$ as follows

$$\partial_t \theta + (\mathbf{u}\cdot\nabla_x)\theta = \mathbf{0}\,, \quad \theta|_{t=0} = \mathbf{x}\,, \tag{4.24}$$

and

$$\partial_\tau \theta^{(1)} + (\mathbf{w}\cdot\nabla_x)\theta\,, \quad \theta^{(1)}|_{\tau=0} = \mathbf{0}\,. \tag{4.25}$$

From θ and $\theta^{(1)}$, we can compute the Jacobian matrix $\mathcal{B}^{(0)}$ as follows:

$$\mathcal{B}^{(0)} = (I - \mathrm{D}_y\theta^{(1)})^{-1}\mathrm{D}_x\theta\,. \tag{4.26}$$

As in the Lagrangian case, we also need to impose a solvability condition for $\theta^{(1)}$. The condition is the same as before; i.e., the τ-average of the cell velocity $\mathbf{w}$ must be zero in order for $\epsilon\theta^{(1)} \to 0$ as $\epsilon \to 0$ (recall $\tau = t/\epsilon$). For this reason, we must project the non-mixable part of the cell velocity to zero; i.e. we apply a projection on $\mathbf{w}$ to filter out the component of $\mathbf{w}$ that has non-zero mean in τ:

$$\mathbf{w} \leftarrow \mathbf{w} - \lim_{T\to\infty}\frac{1}{T}\int_0^T \mathbf{w}\;\mathrm{d}\tau\,.$$

In practice, this projection step can be carried out locally when we integrate the equation from $t_n = n\Delta t$ to $t_{n+1} = (n+1)\Delta t$, with Δt being the coarse-grid time-step. In this case, the time-average window width T should be set to $T = \Delta t/\epsilon$.

The equation for the first-order correction, $\mathbf{u}^{(1)}$, is given by satisfying the Euler equation to $\mathrm{O}(1)$:

$$\begin{aligned}
\partial_\tau \mathbf{u}^{(1)} + \mathcal{B}^{(0)^{\mathrm{T}}}\nabla_y p^{(1)} &= -\Big(\partial_t(\mathbf{u}+\mathbf{w}) \\
&\qquad + (\mathbf{u}+\mathbf{w})\cdot\nabla_x(\mathbf{u}+\mathbf{w}) + \nabla_x(p+q) + \mathcal{B}^{(1)^{\mathrm{T}}}\nabla_y q\Big)\,, \\
\mathbf{u}^{(1)}|_{\tau=0} = \mathbf{0}\,, \\
(\mathcal{B}^{(0)^{\mathrm{T}}}\nabla_y)\cdot\mathbf{u}^{(1)} &= -\nabla_x\cdot\mathbf{w} - (\mathcal{B}^{(1)^{\mathrm{T}}}\nabla_y)\cdot\mathbf{w}\,.
\end{aligned}$$

We have used the necessary condition for the solvability of $\mathbf{u}^{(1)}$ to derive the homogenized equation for $\mathbf{u}$. If $\mathbf{u}^{(1)}$ exists and remains bounded, then it will not affect the homogenized equation for $\mathbf{u}$ and the cell problem for $\mathbf{w}$. To establish the convergence of the multiscale expansion, we need to justify that $\epsilon\mathbf{u}^{(1)} \to 0$ as $\epsilon \to 0$. It is easier to study this issue by transforming the problem into the Lagrangian frame. In the Lagrangian frame, the coefficients in the multiscale expansion of the stream function are governed by elliptic equations (see (4.11)–(4.12)), whose solvability can be analysed more easily. The velocity field in the Eulerian frame can be expressed in terms of the Lagrangian stream function and the flow map. The solvability condition for $\mathbf{u}^{(1)}$ can be derived from the corresponding cell problem for the second-order correction of the flow map, $\mathbf{X}^{(2)}$.

The above multiscale analysis can be generalised to problems with general multiscale initial data without scale separation and periodic structure. In fact, recently we have been developing a multiscale analysis for the incompressible Euler equation with infinitely-many scales that are not separable [32]. For an initial velocity that has an infinite number of scales, the Fourier coefficients of the initial velocity must satisfy a certain decay property in order to have bounded energy. We make only a very mild decay assumption in the Fourier spectrum of the initial velocity field; i.e., $\left|\hat{\mathbf{u}}_k\right| \le C|k|^{-(1+\delta)}$ for large $|k|$, where δ is a small positive coefficient. This decay property is consistent with the Kolmogorov spectrum in the inertial range. The analysis developed for the two-scale velocity field provides us with the critical guideline for this more difficult case.

Another way to generalise the above multiscale analysis for problems with many scales is to develop the discrete homogenization analysis. Let H denote the coarse-grid mesh size, and h denote the fine-grid mesh size. The discrete homogenization is to derive a coarse-grid equation that captures correctly the large-scale behaviour of the well-resolved solution at the fine mesh. By setting ϵ to H, and rescaling the subgrid cell problem by H, we can formally decompose the discrete solution into a large-scale component plus a subgrid-scale component. The large-scale solution corresponds to the 'numerically homogenized' solution, and the local fine-grid problem corresponds to the small-scale cell problem represented by the fine-grid solution within each coarse-grid block. We can carry out a similar multiscale analysis as before and derive essentially the same set of effective equations. Instead of using periodic boundary condition for $\mathbf{w}$ and q as a function of $\mathbf{y}$, we need to develop a microscopic boundary condition at the boundary of a coarse-grid block. Since the cell problem is elliptic, we can apply the over-sampling technique to alleviate the difficulty associated with numerical boundary layer near the edge of the coarse-grid block.

We are currently performing a careful numerical study to validate our multiscale analysis by comparing the large-scale solution obtained by our homogenized equations with that from a well-resolved direct numerical simulation. An important feature of the resulting cell problem for $\mathbf{w}$ is that there is no convection in the fast variable because we treat convection exactly by using the new phase variable. Therefore we can use relatively large time-step in τ when we solve the cell problem. An efficient elliptic solver, such as the multigrid method [53], can be used to solve the cell problem at each time-step. Moreover, when the flow is fully mixed, we expect that the space average of the Reynolds stress term; i.e., $\langle \mathbf{ww} \rangle$, will reach to a statistical equilibrium relatively fast in time. As a consequence, we need only solve for the cell problem in τ for only a small number of fast time-steps to obtain the space-time average of the Reynolds stress term, $\langle \mathbf{ww} \rangle^*$. Moreover, we may express the Reynolds stress term as the product of an eddy diffusivity and the deformation tensor of the averaged velocity field, as in the large-eddy simulation models [45, 36, 12, 25]. For fully mixed homogeneous flow, the eddy diffusivity is supposed to be a constant in space. In this case, we need only to solve one representative cell problem and use the solution of this representative cell problem to evaluate the eddy diffusivity. This would give a self-consistent coarse-grid model that couples the evolution of the small and large scales dynamically.

Bibliography

[1] J. Aarnes and T. Y. Hou, *An efficient domain decomposition preconditioner for multiscale elliptic problems with high aspect ratios*, Acta Mathematicae Applicatae Sinica, **18** (2002), pp. 63–76.

[2] T. Arbogast, *Numerical subgrid upscaling of two-phase flow in porous media*, in Numerical treatment of multiphase flows in porous media, Z. Chen et al., eds., Lecture Notes in Physics 552, Springer, Berlin, 2000, pp. 35–49.

[3] I. Babuska, G. Caloz, and E. Osborn, *Special finite element methods for a class of second-order elliptic problems with rough coefficients*, SIAM J. Numer. Anal., **31** (1994), pp. 945–981.

[4] A. Bensoussan, J. L. Lions, and G. Papanicolaou, *Asymptotic analysis for periodic structures*, Volume 5 of Studies in Mathematics and Its Applications, North-Holland Publ., 1978.

[5] A. Bourgeat, *Homogenized behavior of two-phase flows in naturally fractured reservoirs with uniform fractures distribution*, Comp. Meth. Appl. Mech. Engrg, **47** (1984), pp. 205–216.

[6] F. Brezzi and A. Russo, *Choosing bubbles for advection-diffusion problems*, Math. Models Methods Appl. Sci., bf 4 (1994), pp. 571–587.

[7] F. Brezzi, L. P. Franca, T. J. R. Hughes and A. Russo, $b = \int g$, Comput. Methods in Appl. Mech. and Engrg., **145** (1997), pp. 329–339.

[8] L. Q. Cao, J. Z. Cui, and D. C. Zhu, *Multiscale asymptotic analysis and numerical simulation for the second-order Helmholtz equations with rapidly oscillating coefficients over general convex domains*, SIAM J Numer Anal. (2002), **40**, pp. 543–577.

[9] Z. Chen and T. Y. Hou, *A mixed finite element method for elliptic problems with rapidly oscillating coefficients*, Math. Comput., **72**, No. 242, pp. 541–576, published electronically on June 28, 2002.

[10] J. R. Chen and J. Z. Cui, *A multiscale rectangular element method for elliptic problems with entirely small periodic coefficients*, Applied Math. Comput. (2002), **30**, pp. 39–52.

[11] A. Chorin and J. Marsden, *A Mathematical Introduction to Fluid Mechanics*, Second ed., Springer-Verlag, New York, 1984.

[12] R. Clark, J. H. Ferziger, and W. Reynolds, *Evaluation of subgrid-scale models using an accurately simulated turbulent flow*, J. Fluid Mech., **91** (1979), pp. 1–16.

[13] M. Dorobantu and B. Engquist, *Wavelet-based Numerical Homogenization*, SIAM J.Numer. Anal., **35** (1998), pp. 540–559.

[14] J. Douglas, Jr. and T. F. Russell, *Numerical methods for convection-dominated diffusion problem based on combining the method of characteristics with finite element or finite difference procedures*, SIAM J. Numer. Anal. **19** (1982), pp. 871–885.

[15] L. J. Durlofsky, *Numerical calculation of equivalent grid block permeability tensors for heterogeneous porous media*, Water Resour. Res., **27** (1991), pp. 699–708.

[16] L. J. Durlofsky, R. C. Jones, and W. J. Milliken, *A nonuniform coarsening approach for the scale-up of displacement processes in heterogeneous porous media*, Adv. Water Resources, **20** (1997), pp. 335–347.

[17] W. E and T. Y. Hou, *Homogenization and convergence of the vortex method for 2-D Euler equations with oscillatory vorticity fields*, Comm. Pure and Appl. Math., **43** (1990), pp. 821–855.

[18] Y. R. Efendiev, *Multiscale finite element method (MsFEM) and its applications*, Ph. D. Thesis, Applied Mathematics, Caltech, 1999.

[19] Y. R. Efendiev, L. J. Durlofsky, S. H. Lee, *Modeling of subgrid effects in coarse-scale simulations of transport in heterogeneous porous media*, WATER RESOUR RES, **36** (2000), pp. 2031–2041.

[20] Y.R. Efendiev, T.Y. Hou, and X.H. Wu, *Convergence of a nonconforming multiscale finite element method*, SIAM J. Numer. Anal., **37** (2000), pp. 888–910.

[21] W. E and B. Engquist, *The heterogeneous multi-scale method for homogenization problems*, preprint, 2002, submitted to Multiscale Modeling and Simulation.

[22] A. Fannjiang and G. Papanicolaou, *Convection enhanced diffusion*, SIAM J. Appl. Math. **54** (1994), pp. 333–408.

[23] A. Fannjiang and G. Papanicolaou, *Diffusion in turbulence*, Probab. Theor Relat. Fields, **105** (1996), pp. 279–334.

[24] C.W. Gear, I.G. Kevrekidis, and C. Theodoropoulos, *'Coarse' integration/bifurcation analysis via microscopic simulators: micro-Galerkin methods*, Comput. & Chem Eng. (2002), **26**, pp. 941–963.

[25] M. Germano, U. Pimomelli, P. Moin, and W. Cabot, *A dynamic subgrid-scale eddy viscosity model*, Phys. Fluids A (1991), **3**, pp. 1760–1765.

[26] J. Glimm, H. Kim, D. Sharp, and T. Wallstrom, *A stochastic analysis of the scale up problem for flow in porous media*, Comput. Appl. Math., **17** (1998), pp. 67–79.

[27] T. Y. Hou and X. H. Wu, *A multiscale finite element method for elliptic problems in composite materials and porous media*, J. Comput. Phys., **134** (1997), pp. 169–189.

[28] T. Y. Hou, X. H. Wu, and Z. Cai, *Convergence of a multiscale finite element method for elliptic problems with rapidly oscillating coefficients*, Math. Comput., **68** (1999), pp. 913–943.

[29] T. Y. Hou and X. H. Wu, *A multiscale finite element method for PDEs with oscillatory coefficients*, Proceedings of 13th GAMM-Seminar Kiel on Numerical Treatment of Multi-Scale Problems, Jan 24–26, 1997, Notes on Numerical Fluid Mechanics, Vol. 70, ed. by W. Hackbusch and G. Wittum, Vieweg-Verlag, pp. 58–69, 1999.

[30] T. Y. Hou and D.-P. Yang, *Multiscale analysis for three-dimensional incompressible euler equations*, in preparation, 2003.

[31] T. Y. Hou and D.-P. Yang, *Multiscale analysis for convection dominated transport*, in preparation, 2003.

[32] T. Y. Hou and D.-P. Yang, *Multiscale analysis for incompressible flow with infinite number of scales*, in preparation, 2003.

[33] T. J. R. Hughes, *Multiscale phenomena: Green's functions, the Dirichlet-to-Neumann formulation, subgrid scale models, bubbles and the origins of stabilized methods*, Comput. Methods Appl. Mech Engrg., **127** (1995), pp. 387–401.

[34] P. Jenny, S. H. Lee, and H. Tchelepi, *Multi-scale finite volume method for elliptic problems in subsurface flow simulation*, to appear in J. Comput. Phys., 2003.

[35] P. Langlo and M. S. Espedal, *Macrodispersion for two-phase, immiscible flow in porous media*, Adv. Water Resources **17** (1994), pp. 297–316.

[36] A. Leonard, *Energy cascade in large eddy simulation of turbulent flows*, Adv. in Geophysics, **18**A (1974), pp. 237–248.

[37] A. M. Matache, I. Babuska, and C. Schwab, *Generalized p-FEM in homogenization*, Numer. Math. **86**(2000), pp. 319–375.

[38] A. M. Matache and C. Schwab, *Homogenization via p-FEM for problems with microstructure*, Appl. Numer. Math. **33** (2000), pp. 43–59.

[39] D. W. McLaughlin, G. C. Papanicolaou, and O. Pironneau, *Convection of microstructure and related problems*, SIAM J. Applied Math, **45** (1985), pp. 780–797.

[40] F. Santosa and M. Vogelius, *First-order corrections to the homogenized eigenvalues of a periodic composite medium*, SIAM J. Appl. Math, **53** (1993), pp. 1636–1668.

[41] S. Moskow and M. Vogelius, *First-order corrections to the homogenized eigenvalues of a periodic composite medium: a convergence proof*, Proc. Roy. Soc. Edinburgh, A, **127** (1997), pp. 1263–1299.

[42] P. Park, *Multiscale numerical methods for the singularly perturbed convection–diffusion equation*, Ph.D. Thesis, Applied Mathematics, Caltech, 2001.

[43] O. Pironneau, *On the Transport-diffusion Algorithm and its Application to the Navier–Stokes Equations*, Numer. Math. **38** (1982), pp. 309–332.

[44] G. Sangalli, *Capturing small scales in elliptic problems using a residual-free bubbles finite element method*, to appear in Multiscale Modeling and Simulation.

[45] J. Smogorinsky, *General circulation experiments with the primitive equations*, Mon. Weather Review, **91** (1963), pp. 99–164.

[46] L. Tartar, *Nonlocal effects induced by homogenization*, in PDE and Calculus of Variations, ed by F. Culumbini, et al, Birkhäuser, Boston, pp. 925–938, 1989.

[47] S. Verdiere and M. H. Vignal,*Numerical and theoretical study of a dual mesh method using finite volume schemes for two-phase flow problems in porous media*, Numer. Math. **80** (1998), pp. 601–639.

[48] T. Wallstrom, S. Hou, M. A. Christie, L. J. Durlofsky, and D. Sharp, *Accurate scale up of two-phase flow using renormalization and nonuniform coarsening*, Computational Geoscience, **3** (1999), pp. 69–87.

[49] T. C. Wallstrom, M. A. Christie, L. J. Durlofsky, and D. H. Sharp, *Application of effective flux boundary conditions to two-phase upscaling in porous media*, Transport in Porous Media, **46** (2002), pp. 155–178.

[50] A. Westhead, *Upscaling the two-phase flow in heterogeneous porous media*, Ph. D. Thesis in progress, Applied Mathematics, Caltech, 2003.

[51] X. H. Wu, Y. Efendiev, and T. Y. Hou, *Analysis of upscaling absolute permeability*, Discrete and Continuous Dynamical Systems, Series B, **2** (2002), pp. 185–204.

[52] Y. Zhang and X.-H. Wu, *A Petrov–Galerkin multiscale finite element method*, preprint, 2000, unpublished.

[53] P. M. De Zeeuw, *Matrix-dependent prolongation and restrictions in a blackbox multigrid solver*, J. Comput. Applied Math, **33**(1990), pp. 1–27.

Jonathan Keating is Professor of Mathematical Physics in the School of Mathematics at the University of Bristol. He has also worked at the Basic Research Institute in the Mathematical Sciences (BRIMS) based at the Hewlett–Packard Laboratories, Bristol, and holds a BRIMS Research Fellowship. His interests include semiclassical asymptotics, quantum chaos, random matrix theory, and number theory.

Specifically, Jonathan has contributed to the development of the semiclassical theories of quantum fluctuation statistics and periodic orbit resummation, to eigenvalue statistics for random matrices, and to the theory of the Riemann ζ function. He was educated at the University of Oxford, before taking his PhD at the University of Bristol, where he worked with Sir Michael Berry. He was a Lecturer in Applied Mathematics at Manchester University from 1991 to 1995, and then returned to Bristol.

Chapter 11

Random Matrices and the Riemann ζ-Function: a Review

Jonathan P. Keating[*]

Abstract: The past few years have seen the emergence of compelling evidence for a connection between the zeros of the Riemann ζ-function and the eigenvalues of random matrices. This hints at a link between the distribution of the prime numbers, which is governed by the Riemann zeros, and properties of waves in complex systems (e.g. waves in random media, or in geometries where the ray dynamics is chaotic), which may be modelled using random matrix theory. These developments have led to a significant deepening of our understanding of some of the most important problems relating to the ζ-function and its kin, and have stimulated new avenues of research in random matrix theory. In particular, it would appear that several long-standing questions concerning the distribution of values taken by the ζ-function on the line where the Riemann Hypothesis places its zeros can be answered using techniques developed in the study of random matrices.

Contents

*School of Mathematics, University of Bristol, UK

1 Random matrices and the Riemann zeros

Linear wave theories may be expressed in terms of matrices. Therefore, just as in complex (e.g. chaotic) dynamical systems, where statistical properties of the trajectories may be calculated by averaging with respect to an appropriate measure on phase space, statistical properties of the waves in complex systems may be calculated by averaging over ensembles of random matrices. In this sense, *Random matrix theory* [26] is to wave theories what statistical mechanics is to dynamical systems. In quantum mechanics, where the waves obey the Schrödinger equation, it was developed in the 1950s and 1960s by Wigner, Dyson and others and has been applied to a wide range of problems including systems with a large number of degrees of freedom (e.g. nuclei), systems in which the classical trajectories are chaotic (e.g. atoms, molecules, microelectronic devices), and systems in which the potential is random (e.g. disordered systems). It has found similar applications in acoustics, elasticity, and optics. (For an up-to-date review of the literature, see [15].)

Statistical properties of the eigenvalues and eigenfunctions of self-adjoint operators (e.g. the Schrödinger operator) can be modelled using the corresponding statistics for hermitian matrices, treating the real and imaginary parts of the matrix elements as independently distributed gaussian random variables. In the same way, unitary matrices can be used to model unitary operators (e.g. Green functions). In this case, one can use the fact that $N \times N$ unitary matrices form a compact group — the unitary group $U(N)$ — which comes with a natural invariant (uniform) measure: Haar measure.

My purpose here is to describe some rather surprising connections between the theory of waves in complex systems and number theory. Specifically, these connections concern random matrix theory and the Riemann ζ-function $\zeta(s)$, which is central to the theory of the primes. I will focus on one aspect of this story: the connection between the distribution of values taken by the characteristic polynomials of random unitary matrices and that of the Riemann ζ-function on the critical line, $s = \frac{1}{2} + it$, where the Riemann Hypothesis places its non-trivial (complex) zeros. This has led, conjecturally, as I shall explain, to a general solution to the long-standing problem of determining the moments of $\zeta(\frac{1}{2} + it)$.

The Riemann ζ-function is defined by

$$\zeta(s) = \sum_{n=1}^{\infty} \frac{1}{n^s} = \prod_p \left(1 - \frac{1}{p^s}\right)^{-1}, \tag{1.1}$$

for $\operatorname{Re} s > 1$, where p labels the primes, and then by analytic continuation to the rest of the complex plane. It has a single simple pole at $s = 1$, zeros at $s = -2, -4, -6$, etc., and infinitely many zeros, called the *non-trivial zeros*, in the *critical strip* $0 < \operatorname{Re} s < 1$. The Riemann Hypothesis states that all of the non-trivial zeros lie on the *critical line* $\operatorname{Re} s = \frac{1}{2}$; that is, $\zeta(\frac{1}{2} + it) = 0$ has non-trivial solutions only when $t = t_n \in \mathbb{R}$ [30]. This is known to be true for at least 40% of the non-trivial zeros [8], for the first 100 billion of them [32], and for batches lying much higher [28].

In the following, for ease of presentation, we will assume the Riemann Hypothesis to be true, although this is not strictly necessary.

The mean density of the non-trivial zeros increases logarithmically with height t up the critical line. Specifically, the *unfolded* zeros

$$w_n = t_n \frac{1}{2\pi} \log \frac{t_n}{2\pi} \tag{1.2}$$

satisfy

$$\lim_{W\to\infty} \frac{1}{W} \# \{w_n \in [0, W]\} = 1\,; \tag{1.3}$$

that is, the mean of $w_{n+1} - w_n$ is 1. The question then arises as to the statistical distribution of the unfolded zeros: are they equally spaced, with unit spacing between neighbours, randomly distributed with unit mean spacing, or do they have some other distribution?

It is in this context that the connection with random matrices arises. Let A be an $N \times N$ unitary matrix; that is, $A \in U(N)$. Denote the eigenvalues of A by $\exp(i\theta_n)$, where $1 \le n \le N$ and $\theta_n \in \mathbb{R}$. Clearly the eigenphases θ_n have mean density $N/2\pi$, so the unfolded eigenphases

$$\phi_n = \theta_n \frac{N}{2\pi} \tag{1.4}$$

have unit mean density (i.e. $\phi_n \in [0, N)$). Next, let us define

$$F(\alpha, \beta; A) = \frac{1}{N} \#\{\phi_n, \phi_m : \alpha \le \phi_n - \phi_m < \beta\}\,. \tag{1.5}$$

The key step now is to average $F(\alpha, \beta; A)$ over A, chosen uniformly with respect to (normalized) Haar measure on $U(N)$. This average will be denoted by

$$F_U(\alpha, \beta; N) = \int_{U(N)} F(\alpha, \beta; A)\, \mathrm{d}A\,. \tag{1.6}$$

Dyson proved in 1963 that

$$F_U(\alpha, \beta) = \lim_{N\to\infty} F_U(\alpha, \beta; N) \tag{1.7}$$

exists and takes the form

$$F_U(\alpha, \beta) = \int_\alpha^\beta \Big[1 - \big(\frac{\sin \pi x}{\pi x}\big)^2 + \delta(x)\Big]\, \mathrm{d}x\,, \tag{1.8}$$

where $\delta(x)$ is Dirac's δ-function [26]. The integrand in (1.8) may be thought of as the *two-point correlation function* for the eigenphases of a random unitary matrix, unfolded to have unit mean spacing. The fact that it is a non-trivial function of correlation distance x means that eigenphases are correlated in a non-trivial way.

The connection between the pair correlation of the Riemann zeros, as measured by

$$F_\zeta(\alpha, \beta) = \lim_{W\to\infty} \frac{1}{W} \#\{w_n, w_m \in [0, W] : \alpha \le w_n - w_m < \beta\}\,, \tag{1.9}$$

and that of random matrix eigenvalues was made in 1973 by Montgomery [27], who conjectured that

$$F_\zeta(\alpha,\beta) = F_U(\alpha,\beta)\,. \tag{1.10}$$

This conjecture has turned out to be extremely influential.

In his original paper, Montgomery proved a theorem which provides substantial support for his conjecture (1.10). The fourier transform of the two-point correlation function (i.e. of the integrand in (1.8)) may easily be calculated to be

$$k_U(\tau) = \int_{-\infty}^{\infty} \left[1 - \left(\frac{\sin(\pi x)}{\pi x}\right)^2 + \delta(x)\right] \mathrm{e}^{2\pi i x \tau}\,\mathrm{d}x - \delta(\tau) = \begin{cases} |\tau| & \text{for } |\tau| < 1\,, \\ 1 & \text{for } |\tau| \geq 1\,. \end{cases} \tag{1.11}$$

Montgomery showed that it follows from the prime number theorem (which states that the number of primes less than X grows asymptotically like $X/\log X$ as $X \to \infty$) that the analogue of $k_U(\tau)$ for the Riemann zeros coincides with the expression on the right of (1.11) in the range $|\tau| < 1$. His conjecture thus boils down to the claim that it coincides with the expression on the right in the range $|\tau| \geq 1$ as well.

There is substantial evidence in support of Montgomery's conjecture. First, Odlyzko has computed the two-point correlation function numerically for batches of zeros high up on the critical line (e.g. near to the 10^{20}th zero) and his results are in striking agreement with it [28]; see, for example, Figure 1. Second, the conjectured form for the fourier transform of the two-point correlation function in the range $|\tau| \geq 1$ may be shown to follow, heuristically, from an asymptotic analysis based on a conjecture of Hardy and Littlewood concerning correlations between the primes [22].

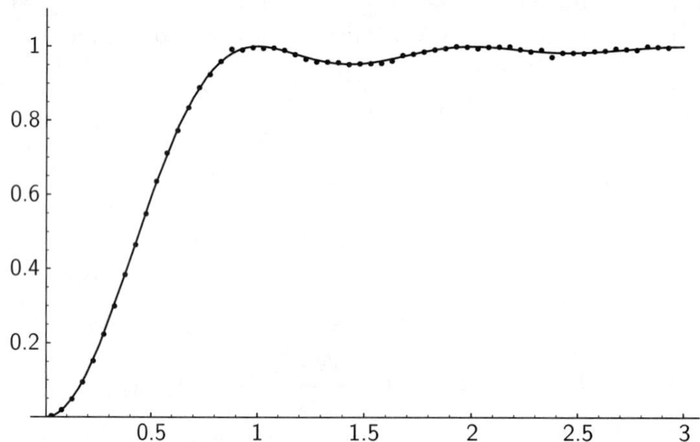

Figure 1. *The two-point correlation function of 10^6 Riemann zeros around the height of the 10^{20}th zero (dots) and of the eigenphases of random unitary matrices in the limit as $N \to \infty$ (smooth curve). (Figure courtesy of AM Odlyzko.)*

Montgomery's conjecture generalizes immediately to relate correlations between n-tuples of zeros and the corresponding correlations between n-tuples of eigenphases. His theorem also generalizes, that is the fourier transforms of the n-point correlation functions for the zeros and random-matrix eigenphases coincide in appropriately restricted ranges [29]. Asymptotic calculations based on prime correlations again support the conjecture outside these ranges [5, 6]. Odlyzko has computed various statistical measures of the zero distribution which depend on the n-point correlation functions for $n > 2$, and all show remarkable agreement with the corresponding random-matrix forms [28]. For example, this is the case for the distribution of spacings between adjacent unfolded zeros, which depends on all of the n-point correlation functions.

The conclusion to be drawn is that the statistical distribution of the Riemann zeros, in the limit as one looks infinitely high up the critical line, coincides with the statistical distribution of the eigenvalues of random unitary matrices, in the limit of large matrix size. (A great deal is also known about the way in which zero statistics asymptotically approach the large-height limit described by random matrix theory — see, for example, [3, 7, 23, 4] — but this will not directly concern us here.)

2 Random matrices and $\log\zeta(\frac{1}{2}+it)$

Having described the connections between the zeros of the Riemann ζ-function and the eigenvalues of random matrices, I now turn to the question of the value distribution of the ζ-function itself on the critical line, or rather, to begin with, the logarithm of the ζ-function on this line.

$\log\zeta(\frac{1}{2}+it)$ is a complex function of the height t up the critical line. An obvious question is: how are the real and imaginary parts of it distributed as t varies? In the limit as $t\to\infty$, the answer to this question is provided by a beautiful theorem due to Selberg [30, 28]: for any rectangle $B\in\mathbb{C}$,

$$\lim_{T\to\infty}\frac{1}{T}\left|\left\{t: T\le t\le 2T\,,\frac{\log\zeta(\frac{1}{2}+it)}{\sqrt{\frac{1}{2}\log\log T}}\in B\right\}\right| = \frac{1}{2\pi}\iint_B \mathrm{e}^{-\frac{1}{2}(x^2+y^2)}\,\mathrm{d}x\,\mathrm{d}y\,; \quad (2.1)$$

that is, in the limit as T, the height up the critical line, tends to infinity, the value distributions of the real and imaginary parts of $\log\zeta(\frac{1}{2}+iT)/\sqrt{\frac{1}{2}\log\log T}$ each tend, independently, to a Gaussian with unit variance and zero mean. Crucially, Odlyzko's computations for these distributions when $T\approx t_{10^{20}}$ show significant systematic deviations from this limiting form [28]. For example, increasing moments of both the real and imaginary parts diverge markedly from the Gaussian values. There is, of course, no contradiction; this merely suggests that the limiting Gaussian distribution is approached rather slowly as $T\to\infty$. It does, though, lead to the question of how to model the statistical properties of $\log\zeta(\frac{1}{2}+it)$ when t is large but finite.

Given its success in describing the statistical properties of the zeros of the ζ-function, it is natural to ask whether random matrix theory might be used as the basis of such a model. The question is, then: what property of a matrix plays the role of the ζ-function? The answer is simple: since the zeros of the ζ-function are distributed like the eigenvalues of a random unitary matrix, the ζ-function might be expected to be similar, in respect of its value distribution, to the function whose zeros are the eigenvalues, that is, to the *characteristic polynomial* of such a matrix. This idea was introduced and investigated in detail in [24]. My aim here is to provide an overview of some of the main results.

The characteristic polynomial of a unitary matrix A may be defined by

$$Z(A,\theta) = \det\bigl(I - A\mathrm{e}^{-i\theta}\bigr)\,. \tag{2.2}$$

The moment generating function for $\mathrm{Re}\log Z$, for example, is thus

$$M_U(s;N) = \int_{U(N)} \exp(s\ \mathrm{Re}\log Z)\,\mathrm{d}A = \int_{U(N)} \bigl|Z(A,\theta)\bigr|^s\,\mathrm{d}A\,, \tag{2.3}$$

where the average over A is, as before, computed with respect to Haar measure on $U(N)$. Obviously Z may be written in terms of the eigenangles of A:

$$Z(A,\theta) = \prod_{n=1}^{N}\bigl(1 - \mathrm{e}^{i(\theta_n-\theta)}\bigr)\,. \tag{2.4}$$

Haar measure on $U(N)$ may also be expressed in terms of these eigenangles [31], allowing one to write

$$\int_{U(N)} \bigl|Z(A,\theta)\bigr|^s\,\mathrm{d}A = \frac{1}{(2\pi)^N N!}\int_0^{2\pi}\cdots\int_0^{2\pi}\prod_{1\le j<m\le N}\bigl|\mathrm{e}^{i\theta_j} - \mathrm{e}^{i\theta_m}\bigr|^2 \times\left|\prod_{n=1}^{N}\bigl(1 - \mathrm{e}^{i(\theta_n-\theta)}\bigr)\right|^s \mathrm{d}\theta_1\cdots\mathrm{d}\theta_N\,. \tag{2.5}$$

This N-dimensional integral may then be computed by relating it to an integral evaluated by Selberg [26], giving

$$M_U(s;N) = \prod_{j=1}^{N}\frac{\Gamma(j)\Gamma(j+s)}{\bigl(\Gamma(j+\frac{1}{2}s)\bigr)^2}\,. \tag{2.6}$$

All information about the value distribution of $\mathrm{Re}\log Z$ is contained within (2.6): moments may be computed in terms of the derivatives of $M_U(s;N)$ at $s = 0$, and the value distribution itself is the fourier transform of $M_U(iy,N)$. In the same way, information about the value distribution of $\mathrm{Im}\log Z$, and the joint value distribution of the real and imaginary parts of $\log Z$ may be computed. This leads to a central limit theorem for $\log Z$ [24] (see also [14, 1]): for any rectangle $B \in \mathbb{C}$

$$\lim_{N\to\infty}\mathrm{meas.}\left\{A \in U(N) : \frac{\log Z}{\sqrt{\frac{1}{2}\log N}} \in B\right\} = \frac{1}{2\pi}\iint_B \mathrm{e}^{-\frac{1}{2}(x^2+y^2)}\,\mathrm{d}x\,\mathrm{d}y\,. \tag{2.7}$$

This theorem corresponds precisely to Selberg's for the value distribution of $\log\zeta(\frac{1}{2}+it)$, suggesting that random matrix theory, in the limit as the matrix-size tends to infinity, can indeed model the value distribution of $\log \zeta(\frac{1}{2}+it)$ as $t \to \infty$. The question that remains is whether it can also model the asymptotic approach to the limit, that is, the value distribution when t is large but finite.

In order to relate the large-t asymptotics for the ζ-function to the large-N asymptotics for the characteristic polynomials we need a connection between t and N. Note that the scaling in Selberg's theorem (2.1) and that in (2.7) coincide if we set

$$N = \log \frac{t}{2\pi}\,. \tag{2.8}$$

Such an identification is natural, because it corresponds to equating the mean density of the Riemann zeros at height t to the mean density of eigenphases for $N \times N$ unitary matrices, and these are the only parameters that appear in the connection between the respective statistics (cf. (1.2) and (1.4)). This therefore prompts the question as to whether the rate of approach to Selberg's theorem as $t \to \infty$ is related to that for (2.7) as $N \to \infty$ (which can be computed straightforwardly using (2.6)) if we make the identification (2.8).

As already noted above, Odlyzko's numerical computations of the value distribution of the ζ-function near to the 10^{20}th zero show significant deviations from the Gaussian limit (2.1). The integer closest to $\log(t_{10^{20}}/2\pi)$ is $N = 42$ ($t_{10^{20}} \approx 1.5202 \times 10^{19}$). Does the value distribution of $\log Z$ for 42×42 random unitary matrices match his data?

Figure 2 shows the value distribution for $\operatorname{Re}\log\zeta(\frac{1}{2}+it)$, scaled as in (2.1), computed by Odlyzko [28], together with the value distribution for $\operatorname{Re}\log Z$, scaled as in (2.7), with respect to matrices taken from $U(42)$. Also shown is the Gaussian with zero mean and unit variance which represents the limit distribution in both cases (as $t \to \infty$ and $N \to \infty$ respectively). The negative logarithm of these curves is plotted in Figure 3, highlighting the behaviour in the tails. In order to quantify the data, the moments of the three distributions are listed in Table 1.

Moment	ζ a)	ζ b)	$U(42)$	Normal
1	0.0	0.0	0.0	0
2	1.0	1.0	1.0	1
3	-0.53625	-0.55069	-0.56544	0
4	3.9233	3.9647	3.89354	3
5	-7.6238	-7.8839	-7.76965	0
6	38.434	39.393	38.0233	15
7	-144.78	-148.77	-145.043	0
8	758.57	765.54	758.036	105
9	-4002.5	-3934.7	-4086.92	0
10	24060.5	22722.9	25347.77	945

Table 1. *Moments of* $\operatorname{Re}\log\zeta(\frac{1}{2}+it)$*, calculated over two ranges (labelled a and b) near the* 10^{20}*th zero (*$t \simeq 1.520 \times 10^{19}$*) (taken from [28]), compared with the moments of* $\operatorname{Re}\log Z$ *for* $U(42)$ *and the Gaussian (normal) moments, all scaled to have unit variance.*

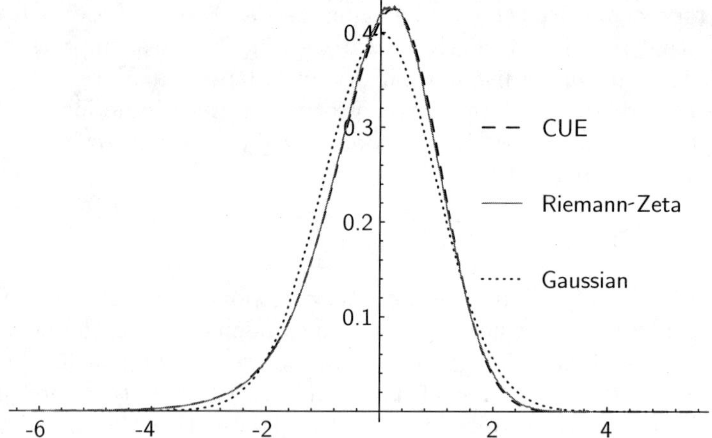

Figure 2. *The value distribution for* $\operatorname{Re}\log Z$ *with respect to matrices taken from* $U(42)$*, Odlyzko's data for the value distribution of* $\operatorname{Re}\log\zeta(\frac{1}{2}+it)$ *near the* 10^{20}*th zero (taken from [28]), and the standard Gaussian, all scaled to have unit variance. (Taken from [24].)*

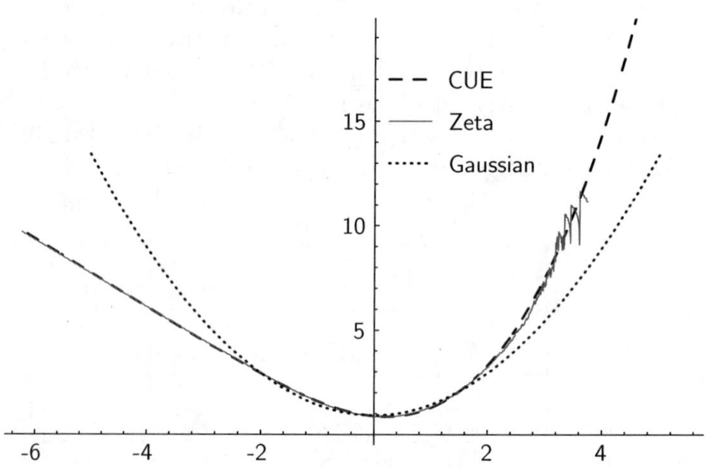

Figure 3. *minus the logarithm of the value distributions plotted in Figure 2. (Taken from [24].)*

It is clear that random matrix theory provides an accurate description of the value distribution of $\operatorname{Re}\log\zeta(\frac{1}{2}+it)$. It also models $\operatorname{Im}\log\zeta(\frac{1}{2}+it)$ equally well [24]. This then suggests that, statistically, the ζ-function at a large height t up the critical line behaves like a polynomial of degree N, where t and N are related by (2.8); and, moreover, that the polynomial in question is the characteristic polynomial of a random unitary matrix.

Of course, specific properties of the ζ-function would be expected to appear in the description of its value distribution. The point is that these contribute at lower order in the asymptotics, with the leading order being given by random matrix theory. For example, it is shown in [24] that as $N \to \infty$

$$\int_{U(N)} (\operatorname{Im} \log Z)^2 \, \mathrm{d}A = \tfrac{1}{2} \log N + \tfrac{1}{2}(\gamma + 1) + o(1) , \tag{2.9}$$

where γ is Euler's constant, while Goldston [16] has proved, under the assumption of the Riemann Hypothesis and Montgomery's conjecture, that as $T \to \infty$

$$\begin{aligned} &\frac{1}{T}\int_0^T \left(\operatorname{Im} \log \zeta(\tfrac{1}{2} + it)\right)^2 \mathrm{d}t \\ &= \tfrac{1}{2} \log\log \frac{T}{2\pi} + \tfrac{1}{2}(\gamma + 1) + \sum_{m=2}^{\infty} \sum_p \frac{(1-m)}{m^2} \frac{1}{p^m} + o(1) . \end{aligned} \tag{2.10}$$

These expression coincide under the identification (2.8), except for the sum over primes in (2.10). Obviously the primes have their origin in number theory, rather than random matrix theory.

3 Random matrices and $\zeta(\frac{1}{2} + it)$

I now turn from the logarithm of the ζ-function to the ζ-function itself. Determining the value distribution of the ζ- function is, it turns out, a significantly harder problem than determining the value distribution of its logarithm. Selberg's theorem completely characterizes the limiting distribution of $\log \zeta(\frac{1}{2} + it)$, while for $\zeta(\frac{1}{2} + it)$ almost nothing is known.

Regarding the moments of $|\zeta(\frac{1}{2} + it)|$, there is a long-standing and important conjecture that $f(\lambda)$ defined by

$$\lim_{T\to\infty} \frac{1}{\log^{\lambda^2}\left(\frac{T}{2\pi}\right)} \frac{1}{T} \int_0^T \left|\zeta(\tfrac{1}{2} + it)\right|^{2\lambda} \mathrm{d}t = f(\lambda)\, a(\lambda) , \tag{3.1}$$

where

$$a(\lambda) = \prod_p \left\{ (1 - 1/p)^{\lambda^2} \left(\sum_{m=0}^{\infty} \left(\frac{\Gamma(\lambda + m)}{m!\Gamma(\lambda)} \right)^2 p^{-m} \right) \right\} , \tag{3.2}$$

exists, and a much-studied problem then to determine the values it takes, in particular for integer λ (see, for example, [30, 19]). Obviously $f(0) = 1$. In 1918, Hardy and Littlewood proved that $f(1) = 1$ [17], and in 1926 Ingham proved that $f(2) = 1/12$ [18]. No other values are known. Based on number-theoretical arguments, Conrey and Ghosh have conjectured that $f(3) = 42/9!$ [12], and Conrey and Gonek that $f(4) = 24024/16!$ [13].

Given the success of random matrix theory in describing the value distribution of $\log \zeta(\frac{1}{2} + it)$, it is natural to ask whether it has anything to contribute on this issue.

Invoking the identification (2.8), the question for the characteristic polynomials that is analogous to (3.1) is whether

$$f_U(\lambda) = \lim_{N\to\infty} \frac{1}{N^{\lambda^2}} \int_{U(N)} |Z(A,\theta)|^{2\lambda} \, \mathrm{d}A \tag{3.3}$$

exists, and, if it does, what values it takes. The answer to this question was given in [24], where it was proved that f_U does indeed exist, that

$$f_U(\lambda) = \frac{G^2(1+\lambda)}{G(1+2\lambda)}, \tag{3.4}$$

where G denotes the Barnes G-function [2], and hence that $f_U(0) = 1$ (trivial) and

$$f_U(k) = \prod_{j=0}^{k-1} \frac{j!}{(j+k)!}, \tag{3.5}$$

for integers $k \geq 1$. Thus, for example, $f_U(1) = 1$, $f_U(2) = 1/12$, $f_U(3) = 42/9!$ and $f_U(4) = 24024/16!$. The fact that these values coincide with those associated, or believe to be associated, with the ζ-function strongly suggests that

$$f(\lambda) = f_U(\lambda), \tag{3.6}$$

for all $\operatorname{Re}\lambda > -\frac{1}{2}$. This conjecture is also supported by Odlyzko's numerical data for non-integer values of λ between zero and two [24]. (Conrey and Gonek's conjecture for $f(4)$ and ours for all integer λ were announced independently at the Erwin Schrödinger Institute in Vienna, in September 1998.)

4 L-functions

The ζ-function is but one example of a more general class of functions known as L-functions. These all satisfy generalizations of the Riemann Hypothesis. For any individual L-function, it is believed that the zeros high up on the critical line are distributed like the eigenvalues of random unitary matrices, that is, exactly like in the case of the Riemann ζ-function [27, 29]. This means that the moment conjecture described above generalizes immediately to all L-functions.

More interesting, however, is the fact that it has been conjectured by Katz and Sarnak [20, 21] that averages over various families of L-functions, with the height up the critical line of each one fixed, are described not only by averages over the unitary group $\mathrm{U}(N)$, but by averages over other classical compact groups, for example the orthogonal group $\mathrm{O}(N)$ or the unitary symplectic group $\mathrm{USp}(2N)$, depending upon the family in question. This raises the important question of whether the moments of L-functions within these families can be determined by random matrix calculations generalizing those described above for the unitary group to the other classical compact groups [9]. The calculations for $\mathrm{O}(N)$ and $\mathrm{USp}(2N)$ were carried out in [25], where it was shown that the results agree with the few moments computed or conjectured using number-theoretical techniques.

5 Asymptotic expansions

The limit (3.1) may be thought of as representing the leading-order asymptotics of the moments of the ζ-function, in that it implies that

$$\frac{1}{T}\int_0^T \left|\zeta(\tfrac{1}{2}+it)\right|^{2\lambda} \mathrm{d}t \sim f(\lambda)\, a(\lambda)\, \log^{\lambda^2}\left(\frac{T}{2\pi}\right), \tag{5.1}$$

as $T \to \infty$. Very little is known about lower order terms (in powers of $\log T$) in the asymptotic expansion of these moments. Does random matrix theory suggest what form these should take?

When λ is an integer, it does. Note first that it follows from (2.3) and (2.6) that

$$\int_{U(N)} |Z(A,\theta)|^{2k}\, \mathrm{d}A = \prod_{j=0}^{k-1} \frac{j!}{(j+k)!} \prod_{i=1}^{k} (N+i+j) = Q_k(N), \tag{5.2}$$

where $Q_k(N)$ is a polynomial in N of degree k^2. This is consistent with the number theorists' guess that the $2k$th moment of the ζ-function should be a polynomial of degree k^2 in $\log(T/2\pi)$ (modulo terms that vanish faster than any inverse power of $\log T/2\pi$ as $T \to \infty$).

Unfortunately it is not easy to see directly how to combine the coefficients in (5.2) with arithmetical information to guess the form of the coefficients of the lower-order terms in the moments of the ζ-function. The expression in (5.2) can, however, be re-expressed in the form [10]

$$\begin{aligned}\int_{U(N)} \left|Z(A,\theta)\right|^{2k} \mathrm{d}A = \frac{(-1)^k}{k!^2 (2\pi i)^{2k}} \oint \cdots \oint \mathrm{e}^{\frac{1}{2}N\sum_{j=1}^{k} z_j - z_{j+k}} \\ \times \frac{G(z_1,\ldots,z_{2k})\Delta^2(z_1,\ldots,z_{2k})}{\prod_{i=1}^{2k} z_i^{2k}}\, \mathrm{d}z_1 \cdots \mathrm{d}z_{2k}\,,\end{aligned} \tag{5.3}$$

where the contours are small circles around the origin,

$$\Delta(z_1,\ldots,z_m) = \prod_{1\le i<j\le m} (z_j - z_i)\,, \tag{5.4}$$

and

$$G(z_1,\ldots,z_{2k}) = \prod_{\substack{1\le \ell\le k \\ k+1\le q\le 2k}} \left(1-\mathrm{e}^{z_q - z_\ell}\right)^{-1}. \tag{5.5}$$

Note that G has simple poles when $z_i = z_j$, $i \neq j$. An evaluation of the contour integral in terms of residues confirms the identity by giving (5.2). This formula has a natural generalization to the ζ-function [11]:

$$\frac{1}{T}\int_0^T \left|\zeta(\tfrac{1}{2}+it)\right|^{2k} \mathrm{d}t = \frac{1}{T}\int_0^T W_k\left(\log\frac{t}{2\pi}\right)\left(1+O(t^{-\frac{1}{2}+\epsilon})\right) \mathrm{d}t\,, \tag{5.6}$$

where

$$W_k(x) = \frac{(-1)^k}{k!^2(2\pi i)^{2k}} \oint \cdots \oint e^{\frac{x}{2}\sum_{j=1}^{k} z_j - z_{j+k}} \times \frac{\tilde{G}(z_1,\ldots,z_{2k})\Delta^2(z_1,\ldots,z_{2k})}{\prod_{i=1}^{2k} z_i^{2k}} \, dz_1 \ldots dz_{2k}\,, \tag{5.7}$$

the path of integration being the same as in (5.3), and

$$\tilde{G}(z_1,\ldots,z_{2k}) = A_k(z_1,\ldots,z_{2k}) \prod_{i=1}^{k}\prod_{j=1}^{k} \zeta(1+z_i-z_{j+k})\,, \tag{5.8}$$

with

$$A_k(z) = \prod_p \prod_{i=1}^{k}\prod_{j=1}^{k}\Big(1-\frac{1}{p^{1+z_i-z_{j+k}}}\Big) \int_0^1 \prod_{j=1}^{k}\Big(1-\frac{e(\theta)}{p^{\frac{1}{2}+z_j}}\Big)^{-1}\Big(1-\frac{e(-\theta)}{p^{\frac{1}{2}-z_{j+k}}}\Big)^{-1} d\theta\,, \tag{5.9}$$

and $e(\theta) = \exp(2\pi i\theta)$. Note that $\tilde{G}$ has the same pole structure as G. An evaluation of this integral in terms of residues shows that W_k is a polynomial of degree k^2 and allows the coefficients to be computed. For example,

$$W_2(x) = 0.0506605918\,x^4 + 0.6988698848\,x^3 + 2.4259621988\,x^2 + 3.2279079649\,x + 1.3124243859 \tag{5.10}$$

and

$$W_3(x) = 0.0000057085\,x^9 + 0.0004050213\,x^8 + 0.0110724552\,x^7 + 0.1484007308\,x^6 + 1.0459251779\,x^5 + 3.9843850948\,x^4 + 8.6073191457\,x^3 + 10.2743308307\,x^2 + 6.5939130206\,x + 0.9165155076 \tag{5.11}$$

(we quote here numerical approximations for the coefficients, rather than the analytical expressions, which are rather cumbersome).

These polynomials describe the moments of the ζ-function to a very high degree of accuracy [11]. For example, when $k = 3$ and $T = 2350000$, the left hand side of (5.6) evaluates to 1411700.43 and the right hand side to 1411675.64. Note that the coefficient of the leading-order term is small. This explains the difficulties, described at length by Odlyzko [28], associated with numerical tests of (3.1).

Alternatively, one can also compare

$$\int_0^\infty \left|\zeta(\tfrac{1}{2}+it)\right|^{2k} \exp(-t/T)\, dt \tag{5.12}$$

with

$$\int_0^\infty W_k\big(\log(t/2\pi)\big) \exp(-t/T)\, dt\,. \tag{5.13}$$

This is done in Table 2.

Similar asymptotic expansions have been derived for the moments of families of L-functions, using expressions analogous to (5.3) [11].

k	(5.12)	(5.13)	difference
1	79499.9312635	79496.7897047	3.14156
2	55088332.55512	55088336.43654	-3.8814
3	708967359.4	708965694.5	1664.9
4	143638308513.0	143628911646.6	9396866.4

Table 2. *Smoothed moments, (5.12) and (5.13), when* T=10000.

6 Conclusions

The conclusion one is led to draw from the results reviewed here is that random matrix theory, specifically results concerning the characteristic polynomials of random unitary matrices, leads to a conjectural solution, supported by all available evidence, to the long-standing problem of calculating the moments of the Riemann ζ-function on its critical line. Moreover, this isn't an accident: the moments of families of other L-functions can be calculated using the same techniques. This opens up several major problems in number theory to investigation using methods developed to understand waves in complex systems (e.g. how the random matrix limit is approached asymptotically).

One obvious question it prompts is: what is the reason for the connection between random matrices and the ζ-function? It has long been imagined there might be a spectral interpretation of the zeros. If the Riemann Hypothesis is true, such an interpretation could be the reason why; for example, if the zeros t_n are the eigenvalues of a self-adjoint operator, or the eigenphases of a unitary operator, then automatically they would all be real. Some speculations along these lines are reviewed in [4], others have been pursued by Connes and co-workers. If the zeros are indeed related to the eigenvalues of a self-adjoint or unitary operator, and if that operator behaves 'typically', this would then suggest that the zeros might be distributed like the eigenvalues of random matrices. The success of random matrix theory in describing properties of the ζ- function might be interpreted as evidence in favour of a spectral interpretation.

Acknowledgements

The programme of research reviewed here was initiated and developed in collaboration with Dr. Nina Snaith.

Bibliography

[1] T. H. Baker and P. J. Forrester; Finite-N fluctuation formulas for random matrices, *J. Stat. Phys.* **88**, 1371–1386 (1997).

[2] E. W. Barnes; The theory of the G-function, *Q. J. Math.* **31**, 264–314 (1900).

[3] M. V. Berry; Semiclassical formula for the number variance of the Riemann zeros, *Nonlinearity* **1**, 399–407 (1988).

[4] M. V. Berry and J. P. Keating; The Riemann zeros and eigenvalue asymptotics, *SIAM Rev.* **41**, 236–266 (1999).

[5] E. B. Bogomolny and J. P. Keating; Random matrix theory and the Riemann zeros I: three- and four-point correlations, *Nonlinearity* **8**, 1115–1131 (1995).

[6] E. B. Bogomolny and J. P. Keating; Random matrix theory and the Riemann zeros II: n-point correlations, *Nonlinearity* **9**, 911–935 (1996).

[7] E. B. Bogomolny and J. P. Keating; Gutzwiller's trace formula and spectral statitics: beyond the diagonal approximation, *Phys. Rev. Lett.* **77**, 1472–1475 (1996).

[8] J. B. Conrey; More than 2/5 of the zeros of the Riemann zeta function are on the critical line, *J. Reine. Ang. Math.* **399**, 1–26 (1989).

[9] J. B. Conrey and D. W. Farmer; Mean values of L-functions and symmetry, *Int. Math. Res. Notices* **17**, 883–908 (2000).

[10] J. B. Conrey, D. W. Farmer, J. P. Keating, M. O. Rubinstein and N. C. Snaith; Autocorrelation of random matrix polynomials, *Commun. Math. Phys.* **237**, 365–395 (2003).

[11] J. B. Conrey, D. W. Farmer, J. P. Keating, M. O. Rubinstein and N. C. Snaith; Integral moments of L-functions, preprint (math.NT/0206018)

[12] J. B. Conrey and A. Ghosh; On mean values of the zeta-function, iii, *Proceedings of the Amalfi Conference on Analytic Number Theory, Università di Salerno* (1992).

[13] J. B. Conrey and S. M. Gonek; High moments of the Riemann zeta-function, *Duke. Math. J.* **107**, 577–604 (2001).

[14] O. Costin and J. L. Lebowitz; Gaussian fluctation in random matrices, *Phys. Rev. Lett.* **75**, 69–72 (1995).

[15] P. J. Forrester, N. C. Snaith and J. J. M. Verbaarschot; Developments in random matrix theory, *J. Phys.* A **36**, R1–R10 (2003).

[16] D. A. Goldston; On the function $S(T)$ in the theory of the Riemann zeta-function, *Journal of Number Theory* **27**, 149–177 (1987).

[17] G. H. Hardy and J. E. Littlewood; Contributions to the theory of the Riemann zeta-function and the theory of the distribution of primes, *Acta Mathematica* **41**, 119–196 (1918).

[18] A. E. Ingham; Mean-value theorems in the theory of the Riemann zeta-function, *Proc. Lond. Math. Soc.* **27**, 273–300 (1926).

[19] A. Ivic; *Mean values of the Riemann zeta function*, Tata Institute of Fundamental Research, Bombay (1991).

[20] N. M. Katz and P. Sarnak; *Random Matrices, Frobenius Eigenvalues and Monodromy*, AMS, Providence, Rhode Island (1999).

[21] N. M. Katz and P. Sarnak; Zeros of zeta functions and symmetry, *Bull. Amer. Math. Soc.* **36**, 1–26 (1999).

[22] J. P. Keating; The Riemann zeta function and quantum chaology, in *Quantum Chaos*; editors, G. Casati, I Guarneri, and U. Smilansky, pages 145–85. North-Holland, Amsterdam (1993).

[23] J. P. Keating; Periodic orbits, spectral statistics, and the Riemann zeros, in *Supersymmetry and trace formulae: chaos and disorder*; editors, I. V. Lerner, J. P. Keating, and D. E. Khmelnitskii, pages 1–15. Plenum, New York (1999).

[24] J. P. Keating and N. C. Snaith; Random matrix theory and $\zeta(1/2+it)$, *Commun. Math. Phys.* **214**, 57–89 (2000).

[25] J. P. Keating and N. C. Snaith; Random matrix theory and L-functions at $s=1/2$, *Commun. Math. Phys.* **214**, 91–110 (2000).

[26] M. L. Mehta; *Random Matrices*, Academic Press, London, second edition (1991).

[27] H. L. Montgomery; The pair correlation of the zeta function, *Proc. Symp. Pure Math* **24**, 181–93 (1973).

[28] A. M. Odlyzko; The 10^{20}th zero of the Riemann zeta function and 70 million of its neighbors, *Preprint* (1989).

[29] Z. Rudnick and P. Sarnak; Zeros of principal L-functions and random-matrix theory, *Duke Math. J.* **81**, 269–322 (1996).

[30] E. C. Titchmarsh; *The Theory of the Riemann Zeta Function*, Clarendon Press, Oxford, second edition (1986).

[31] H. Weyl; *Classical Groups*, Princeton University Press (1946).

[32] See `http://www.zetagrid.net/zeta/rh.html`.

Rupert Klein holds a professorship for "Scientific Computing/Modelling and Simulation of Global Environment Systems" in the Institute for Mathematics and Computer Science at the Freie Universität Berlin. He heads the Data and Computation Department at the Potsdam Institute for Climate Impact Research in Potsdam, Germany and he is an associate member of Konrad-Zuse-Zentrum für Informationstechnik, Berlin (ZIB). His research is characterised by a unique merger of applied mathematical modelling and modern computational techniques. During his 13-year academic career he has addressed problems in theoretical and computational fluid mechanics, ranging from high-speed and low-speed combustion, via the dynamics of slender vortices, to multiplescale phenomena in atmospheric flows.

Professor Klein was born in Wuppertal, Germany, and studied Mechanical Engineering at RWTH Aachen, Germany, where he also received his doctoral degree in 1988. A two-year postdoctoral research fellowship with the Program in Applied and Computational Mathematics at Princeton University, USA, was followed by an assistant professorship with the Department of Mechanical Engineering of RWTH Aachen, Germany. His interest in environmental problems and in man-environment-machine systems led to a professorship with the department of Safety-Technology at Wuppertal University in 1995. Soon afterwards he was appointed to his current position in Potsdam Institute/Freie Universität Berlin, and this has placed him at the interface between climate impact research and modern applied and computational mathematics.

Klein was awarded the Horning Memorial Award and the Arch T Colwell Merit Award from the Society of Automotive Engineers (SAE) in 1990, the Bennigsen–Förder Prize from the state of North-Rhine-Westphalia in 1995, and the International Fellow Award from Johns Hopkins University in 1995/96. More recently he was awarded the Gottfried-Wilhelm-Leibniz-Preis of the Deutscheforschungsgemeinschaft. His list of invited presentations at international conferences includes the International Conference on Numerics in Combustion, the International Conference on Hyperbolic Systems, and the GAMM annual meeting. Currently, he is a member of the editorial boards of *Theoretical and Computational Fluid Dynamics*, the *SIAM Journal of Multiscale Modelling and Simulation* and *Computers and Fluids.*

Chapter 12

An applied mathematical view of meteorological modelling*

Rupert Klein[†]

Abstract: The earth's atmosphere is overwhelmingly complex due to interactions of many different phenomena on widely differing scales. Comprehensive mathematical descriptions covering the entirety of these phenomena do not exist to date. A central theme of atmosphere science is thus the derivation of reduced models that are mathematically tractable, while still representing a relevant subset of the observed phenomena.

This paper first elucidates three simplified models which cover the entire range of meteorological scales from 'kilometers and minutes' to 'planetary lengths and millennia.' Their derivation generally relies on judicious physical reasoning, and intricate mathematical calculations. The former requires intimate knowledge of the scientific field; it is quite hard to follow for the mathematically trained, but meteorologically untrained.

We present a unified mathematical approach to meteorological modelling developed recently by the author. It is based on judiciously chosen coupled asymptotic limits for the Rossby, Froude, Mach, and other non-dimensional parameters, and on specialisations of a very general multiple-scales asymptotic *ansatz*. This scheme allows us to identify a large number of well-known simplified meteorological model equations by specifying (i) some coupled asymptotic limit for the relevant non-dimensional parameters, and (ii) a selection of asymptotically scaled space-time

[1]The research presented here has been funded partially by the Deutsche Forschungsgemeinschaft, Grant KL 611/6. Major parts of this work have been achieved in collaboration with Andrew J. Majda (Courant Institute of Mathematical Sciences, New York, NY, USA), Ann Almgren (Lawrence Berkeley Nat. Lab., Berkeley, CA, USA), and Nicola Botta, Antony Owinoh, Susanne Lützenkirchen (Potsdam Institute for Climate Impact Research). The paper has benefitted greatly from elucidating discussions and extensive helpful comments by U. Achatz, M.J. Cullen, J.C.R. Hunt and V. Petukhov.

[†]FB Mathematik und Informatik, Freie Universität Berlin, Germany, and Potsdam Institut für Klimafolgenforschung (PIK), Potsdam, Germany

coordinates. This multi-scale *ansatz* has proven to be extremely helpful in structuring mathematical discussions on topics of theoretical meteorology. At the same time it opens new routes of exploration for scientific studies by making available a rich applied mathematical toolkit.

Finally, we demonstrate this potential through two examples. The first is joint work with Andrew J. Majda, addressing multi-scale phenomena in the tropics. The second is joint work with Nicola Botta, addressing the construction of asymptotically adaptive 'well balanced' numerical methods for atmospheric flows.

Contents

1 Introduction

The earth's atmosphere is of overwhelming complexity due to a rich interplay between a large number of phenomena interacting on very diverse length and time scales. There are mathematical equation systems which, in principle, provide a comprehensive description of this system. Yet, exact or accurate approximate solutions to these equations covering the full range of complexities they allow for are not available. As a consequence, one of the central themes of theoretical meteorology is the development of simplified model equations that are amenable to analysis and computational approximate solution, while still faithfully representing an important subset of the observed phenomena.

1.1 Governing equations

Throughout this paper we consider the three-dimensional compressible flow equations for an ideal gas with constant specific heat capacities, supplemented with a number of source terms, as the starting point of our derivations.

$$\begin{aligned} \boldsymbol{v}_t + \boldsymbol{v} \cdot \nabla \boldsymbol{v} + \boldsymbol{\Omega} \times \boldsymbol{v} + \rho^{-1} \nabla p &= \boldsymbol{S}_{\boldsymbol{v}} - g\boldsymbol{k} \\ p_t + \boldsymbol{v} \cdot \nabla p + \gamma p \nabla \cdot \boldsymbol{v} &= S_p \\ \theta_t + \boldsymbol{v} \cdot \nabla \theta &= S_\theta , \end{aligned} \tag{1.1}$$

Here $\boldsymbol{v}, p, \theta$ are the fluid flow velocity, the (thermodynamic) pressure, and the fluid's potential temperature. γ is the isentropic exponent, assumed to be constant. $\boldsymbol{\Omega}, g$ are the vector of earth rotation and the acceleration of gravity, $\boldsymbol{k}$ is a radial unit vector, pointing away from the earth's center. The source terms $\boldsymbol{S}_{\boldsymbol{u}}, S_p, S_\theta$ are abbreviations for molecular or turbulent transport terms, for effective energy source terms from radiation, latent heat release from condensation of water vapor, etc.

The potential temperature is a variable, closely related to thermodynamic entropy, and defined by

$$\theta = \frac{p_{\text{ref}}^{\frac{\gamma-1}{\gamma}}}{R} \frac{p^{\frac{1}{\gamma}}}{\rho} , \tag{1.2}$$

where R is the ideal gas constant. This variable is the answer to the following question: Suppose one isolates an infinitesimally small parcel of air at any location in the atmosphere, and lets the parcel's pressure and density be p, ρ, respectively. What would be the parcel's temperature if it were to undergo an adiabatic and quasi-static, i.e., isentropic, process that leads to a final pressure p_{ref}?

The equations in (1.1), (1.2) account for the vapor-water, water-ice, and vapor-ice phase transitions neither through balance equations for the related species densities, nor through the thermodynamic relations for γ and $\theta(p, \rho)$. While this is certainly an over-simplification for realistic meteorological applications, it allows us to present the key ideas of this work in a transparent fashion. The incorporation of moist processes within the present mathematical framework is work in progress.

Similar comments hold for other effects collected in the effective source terms, $\boldsymbol{S}_{\boldsymbol{u}}, S_p, S_\theta$, such as turbulent transport.

1.2 Structure of the rest of the paper

Section 2 will summarise three typical simplified model equation systems that have been developed to describe selected phenomena associated with specific ranges of length and time scales. Models of this type play a central role in meteorology and climate research as they condense the available knowledge regarding the targeted phenomena in a mathematically compact, and computationally tractable fashion.

The derivation of such simplified models generally relies on a combination of judicious physical reasoning and subsequent, sometimes quite intricate, mathematical calculations. The first component, physical reasoning, requires an intimate knowledge of the scientific field, and it is often quite hard to follow for the mathematically trained, but meteorologically untrained. On the other hand, it would be the mathematically trained, who would be in a position to judge, e.g., the well-posedness of the derived reduced model, to show rigorously that solutions of the model equations are somehow “close” to solutions of the original complex equations, etc. Thus it is desirable to bridge between the physics-oriented meteorological viewpoint and the mathematical one.

Section 3 addresses this issue by describing a unified mathematical approach to meteorological modelling developed recently by the author, and anticipated in [26]. The approach is based on a set of carefully chosen distinguished limits for several small non-dimensional parameters, and on specialisations of a very general multiple-scales asymptotic ansatz.

Section 4 will summarise three instructive applications of the approach, ranging from a re-derivation of the well-known semi-geostrophic theory, [18, 39], via one of the recently derived multi-scale models for the tropics, [30], to numerical methods for (1.1), (1.2), that are “well-balanced” with respect to nearly hydrostatic situations. (A numerical method for a complex equation system is called well-balanced w.r.t. some singular limit regime if its accuracy and robustness do not deteriorate as the limit is approached, [8].)

Section 5 draws a few conclusions.

2 Phenomena with widely disparate scales

As mentioned in the introduction, atmospheric flows feature a multitude of different length and time scales. While some of these scales are imposed on the air flow externally; e.g., the characteristic lengths of the bottom topography, others are intrinsic to the atmospheric layer on the rotating earth. An intuitive description of these intrinsic scales may be given by reference to the phase speeds of three physically important phenomena, namely:

$u_{\text{ref}} \approx 10\,\text{m/s}$	characteristic flow velocity;
$c_i \approx 60\,\text{m/s}$	typical propagation speed of internal gravity waves;
$c_e \approx 300\,\text{m/s}$	typical propagation speed of external gravity waves.

The mentioned characteristic length scales are:

$h_{\text{sc}} \approx 10\,\text{km}$	Pressure scale height: vertical distance with significant pressure drop.
$L_1 \approx 70\,\text{km}$	For flows with horizontal characteristic length L_1 the inertial and Coriolis forces are comparable.
$L_{\text{i}} \approx 500\,\text{km}$	Internal Rossby deformation radius: the distance a typical internal gravity wave with speed c_i would have to travel to be affected by Coriolis effects significantly.
$L_{\text{e}} \approx 3\,000\,\text{km}$	Obukhov radius or external deformation radius: analogous to L_{i}, but for the much faster barotropic gravity waves with speed c_e. (see “Lamb waves”, [18])
$L_{\text{p}} \approx 20\,000\,\text{km}$	The planetary scale.

Some physical arguments for the existence of these scales will be given in Section 3 below. The appearance of these separated scales may also be understood, from a mathematical point of view, as being naturally induced by the existence of a single small asymptotic parameter, $\varepsilon \ll 1$, in the sense of multiple scales asymptotics. The section will then "translate" between these two points of view.

Together with the signal speeds u_{ref}, c_i, c_e the length scales just introduced make up an entire grid of different flow regimes as shown in left graph in Figure 1a. Notice, in this context, that

$$t_{\text{ref}} \sim \frac{u_{\text{ref}}}{h_{\text{sc}}} \approx \frac{L_1}{c_i} \approx \frac{L_{\text{i}}}{c_e}. \tag{2.1}$$

Consider, e.g., the left-most three boxes of the grid in the lowest row, which have the time scale t_{ref} in common. The left-most box represents advection over distances comparable to h_{sc}, the middle box covers the regime of internal gravity waves with the same characteristic frequency, while the right-most box represents the regime of external gravity waves that evolve on this time scale. Similarly, the lowest three boxes of the L_{i}-column cover, at successively slower time scales, fast external gravity waves, internal gravity waves, and advection over distances comparable to L_{i}.

Within each of the regimes in the graph, different physical mechanisms dominate, and an associated reduced model equation is conceivable which approximately describes the flow evolution in that regime. Figure 1b shows the regimes to be discussed briefly in this section as examples.

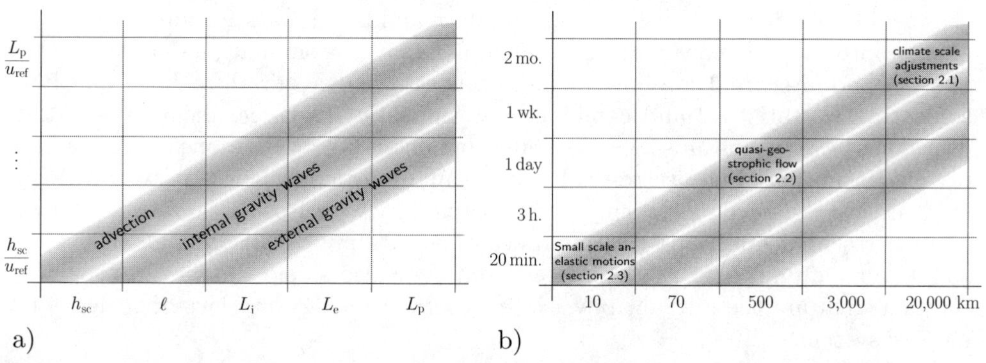

Figure 1. *Regimes of length and time scales for flows involving the bulk of the troposphere in the mid-latitudes. The corresponding regimes for near-equatorial and stratosphere flows involve slightly different scales (see, e.g., Section 4.2).*
a) The backbone of a more involved scale diagram that would include the characteristic scales of turbulence, water phase transitions, radiation, boundary layer processes, atmospheric chemistry, etc.
b) Localisation of the regimes to be discussed in this section as examples.

2.1 Climate scale adjustments

Global climate modelling strategies

Even though they are by far not complete, the scale maps in Figure 1 indicate that the global climate incorporates a very wide range of length and time scales and associated physical processes. Modern climate research proceeds along various pathways in trying to cope with the resulting complexity. One strategy, generally subsumed under the headline of General Circulation Models (GCM), involves the development and numerical solution of a set of model equations that is "as complete as possible". These avoid model reductions based on pre-assumed dominant length and time scales to the widest possible extent. In practice, one aims at model equations that properly describe *all* scales larger than those one can afford to resolve on the most powerful computers available. Today's resolutions involve computational grid sizes of about 100 km in the horizontal, 200 m in the vertical direction, and time steps of several minutes. For all processes taking place on smaller length and time scales, and which can therefore not be described explicitly, one introduces "sub-grid scale parameterisations". These are very similar in spirit to turbulence closures in fluid mechanics, but account for a much larger set of processes, such as small scale (natural) convection due to radiative heating, latent heat release from the condensation of water vapor, interactions of the atmosphere with surface processes near the ground, etc.

Modern GCMs draw their justification and attractiveness from the fact that they directly implement the state-of-the-art knowledge regarding a large number of participating processes. They are "the best climate research can do today" in this sense, and they may be considered as reference and benchmark for alternative modelling approaches. However, the sheer complexity of the simulation data turns their analysis and interpretation into an entire scientific endeavour all by itself, involving sophisticated statistical and combined deterministic/statistical techniques. Moreover, long term, thousands-of-years transients involving interactions of atmosphere, ocean, sea and land ice, terrestrial and oceanic biosphere, land surface processes, etc., are as yet beyond current GCMs' capabilities. Also, for a range of global climate research questions the small scale, short time details provided by GCM simulations are not only not of interest, but they also shield the view, to some extent, on what are the relevant physical interactions on the very largest scales of the climate system.

More recent developments of "Earth system Models of Intermediate Complexity" (EMIC) address these issues, [43, 40, 49]. These models are designed to describe *only* these largest scales and to consider everything below, say, the external Rossby deformation radius $L_{\rm e}$ as "small scale processes" to be parameterised; see, e.g., [49]. In comparison with full-fledged GCMs, such EMICs have the advantage of very high computational efficiency and, by representing only the "climate scales", they provide the desired direct view of the large scale mean variables. Figure 2 shows, as an example, the computed deviations of the large scale, long time temperature mean values between the year 1800 and the year 2100 as obtained with the CLIMBER-2 EMIC in [41]. The figure clearly reveals that the simulations do

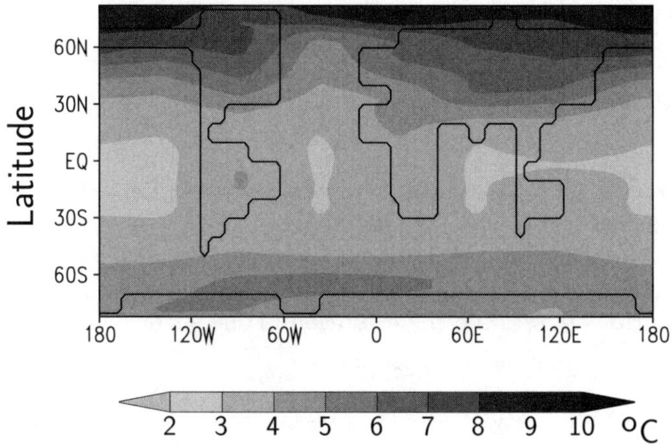

Figure 2. *Map of the simulated annual-mean surface air temperature change in the year 2100 for a standard scenario of future* CO_2 *emissions relative to the year 1800. (Courtesy of Stefan Rahmstorf, Potsdam Institut für Klimafolgenforschung.)*

not display any spacial variations comparable to the typical cyclonic/anti-cyclonic patterns seen on daily weather charts (see Section 2.2 below). In fact, such variability is parameterised within the CLIMBER-2 model by averaging over an assumed Gaussian statistics for these modes.

Structure of an intermediate complexity atmosphere model

Here we report briefly on the structure of POTSDAM-2 (POTsdam Statistical Dynamical Atmosphere Model), the atmosphere component of the CLIMBER-2 EMIC, as described by Petoukhov et al. in [40].

There are only two two-dimensional unsteady evolution equations

$$\frac{\partial Q_T}{\partial t} = \nabla_{\|} \cdot \boldsymbol{F}_T + S_T \tag{2.2}$$

$$\frac{\partial Q_q}{\partial t} = \nabla_{\|} \cdot \boldsymbol{F}_q + S_q \tag{2.3}$$

for depth-averaged thermal energy, Q_t, and water content, Q_q, which are defined as

$$Q_T = \int_{z_s}^{H_a} \rho T \, \mathrm{d}z \qquad \text{and} \qquad Q_q = \int_{z_s}^{H_a} \rho q \, \mathrm{d}z \,, \tag{2.4}$$

respectively. Here T, q, ρ are temperature, water vapor mass fraction, air density, and z_s, H_a are the vertical levels of the bottom topography and of the top of the atmosphere. $\nabla_\parallel$ denotes the horizontal gradient operator. The flux $\boldsymbol{F}_T$ is given by

$$\boldsymbol{F}_T = \int\limits_{z_s}^{H_a} \rho\big(\boldsymbol{u}\theta + \widehat{\boldsymbol{u}'\theta'} + \boldsymbol{D}^\theta\big)\,\mathrm{d}z\,. \tag{2.5}$$

This expression includes advection of potential temperature by the large scale wind field, $\boldsymbol{u}$, the effective transport of potential temperature by synoptic scale fluctuations, $\boldsymbol{u}', \theta'$, and the transport by turbulent motions on even smaller scales subsumed in $\boldsymbol{D}^\theta$. (The humidity flux $\boldsymbol{F}_q$ has a similar representation.)

These equations determine the evolution of the surface fields

$$T_s(t,\boldsymbol{x})\,, \qquad q_s(t,\boldsymbol{x})$$

of temperature and humidity. Closed expressions for the various integrals are obtained by assuming piecewise linear and exponential vertical structures for T, θ, q, ρ, respectively, namely

$$T = \begin{cases} T_s(t,\boldsymbol{x}) - \Gamma(z - z_s) & \text{for} \quad z_s \le z < H_t \\ T_s(t,\boldsymbol{x}) + \Gamma(H_t - z_s) & \text{for} \quad H_t \le z < H_a \end{cases} \tag{2.6}$$

$$\theta = T_s(t,\boldsymbol{x}) - \Gamma_a(z - z_s)\,, \tag{2.7}$$

$$q = q_s(t,\boldsymbol{x})\exp\big(-\frac{z - z_s}{H_e}\big)\,, \tag{2.8}$$

$$\rho = \rho_*\exp\big(-\frac{z}{h_{\text{sc}}}\big)\,. \tag{2.9}$$

Here $\rho_*, h_{\text{sc}}, H_e$ are constants, and the coefficients Γ, Γ_a, H_t are related to T_s, q_s via explicit algebraic functions.

The large scale horizontal wind field is composed of geostrophic and ageostrophic components (see Section 2.2), so that

$$\boldsymbol{u} = \boldsymbol{u}_g + \boldsymbol{u}_a\,, \tag{2.10}$$

where $\boldsymbol{u}_g$ is determined by the current temperature field T and the sea level pressure p_0 via

$$f\,\boldsymbol{k}\times\boldsymbol{u}_g = -\frac{1}{\rho_*}\nabla p_0 - g\,\nabla\int\limits_0^z \frac{T}{T_*}\,\mathrm{d}z' \tag{2.11}$$

with T_* another constant, and f the vertical component of the earth rotation vector $\boldsymbol{\Omega}$. The ageostrophic field $\boldsymbol{u}_a$ is proportional to the gradient of the sea level pressure p_0. Finally, the sea level pressure is expressed in terms of the temperature distribution $T(t,\boldsymbol{x},z)$, using semi-empirical relations for the zonally averaged motions. The reader is referred to [40] for details.

While the relations just described for the large scale wind field can be derived analytically within a rational scaling (asymptotic) analysis, the expressions given in [40] for the turbulent and macro-turbulent fluxes $\boldsymbol{D}^\theta, \widehat{\boldsymbol{u}'\theta'}$, and for the source terms S_T, S_q in (2.2) and (2.3) are semi-empirical closures similar to turbulence closures in fluid mechanics.

For future reference, we summarise the mathematical structure of the large scale, long time evolution equations for the atmosphere: The two-dimensional fields, $T_s(t,\boldsymbol{x}), q_s(t,\boldsymbol{x})$, which represent the surface values of temperature and humidity, are advanced in time by solving depth-averaged balance equations for internal energy and water content. These balance equations involve horizontal fluxes due to large scale advection, and macro- and smaller scale turbulence, as well as other source terms related to radiation, evapo-transpiration, precipitation, etc. The large scale wind field is related algebraically to gradients of the primary fields T_s, q_s as a result of a scale or asymptotic analysis which pre-assumes the characteristic length and time scales of interest. For the various source terms and turbulence effects, semi-empirical parameterisations are introduced that ultimately lead to a closed equation set for the unknown fields $T_s(t,\boldsymbol{x}), q_s(t,\boldsymbol{x})$.

Notes on the closure schemes for fast processes

EMICs are often criticised for over-simplified representations of possibly crucial small scale processes, and there are a number of research activities on the way that aim at (i) an overall model design that still addresses only the large scale, long time climate variables, but which (ii) incorporates more sophisticated closure schemes for the non-resolved processes. For example, Conaty et al., [10], include unsteady evolution equations for the ensemble characteristics of smaller scales instead of the static relations mentioned above. Achatz and Branstator [1] combine a fluid dynamics model for the large scale motions with a closure scheme that directly incorporates the statistics generated by GCM simulations. Their model captures the energetically most important large scale patterns with considerable accuracy and it provides insight into atmospheric variability on time scales that are normally not accessible with statistical dynamical models as described above. This fosters hope that more sophisticated, explicitly time dependent closure schemes might make a much wider range of applications amenable to intermediate complexity models without having to accept the full complexity of a general circulation model.

In this context, a novel approach to obtaining effective *stochastic* equations for the large scale, long time variables has recently been proposed by A.J. Majda and co-workers, [33]. It is assumed, to start with, that a decomposition of the original complex dynamics into equations for the slow and for the fast modes, respectively has been established. The approach then consists of two steps: First, a carefully chosen *stochastic* closure is introduced in the equations for the fast variables so as to make them tractable by the tool kit of stochastic differential equations. No approximations are made at that stage in the equations for the slow modes. Secondly, appropriate stochastic closures for the "slow equations" are *derived systematically* by stochastic projection procedures. The resulting effective closure schemes have rich mathematical structures that are hard to anticipate in an approach that tries to directly guess the functional form of such terms without this two-step procedure.

The next subsection considers in more detail the dynamics on the next smaller "synoptic scales", which appears in low-resolution models and EMICs only through such effective closure schemes.

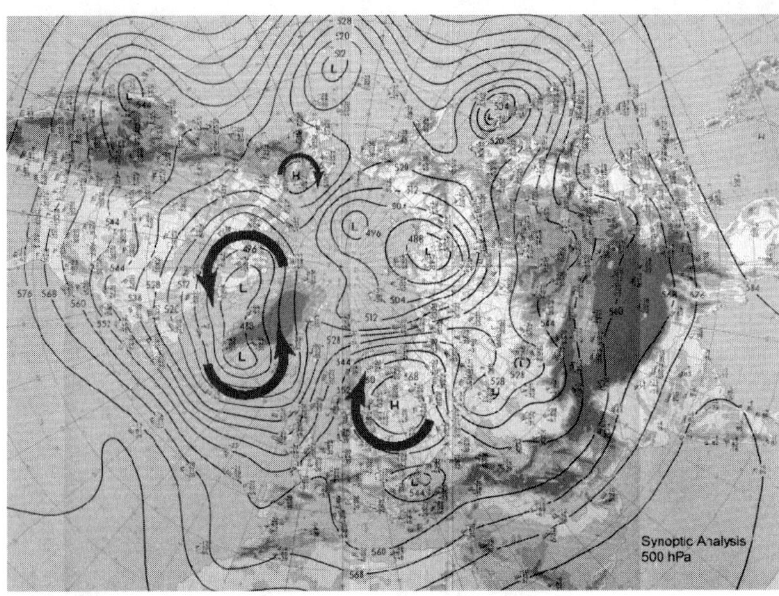

Figure 3. *A typical countour plot of the geopotential height of the* 500 hPa *surface over a large part of the northern hemisphere. (Courtesy of Peter Névir, Institut für Meteorologie, Freie Universität Berlin)*

2.2 Synoptic Scales

Scalings considerations

Figure 3 shows one of the typical results of the "synoptic analysis" of global atmospheric flows. The contours indicate the "geopotential height" $H = \Phi/g_c$ of a surface of constant pressure $p = 500\,\text{hPa}$. Here Φ is the effective geopotential, which is the sum of the earth's gravitational potential and the potential associated with the centripetal forces due to the earth's rotation, and g_c is a constant reference value of the gravitational acceleration which varies by about 0.3% over the globe, [18]. Countour labels show H measured in units of 10 m. The geopotential height approximates the height of a pressure surface above sea level up to deviations of the order of one percent.

The figure exhibits a collection of "synoptic eddies" of high and low pressure around which the main air flow circulates as indicated. The characteristic diameter of these structures is about 500–1 000 km; i.e., comparable to the internal Rossby deformation radius, L_{i}, and their characteristic evolution time scale is about 1–5 days. Figure 1b locates the evolution of these structures in the regime diagram. Assuming a flow field with such characteristic length and time scales, the important "quasi-geostrophic model" is derived (e.g., in [39]) as the leading order approximation in terms of the Rossby number based on the internal deformation radius L_{i},

$$\mathrm{Ro}_{L_i} = \frac{u_{\mathrm{ref}}}{\Omega L_i} \ll 1\,. \tag{2.12}$$

An asymptotic expansion for the solution vector $\mathbf{U}$ for this regime would read

$$\mathbf{U} = \sum_i \mathrm{Ro}_{L_i}^i \mathbf{U}^{(i)}(T, \boldsymbol{X}, Z) \tag{2.13}$$

where

$$T = \frac{tu_{\mathrm{ref}}}{L_i}\,, \qquad \boldsymbol{X} = (X, Y) = \frac{\boldsymbol{x}}{L_i} \qquad Z = \frac{z}{h_{\mathrm{sc}}} \tag{2.14}$$

are non-dimensional time, and horizontal and vertical space coordinates. A detailed derivation is given; e.g., in [39].

The quasi-geostrophic model

The resulting quasi-geostrophic equations are formulated in the β-plane approximation; i.e., they describe the flow of a shallow layer of fluid over a rotating plane with spacially varying Coriolis coefficient

$$f = f_0 + \beta Y\,. \tag{2.15}$$

Here $f = \boldsymbol{\Omega} \cdot \boldsymbol{k}$ is the projection of the earth rotation vector onto the vertical unit vector $\boldsymbol{k}$ in a tangent plane to the earth's surface, f_0 is its value at the location of contact of the plane with the earth's geometry, and β is its variation with distance in the northern direction measured by the cartesian coordinate Y.

The vertical momentum balance reveals that the pressure field is hydrostatic to several orders in terms of the expansion parameter. In particular, there are time independent leading order hydrostatic pressure and density distributions $p_0(z), \rho_0(z)$, and a background stratification of potential temperature characterised by second order deviations, $\Theta_2(z)$, from a constant reference value. The first perturbation pressure with non-trivial horizontal variation, $p' = \rho_0(z)\pi$, still satisfies a hydrostatic relation

$$\frac{\partial \pi}{\partial z} = \theta'\,, \tag{2.16}$$

where θ' denotes the third order perturbation of potential temperature.

Gradients of the perturbation pressure balance the Coriolis forces at leading order in the horizontal direction, so that

$$f_0\, \boldsymbol{k} \times \boldsymbol{u} = -\nabla_{\boldsymbol{X}} \pi\,. \tag{2.17}$$

Here $\boldsymbol{u}$ is the horizontal flow velocity component. This equation is insufficient to determine a flow field *including* its temporal evolution. As usual in singular perturbation analyses, the missing evolution equation is obtained through a consistency condition at the next order. The result is a transport equation for the "potential vorticity", q,

$$\left(\frac{\partial}{\partial T} + \boldsymbol{u} \cdot \nabla_{\boldsymbol{X}}\right) q = 0\,, \tag{2.18}$$

where

$$q = \zeta + \beta Y + \frac{f_0}{\rho_0}\frac{\partial}{\partial Z}\Big(\frac{\rho_0 \theta'}{\mathrm{d}\Theta_2/\mathrm{d}Z}\Big) \tag{2.19}$$

and

$$\zeta = \frac{1}{f_0}\nabla^2\pi \tag{2.20}$$

is the vorticity associated with the horizontal velocity field $\boldsymbol{u}$. This is the appropriate specialisation to quasi-steady flow of a thin fluid layer on a rotating plane of Ertel's general conservation law for potential vorticity; see [47]. Now (2.16)–(2.20) form a closed set of equations for $\zeta, \pi, \theta, \boldsymbol{u}$.

The physical meaning of the transport equation (2.18) is readily understood considering that it is equivalent to the transport equation for vertical vorticity

$$\zeta_T + \boldsymbol{u}\cdot\nabla_{\boldsymbol{X}}\zeta + \beta v + f_0 \nabla_{\boldsymbol{X}}\cdot\boldsymbol{u}' = 0\,, \tag{2.21}$$

where v is the horizontal flow velocity in the direction of the Y-coordinate, and $\boldsymbol{u}'$ is the first order horizontal flow velocity perturbation. The potential vorticity conservation law (2.18) is obtained from (2.21) using the identity $v \equiv (\partial_T + \boldsymbol{u}\cdot\nabla_{\boldsymbol{X}})\,Y$, and by eliminating $\nabla_{\boldsymbol{X}}\cdot\boldsymbol{u}'$ using mass conservation, $\rho_0\nabla_{\boldsymbol{X}}\cdot\boldsymbol{u}' + (\rho_0 w')_Z = 0$, and the transport equation for potential temperature (entropy) $(\partial_T + \boldsymbol{u}\cdot\nabla_{\boldsymbol{X}})\,\theta + w'\dfrac{\mathrm{d}\Theta_2}{\mathrm{d}Z} = 0$.

Equation (2.21) shows that, as the vorticity ζ is advected horizontally, it is augmented by variations of the vertical component of the earth rotation vector, and affected by vortex stretching. These are, in fact, the essential effects determining the evolution of the "synoptic eddies" seen in Figure 3. Notice that the present discussion, which closely follows Pedlosky's classical textbook presentation, [39], does not include source terms from radiative balance, latent heat conversion, boundary layer effects, etc., so that it reveals only the fluid dynamical aspects of these synoptic scale flows.

2.3 Small scale anelastic flows

Here we consider motions on scales comparable to or smaller than the pressure scale height $h_{\rm sc} \approx 10\,\mathrm{km}$, and on the associated advection time scale of $t_{\rm ref} = h_{\rm sc}/u_{\rm ref} \approx 20\,\mathrm{min}$. In this flow regime, one of the prominent issues is the removal of inessential acoustic modes in a reduced model.

The full compressible flow equations from (1.1) feature various flow modes, including "high frequency" acoustic waves. Here "high frequency" denotes acoustic modes with characteristic frequencies higher than 1 min^{-1}. A plausibility argument states that sizeable elastic perturbations cannot establish in the atmosphere, because acoustic waves rapidly redistribute the associated energy and lead to an equilibration void of any acoustic modes. This intuitive explanation has been quantified through asymptotic analysis for vanishing Mach number in non-rotating flows without gravity (see e.g. [45, 31]) and backed up by rigorous justifications; e.g. in [13, 25, 46].

In the absence of energy source terms, such an analysis results in the classical variable density, incompressible flow equations, with zero velocity divergence.

Three-dimensional motions

However, if one includes the effects of gravity and energy source terms from radiation or latent heat conversion, and allows for vertical motions comparable in scale to the pressure scale height, then two distinctly different flow regimes can be identified, [38, 12, 26]. The first regime involves nearly neutral stratification of the atmosphere, so that parcels of air can move freely in the vertical direction without being constrained by buoyancy effects. In this situation, discussed first by Ogura and Phillips, [38], the flow is governed by the anelastic equations in the Boussinesq approximation, so that

$$\begin{aligned} \boldsymbol{u}_t + \boldsymbol{u} \cdot \nabla \boldsymbol{u} + w \boldsymbol{u}_z + \frac{1}{\rho_0} \nabla p' &= \boldsymbol{S}_{\boldsymbol{u}} \\ w_t + \boldsymbol{u} \cdot \nabla w + w w_z + \frac{1}{\rho_0} p'_z &= \theta' + S_w \\ \theta'_t + \boldsymbol{u} \cdot \nabla \theta' + w \theta'_z &= S_{\theta'} \\ \nabla \cdot (\rho_0 \boldsymbol{u}) + \partial_z (\rho_0 w) &= 0 \end{aligned} \tag{2.22}$$

where $\rho_0(z)$ is a time independent, horizontally homogeneous density stratification. Given suitable functional expressions for the source terms $\boldsymbol{S}_{\boldsymbol{u}}, S_w, S_{\theta'}$, these equations govern the temporal evolution of the horizontal and vertical velocity components $\boldsymbol{u}, w$, and the potential temperature perturbation θ'. The perturbation pressure p' is determined, as in the classical case of incompressible flows, by a Poisson-type equation obtained by analyzing the consequences of the divergence constraint $(2.22)_4$ for the two momentum equations $(2.22)_{1,2}$. The complexity and structure of these equations is comparable to that of the classical three-dimensional incompressible flow equations. They are particularly useful in studies of the atmospheric boundary layer, where the assumption of nearly neutral stratification is well justified. (See [12] for an interesting generalisation of (2.22), and the discussions in [2, 26].)

Quasi two-dimensional weak temperature gradient flows

In their original derivation in [38], Ogura and Phillips explicitly state that the stratification of potential temperature in the atmosphere must be nearly neutral to arrive at the reduced equation system from (2.22). Zeytounian [52] pursues an analogous scaling analysis but relaxes that constraint. The result is a system of equations in which the vertical velocity is entirely dominated by buoyancy. In the absence of energy source terms he find that all parcels of air are forced to reside within their respective vertical levels of neutral buoyancy, so that vertical motions are suppressed. This holds at least if the shortest relevant time scale is that of advective motions. Internal gravity waves on shorter time scales *will* induce vertical displacements, but their description would require a multiple time scale analysis; see, e.g., [16] and Section 3 below.

In [26] the author compares both the near-neutral and the buoyancy-controlled regimes and includes energy source terms. For the buoyancy-controlled regime, the following system of equations obtains,

$$\begin{aligned} \boldsymbol{u}_t + \boldsymbol{u} \cdot \nabla \boldsymbol{u} + w \boldsymbol{u}_z + \frac{1}{\rho_0} \nabla p' &= \boldsymbol{S}_{\boldsymbol{u}} \\ w &= \frac{S_\theta}{\mathrm{d}\Theta_2/\mathrm{d}z} \\ \nabla \cdot (\rho_0 \boldsymbol{u}) + \partial_z(\rho_0 w) &= 0 \,. \end{aligned} \tag{2.23}$$

Here $\boldsymbol{u}, w, p'$ are the horizontal and vertical flow velocities and the perturbation pressure as before, while $\Theta_2(z)$ is the background stratification of potential temperature responsible for the deviations from the Ogura and Phillips' regime. These equations are, as far as fluid dynamics is concerned, essentially two-dimensional, whereas the vertical velocity appears as an algebraically determined auxiliary quantity. Similar equation sets have been derived in other contexts including, in particular, near-equatorial motions at much larger horizontal scales, in [9, 20, 6].

While (2.23) has been derived in [26] via low Mach number asymptotics, there is an alternative derivation that is based on the *a priori* assumption that horizontal gradients of temperature are often very weak in practical applications. Referring to this line of thought, models of this class have been labelled "Weak Temperature Gradient" approximations in recent studies of near-equatorial flows, [36, 48, 5, 30].

The reader may wonder, how any reasonable bottom boundary condition, such as $w \equiv 0$ at $z = 0$ could be maintained under (2.23), unless the energy source term represented by S_θ would be unnaturally constrained. The asymptotic limit considered here has obviously removed one degree of freedom in posing boundary conditions that had been present in the original three-dimensional flow equations from (1.1). This is a typical case of a singular limit in which the mathematical type of the equation changes when the limit is approached. In the present case, the failure of solutions of (2.23) to comply with the physically relevant bottom boundary conditions is a hint that a comprehensive analysis will have to include a near-bottom boundary layer, in which different length and/or time scalings have to be assumed. Asymptotic matching of the boundary layer solution to the bulk flow WTG equations will likely lead to the desired physically consistent asymptotic representation all the way down to the bottom of the atmospheric layer.

In the presence of non-trivial orography or surface conditions the near-bottom flow can induce highly non-trivial effects. In this case appropriate boundary layer theories need to be developed ([23, 37]), and their coupling to the bulk of the troposphere in the sense of asymptotic matching needs to be addressed.

2.4 Looking back

Looking back, we have seen in this section that the mathematical structure of reduced models for atmospheric motions depends heavily on the considered length and time scales, and on particular assumptions regarding specifics of the considered flow regime.

In the next section we introduce a unified approach to the analysis of such diverse scaling regimes that lets us get away with a minimal set of a priori assumptions.

The approach provides a systematic basis for structuring further mathematical analyses and discussions. Furthermore, it incorporates multiple scales asymptotic ideas by design, thereby providing a natural framwork for studying interactions between such different phenomena as seen in this section.

3 A Unified Mathematical Modelling Approach

3.1 Overview

The derivation of simplified model equations of theoretical meteorology generally relies on a combination of judicious physical reasoning and subsequent, often intricate, mathematical calculations. The first component, physical reasoning, requires an intimate knowledge of the scientific field, and it is often hard to follow for the mathematically trained, but meteorologically untrained. Readers belonging to this group might benefit from the unified mathematical approach proposed in this paper, which reveals some common underlying structures of meteorological scaling theories. Meteorologists may appreciate the potential of the present approach for systematically addressing multiple scales problems, and for structuring mathematical discussions regarding model derivations and the construction of appropriate numerical techniques.

The approach is based on a set of carefully chosen distinguished limits for several small non-dimensional parameters, and on specialisations of a very general multiple-scales asymptotic *ansatz.* Here is a summary of the key ideas, an early version of which appeared in [26]:

1. *Scale-independent reference quantities*
   First we collect a set of physical quantities that have well-defined characteristic values in the majority of atmosphere flow applications *independently* of the characteristic length and time scales of any particular phenomena.

2. *Universal non-dimensional parameters*
   Next we combine these characteristic values to form non-dimensional parameters, say π_1, π_2, π_3. These parameters will have more or less universally accepted typical values.

3. *Distinguished limits*
   We then group these parameters in a carefully chosen distinguished limit

   $$\begin{aligned} \pi_1 &= \pi_1^*, \\ \pi_2 &= \varepsilon^2\,\pi_2^*\,, \qquad \text{where} \qquad \pi_j^* = \mathrm{O}_s(1) \quad \text{as} \quad \varepsilon \to 0 \quad \text{for} \quad j \in \{1,2,3\}\,. \\ \pi_3 &= \varepsilon^3\,\pi_3^*\,, \end{aligned}$$

   Here we have used the notation $f = \mathrm{O}_s(\varepsilon)$ as $\varepsilon \to 0$ iff $0 < \limsup\limits_{\varepsilon\to 0} \|f\| < \infty$. Subsequently we employ $\varepsilon \ll 1$ as an asymptotic expansion parameter. This step defines a single asymptotic regime for the scale-independent dimensionless parameters π_1, π_2, π_3 which most atmospheric flows belong to.

4. *Dimensionless governing equations*
Using only the scale-independent reference quantities from the first step, we non-dimensionalise the governing equations. Because these reference quantities are related via the distinguished limits from the previous step, the resulting system of dimensionless equations includes the constants π_i^* and the small parameter ε as the only dimensionless characteristic numbers.

5. *Multiple-scales asymptotic expansions*
By definition, reduced model equations of theoretical meteorology obtained through *scale analysis* describe phenomena that are characterised by some typical length and time scales and fluctuation amplitudes. Here we account for such scalings through a general multiple scales asymptotic *ansatz* in terms of ε. The solution $\mathbf{U}(t, \boldsymbol{x}, z; \varepsilon)$ of the multi-dimensional compressible flow equations is expressed as

$$\mathbf{U}(t, \boldsymbol{x}, z; \varepsilon) = \sum_{i \in \mathsf{IN}} \phi^{(i)}(\varepsilon)\, \mathbf{U}^{(i)}\big(\tfrac{t}{\varepsilon}, t, \varepsilon t, \varepsilon^2 t, \ldots, \tfrac{\boldsymbol{x}}{\varepsilon}, \boldsymbol{x}, \varepsilon \boldsymbol{x}, \varepsilon^2 \boldsymbol{x}, \ldots, \tfrac{z}{\varepsilon}, z\big), \quad (3.1)$$

where the $\phi^{(i)}$ form an asymptotic sequence, such as $\phi^{(i)}(\varepsilon) = \varepsilon^i$, and $(t, \boldsymbol{x}, z)$ are time, horizontal space, and vertical coordinates, respectively. More complex asymptotic sequences and non-powerlaw scalings of the coordinates are adopted where necessary. See e.g., [23, 37] for examples necessitating non-powerlaw expansions.

Specialisations of this formulation, obtained by reducing the set of coordinates to merely one time, one horizontal, and one vertical coordinate, have allowed us to recover systematically a large collection of well-known reduced models by applying techniques of formal single-scale asymptotic analysis. Table 1 displays a list of examples (in which $t, \boldsymbol{x}, z$ have been non-dimensionalised with $t_{\text{ref}} = \ell_{\text{ref}}/u_{\text{ref}}, \ell_{\text{ref}} \approx 10\,\text{km}, u_{\text{ref}} \approx 10\,\text{m/s}$; see Tables 2, 3 and equations (3.4)–(3.6)). Notice that throughout this list the parameter ε has exactly the same meaning as established under item 3 above. Thus, even though we address a multitude of different regimes *in terms of length and time scales*, the fundamental distinguished limit for the dimensionless scale-independent parameters π_1, π_2, π_3 remains the same.
The general multiple scales *ansatz* from (3.1) enables us to further study interactions between phenomena acting on separate scales. For example, according to Table 1 we could consider an expansion based on the coordinate set $(\varepsilon t, \varepsilon^2 t, \varepsilon^2 \boldsymbol{x}, z)$ to analyze the interaction of long-wavelength gravity waves that are influenced by Coriolis effects with balanced geostrophic motions. (See Section 4.2 for an example involving multiple scales in the tropics.)

3.2 Asymptotic characterisation of atmosphere flows

Scale-independent reference quantities

There are a few physical quantities that have quite robustly agreed-upon typical values under a wide range of atmospheric flow conditions and independently of the

Table 1. *Coordinate scalings and associated classical models.*

Coordinates	Resulting model describes ...
$\mathbf{U}^{(i)}\left(\frac{t}{\varepsilon}, \boldsymbol{x}, \frac{z}{\varepsilon}\right)$	linear small scale internal gravity waves
$\mathbf{U}^{(i)}\left(t, \boldsymbol{x}, z\right)$	anelastic & pseudo-incompressible flows
$\mathbf{U}^{(i)}\left(\varepsilon t, \varepsilon^2\boldsymbol{x}, z\right)$	gravity waves influenced by Coriolis effects
$\mathbf{U}^{(i)}\left(\varepsilon^2 t, \varepsilon^2\boldsymbol{x}, z\right)$	mid-latitude quasi-geostrophic "QG" flow
$\mathbf{U}^{(i)}\left(\varepsilon^2 t, \varepsilon^2\boldsymbol{x}, z\right)$	equatorial weak temperature gradient flow
$\mathbf{U}^{(i)}\left(\varepsilon^2 t, \varepsilon^{-1}\,\xi(\varepsilon^2\boldsymbol{x}), z\right)$	semi-geostrophic flow
$\mathbf{U}^{(i)}\left(\varepsilon^{\frac{3}{2}} t, \varepsilon^{\frac{5}{2}} x, \varepsilon^{\frac{5}{2}} y, z\right)$	tropical Kelvin, Yanai, and Rossby Waves

particular length and time scales that might otherwise characterise the situation. Table 2 collects quantities related to the rotating earth and Table 3 lists typical reference values for the main fluid dynamical variables; i.e., the thermodynamic pressure and air density near the earth's surface, and a characteristic flow velocity. While there are phenomena where flow velocities exceed the listed 10 m/s considerably, as in high-altitude jet streams and severe storms, the majority of the flow phenomena of interest in meteorology are well characterised by flow velocities of several meters per second.

Table 2. *Properties of the rotating earth*

Earth's radius	a	$\approx 6 \times 10^6\,\mathrm{m}$
rotation frequency	Ω	$\approx 10^{-4}\,\mathrm{s}^{-1}$
acceleration of gravity	g	$\approx 10\,\mathrm{m/s}^2$

Table 3. *Aerothermodynamic conditions*

thermodynamic pressure	p_{ref}	$\approx 10^5\,\mathrm{kg/ms}^2$
air flow velocity	u_{ref}	$\approx 10\,\mathrm{m/s}$
air density	ρ_{ref}	$\approx 1\,\mathrm{kg/m}^3$

Universal non-dimensional parameters

With mass, length, and time these six quantities involve three fundamental physical dimensions, so that Buckingham's theorem, [7], allows us to form three independent dimensionless parameters. Let

$$c_{\text{ref}} = \sqrt{\frac{p_{\text{ref}}}{\rho_{\text{ref}}}} \approx 300\,\text{m/s}\,,$$

denote the speed of long wavelength (barotropic) gravity waves in the atmosphere, which is proportional to the speed of sound sound at reference conditions. Then the following dimensionless combinations turn out to be convenient for the subsequent developments:

$$\begin{aligned} \pi_1 &= \frac{c_{\text{ref}}}{\Omega\, a} &&\approx 0.5 &&\approx 1\,, \\ \pi_2 &= \frac{u_{\text{ref}}}{c_{\text{ref}}} &&\approx 3\times 10^{-2} &&\ll 1\,, \\ \pi_3 &= \frac{\Omega\, c_{\text{ref}}}{g} &&\approx 3\times 10^{-3} &&\ll 1\,. \end{aligned} \tag{3.2}$$

(We discuss the relation of these parameters to more familiar ones, such as the Rossby, Mach, and Froude numbers, in Section 3.3 below.)

These order-of-magnitude estimates are appropriate for a majority of atmospheric flow phenomena, no matter what are their characteristic length and time scales. In constructing a generalised formal approach to atmosphere modelling we are therefore interested in regimes that are compatible with these scalings.

Distinguished limits

When faced with multiple asymptotically small parameters, an important issue is the precise path along which the parameters are to approach their respective limiting values. Consider Figure 4, which displays the space spanned by two parameters ϵ and δ supposed to characterise some physical system. In an actual application these parameters are "small" in analogy with $u_{\text{ref}}/c_{\text{ref}}$ and $\Omega c_{\text{ref}}/g$ in (3.2). Thus we may wish to analyze the system under a limit process that sends $\epsilon \to 0$ and $\delta \to 0$. If the system is nonlinear the result of the analysis will depend on the path in the ε-δ–plane. Possible paths are the sequential limits I or II (see the figure) or any "distinguished limit" such as III, where $\delta = \hat{\delta}(\epsilon) = o(1)$ as $\epsilon \to 0$, and $\hat{\delta}(\,\cdot\,)$ is an appropriate scaling function. Because of the path-dependence of the resulting asymptotic limit equations, the choice of an appropriate distinguished limit is crucial.

Notice also that distinguished asymptotic limits exist under much weaker conditions than multiple parameter expansions that consider each of the small parameters as independent. A multiple parameter expansion in effect seeks to determine the first and higher Frechét derivatives of solutions of the considered equations with respect to the set of parameters, whereas a distinguished limit corresponds to a particular version of a directional or Gateaux derivative. For functions of multiple variables it is known that only if all Gateaux derivatives are continous does the Frechét derivative exist! In this sense, the concept of a distinguished asymptotic limit is more general than that of a multiple parameter asymptotic expansion.

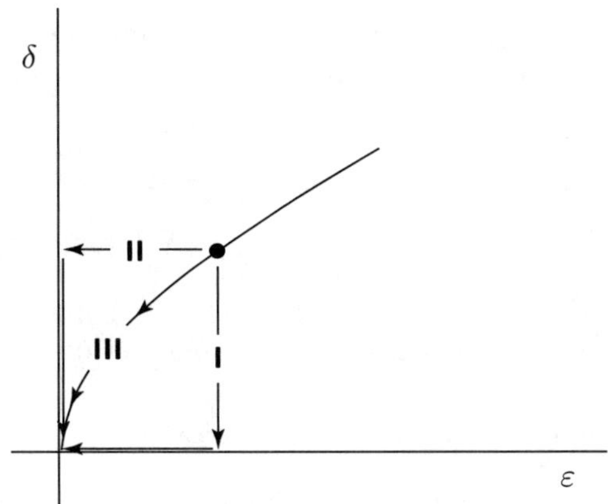

Figure 4. *Limit processes in an asymptotic system with two small parameters*

One of the somewhat surprising conclusions of the present work is that a large collection of simplified models of theoretical meteorology obtains with *one and the same* distinguished limit amongst the non-dimensional parameters from (3.2), namely

$$\pi_1 = \frac{c_{\text{ref}}}{\Omega a} = \pi_1^* \,, \qquad \pi_2 = \frac{u_{\text{ref}}}{c_{\text{ref}}} = \varepsilon^2 \,, \qquad \pi_3 = \frac{\Omega c_{\text{ref}}}{g} = \varepsilon^3 \, \pi_3^* \,, \quad \text{as} \quad \varepsilon \to 0 \,. \tag{3.3}$$

Here π_1^*, π_3^* are independent of ε, which we now adopt as the expansion parameter in subsequent formal asymptotic analyses. (By the way, the specific values shown in (3.2) suggest a range of values $\varepsilon^{-1} \approx 2\pi, \ldots, 7$ in practice.)

Dimensionless governing equations

The aerothermodynamic characteristic values from Table 3 and the ideal gas constant R from (1.2) now serve as reference quantities for non-dimensionalisation. The resulting dimensionless variables read

$$\hat{p} = \frac{p}{p_{\text{ref}}} \,, \qquad \hat{\boldsymbol{v}} = \frac{\boldsymbol{v}}{u_{\text{ref}}} \,, \qquad \hat{\rho} = \frac{\rho}{\rho_{\text{ref}}} \,, \qquad \hat{\theta} = \frac{\theta}{T_{\text{ref}}} \,, \tag{3.4}$$

where $T_{\text{ref}} = p_{\text{ref}}/(R\rho_{\text{ref}})$. Using in addition the acceleration of gravity g, we define reference length and time scales as

$$h_{\text{sc}} = \frac{p_{\text{ref}}}{g\rho_{\text{ref}}} \,, \qquad t_{\text{ref}} = \frac{h_{\text{sc}}}{u_{\text{ref}}} \,, \tag{3.5}$$

and dimensionless space and time coordinates

$$\hat{t} = \frac{t}{t_{\text{ref}}} \,. \qquad \hat{\boldsymbol{x}} = \frac{\boldsymbol{x}}{h_{\text{sc}}} \,, \qquad \hat{z} = \frac{z}{h_{\text{sc}}} \,. \tag{3.6}$$

Dropping the "hat"-indicator for dimensionless variable for convenience again, we obtain the dimensionless governing equations

$$
\begin{aligned}
\boldsymbol{v}_t + \boldsymbol{v}\cdot\nabla\boldsymbol{v} + \varepsilon\,\pi_3^*\,\boldsymbol{\Omega}\times\boldsymbol{v} + \varepsilon^{-4}\,\rho^{-1}\,\nabla p &= \boldsymbol{S}_{\boldsymbol{v}} - \varepsilon^{-4}\,\boldsymbol{k}\,,\\
p_t + \boldsymbol{v}\cdot\nabla p + \gamma p\,\nabla\cdot\boldsymbol{v} &= S_p\,,\\
\theta_t + \boldsymbol{v}\cdot\nabla\theta &= S_\theta\,.
\end{aligned}
\tag{3.7}
$$

Notice that these equations are still *equivalent* to the original dimensional equations from (1.1), since we have merely chosen a particular set of units of measurement.

Multiple scales and related asymptotic expansions

Atmospheric flow phenomena are characterised by a multitude of different length and time scales and by varying typical amplitudes of the variables $\mathbf{U} = (p, \theta, \boldsymbol{v}, \rho)$. We capture this variability through the general multiple scales asymptotic *ansatz*

$$
\mathbf{U}(t, \boldsymbol{x}, z; \varepsilon) = \sum_{i\in\mathsf{IN}} \phi^{(i)}(\varepsilon)\mathbf{U}^{(i)}\big(\tfrac{t}{\varepsilon}, t, \varepsilon t, \varepsilon^2 t, \ldots, \tfrac{\boldsymbol{x}}{\varepsilon}, \boldsymbol{x}, \varepsilon\boldsymbol{x}, \varepsilon^2\boldsymbol{x}, \ldots, \tfrac{z}{\varepsilon}, z\big)\,. \tag{3.8}
$$

Here the $\phi^{(i)}$ are asymptotic scaling functions satisfying

$$
\phi^{(i+1)}(\varepsilon) = \mathrm{o}\big(\phi^{(i)}(\varepsilon)\big) \qquad \text{as} \qquad \varepsilon \to 0\,. \tag{3.9}
$$

In this paper we will consider cases where $\phi^{(i)}(\varepsilon) = \varepsilon^i$ and/or $\phi^{(i)}(\varepsilon) = \varepsilon^{i+\frac{1}{2}}$. More general asymptotic sequences as well as more general coordinate scalings may also arise, [23, 37].

Specialisations of the tuple of arguments in (3.8) as given in Table 1 involving one coordinate each for time, and for the horizontal and vertical directions, respectively, induce standard single-scale asymptotic expansions. They allow us to re-derive a considerable collection of well-known reduced models of theoretical meteorology, and this may be considered as a "validation" of the present approach. If we retain more than one coordinate for either of the independent variables, as in $\mathbf{U}^{(i)}(t, \varepsilon t, \boldsymbol{x}, z)$, multiple scales asymptotic techniques are to be invoked. See, e.g., [45, 24] for textbook material on multi-scale asymptotics, and 4.2 below.

3.3 Physical considerations and scaling arguments

Here we provide first some physical interpretations for the scale-independent non-dimensional parameters introduced in the last section. Next we juxtapose the formal asymptotic *ansatz* proposed here with the physically motivated viewpoint of meteorological scaling analyses, and provide a "translation scheme" to mediate between the respective terminologies.

The universal scale-independent parameters

The first parameter, $\pi_1 = c_{\text{ref}}/(\Omega a)$ in (3.2) may be interpreted as a ratio of two propagation speeds. The quantity c_{ref} is of the order of the speed of sound. At the same time, it is proportional to the speed of long wavelength "barotropic", i.e., vertically homogeneous, gravity waves. The product Ωa is the absolute velocity of points on the earth's surface induced by its rotation. It is also the sun's speed over ground near the equator. The estimate $c_{\text{ref}}/\Omega a < 1$ implies that, near the equator, the sun moves at supersonic speed relative to the earth's surface. In other words: the main energy input into the atmosphere moves faster than the fastest gravity waves near the equator. (If not important, this occurs at least as an amusing observation.)

The second parameter, $\pi_2 = u_{\text{ref}}/c_{\text{ref}} =: M$, is the "Mach number". Small Mach numbers indicate that changes of the fluid density due to compression by inertial forces are small. At the same time, since c_{ref} has also been identified as the speed of barotropic gravity waves, π_2 is also the associated "Froude number".

The third parameter in (3.2) compares two characteristic length scales via the following interpretations

$$\pi_3 = \frac{\Omega c_{\text{ref}}}{g} = \frac{h_{\text{sc}}}{L_{\text{e}}} . \tag{3.10}$$

Here

$$h_{\text{sc}} = \frac{p_{\text{ref}}}{g\rho_{\text{ref}}} \approx 10\,\text{km} \tag{3.11}$$

is the *pressure scale height*; i.e., the characteristic vertical elevation over which the thermodynamic pressure of the atmosphere drops by a factor of order unity. In turn,

$$L_{\text{e}} = \frac{c_{\text{ref}}}{\Omega} \approx 3\,000\,\text{km} \tag{3.12}$$

is the *Obukhov radius* or *external Rossby deformation radius.* It is the horizontal distance that the fastest atmospheric gravity waves will travel before being influenced by the effects of earth rotation.

With these explanations, we obtain a new interpretation of our small expansion parameter ε. Using the distinguished limit from (3.3) we find

$$\frac{1}{\varepsilon} \approx \frac{\pi_2}{\pi_3} = \frac{g u_{\text{ref}}}{\Omega c_{\text{ref}}^2} = \frac{u_{\text{ref}}}{\Omega h_{\text{sc}}} =: \text{Ro}_{h_{\text{sc}}} , \tag{3.13}$$

where $\text{Ro}_{h_{\text{sc}}}$ is called the "Rossby number" based on the pressure scale height. Rossby numbers $\text{Ro}_\ell = u_{\text{ref}}/(\Omega \ell)$ formed with a reference length ℓ generally assess whether fluid motions at speed u_{ref} over distances of ℓ will or will not be affected by Coriolis effects, with small Rossby numbers indicating strong influences.

Scale-dependent non-dimensional parameters

Meteorological scale analysis obtains reduced model equations for specific flow phenomena by focusing on solutions of the full governing equations that are characterised by specific length and time scales. It is good common practice to quantify these scalings through appropriate non-dimensional characteristic numbers. In this section we demonstrate how various classical scaling relations of this type may be expressed within the present asymptotics-based framework.

As an example we choose the family of Rossby numbers with length ℓ as a free parameter; i.e.,

$$\mathrm{Ro}_\ell = \frac{u_{\mathrm{ref}}}{\Omega\,\ell}\,. \tag{3.14}$$

In the discussions of Section 2 we have seen that

$$h_{\mathrm{sc}} \ll L_1 \ll L_{\mathrm{i}} \ll L_{\mathrm{e}}\,, \tag{3.15}$$

and that each of these characteristic scales is associated with its own set of scale-dependent phenomena. In order to preserve this structure of scales and related physical effects in our asymptotic framework, we should preserve the ordering of scales from (3.15) as $\varepsilon \to 0$. In fact, a quick check of equations (3.10) and (3.13) shows that the distinguished limits from (3.3) guarantee

$$L_1 \sim \frac{1}{\varepsilon}\,h_{\mathrm{sc}}\,, \qquad L_{\mathrm{e}} \sim \frac{1}{\varepsilon^3}\,h_{\mathrm{sc}}\,. \tag{3.16}$$

We can ensure in addition that $L_1 \ll L_{\mathrm{i}} \ll L_{\mathrm{e}}$, as required in (3.15), if

$$L_{\mathrm{i}} \sim \frac{1}{\varepsilon^{\alpha+1}}\,h_{\mathrm{sc}} \quad \text{as} \quad \varepsilon \to 0 \qquad \text{with} \qquad 0 < \alpha < 2\,. \tag{3.17}$$

This, however, is not an immediate consequence of the basic order-of-magnitude estimates from Tables 2, 3. Instead, considering the definition of the internal Rossby radius of deformation,

$$L_{\mathrm{i}} \sim \frac{N}{\Omega}\,h_{\mathrm{sc}}\,, \tag{3.18}$$

equation (3.17) amounts to an asymptotic constraint for the Brunt–Väisälä-frequency N and, as a consequence, for the stratification of potential temperaturer; *viz.*,

$$N = \sqrt{\frac{g}{\theta}\frac{\partial\theta}{\partial z}} \sim \frac{1}{\varepsilon^{\alpha+1}}\,\Omega\,. \tag{3.19}$$

This constraint should be observed in subsequent asymptotic expansions schemes. Fortunately, it is also compatible with evidence: Typical values for the Brunt–Väisälä-frequency in the troposphere are $0.01\,\mathrm{s}^{-1}$, and $0.1\,\mathrm{s}^{-1}$ in the stratosphere, which leads to $\alpha \approx 1$ for the troposphere, and $\alpha \approx 2$ for the stratosphere, respectively, when $\varepsilon \approx \frac{1}{7}$. However, up to now we have not found an intrinsic argument associated, e.g., with the radiation balance or with the transport of humidity, that would reveal the mechanisms responsible for establishing these orders of magnitude.

Given these qualifications, the Rossby numbers characterising flow regimes associated with length scales $h_{\mathrm{sc}}, L_1, L_{\mathrm{i}}, L_{\mathrm{e}}$, will have the following asymptotic scalings,

$$\begin{aligned} \mathrm{Ro}_{h_{\mathrm{sc}}} &= O(\varepsilon)\,,\\ \mathrm{Ro}_{L_1} &= O(1)\,,\\ \mathrm{Ro}_{L_{\mathrm{i}}} &= O(\varepsilon^{\alpha})\,,\\ \mathrm{Ro}_{L_{\mathrm{e}}} &= O(\varepsilon^2)\,. \end{aligned} \tag{3.20}$$

In the same fashion, one may relate ε to other scale-dependent dimensionless parameters that characterise particular flow regimes.

Coordinates for specific spatio-temporal scales

Here we explain how one can set up an asymptotic expansion in the present framework to study flows with a given characteristic length scale. As an example, we consider flows with characteristic scale L_{i}, the internal Rossby radius of deformation. The flow shall evolve on the characteristic advection time scale for this length, $t_{\mathrm{ref}} = L_{\mathrm{i}}/u_{\mathrm{ref}}$, and engulf the entire depth, h_{sc}, of the troposphere. In a meteorological scale analysis the appropriate non-dimensionalisation of time and of the horizontal and vertical space coordinates would read

$$\tau = \frac{t' u_{\mathrm{ref}}}{L_{\mathrm{i}}}, \qquad \boldsymbol{\xi} = \frac{\boldsymbol{x}'}{L_{\mathrm{i}}}, \qquad z = \frac{z'}{h_{\mathrm{sc}}}, \tag{3.21}$$

where primes denote the original dimensional coordinates. The small non-dimensional parameter justifying the perturbation analysis would be the Rossby number based on the internal Rossby deformation radius L_{i}; i.e.,

$$\mathrm{Ro}_{L_{\mathrm{i}}} = \frac{u_{\mathrm{ref}}}{\Omega L_{\mathrm{i}}} \approx \tfrac{1}{10}, \ldots, \tfrac{1}{5} \ll 1\,. \tag{3.22}$$

We would then seek solutions of the form

$$\mathbf{U} = \sum_i \mathrm{Ro}_{L_{\mathrm{i}}}^i \mathbf{U}^{(i)}(\tau, \boldsymbol{\xi}, z)\,. \tag{3.23}$$

How would we go about analyzing the same situation within the present asymptotic framework? Suppose we chose the following non-dimensional coordinates in (3.8)

$$t = \frac{t' u_{\mathrm{ref}}}{h_{\mathrm{sc}}}, \qquad \boldsymbol{x} = \frac{\boldsymbol{x}'}{h_{\mathrm{sc}}}, \qquad z = \frac{z'}{h_{\mathrm{sc}}}\,. \tag{3.24}$$

From (3.17) we conclude that the dimensionless time and horizontal coordinates used in the scale analysis expansion (3.23) may be expressed as

$$\tau = \frac{t' u_{\mathrm{ref}}}{h_{\mathrm{sc}}} = \varepsilon^{\alpha+1} t\,, \qquad \boldsymbol{\xi} = \frac{\boldsymbol{x}'}{L_{\mathrm{i}}} = \varepsilon^{\alpha+1} \boldsymbol{x}\,, \tag{3.25}$$

while the vertical coordinate, z, remains unchanged. Furthermore, from (3.20) we have

$$\mathrm{Ro}_{L_{\mathrm{i}}} \sim \varepsilon^{\alpha} \tag{3.26}$$

so that (3.23) may be "translated" to

$$\mathbf{U}(t, \boldsymbol{x}, z; \varepsilon) = \sum_i \left(\varepsilon^{\alpha}\right)^i \mathbf{U}^{(i)} \left(\varepsilon^{\alpha+1} t, \varepsilon^{\alpha+1} \boldsymbol{x}, z\right). \tag{3.27}$$

Analogous "translation schemes" follow immediately when the relevant scales in the scale analysis approach can be expressed in terms of the length, time, and velocity scales considered here; i.e., in terms of $(h_{\mathrm{sc}}, L_1, L_{\mathrm{i}}, L_{\mathrm{e}}, a)$, (N, Ω), and $(u_{\mathrm{ref}}, c_{\mathrm{ref}})$.

There are situations, however, for which this condition is not satisfied. Consider, e.g., the flow over a hill of height $h \approx 500\,\mathrm{m}$, and horizontal extent $L \approx 3\,\mathrm{km}$.

These scales are now imposed by orographical features rather than being directly related to the abovementioned basic scales. As a consequence, two new dimensionless parameters need to be accounted for. Without loss of generality we may choose the ratios

$$\frac{h}{h_{\text{sc}}} \qquad \text{and} \qquad \frac{h}{L} \tag{3.28}$$

for this example. Both these parameters are small and may motivate perturbation analyses.

To cover such a case within the present scheme, one will have to decide upon the path to the origin in the space of parameters $\varepsilon, h/h_{\text{sc}}, L/h_{\text{sc}}$, as discussed in the context of Figure 4. For example, the concrete values for h and L given above would be compatible with

$$\frac{h}{h_{\text{sc}}} \approx 2 \times 10^{-1} = \mathrm{O}(\varepsilon) \qquad \text{and} \qquad \frac{h}{L} \approx 0.6 \times 10^{-1} = \mathrm{O}(\varepsilon)\,. \tag{3.29}$$

With these distinguished limits, we can proceed via asymptotic expansions in ε based on the following set of coordinates for the perturbation functions

$$\mathbf{U}^{(i)}\big(t, \boldsymbol{x}, \tfrac{z}{\varepsilon}, z\big)\,. \tag{3.30}$$

Here the coordinate z/ε will resolve flow structures comparable in height with the hill itself. The coordinates $\boldsymbol{x}, z$ cover structures comparable with the horizontal extension of the hill as well as farfield effects in the vertical direction. A comprehensive discussion of flow effects induced by orography, surface roughness, and surface heat fluxes is work in progress with J. C. R. Hunt and A. Owinoh (see, e.g., [23, 37]).

3.4 "Pro's and Con's" of the present approach

Pro's

1. The proposed approach directly relates any derived reduced models to the three-dimensional compressible flow equations in a transparent fashion.

2. There is a clear-cut mathematical route through each of the derivations. These have a common formal justification from the identification of a small number of universal, scale-independent small parameters and a related distinguished limit.

3. The approach prepares the ground for further studies of multiple scales interactions; see Section 4.2.

4. Item 1 makes the present approach a natural starting point for developments of numerical methods for the compressible flow equations applicable to the various singular limit regimes of atmospheric flows; see Section 4.3.

5. The identification of a nearly universal small parameter $\varepsilon \approx \frac{1}{7}, \dots, \frac{1}{6}$ provides an independent measure, besides empirical tests, for the regimes of validity of reduced model equations. Models derived for a characteristic length and time

scales, L, T, (via scale analysis or via the present approach) loose credibility when the actual scales of a flow field considered cover a range that is comparable or larger than $\varepsilon L, \dots, L/\varepsilon, \varepsilon T, \dots, T/\varepsilon$. In such cases one must, in general, account for the possibility of scale interactions using multple scales techniques. We notice that $\varepsilon \approx \frac{1}{7}, \dots, \frac{1}{6}$ is not extremely small in practice, so that the mentioned scale ranges are limited! On the other hand, reduced models *may* be applicable beyond the assessed ranges, but this requires additional empirical or theoretical justification.

Con's

1. The clear-cut mathematical route provided here may mislead newcomers into neglecting the physical and empirical bases for model reductions. It is relatively easy to come up with formal asymptotic derivations for all sorts of scales; the art is to identify, through careful physical considerations, those regimes that are of practical importance.

2. If we have managed to derive a particular reduced model via the present route, employing some particular distinguished limit and argument scalings. Then by no means does this guarantee that the adopted limit process is the only one leading to this model. In our example from Figure 4 this could mean that the paths II and III, and any in between, yield the same result. Quantification of this degree of freedom requires additional analyses not discussed here.

4 Applications

Here we discuss three instructive applications in which the present approach proved useful. We will re-derive the well-known semi-geostrophic theory using multiple scales techniques in Section 4.1, consider systematic multiple scales models from [30] in Section 4.2, and discuss recent developments of a "well-balanced" numerical method for nearly hydrostatic compressible flows in Section 4.3.

4.1 Semi-geostrophic theory

Weather fronts are examples for atmospheric flow structures that feature anisotropic horizontal scalings. When viewed from above, one may think of a front as being a narrow band of activity centered about some smooth, large scale curve. All flow variables are expected to vary substantially over relatively short distances normal to the front, while they vary on scales comparable to the characteristic length of the curve's geometry in the tangential direction.

An appropriate ansatz to capture this structural behavior in an asymptotic analysis, borrowed from WKB theories, geometrical optics/acoustics, or the theory of thin flames in combustion, could read

$$\mathbf{U}(t, \boldsymbol{x}, z; \varepsilon) = \sum_i \mathbf{U}^{(i)}\left(\varepsilon^2 t, \frac{\Phi(\varepsilon^2 \boldsymbol{x})}{\varepsilon}, \varepsilon^2 \boldsymbol{x}, z\right). \tag{4.1}$$

Here $\Phi(\,\cdot\,)$ is a scalar function, which by the scaling of its argument, $\varepsilon^2\boldsymbol{x}$, in (4.1) will have a characteristic scale comparable to the internal Rossby deformation radius $L_{\rm i}$ (see the previous section). The scaled coordinate $\xi = \Phi/\varepsilon$ will then resolve rapid variations of the solution that occur between levelsets $\Phi = \Phi_0 + {\rm O}(\varepsilon)$. The front will thus be centered about the level set $\Phi(\varepsilon^2\boldsymbol{x}) \equiv \Phi_0$ and have a characteristic thickness of order $\ell = {\rm O}(\varepsilon L_{\rm i})$. According to (3.20), this is the characteristic length L_1 for which the Rossby number would be of order unity.

With the abbreviations

$$\tau = \varepsilon^2 t, \qquad \xi = \frac{\Phi(\varepsilon^2\boldsymbol{x})}{\varepsilon}, \qquad \boldsymbol{X} = \varepsilon^2\boldsymbol{x}\,, \tag{4.2}$$

the partial derivatives in the governing equations (1.1) become

$$\begin{aligned} \partial_t &= \varepsilon^2\partial_\tau \\ \nabla &= \varepsilon\,\sigma\,\boldsymbol{n}\,\partial_\xi + \varepsilon^2\;\nabla_{\boldsymbol{X}} + \boldsymbol{k}\,\partial_z\,. \end{aligned} \tag{4.3}$$

where $\sigma = |\nabla_{\boldsymbol{X}}\Phi|$ and $\boldsymbol{n} = \nabla_{\boldsymbol{X}}\Phi/|\nabla_{\boldsymbol{X}}\Phi|$. (We have left the complications of spherical geometry aside here, as they are not relevant to the main message of this section.)

A somewhat subtle issue in semi-geostrophic theory is the scaling of the velocity components. It is *assumed* that they scale in proportion with the characteristic length for their associated spatial direction. Thus one allows only the flow velocity tangential to the front to be of order $u_{\rm ref}$; i.e., $\boldsymbol{v}\cdot\boldsymbol{t} = {\rm O}(u_{\rm ref})$, where $\boldsymbol{t}(\boldsymbol{X})$ is the tangential unit vector to a level set $\Phi =$ const.. For the normal and vertical components this assumption implies $\boldsymbol{v}\cdot\boldsymbol{n} = {\rm O}(\varepsilon u_{\rm ref})$, and $\boldsymbol{v}\cdot\boldsymbol{k} = {\rm O}(\varepsilon^2 u_{\rm ref})$. Accordingly, the velocity field is represented as

$$\boldsymbol{v} = v^{(0)}\,\boldsymbol{t} + \varepsilon\big(v^{(1)}\,\boldsymbol{t} + u^{(1)}\,\boldsymbol{n}\big) + \varepsilon^2\big(v^{(2)}\,\boldsymbol{t} + u^{(2)}\,\boldsymbol{n} + w^{(2)}\,\boldsymbol{k}\big) + \dots \tag{4.4}$$

It is now straight-forward to insert the (4.1)–(4.4) into the governing equations from (1.1) and to collect the following leading order set of equations. Using the replacements

$$\begin{aligned} \big(\theta^{(4)}, \tfrac{p^{(4)}}{\rho^{(0)}}\big) &\;\rightarrow\; (\theta, \pi) \\ \big(u^{(1)}, v^{(0)}, w^{(2)}\big) &\;\rightarrow\; (u, v, w) \\ \big(S_\theta^{(6)}, S_v^{(2)}\big) &\;\rightarrow\; (S_\theta, S_v) \end{aligned} \tag{4.5}$$

we obtain the semi-geostrophic approximation (see [39], section 8.4 for the case of an incompressible fluid)

$$
\begin{aligned}
-\,f\,v + \sigma\,\frac{\partial \pi}{\partial \xi} &= 0\\
\frac{\mathrm{D}v}{\mathrm{D}\tau} + f\,u^{(1)} + \frac{\partial \pi}{\partial \eta} &= S_v\\
\frac{\partial \pi}{\partial z} - \theta &= 0\\
\frac{\mathrm{D}\theta}{\mathrm{D}\tau} &= S_\theta\\
\frac{\partial u}{\partial \xi} + \frac{\partial v}{\partial \eta} + \frac{1}{\rho_0}\frac{\partial}{\partial z}(\rho_0 w) &= 0\,,
\end{aligned}
\tag{4.6}
$$

Here $\rho_0(z)$ is the background density stratification corresponding to a homentropic atmosphere,

$$
\frac{\mathrm{D}}{\mathrm{D}\tau} = \frac{\partial}{\partial \tau} + v\frac{\partial}{\partial \eta} + u\,\sigma\,\frac{\partial}{\partial \xi} + w\,\frac{\partial}{\partial z}\,,
\tag{4.7}
$$

and

$$
\frac{\partial}{\partial \eta} = \boldsymbol{t}\cdot\nabla_{\boldsymbol{X}}\,.
\tag{4.8}
$$

Lower order expansion functions, such as $\theta^{(2)}, \theta^{(3)}$ and $p^{(2)}, p^{(3)}$, which do not explicitly appear here, can be shown not to participate in the dynamics within a narrow front if the assumed time scaling is to be observed.

The semi-geostrophic equations are used in a range of contexts, including theories for the formation and structure of strong weather fronts. The key difference between these and the quasi-geostrophic equations discussed in Section 2 is that only one of the horizontal velocity components, v, is in geostrophic balance. The velocity component normal to the front is, in contrast, the result of a balance between the Coriolis force, the pressure gradient, and the fluid particle acceleration along the front. The semi-geostrophic equations have, however, several attractive mathematical features, which are reviewed and extensively discussed, e.g., in [22, 11, 42, 27]. One of these is the surprising fact that, by an ingenious nonlinear change of variables developed in [14, 21], and omission of *one* term they can be transformed into the quasi-gestrophic equations discussed in Section 2.2.

Possible extensions Having re-derived another classical result within the present modelling framework, one may ask where to go next in the context of semi-geostrophic flows. Here are two possible lines of thought:

- We observe that analyses involving anisotropic scalings in the vicinity of a narrow front also occur; e.g., in the context of thin-flame fronts in combustion. There, one is interested in (i) approximate equations governing the internal flame structure, and (ii) a flame propagation law that describes the flame's motion in space as time evolves. A slight modification of the asymptotic ansatz in (4.1) would allow us to transfer these ideas into the present context. One would introduce an explicitly time dependent level set function $\Phi(\varepsilon^2 t, \varepsilon^2 \boldsymbol{x})$ to define the front-resolving coordinate, and aim at extracting an evolution equation for this function by asymptotic matching with the surrounding flow.

- Another interesting option is to consider a faster time scale by allowing for an additional time coordinate εt in the asymptotic ansatz. This will lead to a multiple time variable expansion similar to [15, 16, 29], where the authors considered the averaged effects of fast gravity waves passing over quasi-gestrophical mean flows. It has been argued that strong weather fronts in fact undergo fast oscillations that affect their overall behavior non-trivially, and this problem could be investigated with the proposed modification of the analysis.

4.2 Synoptic-planetary interactions in the tropics

Here we summarise recent joint work with A. J. Majda addressing scale interactions in the tropics. A hierarchy of reduced model equations describing a range of possible flow regimes is derived in [30] using systematic multiple scales expansions. One particularly interesting regime involves interactions between the equatorial synoptic and the planetary scales. Here we review the flow regime and the key results of the analysis for this regime. For details the reader may consult the original reference.

Expansion scheme and scaling considerations

The multiple scales expansion scheme for this regime reads

$$\Phi(t, \boldsymbol{x}, z; \varepsilon) = \varepsilon^{\alpha(\Phi)} \sum_i \varepsilon^i \, \Phi^{(i)}(\varepsilon^{\frac{5}{2}} t, \varepsilon^{\frac{5}{2}} \boldsymbol{x}, \varepsilon^{\frac{7}{2}} x, z) \,, \tag{4.9}$$

where $\alpha = 0$ for $\Phi \in \{p, \theta, \rho, \boldsymbol{u}\}$, and $\alpha = \frac{1}{2}$ for $\Phi = w, S_\theta, S_v$. Here $\boldsymbol{u}, w$ are the horizontal and vertical flow velocity components, respectively, and the coordinates $\boldsymbol{x} = (x, y)$ denote the zonal (along the equator) and meridional (north-south) horizontal coordinates. As we will see shortly, this scheme merges the single scale expansion for equatorial geostrophic motions,

$$\Phi(t, \boldsymbol{x}, z; \varepsilon) = \varepsilon^{\alpha(\Phi)} \sum_i \varepsilon^i \, \Phi^{(i)}(\varepsilon^{\frac{5}{2}} t, \varepsilon^{\frac{5}{2}} \boldsymbol{x}, z) \,, \tag{4.10}$$

with a scheme that resolves planetary scale equatorial waves

$$\Phi(t, \boldsymbol{x}, z; \varepsilon) = \varepsilon^{\alpha(\Phi)} \sum_i \varepsilon^i \, \Phi^{(i)}(\varepsilon^{\frac{5}{2}} t, \varepsilon^{\frac{7}{2}} x, \varepsilon^{\frac{5}{2}} y, z) \,. \tag{4.11}$$

The characteristic horizontal length scale accessed by the first scheme in (4.10) is the internal Rossby deformation radius for near-equatorial flows. To verify this, we reconsider the relation $L_{\mathrm{i}} \sim N h_{\mathrm{sc}} / \Omega$ from (3.18). Near the equator, the earth rotation frequency Ω must be replaced with characteristic value of the vertical component $f = \boldsymbol{k} \cdot \boldsymbol{\Omega} = \Omega \sin(\phi)$ of the earth rotation vector, where ϕ is the longitude. In terms of the arclength in the meridional direction, y, we have $\phi = y/a$, where a is the earth's radius. Anticipating that $L_{\rm i} \ll a$, which remains to be verified, we have $\sin(\phi) \sim L_{\rm i}/a$, and $f \sim \beta L_{\rm i}$, where $\beta = \Omega/a$. As a consequence,

$$L_{\rm i} \sim \frac{N}{\beta L_{\rm i}} h_{\rm sc}\,, \qquad \text{or} \qquad \frac{L_{\rm i}}{h_{\rm sc}} \sim \sqrt{\frac{N}{\Omega}\,\frac{a}{h_{\rm sc}}} \sim \varepsilon^{-\frac{5}{2}}\,. \tag{4.12}$$

The last estimate follows from equations (3.19) with $\alpha = 1$, and (3.16), which stated that $N/\Omega \sim \varepsilon^{-2}$ and $a/h_{\rm sc} \sim \varepsilon^{-3}$. We verify that $L_{\rm i} \ll a$, even though the difference is merely of order $\varepsilon^{\frac{1}{2}}$. Equation (4.12) shows that the scaling anticipated in (4.9) and (4.10) in fact accesses the internal Rossby deformation radius $L_{\rm i}$.

The second expansion scheme in (4.11) accesses even larger (planetary) scales via the coordinate

$$X_{\rm P} = \varepsilon^{\frac{7}{2}} x = \frac{x'}{L_{\rm p}} \qquad \text{with} \qquad L_{\rm p} = \varepsilon^{-\frac{7}{2}} h_{\rm sc}\,. \tag{4.13}$$

Notice that this expansion assumes anisotropic scalings along and normal to the equator. This is compatible with theories of equatorial wave motions which reveal confinement of near-tropical dynamics between about $-30°$ and $30°$ degrees latitude; see e.g., [18, 28].

The non-dimensional time coordinate in (4.9)–(4.11) may be re-written in two different ways:

$$\varepsilon^{\frac{5}{2}} t = \varepsilon^{\frac{5}{2}} \frac{t'}{h_{\rm sc}/u_{\rm ref}} = \frac{t'}{L_{\rm i}/u_{\rm ref}} = \frac{t'}{L_{\rm p}/c_{i\,\rm ref}}\,, \tag{4.14}$$

where t' again denotes the dimensional time variable, and $c_{i\,\rm ref} \sim N h_{\rm sc}$ is the characteristic speed of internal gravity waves with vertical scale comparable to $h_{\rm sc}$. This dual representation shows that, on the one hand, the chosen time coordinate resolves advection with the flow velocity $u_{\rm ref}$ over synoptic distances comparable to $L_{\rm i}$. On the other hand, it also resolves internal gravity waves travelling at speeds $c_{i\,\rm ref}$ over planetary distances of the order $L_{\rm p}$. As a consequence, the multiple scales expansion from (4.9) is suited to describe direct interactions of these very different phenomena on one and the same time scale.

Intra-Seasonal Planetary Equatorial Synoptic Dynamics (IPESD)

To simplify the notation, we will use the following replacements in the rest of this section:

$$\begin{aligned}
\left(\theta^{(2)}, \theta^{(3)}, \tfrac{p^{(3)}}{\rho^{(0)}}, \tfrac{p^{(4)}}{\rho^{(0)}}\right) &\;\to\; (\Theta_2, \theta, \pi, \pi') \\
\left(\boldsymbol{u}^{(0)}, \boldsymbol{u}^{(1)}, w^{(2)}, w^{(3)}\right) &\;\to\; (\boldsymbol{u}, \boldsymbol{u}', w, w') \\
\left(S_\theta^{(4)}, S_\theta^{(5)}, \boldsymbol{S}_{\boldsymbol{u}}^{(1)}, \boldsymbol{S}_{\boldsymbol{u}}^{(2)}\right) &\;\to\; (S_\theta, S'_\theta, \boldsymbol{S}_{\boldsymbol{u}}, \boldsymbol{S}'_{\boldsymbol{u}}) \\
\left(\varepsilon^{\frac{5}{2}} t, \varepsilon^{\frac{5}{2}} \boldsymbol{x}, \varepsilon^{\frac{7}{2}} x\right) &\;\to\; (T_{\rm S}, \boldsymbol{X}_{\rm S}, X_{\rm P})\,,
\end{aligned} \tag{4.15}$$

where $\boldsymbol{X}_{\rm S} = \big(X_{\rm S}, Y_{\rm S}\big)$.

Synoptic motions With the abbreviations from (4.15), the leading order set of equations describing motions on the smaller of the considered scales (which is still as large as 2000 km) reads

$$\begin{aligned} \partial_z \pi &= \theta \\ w \frac{\mathrm{d}\Theta_2}{\mathrm{d}z} &= S_\theta \\ -\beta y\, v + \partial_x \pi &= S_u \\ \beta y\, u + \partial_y \pi &= S_v \\ \partial_x(\rho_0 u) + \partial_y(\rho_0 v) + \partial_z(\rho_0 w) &= 0\,. \end{aligned} \tag{4.16}$$

These equations describe, in this sequence, hydrostatic balance in the vertical direction, the generation of vertical motions by heat sources forcing particles to move towards their individual levels of neutral buoyancy, horizontal geostrophic balance in the zonal (x) and meridional (y) directions, and mass conservation.

The equations in (4.16) may be considered as the three-dimensional version of the Matsuno-Webster-Gill type of models [34, 51, 17] (see also [18], section 11.14, [35]), who derived models for steady forced synoptic scale motions near the equator in the context of the shallow water approximation (see also section 2 in [30]). Gill demonstrates in [17, 18] that these equations reproduce typical quasi-steady large scale near-equatorial flow patterns when some physically reasonable closures for the source terms are assumed. His examples include qualitative models for the important "Hadley" and "Walker" circulations.

For given source terms S_θ, S_u, S_v the system in (4.16) is linear in u, v, w, π, θ. General solutions, in this case, are superpositions of particular and homogeneous solutions. One particular solution $u^p, v^p, w^p, \pi^p, \theta^p$ is determined by, [30],

$$w^p = \frac{S_\theta}{\mathrm{d}\Theta_2/\mathrm{d}z}\,, \qquad v^p = \frac{1}{\beta}\left(S_{v,x} - S_{u,y}\right) + \frac{y}{\rho_0}\left(\frac{\rho_0 S_\theta}{\mathrm{d}\Theta_2/\mathrm{d}z}\right)_z\,, \tag{4.17}$$

$$\partial_x u^p = -\partial_y v^p - \frac{1}{\rho_0}\,\partial_z\left(\rho_0 w^p\right) \qquad \text{with} \qquad \beta y\, \overline{u^p} = \overline{S_v}\,, \tag{4.18}$$

$$\left.\begin{aligned} \partial_x \pi^p &= S_u - \beta y\, v^p \\ \partial_y \pi^p &= S_v + \beta y\, u^p \end{aligned}\right\} \qquad \text{with} \qquad \overline{\pi^p}(t, 0, z) \equiv 0\,, \tag{4.19}$$

$$\theta^p = \partial_z \pi^p\,. \tag{4.20}$$

Homogeneous solutions to (4.16) satisfy,

$$v = w \equiv 0, \qquad (\theta, u, \pi) = \big(\Theta, U, P\big)(t, y, z)\,, \tag{4.21}$$

where Θ, U, P are arbitrary except for the constraints

$$\partial_y P = -\beta y\, U\,, \qquad \partial_z P = \Theta\,. \tag{4.22}$$

A few remarks regarding this synoptic scale model and the above particular solution are in order:

- With the present results we could extend Table 1 by adding the line

  $\mathbf{U}^{(i)}\big(\varepsilon^{\frac{5}{2}}t, \varepsilon^{\frac{5}{2}}x, \varepsilon^{\frac{5}{2}}y, z\big)$ Forced quasi-steady equatorial synoptic motions

  described by quasi-steady "Matsuno-Webster-Gill"-models, [34, 51, 17, 18].

- The equations incorporate a mechanism for the generation of large scale horizontal flow divergence from energy source terms. Heat addition through S_θ induces non-zero vertical velocities, which in turn drive horizontal flow divergences through the continuity of mass. In fact,

$$u_x + v_y = -\frac{1}{\rho_0}\left(\frac{\rho_0 S_\theta}{\mathrm{d}\Theta_2/\mathrm{d}z}\right)_z .$$

  This mechanism is common to the family of "WTG" models, [9, 20, 6, 5, 48, 30], which are currently being used extensively in the context of tropical meteorology. These models are often derived on the basis of the assumption of **W**eak horizontal **T**emperature **G**radients, and hence the acronym. Often, these models do not explicitly resolve all three space dimensions, but rather utilise Galerkin-type projections onto the dominant vertical flow modes to obtain reduced, effectively two-dimensional models, [17, 50, 36, 32]. This kind of further reduction could, of course, be applied here as well. We do not discuss this aspect here, as this step has no direct justification within the present asymptotic framework.

- There is a consistency condition on the spatial structure of the source terms. It is obtained by averaging the first equation in (4.19) with respect to x, and using the expression for v^p from (4.17),

$$\left(\tfrac{1}{y}\,\overline{S_u}\right)_y = \frac{\beta}{\rho_0}\left(\frac{\rho_0 \overline{S_\theta}}{\mathrm{d}\Theta_2/\mathrm{d}z}\right)_z . \tag{4.23}$$

  When the source terms S_u, S_θ are considered to be given explicitly, then this is merely a constraint to be be observed. If these source terms are, however, replaced with "parameterisations" of various net effects induced by small scale motions not resolved by the present model, the above constraint attains a quite different meaning. Generally, the source terms will then be functions of (π, θ, u, v, w), and the consistency condition will introduce a functional relationship between these variables *that is entirely determined by the chosen parameterisations*! This corroborates the often-voiced warning that adequate parameterisations are at least as important for the success of a model as its fluid dynamical consistency.

From here on $u^p, v^p, w^p, \pi^p, \theta^p$ denote the particular solutions (4.17) to (4.20), and the source terms S_u, S_v, S_θ are assumed to satisfy (4.23).

Planetary waves As pointed out in conjunction with (4.21), the equation system for the synoptic scales determines solutions only up to a zonal (along the equator) shear flow. In the present context of a multiple scales expansion, this zonal shear flow may still not depend on x, but it may well depend on the planetary scale coordinate X_{P}, and multiple scales asymptotic techniques should allow us to derive evolution equations for these large scale averaged mean motions. In fact, the following system is derived in [30],

$$\begin{aligned}
\overline{\mathrm{D}_t^p}\, U + \partial_X P - \beta\, y\, \overline{v'} &= \overline{S'_u} - \overline{\mathrm{D}_t^p\, u^p} \\
\overline{\mathrm{D}_t^p}\, \Theta + \overline{w'}\, \frac{\mathrm{d}\Theta_2}{\mathrm{d}z} &= \overline{S'_\theta} - \overline{\mathrm{D}_t^p\, \theta^p} \\
\beta\, y\, U + \partial_y P &= 0 \\
\partial_z P &= \Theta \\
\partial_X U + \partial_y \overline{v'} + \frac{1}{\rho_0}\, \partial_z (\rho_0 \overline{w'}) &= 0\,,
\end{aligned} \tag{4.24}$$

where

$$\begin{aligned}
\mathrm{D}_t^p &= \partial_t + u^p\, \partial_x + v^p\, \partial_y + w^p\, \partial_z \\
\overline{\mathrm{D}_t^p} &= \partial_t + \overline{v^p}\, \partial_y + \overline{w^p}\, \partial_z\,.
\end{aligned} \tag{4.25}$$

These equations are the three-dimensional analogue of the linear equatorial long wave equations, [19], supplemented with the net large scale effects of the synoptic scale transport as represented by the terms, $\overline{\mathrm{D}_t^p\, u^p}, \overline{\mathrm{D}_t^p\, \theta^p}$, and with advection by the mean synoptic meridional velocity $\overline{v^p}$, see also [30]. Again, a few general remarks are in order:

- If we restrict here to very weak source terms, so that $S_u = S_v = S_\theta \equiv 0$ and, as a consequence, $(u^p, v^p, w^p, \theta^p, \pi^p) \equiv 0$, then the operators $\mathrm{D}_t^p, \overline{\mathrm{D}_t^p}$ reduce to the partial time derivative ∂_t, and the only source terms left in (4.24) are the perturbation source terms $\overline{S'_u}$, $\overline{S'_\theta}$.

  The equations then describe the generation of linear equatorial long waves by weak momentum and energy sources, as in an inhomogeneous Heckley–Gill model, [19]. We have thus successfully added another line to Table 1,

  $$\mathbf{U}^{(i)}\big(\varepsilon^{\frac{5}{2}} t, \varepsilon^{\frac{7}{2}} x, \varepsilon^{\frac{5}{2}} y, z\big)\,. \quad \text{Linear equatorial long-wave motions}$$

- The equatorial synoptic scale may be estimated as $L_{\text{i}} \approx 2000\,\text{km}$. Clearly, with $\varepsilon \approx \frac{1}{7}$, a length scale of order L_{i}/ε must be expected to participate in the equatorial dynamics. Single scale expansions that do not acknowledge the presence of these asymptotically separated scales, will not be able to describe the scale interactions revealed in this section.

4.3 Balancing numerical methods for nearly hydrostatic motions

This section summarises a quite different development that was motivated by our asymptotic considerations for atmospheric flows. In [3] we address a nagging numerical issue associated with the (asymptotically) dominant balance between pressure forces and the gravitational acceleration in most realistic atmosphere flow regimes.

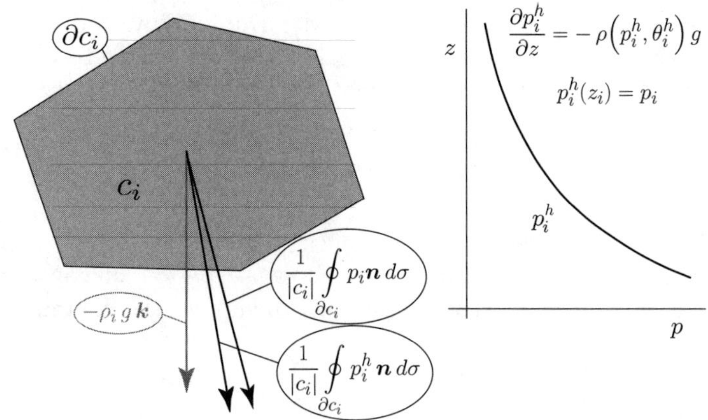

Figure 5. *Archimedes' principle for the gravity source term.*

Consider the vertical component of the momentum balance in the dimensionless governing equations from (3.7); i.e.,

$$w_t + \boldsymbol{u} \cdot \nabla w + w w_z + \varepsilon\, \pi_3^*\, \boldsymbol{k} \cdot (\boldsymbol{\Omega} \times \boldsymbol{v}) + \varepsilon^{-4} \left(\rho^{-1} \frac{\partial p}{\partial z} + 1\right) = S_w\,. \tag{4.26}$$

Suppose further that we had adopted a numerical discretisation of all terms that is second order accurate in terms of the (vertical) space discretisation parameter $h = \Delta z / h_{\text{sc}}$. Then the numerical truncation error induced by the two terms in (4.26) multiplied by $1/\varepsilon^4$ would read

$$\delta_{\text{num.}} = \mathrm{O}\Big(\frac{h^2}{\varepsilon^4}\Big)\,. \tag{4.27}$$

Current production runs with numerical weather forecasting or climate simulation codes use around 30 grid layers to resolve the vertical direction, so that $h \approx 1/30$. Our earlier estimates indicate that $\varepsilon \approx \frac{1}{7}, \dots, \frac{1}{6}$, so that

$$\frac{h^2}{\varepsilon^4} = \mathrm{O}(1) \tag{4.28}$$

under realistic conditions for up-to-date computational simulations of atmospheric flows. We conclude that the computed vertical accelerations will be highly inaccurate, unless special measures are taken to ensure that the limiting situation of hydrostatic balance is adequately captured by the numerical discretisation.

Construction principle for a well-balanced scheme

In [3] the authors propose a quite generally applicable remedy for this problem, and specifically describe an implementation in the context of conservative finite volume compressible flow solvers. The key ideas involved are as follows.

Archimedes' principle for the gravity source term: Figure 5 displays a general control volume that might serve as the ith grid cell c_i for our finite volume method. A straight-forward second order approximation of the gravity source term for such a volume would read

$$\int_{c_i} \rho \boldsymbol{k} \, \mathrm{d}^2 x = |c_i| \, \rho_i \boldsymbol{k} + \mathrm{O}(\delta^2) \tag{4.29}$$

where ρ_i is the computed cell-averaged density and $|c_i|$ is the cell's volume, and $\delta = \operatorname{diam} c_i$ is its characteristic diameter. In the context of conservative finite volume methods the integral of the pressure gradient over the volume is discretised as

$$\int_{c_i} \nabla p \, \mathrm{d}^2 x = \oint_{\partial c_i} p \boldsymbol{n} \, \mathrm{d}\sigma = \sum_j A_j p_j \boldsymbol{n}_j + \mathrm{O}(\delta^2) \tag{4.30}$$

where A_j, p_j, $\boldsymbol{n}_j$ are the length (area) of the jth section of the control volume's boundary, and approximations to the pressure and outward unit normal vector on that cell interface section. As indicated, these approximations are second order accurate in terms of δ. But this alone is insufficient to properly capture the hydrostatic limit, because the magnitude of the truncation errors is independent of whether the flow state is or is not close to hydrostatic.

To overcome this difficulty, see [3], we observe that the vector $-\rho \boldsymbol{k}$ may be interpreted as the gradient of a virtual hydrostatic pressure distribution; i.e.,

$$\rho \boldsymbol{k} = -\nabla p^h \, . \tag{4.31}$$

The gravity source term may then be rewritten as

$$\int_{c_i} \rho \boldsymbol{k} \, \mathrm{d}^2 x = -\oint_{\partial c_i} p^h \boldsymbol{n} \, \mathrm{d}\sigma = -\sum_j A_j p_j^h \boldsymbol{n}_j + \mathrm{O}(\delta^2) \, . \tag{4.32}$$

As a consequence, the sum of the gravity source term and the pressure gradient is discretised as

$$\int_{c_i} (\nabla p + \rho \boldsymbol{k}) \, \mathrm{d}^2 x = \sum_j A_j (p_j - p_{i,j}^h) \boldsymbol{n}_j + \mathrm{O}(\delta^2 \|p - p_i^h\|) \, , \tag{4.33}$$

where the $p_{i,j}^h = p_i^h(z_j)$ are values of a locally reconstructed virtual hydrostatic pressure distribution evaluated at the gric cell interface centers. (For the cell-wise construction of $p_i^h(z)$ see below.) This modification will not change the approximation order of the scheme, which is still of second order for suitable formulations of the hydrostatic pressure distribution $p_i^h(z)$. Yet, we observe that the numerical approximate expression for the sum of the two terms now vanishes identically when the p_j match the $p_{i,j}^h$; i.e., when the pressure is hydrostatic in the sense of p_i^h. Even if the approximate construction of p_i^h is inexact, so that $\|p - p_i^h\| \not\to 0$ as the hydrostatic limit is approached, the truncation error will be $\mathrm{O}(\delta^2 \|p - p_i^h\|)$ instead of $\mathrm{O}(\delta^2)$, and thus reduced as much as we manage to reproduce the local vertical balance.

Local hydrostatic reconstructions: The local hydrostatic pressure distributions $p_i^h(z)$ obey

$$\frac{\mathrm{d}p_i^h}{\mathrm{d}z} = -\rho\left(p_i^h, \theta_i^h\right) g \tag{4.34}$$

with initial data

$$p_i^h(z_i) = p_i \tag{4.35}$$

where p_i is the pressure of the $i$the cell evaluated using the available cell averages of mass, momentum, and energy. The function $\theta_i^h(z)$ is a local approximate distribution of potential temperature as constructed from the cell-centered data via standard first or second order reconstruction formulae. In practice we have obtained very good results by using piecewise constant distributions $\theta_i^h(z) \equiv \theta_i$, while piecewise linear reconstructions yield further improvement.

Piecewise hydrostatic first order scheme: Standard first order finite volume schemes for hyperbolic conservation laws *without source terms* rely on an interpretation of the grid cell averages u_i of some conserved quantity u as providing cell-by-cell piecewise constant reconstructions

$$u^{\text{rec.}}(\boldsymbol{x}) = u_i \quad \text{for} \quad \boldsymbol{x} \in c_i\,. \tag{4.36}$$

Second order schemes are then designed by improving these reconstructions, and to adopt piecewise linear or higher order polynomial distributions which are compatible with the grid cell averages at the cell centers. This approach satisfies the minimal requirement that spatially constant initial data, which constitute a stationary state for compatible boundary conditions, are also stationary in the discrete sense.

In the presence of, e.g., the gravity source term, this latter requirement is no longer satisfied, as stationary states feature non-trivial hydrostatic distributions of pressure and density. This incompatibility is avoided by adopting piecewise hydrostatic distributions as discussed in the last paragraph as the first order reference states. Higher order schemes are then constructed, see [3], by piecewise polynomial reconstructions of *deviations* from these local hydrostatic states.

Discussion and results

Properties of the well-balanced scheme The improved scheme just discussed has several desirable features.

- *Straight-forward implementation:*
  The implementation of the balancing approach merely requires re-formulation of the gravity source term and application of discretisation stencils to deviations from a hydrostatic state instead of deviations from piecewise constant distributions.

- *Flexibility w.r.t. the underlying base scheme:*
  The design idea carries over to finite difference methods and finite element schemes. In fact, we have successfully implemented the approach in a test

version of the local model "LM" of the Deutscher Wetterdienst (DWD), which uses finite difference discretisations.

- *Improved accuracy on a given grid:*
  The scheme provides considerably improved accuracy *on a given grid* in comparison with the underlying non-balanced base scheme. In [3] it is demonstrated that for some test cases one obtains the same accuracy with (i) the standard scheme using 80 vertical discretisation layers and (ii) the well-balanced scheme using just 32 vertical layers. As compute time is a limiting factor in meteorological applications, this is of considerable interest in practice.

- *Robustness w.r.t. inessential details of a discretisation:*
  Non-balanced schemes are much more sensitive w.r.t. details of the numerical discretisation scheme. For instance, we have found that the choice of a slope limiting function, used to avoid unphysical overshoots in the piecewise linear reconstructions of a second order scheme, will non-trivially affect the computational results. This is understood by comparing the limited slope for some quantity u within the ith grid cell obtained with the standard approach,

$$\left(\frac{\partial u}{\partial z}\right)_i \approx \mathrm{Lim}\Big(\Big[\frac{\partial u^h}{\partial z} + \Big(\frac{\partial \delta u}{\partial z}\Big)^l\,\Big], \Big[\frac{\partial u^h}{\partial z} + \Big(\frac{\partial \delta u}{\partial z}\Big)^r\,\Big]\Big)_i \tag{4.37}$$

  and the version from the well-balanced scheme

$$\left(\frac{\partial u}{\partial z}\right)_i \approx \left(\frac{\partial u^h}{\partial z}\right)_i + \mathrm{Lim}\Big(\Big(\frac{\partial \delta u}{\partial z}\Big)^l, \Big(\frac{\partial \delta u}{\partial z}\Big)^r\Big)_i. \tag{4.38}$$

  Here $\mathrm{Lim}(\,\cdot\,,\,\cdot\,)$ is some nonlinear limiting function that guarantees second order accuracy in the end, but avoids the generation of new extrema in the reconstructed distributions. The superscripts l, r indicate slope approximations using data from grid cells $\{i-1, i\}$ and $\{i, i+1\}$, respectively.

  For small deviations from a hydrostatic state, $\delta u \ll 1$, the standard version from (4.37) will yield a result that depends strongly on the choice of the limiter function and on the hydrostatic background distributions, $u^h(z)$, and it will generally *not* reproduce $(\partial u^h/\partial z)_i$. In contrast, the version from (4.38) will reduce properly to the desired hydrostatic slope.

Sample results Unless the problem of unbalanced truncation errors explained above is addressed, numerical methods for flows under the influence of gravity will generate spurious motions under nearly hydrostatic conditions. This often becomes particularly prominent in the vicinity of steep orography as seen in Figure 6.

There are classes of vertical distributions of potential temperature (entropy) for which the well-balanced scheme described above is able to reproduce the steady state up to machine accuracy by construction. These involve piecewise constant and piecewise linear distributions, in particular. For more general distributions, the truncation error is still reduced considerably as explained in the context of equation (4.33) above. An important example is an "inversion situation", for which the

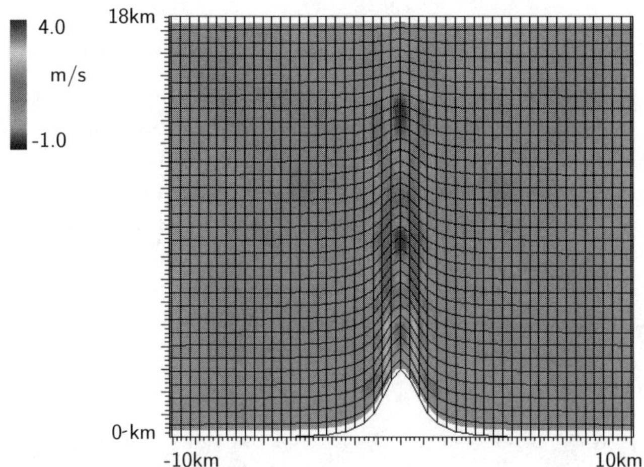

Figure 6. *Spurious vertical winds in the vicinity of steep orography generated by an unbalanced conservative finite volume method, [3]. An* 18 km *high layer of air at rest is placed near a* 3 km *high "mountain". Vertical winds of up to* 4 m/s *are observed within about* 60 min *physical time on a computational grid with* 32×128 *grid points. The graph shows a color-plot of vertical velocity in a* $-10, \ldots, 10$ km *vicinity of the orographical feature, involving about 50 gridpoints in the horizontal direction.*

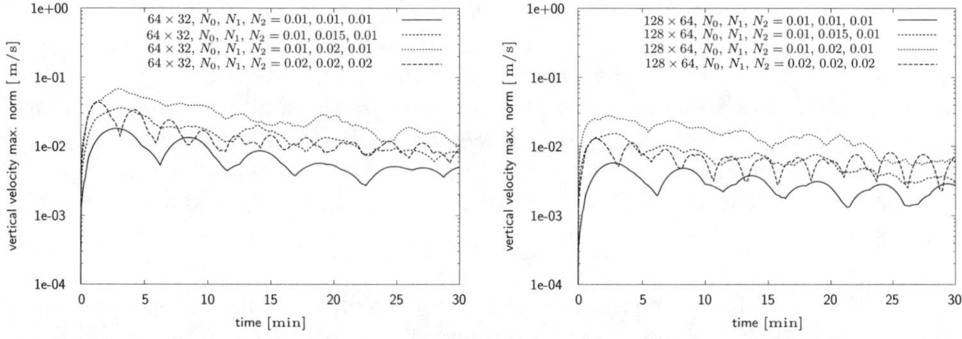

Figure 7. *Remaining spurious motions for an inversion situation with piecewise constant vertical distributions of the Brunt–Väisälä-frequency* N*; see (3.19).*

vertical gradient of potential temperature changes abruptly between several limited vertical layers of air. Figure 7 shows the time evolution of vertical velocity for a test case involving hydrostatic initial conditions in the vicinity of a 2 km high "mountain" and with the following vertical distributions of the Brunt–Väisälä-frequency.

$$N = N_i \qquad \text{for} \qquad z_i \leq z < z_{i+1} \quad (i \in \{0, 1, 2\}) \tag{4.39}$$

where

$$z_0 = 0\,\mathrm{m}\,, \quad z_1 = 750\,\mathrm{m}\,, \quad z_2 = 1\,200\,\mathrm{m}\,, \quad z_3 = 8\,000\,\mathrm{m}\,. \tag{4.40}$$

Here z_3 indicates the top of the computational domain, where the initial state is imposed as a one-sided boundary condition in the approximate Riemann solutions solved at the domain boundaries. This formulation is also used at the lateral boundaries. (See the legends for values of N.) The graph shows that the maximum (spurious) vertical velocities for all cases considered do not exceed 0.1 m/s, and remain as small as a few centimeters per second in general. This is in sharp contrast with the situation displayed in Figure 6, obtained with the standard scheme, where vertical velocities reached levels of 4 m/s, a level that is entirely unacceptable for meteorological applications. For details of this test case, the reader is referred to the original reference [3].

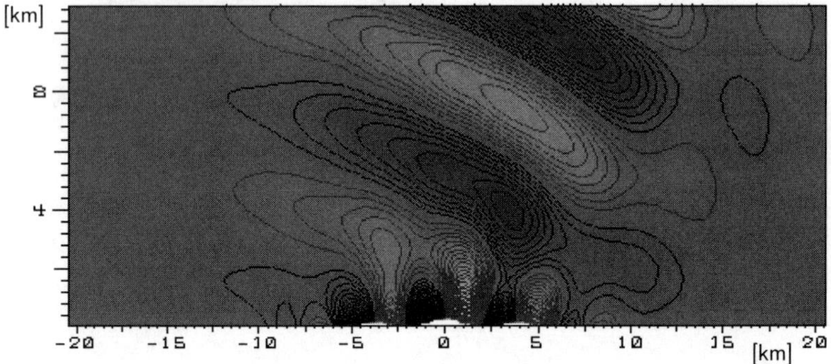

Figure 8. *Steady state lee waves over complex terrain in a test case taken from Schaer et al., [44], represented by countours of the vertical velocity. Computations are based on the present well-balanced conservative finite volume scheme, and a grid resolution of* 800×128 *nearly equally spaced grid cells for the discretisation of the computational domain covering* $19.5\,\mathrm{km} \times 200\,\mathrm{km}$. *A single "bump" is thus resolved by about* 10 *grid cells. Only a* $9.5\,\mathrm{km} \times 40\,\mathrm{km}$ *near-topography subdomain, resolved by about* 160×64 *grid cells, is shown.*

With the well-balancing approach in place, the finite volume compressible flow solver is now competitive in terms of accuracy with other well-established numerical techniques used in numerical weather forecasting. Its advantage is that it implements conservation of mass and energy up to machine accuracy, and momentum conservation in cases where the momentum source terms have a divergence form representation. Results for a challenging test case involving lee wave generation over non-trivial terrain from Schaer et al., [44], are shown in Figure 8.

The test case involves dry air flow past the "wavy" topography

$$z_b(x) = h\,\exp\bigl(-\frac{x^2}{a^2}\bigr)\,\cos^2\bigl(\frac{\pi\,x}{\lambda}\bigr) \tag{4.41}$$

where $h = 250\,\mathrm{m}$, $a = 5\,\mathrm{km}$, and $\lambda = 4\,\mathrm{km}$. The Brunt–Väisälä-frequency is assumed constant at $N = 0.01\,\mathrm{s}^{-1}$, and the bottom pressure and temperature are set to $p_0 =$

$10^5\,\mathrm{kg/ms^2}$ and $T_0 = 273,16\,\mathrm{K}$, respectively. The horizontal flow velocity imposed at the upstream (left) entrance to the domain is $10\,\mathrm{m/s}$ for $0 \leq z \leq 10.395\,\mathrm{km}$, and decreases linearly to zero from there to the top of the domain at $19.5\,\mathrm{km}$. The vertical velocity is set to zero throughout the domain initially. The boundary condition formulation is the same as that explained above in the context of the static cases without mean flow.

Figure 8 shows contours of the vertical velocity after the solution has reached near steady state conditions. These results are consistent with those obtained in [44] as a result of linear perturbation theory. At the same time they are comparable in terms of quality with simulations in [44] using the canadian MC2-model, [4], and a newly proposed computational grid structure that drastically reduces grid distortions away from the bottom topography. Our results have been computed using the well-balanced conservative finite volume scheme on a standard terrain-following grid without such modifications, albeit with a 20% higher spatial resolution. For further details the reader is referred once more to the original references.

5 Conclusions

The present paper proposes a systematic approach, based on techniques of multiple scales asymptotics, for structuring the vast variety of reduced model equations of theoretical meteorology. The bottom line of the approach has been summarised focusing on the fluid mechanics of atmospheric motions. "Diabatics" (i.e., energetic processes), boundary layer problems, and many other meteorological processes of interest are not addressed here, due to both a lack of space and the fact that further work is still needed in this direction.

The concrete applications of the technique may have convinced the reader that it gives rise to interesting extensions of existing theories, that it leads to novel multi-scale models, and that it is quite useful also in the analysis of numerical methods.

Bibliography

[1] Achatz U., and G. Branstator, 1999: A two-layer model with empirical linear corrections and reduced order for studies of internal climate variability, *J. Atmos. Sci.*, **56**, 3140–3160

[2] Botta N., R. Klein, A. Almgren, 1999: Asymptotic Analysis of a Dry Atmosphere. ENUMATH, Jyväskylä, Finnland,

[3] Botta N., R. Klein, S. Langenberg, S. Lützenkirchen, 2002: *Well-Balanced Finite Volume Methods for Near-Hydrostatic Flows*, under revision for *J. Comp. Phys.* (Preprint: PIK–Report No. 84, Potsdam Institute for Climate Impact Research.)

[4] Benoit R., M. Desgagne, P. Pellerin, S. Pellerin, Y. Chartier, and S. Desjardin, 1997: The Canadian MC2: A semi-Lagrangian, semi-implicit wideband atmospheric model suited for finescale process studies and simulation. *Mon. Wea. Rev.*, **125**, 2382–2415.

[5] Bretherton C., and A. Sobel, 2001: The Gill model and the weak temperature gradient (WTG) approximation, submitted to *J. Atmos. Sci.* on 3/7/01.

[6] Browning G. L., H. O. Kreiss, and W. H. Schubert, 2000: The Role of Gravity Waves in Slowly Varying in Time Tropospheric Motions near the Equator. *J. Atmos. Sci.*, **57**, 4008–4019.

[7] Buckingham E., 1915: Model experiments and the forms of empirical equations. *Trans. Am. Soc. Mech. Eng.*, **37**, 263–296.

[8] Cargo P., and A. Y. Le Roux, 1994: Un schéma équilibre adapté au modéle d' atmosphère avec termes de gravité. *C. R. Acad. Sci. Paris*, **318** Série 1, 73–76.

[9] Charney J. G., 1963: A note on large-scale motions in the tropics. *J. Atmos. Sci.*, **20**, 607–609.

[10] Conaty A., J. Jusem, L. Takacs, D. Keyser, and R. Atlas, 2001: The structure and evolution of extratropical cyclones, fronts, jet streams, and the tropopause in the geos general circulation model. *Bull. Amer. Meteorol. Soc.*, **82**, 1853–1867.

[11] Cullen M. J. P., J. Norbury, R. J. Purser, 1991: Generalized Lagrangian solutions for atmospheric and oceanic flows. *SIAM J. Appl. Math.*, **51**, 20–31.

[12] Durran D. R., 1989: Improving the anelastic approximation. *J. Atmos. Sci.*, **46**, 1453–1461.

[13] Ebin D. G., 1982: Motion of slightly compressible fluids in a bounded domain. *Comm. Pure Appl. Math.*, **35**, 451–485.

[14] Eliassen A., 1962: On the vertical circulatin in frontal zones. *Geophys. Publ.*, **24**, 147–160.

[15] Embid P., and A. Majda, 1996: Averaging over fast gravity waves for geophysical flows with arbitrary potential vorticity. *Comm. Partial Diff. Eqns.*, **21**, 619–658.

[16] Embid P., and A. Majda, 1998: Low Froude number limiting Dynamics for stably stratified flow with small or finite Rossby numbers. *Geophys. Astrophys. Fluid Dynamics* **87**, 1–50.

[17] Gill A. E., 1980: Some simple solutions for heat-induced tropical circulation. *Quart. J. Roy. Meteor. Soc.*, **106**, 447–462.

[18] Gill A. E., 1982: *Atmosphere-Ocean Dynamics* Intl. Geophysics Series **30**, Academic Press, San Diego.

[19] Heckley W. A., and A. E. Gill, 1984: Some simple analytical solutions to the problem of forced equatorial long waves. *Quart. J. Roy. Meteor. Soc.*, **110**, 203–217.

[20] Held I. M., and B. J. Hoskins, 1985: Large–scale eddies and the general circulation of the troposphere. *Advances in Geophysics*, **28**, Academic Press, 3–31.

[21] Hoskins B. J., and F. P. Bretherton, 1972: Atmospheric frontogenesis models: mathematical formulation and solution. *J. Atmos. Sci.*, **29**, 11–37.

[22] Hoskins B. J., 1975: The geostrophic momentum approximation and the semi-geostrophic equations. *J. Atmos. Sci.*, **32**, 233–242.

[23] Hunt J. C. R., K. J. Richards and P. W. M. Brighton, 1988: Stably Stratified Shear Flow over Low Hills. *Q.J.R. Met. Soc.*, **114**, 859–886.

[24] Kevorkian J., and J. D. Cole, 1981: *Perturbation methods in Applied Mathematics*, Applied Mathematics Sciences **34**, Springer-Verlag, New York.

[25] Klainerman S. and A. J. Majda, 1982: Compressible and incompressible fluids. *Comm. Pure Appl. Math.*, **35**, 629–656.

[26] Klein R., 2000: Asymptotic analyses for atmospherics flows and the construction of asymptotically adaptive numerical methods. *Zeitschr. Angew. Math. Mech.*, **80**, 765–777.

[27] McIntyre M. E., and I. Roulstone, 2002: Are there higher-accuracy analogues of semi-geostrophic theory? in: *Large-scale Atmospere-Ocean Dynamics II: Geometric Methods and Models Cambridge Univ. Press.*

[28] Majda A., 2001: Introduction to PDE's and Waves for the Atmosphere and Ocean. Lecture Notes from Courant Institute 2000–2001, published by *American Mathematical Society* Spring 2003, 189 pages.

[29] Majda A., and P. Embid, 1998: Averaging over fast gravity waves for geophysical flows with unbalanced initial data. *Theoret. Comput. Fluid Dynamics*, **11**, 155–169.

[30] Majda A., and R. Klein, 2003: Systematic multi-scale models for the tropics. *J. Atmos. Sci.*, **60**, 393–408.

[31] Majda A. and J. Sethian, (1985): The derivation and numerical solution of the equations for zero Mach number combustion. *Combustion Science and Technology*, **42**, 185–205.

[32] Majda A., and M. Shefter, 2001: Models for stratiform instability and convectively coupled waves. *J. Atmos. Sci.*, **58**, 1567–1584.

[33] Majda A., I. Timofeyev, and E. Vanden-Eijnden, 2003: Systematic Strategies for Stochastic Mode Reduction in Climate, *J. Atmos. Sci.*, **60**, 1705–1722.

[34] Matsuno T., 1966: Quasi-geostrophic motions in the equatorial area. *J. Meteor. Soc. Japan*, Ser II, **44**, 25–43.

[35] Neelin J. D., 1989: On the integration of the Gill model. *J. Atmos. Sci.*, **46**, 2466–2468.

[36] Neelin J. D., and N. Zheng, 2000: A quasi-equilibrium tropical circulation model–Formulation. *J. Atmos. Sci.*, **57**, 1741–1766.

[37] Newley T. M. J., H. J. Pearson and J. C. R. Hunt, 1991: Stably Stratified Rotating Flow through a Group of Obstacles. *Geophys. Astrophys. Fluid Dynamics*, **58**, 147–171.

[38] Ogura Y. and N. A. Phillips, 1962: Scale analysis of deep moist convection and some related numerical calculations. *J. Atmos. Sci.*, **19**, 173–179.

[39] Pedlosky J., 1987: *Geophysical Fluid Dynamics*, Springer-Verlag, New York, Second edition.

[40] Petoukhov V., A. Ganopolski, V. Brovkin, M. Claussen, A. Eliseev, C. Kubatzki, S. Rahmstorf, 2000: CLIMBER-2: A climatge system model of intermediate complexity. Part I: Model description and performance for the present climate. *Climate Dynamics*, **16**, 1–17.

[41] Ganopolski A., and S. Rahmstorf, 1999: Long-term global warming scenarios computed with an efficient coupled climate model. *Climatic Change*, **43**, 353–367.

[42] Roulstone I., and M. J. Sewell, 1997: The mathematical structure of theories of semi-geostrophic type. *Phil. Trans. R. Soc. London*, **A 355**, 2489–2517.

[43] Saltzmann B., 1978: A survey of statistical-dynamical models of the terrestrial climate. *Adv. Geophysics*, **20**, 183–304.

[44] Schär C., D. Leuenberger, O. Fuhrer, D. Lüthi, and C. Girard, 2002: A new terrain-following vertical coordinate formulation for atmospheric prediction models. *Mon. Wea. Rev.*, **130**, 2459–2480.

[45] Schneider W., 1978: *Mathematische Methoden in der Strömungsmechanik.* Vieweg.

[46] Schochet S, 1994: Fast singular limits of hyperbolic pdes. *Journal of Differential Equations*, **114**, 476–512.

[47] Schroeder, W., 1997: *Ertel's potential vorticity.* (Ertel's collected papers Vol. III) Interdivisional Commission on History of IAGA / History Commission of the German Geophyical Society, Mitteilungen **16**

[48] Sobel A., J. Nilsson, and L. Polvani, 2001: The weak temperature gradient approximation and balanced tropical moisture waves. *J. Atmos. Sci.*, **58**, 3650–3665.

[49] Stone P., and M.-S. Yao, 1990: Development of a two-dimensional zonally averaged statistical–dynamical model. Part III: The parameterization of the eddy fluxes of heat and moisture. *J. Climate*, **3**, 726–740.

[50] Wang B., and T. Li, 1993: A simple tropical atmosphere model of relevance to short-term climate variations. *J. Atmos. Sci.*, **50**, 260–284.

[51] Webster P. J., 1972: Response of the tropical atmosphere to local steady forcing. *Mon. Wea. Rev.*, **100**, 518–541.

[52] Zeytounian R. K., 1991: *Meteorological Fluid Dynamics.* Lecture Notes in Physics, m5, Springer, 1991.

At Boston University, **Nancy Kopell** is co-director of the Center for BioDynamics (CBD), whose mission is the training of undergraduates, graduate students, postdoctoral students and faculty to work at the interfaces among dynamical systems, biology and engineering. She is also the codirector of the Burroughs Wellcome Fund Training Program in Mathematical and Computational Neuroscience, designed to facilitate the transition of those trained in mathematics, physics and computer science to work on problems of neuroscience.

Professor Kopell received her AB from Cornell University in 1963 and her PhD from the University of California at Berkeley in 1967. She was a CLE Moore Instructor at Massachusetts Institute of Technology from 1967–69, then taught at Northeastern University until 1986. She is now at Boston University as the first WG Aurelio Professor of Mathematics and Science. Her research currently focuses on dynamics of the nervous system, especially rhythmic dynamics associated with motor control and cognition. She is also very interested in geometric methods for analysis of systems with many timescales. She is a member of the National Academy of Sciences and the American Academy of Arts and Sciences; she has held fellowships from the John D and Catherine T MacArthur Foundation, the Guggenheim Foundation, and the Sloan Foundation. In 2001 she was the recipient of the Wright Prize from Harvey Mudd College.

Chapter 13

Rhythms in the Nervous System: from Cells to Behavior via Dynamics

Nancy Kopell

Abstract: The nervous system produces rhythmic electrical activity in many frequency ranges, and the rhythms displayed during waking are tightly tied to cognitive state. This talk describes ongoing work whose ultimate aim is to understand the uses of these rhythms in sensory processing, cognition and motor control. The method used is to address the biophysical underpinnings of the different rhythms and transitions among them, to get clues to how specific important subsets of the cortex and hippocampus process and transform spatio-temporal input. We focus on the gamma rhythm (30–80 Hz), which is associated with attention and awareness, and the theta (4–12 Hz), associated with active exploration and learning of sequences. Via case studies, we show that different biophysics corresponds to different dynamical structure in the rhythms, with implications for function. The mathematical tools come from dynamical systems, and include the use of low-dimensional maps, probability and geometric singular perturbations.

Contents

1 Introduction

It has been known for more than half a century that the brain produces rhythms in its electrical activity, day and night. Due to increasingly better technology and signal processing, we now know far more than fifty years ago, with many different tools for measuring and interpreting rhythms.

In awake humans, EEG and MEG recordings can now pick up these rhythms at dozens of points on the scalp and record the changes in the rhythms as people do various tasks [48, 10, 16, 37]. Since the EEG and MEG do not have adequate spatial resolution, related work is also done with more invasive experiments on animals, recordings that can measure collections of cells (field recording or local EEG) or even single cell activity [9, 6, 19]. This is further supplemented by work in "slices" of tissue, where it is possible to get detailed information about the physiology and pharmocological sensitivities of the tissue [52, 20, 3].

There is a large and growing literature suggesting that different rhythms or combinations of them are associated with different cognitive states. Gamma has become famous as the rhythm associated with "binding" or distributed processing [23, 47, 24]. It appears to be connected with the cognitive states of attention, perception, awareness, and early sensory processing [19, 42, 43, 30]. Beta often forms after gamma [25], is more widespread, appears to be associated with higher-order processing, and be involved in motor planning [38, 16, 12]. Alpha appears in the visual area when one's eyes are closed, but versions of alpha appear in other parts of the nervous system at other times of relaxation or end of concentration [37, 44]. Theta appears when an animal is actively exploring, and is believed by many to be important for the learning of associations [26, 14].

In addition to a growing literature tying normal cognitive and motor behavior with these and other rhythms, there is another line of evidence suggesting a relationship between these rhythms and cognition: Various mental diseases, including schizophrenia, Alzheimer's and the cognitive aspects of Parkinson's disease are associated with pathologies in these rhythms [35, 40, 18, 2]. Furthermore, a variety of anesthetics/amnesiacs affect neural rhythms [17].

The main and contentious question associated with all this is: Is there a connection between these rhythms and cognitive function? Many of us would be astonished if the answer were no, since there seems to be so much evidence linking them. But how could one prove it? How does one even get provable conjectures about what exactly the rhythms might be doing?

As mentioned above, dynamics in the nervous system are studied at many different levels, so there is a range of data relevant to the issue. What is not obvious from the data is how they fit together — how dynamics of individual cells effects that of networks, and how many different networks of neurons in different parts of the brain work together to do sensory processing, cognition and motor activity. This is the niche where math and modeling have found a home, and where I believe they are becoming indispensible.

The levels of experimental work are reflected in levels at which models and analysis can be done, from single cell through complex functional networks representing many different substructures and many different types of cells. The question we want to understand is: with complicated and real networks of subsets of the nervous system, what is the relevance of rhythms? However, we need the kind of insights that we get from single cells and small networks to start parsing what happens in the complicated networks. So most of this paper refers to single cells and quite simple networks, and the insights we get from studying them. At the end, I'll discuss a much more complex problem that makes use of the insights gathered from the smaller case studies.

2 The Hodgkin–Huxley Equations

The mathematical framework for this program was proposed by Hodgkin and Huxley [29], based on difficult and ground breaking experiments, and earned them a Nobel Prize. These equations play the same role in neural dynamics that Navier–Stokes play in fluids. In both cases, they cover an enormous range of different concrete possibilities, and analysis is usually in the context of particular examples.

The main equation is an elaborate analogue of electric circuit theory. Currents carried by charged ions move across the cell membrane changing its voltage; the main equation is conservation of current, where v is the voltage across the membrane.

$$C\frac{dv}{dt} = -\left[\sum I_{\text{ion}} + \nabla^2 v + \sum I_{\text{syn}}\right] + I_{\text{appl}}\,. \tag{2.1}$$

Here C is the capacitance of the membrane and I_{ion} is any one of the intrinsic currents, most carried by Na^+, K^+ or Ca^{++}. The $\nabla^2 v$ term corresponds to flows of current through the cell, which has axons and dendrites spread out in space. Thus this equation, for even a single neuron, is really a PDE, but I will ignore that for most of this paper. I_{syn} is a current through a synapse which connects the cell to others, and I_{appl} is current applied from the outside (e.g., by an experimentalist).

Each current is given by the standard Ohm's law: electromotive force is current times resistance. In neurobiology, one uses conductance = 1/resistance instead of resistance, so each current is conductance times electromotive force:

$$I_{\text{ion}} = \text{conductance} \times [v - V_R] \tag{2.2}$$

Each different current is associated with a different "battery", so each one has a different reversal potential V_R. (For more about the basics of computational neuroscience, see [31, 13, 33].)

What makes the mathematics so challenging is the nature of the conductances, which are not constant. Each conductance is the product of a maximal conductance $\bar{g}$ times a power of one or more gating variables, each of which is the fraction of relevant ionic channels open at a given moment. The opening and closing of the channel is voltage dependent (and possibly dependent on other quantities, such as calcium, though that is not reflected in the simplest versions of the equations discussed here). These changes in fraction of channels open is often modeled by first order kinetics

$$\frac{\mathrm{d}x}{\mathrm{d}t} = \frac{[x_\infty(v) - x]}{\tau_x(v)}, \tag{2.3}$$

where each gating variable x has its own kinetics (2.3). Here $x_\infty(v)$ is the fraction of channels open at voltage v if the voltage is held fixed for sufficiently long, and $\tau_x(v)$ is the voltage-dependent time constant. The classic Hodgkin–Huxley (H–H) equations have a Na^+ conductance that is $\bar{g}_{\mathrm{Na}}m^3h$ (with two gating variables m and h), a K^+ conductance usually written $\bar{g}_{\mathrm{K}}n^4$, and a constant leak conductance $\bar{g}_{\mathrm{leak}}$.

Since the cell may have many ionic currents, even a single model cell may have many time constants, and the currents associated with the synapses add still more. What we have from this is a very large-dimensional and highly nonlinear system. However, in many examples, they appear to behave like much lower dimensional systems, as I'll discuss below.

3 Single cells

In general, the mathematical description of even single cells can be very complicated, even ignoring the spatial distribution. But some cells are fairly well modeled by using only the ionic currents needed to produce the action potential, or spike, and a so-called leak current, which is not voltage-gated. Cells like these are sometimes simplified further into what is known as "integrate and fire" (I&F) neurons. In the simplest version, the cells behave in a linear way between spikes, charging up, and are reset (spike ignored) when they reach some threshold. The equations are

$$v' = I - v + \text{ spike } + \text{ reset at a threshold }. \tag{3.1}$$

There are various versions of this, such as the so-called "quadratic integrate and fire" [36] with somewhat different properties. But the crucial point is that they are one-dimensional, and are meant to capture the behavior of voltage between spikes.

Though the I&F cell has been the subject of a seemingly endless stream of papers, most of the neurons that concern us have other nonlinear conductances that are on between spikes, and which make the neuron fundamentally higher dimensional in its description. A set of cells that will be part of a running description throughout the paper are ones that appear to be important in the creation of theta rhythms in different structures of the brain.

In the right conditions, these cells are oscillators, with a range of preferred frequencies. Unlike the I&F cell, which can be driven to very high frequencies by increasing applied current I, these have their own time-scales from the kinetics of some important ionic currents. These include a sodium current that is "persistent" (on even between spikes) and peculiar currents belonging to a class known as h-currents. I'll describe later what makes them weird, and the implications of that weirdness for how the cells interact with others.

4 Structure of H–H equations and reduction to lower dimensions

The details of how these kinds of cells oscillate at first seems like an arcane matter irrelevant to network or animal behavior, but it turns out that details like this can be highly significant for behavior at much higher levels. I'm going to make a brief digression here that is relevant to much of the rest of this paper, as well as to understanding how these cells oscillate. This digression, about structure, is at the heart of why the H–H equations often behave like low dimensional equations.

The central idea comes from the structure of the main equation (2.1) of the H–H set, the current conservation equation. It is possible to rewrite (2.1) as

$$C\frac{\mathrm{d}v}{\mathrm{d}t} = -G(v - V_T)\ , \tag{4.1}$$

where G is the (time and voltage dependent) sum of the conductances and V_T is a weighted sum of all the reversal potentials, weighted by the conductances. Thus, instantaneously, the voltage equation behaves like a linear equation with a target voltage V_T. Furthermore, between spikes, V_T changes relatively slowly so that the voltage tracks V_T. Even at times when V_T changes quickly, the voltage stays fairly close to V_T because at those times G is generally very large.

It is often true that, even though there is a large number of conductances, there are only a few that are large at a given time, and thus have a large effect on V_T. By using the gating variables of only those, we have a projection to a lower dimensional space, and estimates on errors; we take different projections along different parts of the trajectory, changing the variables as different ones become relevant. Within that smaller dimensional system, we can do more using differences in time-scales [11].

This method works in principle for networks and single cells. An example is a model of a single O–LM cell, which has 6 dimensions before reduction [45]. This method helps us to understand where the oscillations come from in the O–LM cells by dividing the oscillation into different epochs in which different currents are important. When the cells are spiking, the standard currents are important, and the equations turn out to be just a perturbation of the simplest H–H equation, as caricatured by the integrate and fire equation. Things get more interesting between spikes, where the currents important for spikes are small, and the persistent Na and two h-currents are more important. Ignoring the spike currents, we are then down to four dimensions. However, one of these adapts almost instantly to the voltage, so is not a dynamic variable. One of the remaining ones changes quite slowly, so its gating variable is essentially a slowly moving parameter. Hence, we can think of the dynamics between spikes as a family of 2-D systems. As the slow gating variable increases in time, the nullclines pull off one another, eliminating a quasi-steady-state and allowing the trajectory to escape to a different part of phase space (Figure 1a). Trajectories coming from the spike represent a small part of phase space we can track, and see what happens when the quasi-steady state is lost (Figure 1b). This kind of analysis allows us to see exactly what the different currents are doing in creating the dynamics.

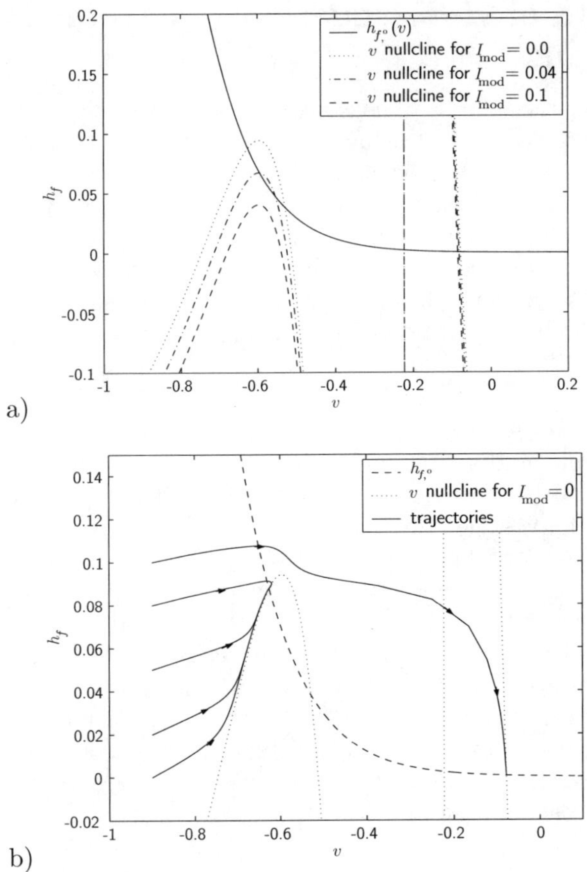

Figure 1. *a) Nullclines for different parameter values, corresponding to slowly changing gating variable. b) Phase-plane diagram for one set of parameter values.*

5 Coupling cells into networks: inhibition, excitation and electrical coupling

It is networks rather than individual cells that we are mainly concerned about, and they have properties one could not necessarily guess from the individual cells. The cells are connected primarily by chemical or electrical synapses.

In detail, the chemical synapses are complicated, but models of some of them (the so-called "ionotropic", or fast synapses) can be made that have the same form as the other (intrinsic) ionic currents: they are a conductance times a driving force. The difference is that the conductance depends on the voltage of the presynaptic cell, and the driving force depends on the voltage of the postsynaptic cell.

$$I_{\text{syn}} = g_{\text{syn}}(v_{\text{pre}}) \times \left[v_{\text{post}} - V_{\text{syn}}\right] \tag{5.1}$$

For example, if an excitatory, or E-cell, is sending a pulse to an inhibitory, or I-cell, the E-cell is pre-synaptic and the I-cell is post-synaptic. Whether the synapse is excitatory or inhibitory depends on the reversal potential $V_{\rm syn}$ for the synapse; an excitatory synapse creates a current that drives the cell toward its threshold for firing, and an inhibitory one gives a current that moves the voltage away from threshold or shunts out other currents. With chemical synapses, the structure of the equations is as in (4.1), with the voltage following a target voltage. There are also electrical synapses, which change the structure of the equation. For such connections (also known as gap junctions), the standard synapses can be modeled in a very simple way as discrete diffusion; i.e., an extra current of the form

$$g_{\rm el} \times [\hat{v} - v] \tag{5.2}$$

is added to the RHS of (2.1), where $g_{\rm el}$ is the (constant) conductance of the synapse, v is the voltage of the cell receiving this current, and $\hat{v}$ is the voltage of another cell to which it is connected. Though the mathematical description is simpler, the resulting network dynamics is harder to analyze, because the equations no longer have the form (2.2). (A paper showing the complications of nonlinear dynamics with electrical coupling is [32].)

6 Gamma, a "simple" rhythm.

The simplest networks consists of cells for which only the standard Hodgkin–Huxley currents, important for spiking, are involved. Even these simplest networks keep inspiring mathematical work. It turns out that, for cells without voltage-gated currents between spikes, inhibition between cells is more effective at creating synchrony than mutual excitation. This synchronization is not intuitive, because one would think that inhibition between cells trying to step on one another would drive them apart. There are many ways one can analyze such a network, and work has been done by a variety of people to understand how it can happen. (Many references are given in [33].) The key point is that the inhibitory current is on-going. In terms of the tracking theory discussed above, the inhibitory conductance is the main one between spikes, and the cells track the voltage determined by the inhibition as the conductance declines. After a short time, all history of initial conditions are gone. Thus, if identical cells receive common inhibition, they synchronize [4]. Note that this is not true with (fast) excitation, which comes and goes very quickly, and indeed, does not synchronize the cells. Other analysis uses the so-called spike-response method [21, 33] and weak coupling methods [50].

When simple cells are connected by standard inhibition, the rhythm they create is the so-called "gamma" rhythm of 30–80 Hz, depending on the drive to the cells [7, 52]. Although individual cells have no natural time constant, the decay time of the synapse provides the periodicity of the rhythm, and it can be shown, both mathematically [8] and experimentally [51], that the period is proportional to the decay time, with the proportionality constant depending on the drive. For completely homogeneous networks, this is not necessarily true, but in the presence of even a small amount of difference among the cells, the system stays coherent only in the gamma range. This form of gamma is called "ING" for inhibitory (or interneuron) gamma.

Another version of gamma in which the period is also related to the decay time of inhibition is "PING", for pyramidal-interneuron gamma. The pyramidal cell is excitatory, the interneuron inhibitory. In this case, the I-cell does not have enough drive to fire by itself with input. The E-cell fires, stimulates the I to fire, which inhibits the E-cell until the inhibition wears off. Again, voltages track the decay of the inhibition. This network is simple to understand for two cells, but the behavior is very subtle when there are many cells, heterogeneity, sparse coupling, noise etc., as discussed below.

7 Interaction of intrinsic and synaptic currents

In the previous section, the only really important time-scale was the decay of the inhibition, and it determined the period of the network oscillation. For more complicated cells, having nonlinear conductances between the spikes, adding excitation and inhibition produces new nonlinear effects. These effects have to do with the how the conductances depend on the voltage of the cell.

For most conductances, increasing the voltage increases the conductance; i.e., opens a gate more for more current to flow. However, there are some conductances that act in an anti-intuitive manner: the higher the voltage, the smaller the conductance. This is true of a class of currents known as "hyperpolarization-activated currents" or I_h, mentioned above. These currents turn up in many cells in the nervous system, and in at least some of the cells associated with theta rhythms, is important in determining the voltage between spikes for those cells. When inhibition or excitation is added to a cell that has nonlinear conductances between spikes, it doesn't just change the voltage by adding a new current; it changes the other currents that are sensitive to the voltage. Adding inhibition to the O–LM cell can initially lower the voltage, but then the h-current turns on and makes the voltage go back up. So inhibition in such a cell can actually make the next spike come faster.

This current also totally reverses the synchronizing properties of excitation and inhibition ([46], in preparation; see also [1, 15] for related work). O–LM cells are inhibitory, and model O–LM cells, when connected with inhibition, do not synchronize for most initial conditions; this can be traced to the effects of inhibition on this current. It can be understood from maps constructed using the projection techniques I mentioned earlier, and builds on the dynamics of the individual cell. With excitation, such model cells do synchronize [1]; this prediction was recently confirmed by the lab of M. Banks [39].

This kind of understanding can help to make predictions about more complicated networks. For example, hippocampal slices (in the CA1 region) can produce a theta rhythm with no excitatory cells taking part. This network contains the O–LM cells, which have been shown to be critical for the theta [22]. The above result, from both the math and simulations, seems to contradict this: it says that individual O–LM cells can produce theta, but together they should not be coherent. But this raises a question, since the experimental data do show coherence.

The mathematics and simulations make the prediction that the network requires not just the O–LM cells, but other inhibitory neurons as well in order to get the coherent theta. A network with both standard I cells (as described above, and known as "basket cells") and O–LM cells are able to produce a coherent theta rhythm, with the two types of cells firing at different times in the cycle; the common inhibition from the I-cells is part of the mechanism causing coherence in the O–LM cells. The different time-scales of the inhibitory synaptic currents induced by the two kinds of cells also plays a role in the ability to get a theta rhythm (The O–LM induces a current that is much longer than that of the standard I-cell.)

8 Larger networks: heterogeneity and noise

The phenomena described in the smaller networks persist when the networks are scaled up. But larger networks with the same elements can have other behaviors as well. Indeed, there are some kinds of influences, such as sparse coupling, noise and totally asynchronous behavior that cannot be investigated in small networks. At the moment, these kinds of issues have been explored mathematically only in the simpler kinds of networks in which the main nonlinearities are in the spiking currents and the synapses, not the other fancy intrinsic currents. For example, C. Borgers and I have shown that the PING network is very robust to sparse coupling, and explained why, using combinations of probability theory and dynamical systems [4].

Another example is recent work with Borgers showing the interplay of noise and frequency, where the interplay depends on having a large network and possibly asynchronous behavior. The motivating question is why this network of E-cells and I-cells have a frequency in the gamma range. In the case of a purely inhibitory network, if there is even a small amount of heterogeneity, the frequency is tightly tied to the decay time of the inhibition, and one can analyze this from a 2-cell network. In the case of PING, one can make the 2-cell E–I network fire at a huge range of frequencies, so the control of frequency is mysterious.

That question involves looking at what happens in much larger networks. Focusing on the lower boundary of frequency, it turns out that there are at least two different mechanisms by which coherence can be lost in PING networks as the drive to the I-cells is increased (via constant drive or phasic E–I synapses) compared with drive to the E-cells. In one, which happens when the frequencies are below gamma, the I-cells may become incoherent, and the effect on the E-cells is to totally suppress them. It can be shown that inhibition is more efficient at suppression of cells if the inhibition is spread out in time [5]. There can also be synchrony at such lower frequencies, with different initial conditions. There is another mechanism when the inhibition is too strong, in which the I-cells do not wait to be fired by the E-cells, and precess them. The interplay of these creates a region in parameter space of bistability between synchrony and complete incoherence of the I-cells, with the E-cells totally suppressed — if something can make the I-cells incoherent, the synchrony falls apart. This happens only when the drive to the E-cells is low; i.e. in low frequency ranges.

Here is where noise comes in: unlike inhibitory networks, PING is much more tolerant of heterogeneity, and even with significant heterogeneity in the natural

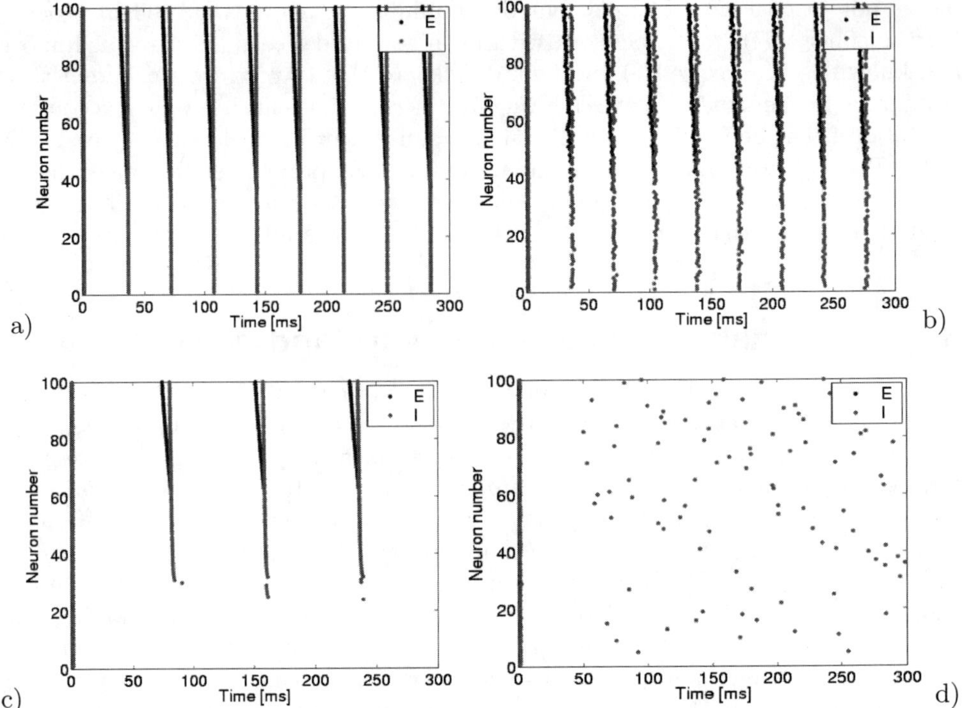

Figure 2. *a) Raster diagram for E/I network with gamma frequency rhythm. The connections are* E $\to$ I *and* I $\to$ E. *The E-cells have a range of intrinsic drives, accounting for the lack of exact synchrony (vertical alignment) of those cells.*
b) Adding noise creates raggedness, but does not eliminate coherence. Noise is added as Poisson-distributed current pulses to the I and E cells.
c) Raster diagram as in (a), with lower levels of drive to the E and I cells. Even with heterogeneity, there is coherence, now at a lower frequency.
d) Adding noise to the network displaying the low-frequency gamma makes the I-cells incoherent and suppresses the E-cells.

frequency of the cells, the network can fire at a large range of frequencies, including low ones. But if a little noise is added, the synchrony disappears at low frequencies, but not at high ones (see Figure 2). We can understand this from the previous analysis — at the low frequency, there is synchrony, but it is bistable with the complete incoherence, and noise knocks it from one basin to another. The above gets rid of sub-gamma frequencies when there are no I–I connections; when there are, it takes significant noise to knock out the low-frequency coherence. Thus, I–I connections are very relevant in the gamma frequency range.

There are also mathematical issues associated with large networks of inhibitory neurons, in which the cells are connected not just by inhibitory synapses but also by gap junctions (electrical synapses). It has been shown in distributed networks that blocking the gap junctions disrupts the coherence of the inhibitory gamma rhythm, and this was reproduced in large biophysically detailed simulations [49]. In recent work, Ermentrout and I [34] have shown why electrical coupling can have a powerful effect on coherence, even though the coupling conductance is small relative to that of inhibition, and each cell is coupled to many fewer cells by gap junctions than by inhibitory synapses. Mathematics using modified integrate-and-fire neurons and techniques from dynamical systems show that the electrical coupling provides a dynamical mechanism complementary to that of the inhibition: the inhibition acts to remove the effects of initial conditions (as described above), while the electrical coupling removes effects of heterogeneity in intrinsic or synaptic currents.

9 Functional networks, functional questions and a biophysical approach

All the networks I discussed above are bits and pieces of real networks. As a last topic, I'd like to give a sense of what it is like to put the bits together to address a functional question. This is still a small portion of "cognition", but an order of magnitude more complex than the pieces I've discussed. The example I have in mind involves the hippocampal formation, a series of substructures that form a loop, with extra connections, deep inside the brain. The hippocampus and its input/output structure, the entorhinal cortex, is thought to be the place where new memories are encoded, especially memories of sequential events [14].

The issue here is about the creation of cell assemblies, or synchronous subsets of cells, which are thought to carry information about which cells are engaged with which at a given time. Cell assemblies are thought to be very important to "plasticity", the ability of the synapses to increase or decrease their potency with behavior [28]. Though synchrony does not absolutely depend on rhythms, there is evidence [44] that it may be critical when there are conduction delays between cells that are to be part of the assembly. There is also evidence [41] that gamma rhythms are especially important for the creation of cell assemblies.

The issue for this case study is the involvement of cell assemblies in both encoding of memories and their recall; as has been pointed out (e.g., [27]), there are difficulties doing both at the same time, because plasticity is required in one and has potentially bad effects in the other. Hasselmo suggested that the separation is done during different parts of the theta cycle, and showed with high-level models that if this is done, it is possible to account for how some features of how animals learn and extinguish learning in order to learn new things. The project was to show that what we know about anatomy and physiology of the cells makes this plausible, and suggest biophysical tests of the idea.

The model (Kunec, Hasselmo, Kopell, in preparation) focuses on CA3, a part of the hippocampus that gets input from several places. The model and the story are complicated, and will not be spelled out here. But there are a few points to emphasize about this kind of work. First, the model involves all the types of cell mentioned above, and may eventually need more. This is in contrast to the simple models above with at most two kinds of cells. There are also inputs from several outside structures (entorhinal cortex and septum), and inputs themselves are temporally complex, with gamma and theta rhythms. We also must now take into account the fact that cells are not point neurons, but are distributed in space. This is especially important for the E-cells, which get inputs at different parts of its dendritic structure; it matters where the inputs come in, since inhibition near the cell body can shunt out the excitation arriving further away from the cell body. (It does not require PDEs however; most investigators divide up the dendrites into "compartments" that are electrically coupled.) The properties of the O–LM cells are especially important, including the ways that they react to inhibition from the septum and shunt out some of the excitation from the entorhinal cortex. The ability of the E-cells and I-cells to together form a gamma also contributes to the network behavior. The small models discussed above provides the heuristics for a complex one such as this. From the heuristics and the simulations, one can start to simplify and get new mathematical issues, so there is a learning spiral.

10 Rhythms, Past and Future

I'd like to end by putting the case-studies I mentioned into a broader perspective. As technology becomes more advanced, and more and more data become available, it is becoming easier and easier to get "facts" about neuroscience, but no easier to understand their significance. At least among the enlightened, it is becoming clearer that mathematics and computation are going to be playing increasingly important roles. But in what way?

The scientific program that I discussed today is a focused but broad-reaching way of putting together information available, or potentially available, at different levels of organization. For more than a decade, mathematicians have been studying the properties of single cells and small networks and their relationship to biophysics. Physicists have been especially interested in large but fairly homogeneous networks, for which techniques and ideas of statistical physics could be applied. What is much newer is the possibility of bridging the entire scale. We are beginning to see how to make use of the insights from small, and large but relatively simple, networks to address functional questions at a biophysical, rather than abstract, level. Furthermore, advances in molecular biology are bringing into this research new ways to manipulate biophysics to test ideas worked out at the higher levels (e.g., [20]). At the other end, new imaging techniques are allowing researchers to see more deeply what is happening during animal behavior, placing better constraints on the theories one develops for the functional networks.

Rhythms turn out to be an excellent lens through which to view the integration of these different levels. With the right mathematics, some of which will be new, this point of view is potentially a powerful way of investigating the workings of the brain.

Bibliography

[1] C. Acker, N. Kopell and J. White (2003); *Synchronization of strongly coupled excitatory neurons: relating network behavior to biophysics*, J. Comput. Neurosci 15, pp. 71–90.

[2] T. Baldeweg, S. Spence, S. Hirsch and J. Gruzelier (1998); *Gamma-band electroencephalographic oscillations in a patient with somatic hallucinations*, Lancet 352, pp. 620–621.

[3] M. Beierlein, J. R. Gibson and B. Connors (2000); *A network of electrically interneurons drives synchronized inhibition in the neocortex*, Nature Neurosci. 3, pp. 904–910.

[4] C. Borgers and N. Kopell; *Synchronization in network of excitatory and inhibitory neurons with sparse, random connectivity*, Neural Computation 15 (2003), pp. 509–538.

[5] C. Borgers and N. Kopell; *Effects of noisy drive on rhythms in networks of excitatory and inhibitory neurons*, in preparation.

[6] A. Bragin, G. Jando, Z. Nadasdy, J. Heike, K. Wise and G. Buzsáki (1995); *Gamma (40–100 Hz) oscillation in the hippocampus of the behaving rat*, J. Neurosci. 15, pp. 47–60.

[7] G. Buzsáki and J. J. Chrobak (1995); *Temporal structure in spatially organized neuronal ensembles: a role for interneuronal networks*, Current Opinion in Neurobiology 5, pp. 504–510.

[8] C. Chow, J. White, J. Ritt, and N. Kopell (1998); *Frequency control in synchronous networks of inhibitory neurons*, J. Comput. Neurosci., 5, pp. 407–420.

[9] J. J. Chrobak and G. Buzsáki (1998); *Gamma oscillations in the entorhinal cortex of the freely behaving rat*, J. Neurosci 18, pp. 388–98.

[10] J. Classen, C. Gerloff, M. Honda and M. Hallett (1998); *Integrative visuomotor behavior is associated with interregionally coherent oscillations in the human brain*, J. Neurophysiol. 79, pp. 1567–1573.

[11] R. Clewley, N. Kopell; *Reduction of dimensions in networks of neurons*, in preparation.

[12] J. P. Donoghue, J. N. Sanes, N. G. Hatsopoulos, G. Gaal (1998); *Neural discharge and local field potential oscillations in primate motor cortex during voluntary movements*, J. Neurophysiol. 79, pp. 159–73.

[13] P. Dayan and L. F. Abbott; Theoretical Neuroscience, MIT Press, Cambridge MA 2001.

[14] H. Eichenbaum, (1996); *Is the rodent hippocampus just for "place"?*, Curr. Opin. Neurobiol. 6, pp. 187–195.

[15] G. B. Ermentrout, M. Pascal and B. Gutkin (2001); *The effects of spike frequency adaptation and negative feedback on the synchronization of neural oscillators*, Neural Comput. 13, pp. 1285–1310.

[16] S. F. Farmer (1998); *Rhythmicity, synchronization and binding in human and primate motor systems*, J. Physiol. 509 1, pp. 3–14.

[17] H. J. Faulkner, R. D. Traub and M. A. Whittington (1999); *Anaesthetic/amnesic agents disrupt beta frequency oscillations associated with potentiation of excitatory synaptic potentials in the rat hippocampal slice*, Brit. J. of Pharmacol. 128, pp. 1813–1825.

[18] H. Forstland, P. Fischer (1994); *Diagnostic confirmation, severity, and subtypes of Alzheimer's disease. A short review on clinico-pathological correlations*, Eur. Arch. Psychiatry Clin. Neurosci. 244, pp. 252–60.

[19] P. Fries, P. R. Roelfsema, A. K. Engel, P. Konig and W. Singer (1997); *Synchronization of oscillatory responses in visual cortex correlates with perception in interocular rivalry*, Proc. Nat. Acad. Sci USA 94, pp. 12699–12704.

[20] E. C. Fuchs, H. Doheny, H. Faulkner, A. Caputi, R. D. Traub, A. Bibbig, N. Kopell, M. Whittington, and H. Monyer, (2001); *Genetically altered AMPA-type glutamate receptor kinetics in interneurons disrupt long-range synchrony of gamma neurons*, Proc. Nat. Acad. Sci USA 98, pp. 3571–3576.

[21] W. Gerstner, J. L van Hemmen and J. Cowan (1996); *What matters in neuronal locking?*, Neural Comp. 8, pp. 1653–1676.

[22] M. J. Gillies, R. D. Traub, F. E. N. LeBeau, C. H. Davies, T. Gloveli and E. H. Buhl (2002); *A model of atropine-resistent theta oscillations in rat hippocampal area CA1*, J. Physiol. 543, pp. 779–793.

[23] C. M. Gray, P. Konig, A. K. Engel and W. Singer (1989); *Oscillatory responses in cat visual cortex exhibit intercolumnar synchronization which reflects global synchronization properties*, Nature 338, pp. 157–161.

[24] C. M. Gray, (1999); *The temporal correlation hypothesis of visual feature integration: still alive and well*, Neuron 24, pp. 31–47.

[25] C. Haenschel, T. Baldewig, R. Croft, M. Whittington and J. Gruzelier (2000); *Gamma and beta frequency oscillations in response to novel auditory stimuli: A comparison of human electroencephalogram (EEG) data with in vitro models*, Proc. Nat. Acad. Sci USA 97, pp. 7645–7650.

[26] M. E. Hasselmo, (1999); *Neuromodulation and the hippocampus: memory function and dysfunction in a network simulation*, Prog. Brain Res. 131, pp. 13–18.

[27] M. E. Hasselmo, C. Bodelson, B. Wyble (2002); *A proposed function for hippocampal theta rhythm: separate phases of encoding and retrieval enhance reversal of prior learning*, Neural Comput. 14, pp. 793–817.

[28] O. D. Hebb, (1948); The Organization of Behavior: A Neuropsychological Theory, John Wiley NY.

[29] A. L. Hodgkin and A. F. Huxley (1952); *A quantitative description of membrane current and its application to conduction and excitation in nerves*, J. Physiol (London) 117, pp. 500–544.

[30] M. Joliot, U. Ribary, R. Llinas (1994); *Human oscillatory brain activity near 40 Hz coexists with cognitive temporal binding*, Proc. Nat. Acad. Sci. USA 91, pp. 11748–11751.

[31] C. Koch and I. Segev, (1999); Methods of Neuronal Modeling, MIT Press (Bradford Book), Cambridge MA.

[32] N. Kopell, L. Abbott and C. Soto-Trevino, (1998); *On the behavior of a neural oscillator electrically coupled to a bistable element*, Physica D, 121, pp. 367–395.

[33] N. Kopell and G. B. Ermentrout, (2002); *Mechanisms of phase-locking and frequency control in pairs of coupled neural oscillators*, for Handbook on Dynamical Systems, Volume 2, Toward Applications. Ed. B. Fiedler, Elsevier, pp. 3–54.

[34] N. Kopell and G. B. Ermentrout; *Electrical coupling and chemical synapses have complementary roles in synchronization of gamma rhythms*, in preparation.

[35] J. S. Kwon, B. F. O'Donnell, G. V. Wallenstein, R. W. Greene, Y. Hirayasu, P. G. Nestor, M. E. Hasselmo, G. F. Potts, M. E. Shenton, R. W. McCarley (1999); *Gamma frequency-range abnormalities to auditory stimulation in schizophrenia*, Arch Gen Psychiatry 56, pp. 1001–1005.

[36] P. E. Latham, B. J. Richmond, P. G. Nelson and S. Nirenberg, (2000); *Intrinsic dynamics in neuronal networks, I. Theory*, J. Neurophysiol. 83, pp. 808–827.

[37] F. Lopez da Silva, (1991); *Neural mechanisms underlying brain waves: from neural membranes to networks*, Electroencephalography and clinical Neurophysiol. 79, pp. 81–93.

[38] V. N. Murthy, and E. E. Fetz (1992); *Coherent 25–35 Hz oscillations in the sensorimotor cortex of awake behaving monkeys*, Proc. Nat. Acad. Sci USA 89, pp. 5670–5674.

[39] T. Netoff, M. Banks, A. Dorval. C. Acker, J. Haas, N. Kopell and J. White; *Synchronization and phase locking in hybrid neuronal networks of the hippocampal formation*, submitted.

[40] M. Y. Neufeld, S. Blumen, I. Aitkin, Y. Parmet, A. D. Korczyn (1994); *EEG frequency analysis in demented and nondemented parkinsonian patients*, Dementia 5, pp. 23–28.

[41] M. Olufsen, M. A. Whittington, M. Camperi and N. Kopell; *New functions for the gamma rhythm: population tuning and preprocessing*, in J. Comput. Neurosci. 14 (2003).

[42] F. Pulvermuller, N. Birbaumer, W. Lutzenberger and B. Mohr (1997); *High-frequency brain activity: its possible role in attention, perception and language processing*, Progress in Neurobiology 52, pp. 427–425.

[43] E. Rodriguez, N. George, J.-P. Lachaux, J. Martinerie, B. Renault and F. Varela (1999); *Perception's shadow: long-distance synchronization of human brain activity*, Nature, 397, pp. 430–433.

[44] P. R. Roelfsema, A. K. Engel, P. Konig and W. Singer (1997); *Visuomotor integration is associated with the zero time-lag synchronization among cortical areas*, Nature 385, pp. 157–161.

[45] H. Rotstein, N. Kopell; *Analysis of a hippocampal oscillator via dimension reduction*, in preparation.

[46] H. Rotstein, M. Gillies, C. Acker, J. A. White, E. Buhl, M. Whittington and N. Kopell; *Slow and fast inhibition interact to create a theta rhythm in a model of hippocampal CA1*, in preparation.

[47] W. Singer and C. Gray, (1995); *Visual feature integration and the temporal correlation hypothesis*, Ann. Rev. Neurosc 18, pp. 555–586.

[48] C. Tallon-Baudry and O. Bertrand (1999); *Oscillatory gamma activity in humans and its role in object representation*, Trends in Cognitive Neurosci 3, pp. 151–162

[49] R. D. Traub, N. Kopell, A. Bibbig, E. H. Buhl, F. E. N. LeBeau, and M. A. Whittington, (2001); *Gap junctions between interneuron dendrites can enhance synchrony of gamma oscillations in distributed networks*, J. Neurosci. 21 (23), pp. 9478–9486.

[50] C. van Vreeswijk, L. L. Abbott and G. B. Ermentrout (1994); *When inhibition, not excitation synchronizes neural firing*, J. Comput. Neurosci., pp. 313–322.

[51] M. A. Whittington, R. D. Traub and J. G. R. Jefferys (1995); *Synchronized oscillations in interneuron networks driven by metabotropic glutamate receptor activation*, Nature 373, pp. 612–615.

[52] M. A. Whittington, R. D. Traub, N. Kopell, G. B. Ermentrout and E. H. Buhl, (2000); *Inhibition-based rhythms: Experimental and mathematical observation on network dynamics*, Int. J. of Psychophysiology 38, pp. 315–336.

Invited speaker Tom Leighton (Akamai, USA). photography by *Happy Medium Photo Co.*

Peter Markowich currently holds the chair for Applied Analysis at the Institute for Mathematics of the University of Vienna. His research interests gravitate around differential equations, where he has contributed to modelling, analytical and numerical analysis problems. Markowich has held Professorships at the Technical University of Berlin, Purdue University and the Johannes-Kepler-University in Linz. In the year 2000 he received the prestigious Wittgenstein Award of the Austrian Science Fund FWF. Moreover, he is co-directing the Graduate School in Differential Equations, also funded by the FWF. Recently, he has become a founding member of the Wolfgang Pauli Institute (WPI) in Vienna.

Professor Markowich received his PhD in 1980 from the Technical University of Vienna with a thesis on asymptotics of singular ordinary differential equations. In his postdoctoral years (Mathematics Research Center at Madison, Wisconsin and University of Texas at Austin), he started to work on partial differential equations, focusing on models in solid state physics. The two books he has authored on semiconductor models have long since become classical reference books in this very active research area. Meanwhile he has contributed to kinetic equations, analysis and numerics of quantum transport problems, theory of Wigner transforms, homogenisation of dispersive equations, entropy techniques for nonlinear diffusion equations and generalised Sobolev inequalities.

Chapter 14

Highly Oscillatory Partial Differential Equations*

Peter A. Markowich[†]
Christof Sparber[†]

Abstract: Partial differential equations with highly oscillatory solutions occur in many areas of science like quantum mechanics and acoustics, with important spin-offs to semiconductors, nanotechnology and low-temperature physics.

These equations pose a great challenge to mathematical and numerical analysis. Recently a new mathematical technique has been developed to treat highly oscillatory PDEs, namely Wigner transforms. They allow deep new insights into high-frequency asymptotics. The state-of-the-art is reported here, with emphasis on applications like quantum semiconductor devices and Bose–Einstein condensation.

Contents

[1]This work has also been supported by the Austrian Science Foundation FWF through grant no. W8 and the Wittgenstein Award 2000 of P. M. Additional financial sponsorship has been given by the EU research network *HYKE*.

[†]Wolfgang Pauli Institute Vienna and Department of Mathematics, University of Vienna.

1 Introduction

Highly oscillatory problems frequently appear in the study and application of various *linear and nonlinear* partial differential equations (PDEs), most of them being either *hyperbolic* or *dispersive.* A basic example is the free *semi-classical scaled Schrödinger equation* with plane-wave initial data; i.e.

$$-i\varepsilon\,\partial_t u^\varepsilon - \tfrac{1}{2}\varepsilon^2\Delta u^\varepsilon = 0\,, \quad x\in\mathbb{R}^d\,,\ t\in\mathbb{R}\,, \tag{1.1}$$

$$u^\varepsilon(x,0) = e^{ik\cdot x/\varepsilon}\,, \tag{1.2}$$

where $\varepsilon \sim \hbar$ (the scaled Planck's constant), represents the *small semi-classical parameter*; i.e.the *microscopic/macroscopic scale ratio*: $0 < \varepsilon \ll 1$. Here and in the following ε-dependence is denoted by superscripts. The solution of (1.1), (1.2) is explicitly given by

$$u^\varepsilon(x,t) = \exp\Big(i\big(k\cdot\frac{x}{\varepsilon} - |k|^2\frac{t}{2\varepsilon}\big)\Big) \tag{1.3}$$

and hence, we observe that (1.1) propagates *oscillations* with $\mathrm{O}(\varepsilon)$-wave length in *time and space.* In the following we deal with such kind of oscillations and we review (a highly biased selection of) problems, techniques and results in their asymptotic description. In particular, we shall focus on linear and nonlinear *dispersive limits, semi-classical asymptotics, homogenization problems* and *Wigner transformation techniques.*

To this end, we restrict ourselves to PDEs, in which a *small* (dimensionless) parameter ε is present already in the equation itself and ignore cases where ε-oscillations are induced only by the initial-data, since the latter case sometimes is different in several aspects from the former and also, since the former is more frequently encountered in applications. Also we shall not include stochastic models, since their description is beyond the scope of this article. From a mathematical point of view, it is essential for all following examples and methods that the considered systems are *conservative* in the sense that they allow for an *a priori L^2-estimate* on their respective solution.

2 Oscillations in linear PDEs

2.1 The WKB-Approach for Schrödinger-type equations

In this section we consider a class of scalar IVP's for linear dispersive PDEs with *fast* temporal and spatial scales subject to *highly oscillatory initial data.* We are then interested in the *high frequency asymptotics*; i.e.the *geometrical-optics limit* of these equations. The Cauchy problem of the semi-classical Schrödinger equation serves as a typical example

$$-\,i\varepsilon\,\partial_t u^\varepsilon - \tfrac{1}{2}\varepsilon^2 \Delta u^\varepsilon + V(x)u^\varepsilon = 0\,, \quad x \in \mathbb{R}^d\,,\ t \in \mathbb{R}\,, \tag{2.1}$$

$$u^\varepsilon(x,0) = a_I(x)\, e^{i\varphi_I(x)/\varepsilon}\,. \tag{2.2}$$

In the particular case of the Schrödinger equation, the asymptotic regime $\varepsilon \to 0$ is called the *semi-classical limit*; *cf.*[78, 85]. We have seen in the introduction that equations of the form (2.1) propagate oscillations of wave lengths ε. These oscillations inhibit u^ε from converging strongly in a suitable sense and hence the limit $\varepsilon \to 0$ is by no means straightforward, in particular for physical observables, which typically are quadratic in u^ε. The traditional way to deal with this problem is the so-called *WKB-Approach*, named after Wentzel [95], Kramers [63] and Brillouin [13], where one seeks an asymptotic description for the solution u^ε in the following form:

$$u^\varepsilon(x,t) \sim e^{i\varphi(x,t)/\varepsilon}\big(a_0(x,t) + \mathrm{O}(\varepsilon)\big)\,. \tag{2.3}$$

with *real-valued principal amplitude* a_0 and *real-valued phase-function* φ. Formally, plugging (2.3) into (2.1), one obtains in lowest order the *Hamilton–Jacobi equation* (HJ) or *eikonal equation* for the phase:

$$\partial_t\varphi + \tfrac{1}{2}\big|\nabla\varphi\big|^2 + V(x) = 0\,, \tag{2.4}$$

$$\varphi(x,0) = \varphi_I(x)\,. \tag{2.5}$$

Here and in the following "∇" denotes the d-dimensional gradient w.r.t. $x \in \mathbb{R}^d$, unless otherwise indicated. In terms of the *Hamiltonian function*

$$H(x,\xi) := \tfrac{1}{2}|\xi|^2 + V(x) \tag{2.6}$$

this equivalently can be written as

$$\partial_t\varphi + H(x,\nabla\varphi) = 0\,, \tag{2.7}$$

$$\varphi(x,0) = \varphi_I(x)\,. \tag{2.8}$$

We remark that (2.7) is valid also for more general (scalar) dispersive PDEs than the Schrödinger equation; *cf.*[88]. As a next step in the asymptotic expansion, the terms of first order in ε lead to a transport equation for the principal amplitude

$$\partial_t a_0 + \nabla a_0 \cdot \nabla\varphi + \tfrac{1}{2}a_0\Delta\varphi = 0\,, \tag{2.9}$$

$$a_0(x,0) = a_I(x)\,, \tag{2.10}$$

which can be rewritten as a conservation law for the *position density* $\rho := |a_0|^2$:

$$\partial_t\rho + \mathrm{div}(\rho\nabla\varphi) = 0\,, \tag{2.11}$$

$$\rho(x,0) = |a_I(x)|^2. \tag{2.12}$$

For more general PDEs (defined in (2.15 below) one obtains (*cf.*[35, 88])

$$\partial_t \rho + \operatorname{div}\big(\rho \nabla_\xi H(x, \nabla\varphi)\big) = 0\,, \tag{2.13}$$

$$\rho(x,0) = |a_I(x)|^2. \tag{2.14}$$

The system (2.7), (2.13) is called the *WKB-System.* It determines the leading order terms in the expansion (2.3), where the scalar-valued function u^ε is a solution of a general scalar dispersive PDE of the form

$$-i\varepsilon\, \partial_t u^\varepsilon + H^W(x, \varepsilon D) u^\varepsilon = 0\,, \quad x \in \mathbb{R}^d\,,\ t \in \mathbb{R}\,. \tag{2.15}$$

Here $D := -i\nabla_x$ and $H^W(x, \varepsilon D)$ is the *Hamiltonian operator* (assumed to be self-adjoint on $L^2(\mathbb{R}^d)$) obtained by *Weyl-quantising* the symbol $H(x,\xi)$; i.e.

$$H^W(x,\varepsilon D)\, f(x) := \frac{1}{(2\pi)^d} \int_{\mathbb{R}^d}\int_{\mathbb{R}^d} H\big(\tfrac{1}{2}(x+y)\,, \varepsilon\xi\big)\, f(y) e^{i(x-y)\xi}\, \mathrm{d}\xi\, \mathrm{d}y\,. \tag{2.16}$$

Here and in the following, the Weyl-quantisation is used because it is closely connected [39] with the Wigner transform defined in section 5 below. Other quantisation rules could be used as well.

In general, (i.e.except for some special examples of initial conditions) one cannot expect the phase function φ to be smooth for all $t \in \mathbb{R}$, due to the fully nonlinear structure of (2.7). The appearing singularities are usually called *caustics* and it is clear that one can rigorously justify the asymptotic expansion (2.3) only up to times $|t| < t_c$, where $\pm t_c$ denotes the time at which the first caustic appears. More precisely, one can prove the following local-in-time result:

Theorem 1. *Let $u^\varepsilon(t)$ be the solution of (2.15), (2.2) and let a_0, φ be the sufficiently smooth solutions of (2.7), (2.8) and (2.13), (2.14). Then it holds*

$$\sup_{|t|<t_c} \big\| u^\varepsilon(\,\cdot\,,t) - e^{i\varphi(\,\cdot\,,t)/\varepsilon} a_0(\,\cdot\,,t)\big\|_2 \le \mathrm{O}(\varepsilon)\,, \tag{2.17}$$

where $\|\cdot\|_2$ denotes the $L^2(\mathbb{R}^d)$-norm.

The first rigorous result in this direction has been obtained by P. Lax [65] and we refer to [35, 53, 78] and [83] for more details on the WKB-Approach.

An alternative point of view on this problem is given by considering the, so called, *Quantum Hydrodynamic System* (QHD): One represents the solution of Schrödinger's equation in the form

$$u^\varepsilon(x,t) = e^{i\varphi^\varepsilon(x,t)/\varepsilon}\, a^\varepsilon(x,t)\,. \tag{2.18}$$

Inserting this *ansatz* into (2.1), defining $\rho^\varepsilon := |a^\varepsilon|^2$ and $j^\varepsilon := \rho^\varepsilon \nabla\varphi^\varepsilon$, one arrives, after separating real and imaginary parts, at a closed hydrodynamic-type system for ρ^ε and j^ε:

$$\partial_t \rho^\varepsilon + \operatorname{div} j^\varepsilon = 0\,, \tag{2.19}$$

$$\partial_t j^\varepsilon + \operatorname{div}\Big(\frac{j^\varepsilon \otimes j^\varepsilon}{\rho^\varepsilon}\Big) + \rho^\varepsilon \nabla V = \tfrac{1}{2}\varepsilon^2 \rho^\varepsilon \nabla\Big(\frac{1}{\sqrt{\rho^\varepsilon}}\Delta\sqrt{\rho^\varepsilon}\Big)\,. \tag{2.20}$$

This system is exact; i.e. *equivalent* to the Schrödinger equation (as proved in [42]) and well posed for all $t \in \mathbb{R}$, due to the third order dispersive regularisation term. For $\varepsilon = 0$ the system (at least formally) simplifies to the *zero temperature Euler equations* of gas dynamics, for which it is known that in general singularities will appear in finite time. From this point of view, the semi-classical limit is equivalent to a *zero-dispersion limit* [43], a problem which will be again encountered in section 3.1.

2.2 The stationary phase method

The results of section 2.1 show that the WKB-Method gives only local-in-time results, due to the appearance of caustics. Geometrically this corresponds to the crossing of rays, which are the characteristics corresponding to the HJ equation (2.7). After the first caustic and since the underlying dispersive PDE is indeed *linear* we expect the superposition of such rays; i.e. a superposition of WKB-modes.

Let us illustrate this by the following example: Once again, we consider the free Schrödinger equation in $\mathbb{R}^d$, i.e.

$$-i\varepsilon\,\partial_t u^\varepsilon - \tfrac{1}{2}\varepsilon^2 \Delta u^\varepsilon = 0\,, \quad x \in \mathbb{R}^d\,,\ t \in \mathbb{R}\,, \tag{2.21}$$

$$u^\varepsilon(x,0) = a_I(x) e^{i\varphi_I(x)/\varepsilon}\,. \tag{2.22}$$

Its solution is explicitly given by an *oscillatory integral*:

$$u^\varepsilon(x,t) = (2\pi\varepsilon)^{-d} \int_{\mathbb{R}^d}\int_{\mathbb{R}^d} a_I(y) e^{i\varphi(x,y,\xi,t)/\varepsilon}\, \mathrm{d}y\, \mathrm{d}\xi\,, \tag{2.23}$$

where

$$\varphi(x,y,\xi,t) := (x-y)\cdot\xi + \tfrac{1}{2}t|\xi|^2 + \varphi_I(y)\,. \tag{2.24}$$

Next, we recall the theorem of stationary phase [54], where sgn A denotes the number of positive eigenvalues of A minus the number of negative eigenvalues.

Theorem 2. *Let $a \in C_0^\infty(\mathbb{R}^d)$, $\Phi \in C^\infty(\mathbb{R}^d)$ and assume that the set $\{y : \nabla\Phi(y) = 0, y \in \mathrm{supp}(a)\}$ consists of finitely many points y_j, with $j = 1,\dots,M \in \mathbb{N}$. If the Hessians $D^2\Phi(y_j)$ are non-singular, then for $\varepsilon \ll 1$ it holds:*

$$(2\pi\varepsilon)^{-d} \int_{\mathbb{R}^d} a(y) e^{i\Phi(y)/\varepsilon}\, \mathrm{d}y \sim$$

$$\sum_{j=1}^{M} \frac{1}{\sqrt{|\det D^2\Phi(y_j)|}} \exp\Big(\frac{i}{\varepsilon}\Phi(y_j) + \frac{i\pi}{4} m_j\Big)\big(a(y_j) + \mathrm{O}(\varepsilon)\big)\,,$$

where y_j and $m_j := \mathrm{sgn}\, D^2\Phi(y_j)$ is the, so called, Maslov index *of the the j-th ray.*

This implies that locally the main contribution to the solution of the Schrödinger equation stems from stationary points w.r.t. y and ξ, i.e. points at which $\nabla_{y,\xi}\varphi = 0$, which gives

$$\xi = \nabla\varphi_I(y)\,, \quad x = y + t\xi\,. \tag{2.25}$$

The problem of the missing compactness of $\text{supp}(a_I)$ w.r.t. $\xi \in \mathbb{R}^d$ can easily be overcome by an approximation argument, since the number of stationary points only depends on the phase. Hence, we get a ray-map, defined by the following relation

$$x = \hat{x}(t, y) = y + t\nabla\varphi_I(y) \,. \tag{2.26}$$

For small t the map $y \mapsto \hat{x}(t, y)$ is *single-valued.* In general however, there exist (maybe infinitely) many $y_j = y_j(x, t)$, which obey the equation (2.26). Note that the functions $\varphi_I\big(y_j(x,t)\big)$ are local solutions of the Hamilton–Jacobi equation corresponding to (2.21)

$$\partial_t\varphi + \tfrac{1}{2}|\nabla\varphi|^2 = 0 \,, \tag{2.27}$$

called *multi-valued solutions.* The search for an efficient numerical algorithm to compute such solutions has been intensively studied in recent years; *cf.*[7, 48, 86].) Hence, if the assumptions of theorem 2 are satisfied, we obtain (locally away from caustics), the following *multi-valued WKB-approximation* of u^ε:

$$u^\varepsilon(x,t) \sim \sum_{j=1}^{M} \frac{a_I\big(y_j(x,t)\big)}{\big|1 + tD^2\varphi_I\big(y_j(x,t)\big)\big|} \exp\Big(\frac{i}{\varepsilon}\varphi_I\big(y_j(x,t)\big) + \frac{i\pi}{4}m_j\Big) + \mathrm{O}(\varepsilon) \,. \tag{2.28}$$

For equations with variable coefficients, the above concepts have to be generalised, leading to the definition of *Fourier-integral operators* (FIO) and for more details in this direction, we refer to [3, 24, 30, 31, 35, 54] and the references given therein. At caustic-points the FIO is an integral which can be brought into a canonical form and evaluated in terms of special integral functions, *cf.*[1, 30, 72], hence providing a complete description of the high frequency waves. We finally remark that the above theory can be generalised to the case of *matrix-valued equations*; i.e.linear systems of PDEs, though the analysis becomes more involved and the geometric content of the obtained transport equations plays an important role, *cf.*[32]. A particular application from quantum mechanics is the *Dirac equation* [11].

Nowadays the geometrical optics limit for solution of linear PDEs is well understood and there are powerful and very precise analytical machineries available to deal with these kind of problems; *cf.*[35]. As we shall see, the situation is quite different in the case of nonlinear PDEs, where up to now only partial results for a few types of equations are available.

Before we consider nonlinear problems we shall briefly discuss a different source of ε-oscillations in the solution of linear PDEs in the next section.

2.3 Linear Homogenization

In the mathematical modeling of microscopic effects in solid-state physics, various local characteristics, which henceforth appear as coefficients in certain PDEs, are described by functions of the form $a(x, x/\varepsilon, t)$. Here, again $0 < \varepsilon \ll 1$ represents the ratio of microscopic *vs* macroscopic scales and $a(x, y, t)$ is Y-*periodic* w.r.t. $y \in \mathbb{R}^d$, where Y denotes the unit cube in R^d (to keep the discussion simple). The study of the expected high-frequency effects in these coefficients is called *homogenization theory* and a basic reference for it is [9].

In the setting of classical physics for crystals, temperature distributions for example are often described by *elliptic* PDEs of the form

$$-\operatorname{div}\big(a(\tfrac{x}{\varepsilon})\nabla u^\varepsilon\big) = f(x)\,, \quad x \in \Omega \subset \mathbb{R}^d\,, \tag{2.29}$$

$$u^\varepsilon = 0\,, \qquad \text{on } \partial\Omega\,, \tag{2.30}$$

where $f : U \to \mathbb{R}$ is a given function, as is the periodic coefficients matrix $a = \big(a_{i,j}(y)\big) \in \mathbb{R}$, with $i, j = 1, \ldots, d$. We assume uniform ellipticity of the equation and suppose that (at least formally) u^ε admits a *two-scale asymptotic expansion* of the form

$$u^\varepsilon(x) = u_0\big(x, \tfrac{x}{\varepsilon}\big) + \varepsilon u_1\big(x, \tfrac{x}{\varepsilon}\big) + \mathrm{O}(\varepsilon^2)\,. \tag{2.31}$$

Then it can be shown [9] that the leading order term u_0 only depends on x; i.e.$u_0 = v(x)$ and solves the *homogenized problem*:

$$-\operatorname{div}\big(\bar{a}\nabla v\big) = f(x)\,, \quad x \in \Omega \subset \mathbb{R}^d\,, \tag{2.32}$$

$$u = 0\,, \qquad \text{on } \partial\Omega\,, \tag{2.33}$$

in the weak sense. In (2.32), the *homogenized coefficient matrix* $\bar{a} = \bar{a}_{i,j} \in \mathbb{R}$ is given by

$$\bar{a}_{i,j} := \int_Y a_{i,j}(y) - \sum_{k=1}^{d} a_{j,k}(y)\partial_k \chi_i(y)\,\mathrm{d}y\,, \tag{2.34}$$

where χ_i is the solution of the *corrector problem*, stated on the unit cell Y:

$$\operatorname{div}\big(a(y)\nabla\chi_i\big) = \sum_{j=1}^{d} \partial_j a_{i,j}(y)\,, \quad y \in Y\,, \tag{2.35}$$

$$\chi_i \text{ being } Y\text{-periodic, } \forall i = 1, \ldots, d. \tag{2.36}$$

More precisely, one obtains the following convergence result, proved in [9]:

Theorem 3. *Let $u^\varepsilon \in H_0^1(\Omega)$ be the solution of (2.29). Then there exist* a subsequence *ε_k, with $\varepsilon_k \to 0$, as $k \to \infty$, and $u \in H_0^1(\Omega)$, satisfying (2.32), such that*

$$u^\varepsilon \overset{k\to\infty}{\longrightarrow} u \quad \textit{weakly in } H_0^1(\Omega)\,. \tag{2.37}$$

Similar results hold for coefficients including an additional *slow scale*; i.e.$a = a(x, x/\varepsilon)$, for *almost periodic operators* and also for certain classes of integral equations with highly oscillatory coefficients; *cf.*[9, 21, 58, 94].

The situation gets much more complicated when one passes to *non-elliptic problems*. A typical example is the Schrödinger equation for the motion of an electron in a crystal:

$$-i\varepsilon\,\partial_t u^\varepsilon - \tfrac{1}{2}\varepsilon^2\Delta u^\varepsilon + V\Big(\frac{x}{\varepsilon}\Big)u^\varepsilon = 0\,, \quad x \in \mathbb{R}^d\,,\ t \in \mathbb{R}\,, \tag{2.38}$$

$$u^\varepsilon(x, 0) = u_I^\varepsilon(x)\,. \tag{2.39}$$

Here $V(y) = V(y+\gamma)$, $\gamma \in \Gamma$, typically models the periodic potential of the *crystal-lattice ions*. In the case of one spatial dimension, $\Gamma = 2\pi Z$ and the *first Brillouin-Zone* reads $Y = [-\frac{1}{2}, \frac{1}{2}]$, which is nothing but the *Wigner–Seitz cell* of the *dual lattice* $\Gamma^* = Z$; *cf.*[9].

In (2.38), $\varepsilon \to 0$ corresponds to the *simultaneous* semi-classical and homogenization limit. (See also [82] for a more detailed study on the connection of these limiting regimes.) A naive WKB-Approach along the lines of section 2.1 gives a weakly coupled system

$$\partial_t \varphi + \tfrac{1}{2}|\nabla\varphi|^2 + V\Big(\frac{x}{\varepsilon}\Big) = 0\,, \tag{2.40}$$

$$\partial_t \rho + \operatorname{div}(\rho\nabla\varphi) = 0\,, \tag{2.41}$$

which still carries the small parameter ε. Hence, one would be forced to homogenize the HJ equation (2.40), a procedure, which will be discussed in section 4 below and which is known to neglect important dispersive effects of the original Schrödinger equation (2.38). Numerical evidence for this fact can be found in [50]. To overcome this difficulty, a generalised *two-scale WKB-expansion* has been introduced in [52]

$$u^\varepsilon(x,t) \sim e^{i\varphi(x,t)/\varepsilon}\big(a_0(x,t,x/\varepsilon) + \mathrm{O}(\varepsilon)\big)\,, \tag{2.42}$$

where a_0 is *lattice-periodic* w.r.t. the fast variable; i.e.$a_0(x,t,y) = a_0(x,t,y+\gamma)$, $\gamma \in \Gamma$. At least formally [52, 50], this leads to the following *Bloch-band Hamilton–Jacobi equation*:

$$\partial_t \varphi + E_n(\nabla\varphi) = 0\,, \tag{2.43}$$

where the *n-th band energy* $E_n = E_n(k)$, $n \in \mathbb{N}$, is obtained from the classical *Bloch eigenvalue-problem* [10] (see also [84]):

$$\begin{aligned} -\tfrac{1}{2}\Delta\psi_n(y,k) + V(y)\psi_n(y,k) = E_n(k)\psi_n(y,k)\,, \quad y \in \mathbb{R}\,,\ k \in Y\,, \\ \psi_n \text{ and } \nabla\psi_n \text{ being } Y\text{-quasi-periodic.} \end{aligned} \tag{2.44}$$

Here, *quasi-periodic* means that for some $k \in Y$, the eigenfunctions ψ_n satisfy

$$\psi_n(y+\gamma,k) = e^{ik\cdot y}\psi_n(y,k)\,, \quad \forall y \in \mathbb{R}\,,\ n \in \mathbb{N}\,. \tag{2.45}$$

Now, writing $a_0(x,t,x/\varepsilon) = \tilde{a}(x,t)\sigma_n(x/\varepsilon,k)$, with $k = \nabla\varphi(x,t)$ and $\sigma_n(y+\gamma,k) = \sigma_n(y,k)$ one gets a conservation law for the n-th band density

$$\partial_t \tilde{\rho} + \operatorname{div}\big(\tilde{\rho}\nabla_k E_n(\nabla\varphi)\big) = 0\,, \quad \tilde{\rho} := |\tilde{a}|^2. \tag{2.46}$$

A more rigorous approach to this is via Bloch's theorem [10, 84], which asserts a decomposition of the original Hilbert-space into *countable many orthogonal (spectral) subspaces* $L^2(\mathbb{R}^d) = \oplus_{n=1}^{\infty} S_n^\varepsilon$, each of which is invariant under the lattice Hamiltonian. This leads to more sophisticated treatments of this problem, using either Wigner transforms; *cf.*the references in section 5.3, or *adiabatic decoupling theory* [55, 81]. Both of these methods do not suffer from the occurrence of caustics, in contrast to the two-scale WKB-approach above. Nevertheless, even in these more advanced approaches the problem of *band-crossings* is encountered and up to now resolved for some special cases only; *cf.*[37, 38] and the references given therein. In this case, the main obstruction for a rigorous treatment is the possible energy-exchange between different Bloch-bands, the so-called *Landau–Zener effect*, [64, 98].

We finally remark that there are several recent papers in which the connection between the Bloch-decomposition and traditional homogenization theory is studied in more detail; *cf.*[21, 34, 50].

3 Oscillation in nonlinear PDEs

The theory of oscillations in nonlinear PDEs is far from being completed. The appearing high-frequency phenomena depend on the considered equation, the size of the inital data, the structure and the scaling of the nonlinearities. In the following subsections we shall consider two important research directions.

3.1 Nonlinear Dispersive limits

We have seen in section 2.1 that high-frequency asymptotics are closely related to *small dispersion limits.* One of the best studied examples in this field is the quasi-linear *Korteweg–de Vries equation* (KdV):

$$\partial_t u^\varepsilon - u^\varepsilon \partial_x u^\varepsilon = \tfrac{1}{3}\varepsilon^2\, \partial^3_{xxx} u^\varepsilon\,, \quad x\,,\ t \in \mathbb{R}\,, \tag{3.1}$$

$$u^\varepsilon(x,0) = u_I(x)\,. \tag{3.2}$$

This equation appears in the descriptions of shallow water-waves [96] and can be seen as a regularised version (due to the third order dispersive term) of *Burgers' equation*:

$$\partial_t u - u\,\partial_x u = 0\,, \quad x\,,\ t \in \mathbb{R}\,, \tag{3.3}$$

$$u^\varepsilon(x,0) = u_I(x)\,. \tag{3.4}$$

Similar to the semi-classical limit for the QHD-system, the formal limit from KdV to Burgers, as $\varepsilon \to 0$, holds only up to caustics; i.e.for $|t| < t_c$, where $\pm t_c$ is the time of formation of the first shock in Burgers' equation. However, due to fact that the KdV-equation represents a *fully integrable infinite dimensional dynamical system*, one can indeed pass over the caustic and give a description of the zero-dispersion limit also for $|t| \geq t_c$; *cf.*[66, 67, 68]. However, these results heavily rely on an inverse-scattering-approach, which is beyond the scope of this article and which we shall not discuss in detail, also, since this technique is precisely tuned to the fully integrable case and can not be generalised to non-integrable PDEs.

A closely related problem is the semi-classical limit for *nonlinear Schrödinger-type equations* (NLS):

$$-\, i\varepsilon\, \partial_t u^\varepsilon - \tfrac{1}{2}\varepsilon^2 \Delta u^\varepsilon + V\big(|u^\varepsilon|^2\big) u^\varepsilon = 0\,, \quad x \in \mathbb{R}^d\,,\ t \in \mathbb{R}\,, \tag{3.5}$$

$$u^\varepsilon(x,0) = a_I(x) e^{i\varphi_I(x)/\varepsilon}\,. \tag{3.6}$$

Equations of NLS-type arise in various branches of theoretical physics, like water-wave dynamics, plasma physics and Bose–Einstein condensation; *cf.*[93]. Locally in time, the limit $\varepsilon \to 0$ can be rigorously justified by using the WKB-ansatz (2.18), rewriting (3.5) as a corresponding QHD-system

$$\partial_t \rho^\varepsilon + \operatorname{div} j^\varepsilon = 0\,, \tag{3.7}$$

$$\partial_t j^\varepsilon + \operatorname{div}\Big(\frac{j^\varepsilon \otimes j^\varepsilon}{\rho^\varepsilon}\Big) + \rho^\varepsilon \nabla V(\rho^\varepsilon) = \tfrac{1}{2}\varepsilon^2 \rho^\varepsilon \nabla\Big(\frac{1}{\sqrt{\rho^\varepsilon}}\Delta\sqrt{\rho^\varepsilon}\Big) \tag{3.8}$$

and studying its zero-dispersion limit: The similarity to (3.1) becomes more clear if we define $\nabla\varphi^\varepsilon =: u^\varepsilon$. Then, since $j^\varepsilon = \rho^\varepsilon u^\varepsilon$, (3.8) can be formally rewritten as

$$\partial_t u^\varepsilon - (u^\varepsilon\cdot\nabla)u^\varepsilon + \nabla V(\rho^\varepsilon) = \tfrac{1}{2}\varepsilon^2 \nabla\Big(\frac{1}{\sqrt{\rho^\varepsilon}}\Delta\sqrt{\rho^\varepsilon}\Big)\,, \quad x\,,\ t \in \mathbb{R}^d\,, \tag{3.9}$$

which again furnishes a Burgers' equation with third order dispersive regularisation. The first rigorous results as $\varepsilon \to 0$ were given in [45, 51], where the nonlinearity is assumed to be such that: $V \in C^\infty(\mathbb{R}_+;\mathbb{R})$ and $V' > 0$; i.e.a so called *defocusing nonlinearity.* For more general defocusing nonlinearities (involving also derivatives of u^ε) we refer to [43] and the references given therein. We remark that in the *focusing case* (i.e.$V' < 0$) the limiting hydrodynamical system is *not* hyperbolic and the rigorous zero-dispersion limit of (3.7), (3.8) has not yet been found.

To make things even worse, there are only a few global-in-time results for small ε available, since, in contrast to the linear case, one can *not* expect a superposition of WKB-modes to be a valid description after the caustic. Also, the highly advanced machinery of FIO's in general does not apply to nonlinear PDEs (for a notable exception see [61]).

For the special case $d = 1$, $V = \mathrm{id}$, (i.e.the so called *cubic defocusing NLS*) the equation (3.5) is again fully integrable leading to a similar global-in-time inverse scattering approach as in the KdV-case; *cf.*[69]. For the much more often encountered *non*-fully integrable NLS equations, the only global-in-time results are given in [16, 18] (using scattering theory), [62] and we also refer to the paper [19], which examines in more detail the role of quadratic oscillations in nonlinear Schrödinger equations.

We conclude from the above that so far there exists only a weak mathematical theory for the description of nonlinear high frequency limits of dispersive equations and we refer the interested reader to [2, 46, 57, 91] for further developments.

3.2 Weakly nonlinear dispersive geometrical optics for first order hyperbolic systems

In the examples of the previous section, regularising higher order (dispersive) effects are lost in the limit $\varepsilon \to 0$. On the other hand it is well known that ε-oscillations are also present in *first order hyperbolic systems*, which do not include any higher order terms. The first rigorous work on the corresponding geometrical optics approximation is [65] (see also [83]), which deals with linear equations. For nonlinear systems the theory has been developed in [29, 56].

Consider the following class of oscillatory IVP:

$$A_0(u^\varepsilon)\varepsilon\,\partial_t u^\varepsilon - \sum_{k=1}^{d} A_k(u^\varepsilon)\varepsilon\,\partial_k u^\varepsilon + g(x) = f(u^\varepsilon)\,, \quad x \in \mathbb{R}^d\,, \quad t \in \mathbb{R}\,, \tag{3.10}$$

$$u^\varepsilon(x,0) = \varepsilon^p a_I(x) e^{i\varphi(x)/\varepsilon}\,, \; p \in Q_+\,, \tag{3.11}$$

where either $u^\varepsilon \in \mathbb{R}^d$ or $u^\varepsilon \in C^d$ and the matrices A_k are such that (3.10) constitutes a hyperbolic system; i.e.

$$A_k^*(u^\varepsilon) = A_k(u^\varepsilon)\,, \quad A_0(0) > 0\,. \tag{3.12}$$

The exponent p, determing the amplitude in (3.11), will be fixed below. Further assume on the nonlinearities that $A(u^\varepsilon)$, $f(u^\varepsilon)$ are sufficiently smooth and homogeneous of order $O(u^J)$, with $J \geq 2$. Then, it is shown in [29] that, locally in time, it holds

$$u^\varepsilon(x,t) \sim v^\varepsilon(t,x) := \varepsilon^p v_0(x,t,\varphi(x,t)/\varepsilon) + O(\varepsilon^{p+1})\,, \quad |t| < t_c\,, \tag{3.13}$$

where $\varphi(x,t) \in \mathbb{R}$ is a real-valued phase-function, the principal amplitude $v_0(x,t,\theta)$ is 2π-periodic w.r.t. θ and the *critical exponent* p is determined such that the *normalisation condition*:

$$pJ = p + 1 \tag{3.14}$$

holds. With this normalisation, the nonlinearity does *not* enter in lowest order; i.e.in the Hamilton–Jacobi equation for the phase

$$\det\Big(iA_0(0)\,\partial_t\varphi - i\sum_{k=1}^{d} A_k(0)\,\partial_k\varphi + g(x)\Big) = 0\,, \tag{3.15}$$

$$\varphi(x,0) = \varphi_I(x)\,, \tag{3.16}$$

but only in the (nonlinear) transport-equation for the principal amplitude v_0. Hence, one usually refers to this asymptotic regime as *weakly nonlinear geometrical optics*. It can be shown that p is critical in the sense that for amplitudes $O(\varepsilon^p)$ one can prove simultaneously existence of the approximate smooth solution for times $t = O(1)$, i.e.on a time-scale *independent* of ε, and nontrivial nonlinear behavior in the principal term of the approximation. More precisely, we have for all multi-indices α

$$\partial_x^\alpha(u^\varepsilon - v^\varepsilon) = O(\varepsilon^n), \quad \forall n \in \mathbb{N}\,, \tag{3.17}$$

uniformly on $[0,T] \times \mathbb{R}^d$, where $T \in (0, t_c)$. In general the nonlinearity will produce *higher order harmonics*; i.e.we obtain

$$v_0(x,t,\theta) = \sum_{m\in Z} v_{0,m}(x,t)e^{im\theta}\,, \tag{3.18}$$

where for every relevant physical model (considered so far), one can prove that only a finite number of these harmonics are non-zero. For a particular class of first-order systems which in addition satisfy the so-called *transparency condition*, it is even possible to conclude the existence of O(1)-asymptotic solutions, using a tricky change of unknowns; *cf.*[60]. However, global-in-time results in general are out of reach (as for nonlinear higher order dispersive equations).

Recent applications of this method include the semi-classical asymptotic description of the *Dirac–Maxwell system* [89] and the rigorous derivation of nonlinear Schrödinger equations from quadratic hyperbolic systems [20]. Finally, the validity of geometrical optics for first order hyperbolic systems can be extended to more general *oscillating solutions* u^ε; *cf.*[27, 28, 91], which also provides a link between geometrical optics and *compensated compactness.*

4 Homogenization of nonlinear PDEs

Homogenization theory has been successfully adapted to various nonlinear PDEs. For nonlinear elliptic problems (i.e.generalisations of 2.29) the results are similar to the ones obtained in the linear case, assuming that certain technical conditions on the considered operators hold (monotonicity for example). We therefore omit more details on this field of research and refer the interested reader to [58, 41, 94] and the references given therein.

Rather we shall turn our attention to the homogenization problem of fully nonlinear first order PDEs, most prominently *periodic Hamilton–Jacobi equations.* Hence, we consider

$$\partial_t u^\varepsilon + H\Big(\frac{x}{\varepsilon}, \nabla u^\varepsilon\Big) = 0\,, \quad x \in \mathbb{R}^d\,,\ t \in \mathbb{R}\,, \tag{4.1}$$

$$u^\varepsilon(x,0) = u_I(x)\,, \tag{4.2}$$

where the Hamiltonian $H(y,\xi)$ is assumed to be (uniformly) *convex* w.r.t $\xi \in \mathbb{R}^d$ and Y-periodic w.r.t. $y \in \mathbb{R}^d$. The first result in this direction is given in an unpublished but nevertheless famous paper by P. L. Lions, G. Papanicolaou and S. R. Varadhan in 1988. More recent references are [33, 22, 23] and [80].

The homogenization results are based on the notion of *viscosity solutions* [25], which proved to be the right concept of generalised solutions for fully nonlinear PDEs of first and second order [26]. One can show that in the sense of viscosity solutions $u^\varepsilon \to u$, as $\varepsilon \to 0$, uniformly on compact subsets of $[0,\infty) \times \mathbb{R}^d$, where u solves a *homogenized Hamilton–Jacobi equation* of the form

$$\partial_t u + \overline{H}(\nabla u) = 0\,, \quad x \in \mathbb{R}^d\,,\ t \in \mathbb{R}\,, \tag{4.3}$$

$$u(x,0) = u_I(x)\,. \tag{4.4}$$

Here, $\overline{H} : \mathbb{R}^d \to \mathbb{R}$ is the *effective Hamiltonian*, which is nothing but the *unique real number*, such that the following *corrector problem*

$$H(P + \nabla v, y) = \overline{H}(P)\,, \quad P \in \mathbb{R}^d \text{ fixed}\,, \tag{4.5}$$

$$v(P,y) \text{is } Y\text{-periodic w.r.t. } y \in \mathbb{R}^d \tag{4.6}$$

admits a viscosity solution $v = v(P,y)$, which, however, may not be unique; *cf.*[33].

We remark that a recent paper of L. C. Evans [34] links the eigenvalue-problem (2.44), HJ homogenization theory and *Mather's theory of action minimising measures.* Using this connection, an alternative approach to the WKB-approximation for time-independent Schrödinger equations is given. We further remark that the homogenization problem for HJ equations still is an intense field of ongoing research, in particular by including stochasticity; *cf.*[15, 71, 87].

5 Wigner transformation techniques

The Wigner transformation approach provides a rigorous tool for the description of high-frequency and homogenization-limits of physical observables; i.e. *quadratic quantities* build out of the solution u^ε of some linear dispersive PDE. The foundation for this theory is laid in the early papers on *H-measures* and *microlocal defect-measures*; *cf.*[44, 92] (see also [14]). In the following, the basic idea is to pass from PDEs in physical space R^d_x to an equivalent description in *phase-space* $R^d_x \times \mathbb{R}^d_\xi$, where the problem of caustics no longer appears.

5.1 Phase-space description for dispersive equations

We again consider a linear scalar dispersive equation of the the form (2.15)

$$- i\varepsilon\, \partial_t u^\varepsilon + H^W(x, \varepsilon D) u^\varepsilon = 0\,, \quad x \in \mathbb{R}^d\,,\ t \in \mathbb{R}\,, \tag{5.1}$$

$$u^\varepsilon(x,0) = u_I^\varepsilon(x)\,, \tag{5.2}$$

where $u^\varepsilon(x,t) \in C$ and the Weyl-quantised operator $H^W(x,\varepsilon D)$ is assumed to be a *self-adjoint (pseudo-) differential operator* on $L^2(\mathbb{R}^d)$. Hence, from Stone's theorem we obtain

$$\left\|u^\varepsilon(\,\cdot\,,t)\right\|_2 = \left\|u_I^\varepsilon(\,\cdot\,)\right\|_2 . \tag{5.3}$$

We can now define the *Wigner transform* [97] corresponding to u^ε:

$$w^\varepsilon(x,\xi,t) := \frac{1}{(2\pi)^d} \int_{\mathbb{R}^d} u^\varepsilon\big(x - \tfrac{1}{2}\varepsilon\eta\,, t\big)\overline{u^\varepsilon}\big(x - \tfrac{1}{2}\varepsilon\eta, t\big) e^{i\xi\cdot\eta}\, \mathrm{d}\eta\,. \tag{5.4}$$

In the context of quantum mechanics, the *real-valued* Wigner transform is interpreted as a *phase-space description* of the quantum state u^ε, although in general w^ε is *not positive a.e.*; *cf.*[39].

As an example, we apply the Wigner transform to the Schrödinger equation (2.1), which yields [70]

$$\partial_t w^\varepsilon + \xi \cdot \nabla_x w^\varepsilon - \theta^\varepsilon[V] w^\varepsilon = 0\,, \quad x\,, \xi \in \mathbb{R}^d\,,\ t \in \mathbb{R}\,, \tag{5.5}$$

$$w^\varepsilon(x,\xi,0) = w_I(x,\xi)\,, \tag{5.6}$$

where $\theta^\varepsilon[V]$ is a pseudo-differential operator, defined by

$$\theta^\varepsilon[V] w^\varepsilon(x,\xi,t) := \frac{i}{\varepsilon(2\pi)^d} \int_{\mathbb{R}^{2d}} \big[V(x - \tfrac{1}{2}\varepsilon) - V(x + \tfrac{1}{2}\varepsilon)\big] w^\varepsilon(x,\xi',t)\, e^{i\eta(\xi-\xi')}\, \mathrm{d}\eta\, \mathrm{d}\xi'\,. \tag{5.7}$$

Note that in the free motion case (i.e.$V(x) \equiv 0$) the Wigner equation becomes the free transport equation of classical statistical mechanics. The important feature of the Wigner transform is that it facilitates a *classical* computation of expectation values (mean values) for *quantum mechanical observables* $A^W(x, \varepsilon D)$ in a state u^ε, namely

$$\big\langle u^\varepsilon(\,\cdot\,,t)\,, A^W(x,\varepsilon D) u^\varepsilon(\,\cdot\,,t)\big\rangle_2 = \int_{\mathbb{R}^d}\int_{\mathbb{R}^d} A(x,\xi) w^\varepsilon(x,\xi,t)\, \mathrm{d}x\, \mathrm{d}\xi\,, \tag{5.8}$$

where the corresponding *classical observable* is represented by a real-valued symbol $A(x,\xi) \in \mathcal{S}(\mathbb{R}^{2d})$. The Wigner transform can be generalised to the, so called, *Wigner matrix* [47]

$$W^\varepsilon(x,\xi,t) := \frac{1}{(2\pi)^d}\int_{\mathbb{R}^d} u^\varepsilon(x - \tfrac{1}{2}\varepsilon\eta\,,t) \otimes \overline{u^\varepsilon}(x - \tfrac{1}{2}\varepsilon\eta\,,t)\, e^{i\xi\cdot\eta}\, \mathrm{d}\eta\,, \tag{5.9}$$

where u^ε is a n-vector valued function and $\otimes$ denotes the tensor product of vectors. This allows one to treat matrix-valued *first-order systems, wave equations* as well as *matrix-valued observables.* For more details, examples and applications we refer to [14, 39, 70] and [88].

5.2 Semi-classical limit of physical observables

The following main theorem on Wigner transforms is proved in [47]:

Theorem 1. *If $u^\varepsilon \in L^\infty\big(\mathbb{R}\,; L^2(\mathbb{R}^d)\big)$, then its Wigner transform w^ε is uniformly bounded in $L^\infty\big(\mathbb{R}\,;\mathcal{S}'(\mathbb{R}^d_x\times\mathbb{R}^d_\xi)\big)$ as $\varepsilon\to 0$. Thus, by compactness, there exists a sub-sequence ε_k, with $\varepsilon_k \to 0$, as $k\to\infty$, and a distribution $w\in\mathcal{S}'(\mathbb{R}^d_x\times\mathbb{R}^d_\xi)$, such that*

$$w^{\varepsilon_k} \stackrel{k\to\infty}{\longrightarrow} w \ \text{in}\ L^\infty\big(\mathbb{R}\,;\mathcal{S}'(\mathbb{R}^d_x\times\mathbb{R}^d_\xi)\big)\ \text{weak-}\star. \tag{5.10}$$

Moreover $w(t)\in\mathcal{M}^+(\mathbb{R}^d_x\times\mathbb{R}^d_\xi)$; i.e.it is a bounded positive Borel-measure on phase space.

Hence, w indeed can be interpreted as a classical phase-space measure, called the *Wigner measure* of u^{ε_k}. In particular, if u^ε is given in WKB-form (2.3), then (up to extraction of subsequences)

$$w^\varepsilon \stackrel{\varepsilon\to 0}{\longrightarrow} \rho(x,t)\,\delta\big(\xi - \nabla\varphi(x,t)\big)\,, \quad \rho(x) := |a_0(x,t)|^2. \tag{5.11}$$

An evolution equation for the limiting Wigner measure can be derived, too.

Theorem 2. *Let $u^\varepsilon\in L^\infty\big(\mathbb{R}\,;L^2(\mathbb{R}^d)\big)$ be the solution of equation (5.1), then the corresponding Wigner measure is a weak solution of the classical* Liouville equation

$$\partial_t w + \{H, w\} = 0, \quad x\in\mathbb{R}^d\,,\ t\in\mathbb{R}\,, \tag{5.12}$$

$$w(x,\xi,0) = w_I(x)\,, \tag{5.13}$$

where $\{f, g\} := \nabla_x f \cdot \nabla_\xi g - \nabla_\xi f \cdot \nabla_x g$.

Note that for these results, we do not need that u_I^ε is of WKB-type and moreover we do not assume u^ε to be very regular (which is required for example in the stationary phase theorem).

In comparison to WKB-methods, the great advantage of the Wigner formalism is the fact that no singularities (i.e.caustics) appear in the solution of (5.12). More precisely, one can prove [88], that the solution of (5.12) is a so called *mono-kinetic distribution* (5.11), if and only if ρ, φ solve the WKB-system (2.7), (2.13). Moreover, assuming that only a *finite* number of branches appear in the multi-valued Hamilton–Jacobi solution, one can show, that locally away from caustics, w can always be decomposed into a finite sum of mono-kinetic distributions. This statement in some sense is the Wigner measure analogue of the stationary phase theorem.

The above results imply that the Wigner transform allows a rigorous and quite straightforward treatment of the semi-classical limit for physical observables, since from (5.8) and theorem 1 we obtain

$$\big\langle u^\varepsilon(\,\cdot\,,t), A(x,\varepsilon D)\, u^\varepsilon(\,\cdot\,,t)\big\rangle_2 \overset{\varepsilon\to 0}{\longrightarrow} \int_{\mathbb{R}^d}\int_{\mathbb{R}^d} A(x,\xi) w(x,\xi,t)\, \mathrm{d}x\, \mathrm{d}\xi\,, \tag{5.14}$$

at least for $A(x,\xi) \in \mathcal{S}$, but this also holds for more general symbols A, under the additional assumption that u_I^ε is *ε-oscillatory* and *compact at infinity*; *cf.*[47, 70]. A function u_I^ε satisfying these properties can be seen as generalised WKB-initial data.

5.3 Quadratic Homogenization and Wigner transforms

A particularly important application of Wigner measures can be found in *quadratic homogenization problems*; i.e.*homogenization of physical observables*. This has been done for (acoustic) wave equations in [73] (see also [40] for a H-measure treatment) and for Schrödinger equations in [75, 82] (see also [50] for a numerical study). To this end, the more adapted concept of *Wigner series* has been developed; *cf.*[47]:

$$w_s^\varepsilon(x,k,t) := \frac{1}{|Y|}\sum_{\gamma\in\Gamma} u^\varepsilon\big(x - \tfrac{1}{2}\varepsilon\gamma\,,t\big)\overline{u^\varepsilon}\big(x + \tfrac{1}{2}\varepsilon\gamma\,,t\big)\, e^{ik\cdot\gamma}\,, \quad x \in \mathbb{R}^d\,,\ k \in Y\,, \tag{5.15}$$

where Y again denotes the first Brillouin Zone of the dual lattice Γ^*. Having in mind the example of a periodic Schrödinger equation (2.38) we start from a scalar periodic IVP

$$-\,i\varepsilon\,\partial_t u^\varepsilon + H^W\Big(\frac{x}{\varepsilon}\,,\varepsilon D\Big)u^\varepsilon = 0\,, \quad x \in \mathbb{R}^d\,,\ t \in \mathbb{R}\,, \tag{5.16}$$

$$u^\varepsilon(x,0) = u_I^\varepsilon(x) \tag{5.17}$$

and, by using a Bloch-decomposition [84] $L^2(\mathbb{R}^d) = \oplus_{n=1}^\infty S_n^\varepsilon$; i.e.writing

$$u^\varepsilon(x,t) = \sum_{n\in\mathbb{N}} u_n^\varepsilon(x,t)\,, \tag{5.18}$$

we arrive at a IVP within each Bloch-band space

$$-i\varepsilon\,\partial_t u_n^\varepsilon + E_n(\varepsilon D)u_n^\varepsilon = 0\,, \quad x\in\mathbb{R}^d\,,\ t\in\mathbb{R}\,, \tag{5.19}$$

$$u_n^\varepsilon(x,0) = u_{n,I}^\varepsilon(x)\,. \tag{5.20}$$

Here E_n is the Fourier-multiplier pseudo-differential operator corresponding to the Bloch-eigenvalue $E_n(k)$, obtained in (2.44). One can then pass to the limit $\varepsilon\to 0$ within the corresponding Wigner series and since the $u_n^\varepsilon \perp u_m^\varepsilon$, for $n\neq m$, we can decompose the limiting measure in its Bloch-band representation [47]

$$w_s(x,k,t) = \sum_{n\in\mathbb{N}} w_n(x,k,t)\,. \tag{5.21}$$

In each band we henceforth derive the following homogenized IVP

$$\partial_t w_n + \nabla_k E_n(k)\cdot\nabla_x w_n = 0\,, \quad x\in\mathbb{R}^d\,,\ t\in\mathbb{R}\,, \tag{5.22}$$

$$w_n(x,k,0) = w_{n,I}(x,k)\,, \tag{5.23}$$

which again can be connected with the corresponding two-scale WKB-approach, iff w_n is in mono-kinetic form

$$w_n(x,k,t) = \rho(x,t)\,\delta_\sharp\big(k-\nabla\varphi(x,t)\big)\,, \tag{5.24}$$

where $\delta_\sharp$ denotes the Γ-periodic Delta-distribution. The homogenization of physical observables then follows as in section 5.2 above. Of course, the above procedure can only be applied if the initial Wigner measure vanishes on sets of band-crossings. To study energy/charge-transitions at crossings a more refined analysis must be performed; *cf.*[38]. For completeness, we remark that the long standing problem of including simultaneously periodic *and* external fields (which are neglected in [47] and [75]) has been resolved recently in [81], by using adiabatic decoupling theory.

In recent years, the Wigner transforms have been successfully applied to more advanced problems, such as the semi-classical limit of the *Dirac equation* [90], the derivation of effective mass theorems in crystals [82], the semi-classical limit for *weakly nonlinear Schrödinger–Poisson systems* [74, 99], the refraction of high-frequency waves [79], the high-frequency analysis of the *Helmholtz equation* [8] and the study Landau–Zener phenomena; *cf.*[36, 37, 38].

A drawback of the Wigner measure approach in comparison to WKB-type methods is the *loss of phase-information.* For example, the change of the Maslov indices, experienced when passing through the caustic in the multi-valued WKB-representation (2.28), is not tractable by using Wigner transforms; *cf.*[88]. Another, even more severe problem is that Wigner transforms so far can not be applied to strongly nonlinear problems, like for example semi-classical nonlinear Schrödinger equations. In this context, the results in [17] show that the semi-classical limit for a class defocusing nonlinear Schrödinger equations is *ill-posed* in terms of Wigner measures.

6 Numerical challenges

The numerical solution of highly oscillatory PDEs poses formidable challenges to numerical analysis. Typical questions are:

- Is it necessary to resolve all the fine-structure (oscillations) accurately by the discretisation or can a numerical scheme be devised on a relatively coarse grid such that (at least) weak limits of solutions or observables, resp. are well approximated? How coarse can this grid be (in terms of the small parameter characterising the oscillation wave length)?
- If there are no analytical results available on the accuracy of discretisations or/and on the limiting behavior of solutions or observables resp. (as is often the case for nonlinear dispersive IVPs), then how can the numerically obtained results be trusted, particularly in $d = 3$ spatial dimensions when, as is often the case, another grid refinement is *prohibitive* due to limited computer resources/time?
- How can analytical information on the continuous problem and on the limiting behavior of solutions be efficiently incorporated in the numerical technique?

Let us start the discussion with:

1. *Linear Schrödinger-type equations*:
It was shown in [76, 77], that finite difference discretisations have significant drawbacks in semi-classical limit regimes for linear Schrödinger equations. In the best of all cases (i.e.in the *Crank–Nicolson scheme* and *leap-frog scheme*) they require the temporal and spatial grid sizes to be much smaller than the semi-classical parameter $\Delta t = \mathrm{o}(\varepsilon)$, $\Delta x = \mathrm{o}(\varepsilon)$, in order to give accurate observables. Much more stringent mesh size constraints are required for accurate wave-functions and for non-time reversible discretisations like *Euler schemes.* These serious meshing restrictions of finite difference schemes is intimately connected with the fact that they are *not time-transverse-invariant*; i.e.contrary to the continuous Schrödinger equation a constant shift of the potential changes the numerically obtained observables.

    In comparison, the *Time-Splitting-Trigonometric-Spectral Method* (TSTSM), based on splitting the kinetic term from the potential term and trigonometric spectral discretisation of the Laplacian (which allows for explicit time-integration of both splitting steps in every time step), behaves significantly better. It was shown in [5] that the TSTSM conserves the total charge, is time-transverse-invariant, reversible and requires only $\Delta x = \mathrm{O}(\varepsilon)$ and $\Delta t = \mathrm{O}(1)$ (but dependent on the prescribed error tolerance), for giving accurate observables. Again, more restrictive meshes are required in order to obtain accurate wave-functions. We remark that the analysis of the approximation of observables obtained by discretisations of linear Schrödinger-type equations in the semi-classical regime was based on Wigner-measure techniques and not on the classical consistency-stability concept of numerical analysis, which gives non-sharp results.

    Recently, *hyperbolic technology* based on the concept of *K-multivalued solutions* of conservation laws [12] was used to solve the WKB-system; *cf.*[48, 49]

and a level-set method was presented in [59]. These methods have the advantage that they give more direct information on the caustic structure of the underlying Schrödinger equation since they are based on solving directly for macroscopic semi-classical quantities. Their disadvantage is the bad accuracy in the reconstructed wave-function at caustic manifolds (*phase transitions*).

2. *Quadratic homogenization for Schrödinger equations with periodic potentials*:
The main numerical difficulty is that, as $\varepsilon \to 0$, higher and higher *Fourier-modes* of the lattice periodic potential affect the wave-function. A systematic approach, bypassing this difficulty, was presented in [50]. At first the Bloch (spectral) decomposition is carried out numerically by a spectral technique, then the (differentiated) Hamilton–Jacobi equation (2.43) (in the n-th energy band, $n \in \mathbb{N}$ fixed) is solved for the multivalued n-th band particle velocity by the method of K-multivalued solutions of conservation laws and, as postprocessing, the transport equation (2.46) for the wave-function amplitudes is solved numerically. Then the wave-function is reconstructed by following the two-scale WKB technique. So far, the method of K-multivalued solutions of conservation laws is restricted to one-dimensional problems, multi-dimensional versions are being investigated at the time being.

3. *Nonlinear Schrödinger equations*:
The main numerical obstacles are the formation of singularities in focusing nonlinear Schrödinger equations and the creation of new scales at caustics for focusing and defocusing NLS. Typically, *Krasny filters* (high Fourier-mode cut-off) are needed in order to avoid artifacts (like symmetry breaking) in focusing NLS computations. Finite difference methods typically require prohibitively fine meshes to even approximate observables well in semi-classical defocusing and focusing NLS computations. However, due to the creation of new oscillation scales, also the TSTSM needs more severe meshing restrictions for post-caustic NLS than for linear SE computations (though less severe than in finite difference methods). For a description of TSTSM in NLS computations, an exhaustive set of test runs and a collection of references we refer to [6]. Also we remark that the TSTSM has recently been successfully applied in [4] to the numerical simulation of the Gross–Pitaevskii (quadratically confined) NLS modeling Bose–Einstein condensation (BEC).

Bibliography

[1] G. B. Airy; *On the intensity of light in a neighborhood of a caustic*, Trans. Cambr. Philos. Soc. 6 (1938), pp. 379–403.

[2] H. Bahouri, P. Gérard; *High frequency approximation of solutions to critical nonlinear wave eqautions*, Amer. J. Math. 121 (1999), no. 1, pp. 131–175.

[3] C. Bardos, L. Boutet de Monvel; *From the atomic hypothesis to microlocal analysis*, preprint (2002), available at `http://mapage.noos.fr/bardos`.

[4] W. Bao, D. Jaksch, P. A. Markowich; *Numerical solution of the Gross–Pitaevskii Equation for Bose–Einstein condensation*, to appear in J. Comput. Phys. (2003).

[5] W. Bao, S. Jin, P. A. Markowich; *On time-splitting spectral approximations for the Schrödinger equation in the semi-classical regime*, J. Comput. Phys. 175 (2002), no. 2, pp. 487–524.

[6] W. Bao, S. Jin, P. A. Markowich; *Numerical study of time-splitting spectral discretizations of nonlinear Schrodinger equations in the semi-classical regimes*, to appear in SIAM J. of Comp. (2003).

[7] J. D. Benamou; *Direct computation of multivalued phase space solutions for Hamilton–Jacobi equations*, Comm. Pure Appl. Math. 52 (1999), pp. 1443–1475.

[8] J. D. Benamou, F. Castella, T. Katsaounis, B. Perthame; *High frequency limit of the Helmholtz equation*, Rev. Mat. Iberoamericana 18 (2002), no. 1, pp. 187–209.

[9] A. Bensoussan, J. L. Lions, G. Papanicolaou; *Asymptotic analysis for periodic structures*, Studies in Mathematics and its Applications 5, North-Holland Publishing Co. (1978).

[10] F. Bloch; *Über die Quantenmechanik der Elektronen in Kristallgittern*, Z. Phys. 52 (1928), pp. 555–600.

[11] J. Bolte, S. Keppeler; *A semiclassical approach to the Dirac equation*, Ann. Phys. 274 (1999), pp. 125–162.

[12] Y. Brenier, L. Corrias; *A kinetic formulation for multibranch entropy solutions of scalar conservation laws*, Henri Poincaré, Analyse non linéare 15 (1998), pp. 169–190.

[13] L. Brillouin; *Notes on undulatory mechanics*, J. Phys. 7 (1926), pp. 353–365.

[14] N. Burq; *Mesures Semi-Classique et Mesures de Défaut*, Séminaire Bourbaki no. 826 (1997).

[15] L. Caffarelli, P. E. Souganidis, L. Wang; *Homogenization of fully nonlinear, second order, elliptic partial differential equations in stationary ergodic random environment*, preprint (2003).

[16] R. Carles; *Geometric optics with caustic crossing for some nonlinear Schrödinger equations*, Indiana Univ. Math. J. 49, (2000), no. 2, pp. 475–551.

[17] R. Carles; *Remarques sur les mesures de Wigner*, C. R. Acad. Sci. Paris 332 (2001), pp. 981–984.

[18] R. Carles; *Semi-classical Schrödinger equations with harmonic potential and nonlinear perturbation*, Ann. Henri Poincaré, Analyse non linéare 20 (2003), no. 3, pp. 501–542.

[19] R. Carles, C. Fermanian–Kammerer, I. Gallagher; *On the role of quadratic oscillations in nonlinear Schrödinger equations*, to appear in J. Funct. Anal (2003).

[20] T. Colin; *Rigorous derivation of the nonlinear Schrödinger equation and Davey–Stewartson systems from quadratic hyperbolic systems*, Asympt. Anal. 31 (2002), pp. 69–91.

[21] C. Conca, M. Vanninathan; *Homogenization of periodic structures via Bloch decomposition*, SIAM J. Appl. Math. 57 (1997), no. 6, pp. 1639–1659.

[22] M. Concordel; *Periodic homogenization of Hamilton–Jacobi equations I: additive eigenvalues and variational formula*, Indiana Univ. Math. J. 45 (1996), pp. 1095–1117.

[23] M. Concordel; *Periodic homogenization of Hamilton–Jacobi equations II: eikonal equations*, Proc. Roy. Soc. Edinburgh 127 (1997), pp. 665–689.

[24] A. Cordoba, C. Fefferman; *Wave packets and Fourier integral operators*, Comm. Part. Diff. Equ. 3 (1978), pp. 979–1005.

[25] M. Crandall, P. L. Lions; *Viscosity solutions of Hamilton–Jacobi equations*, Trans. Am. Math. Soc. 277 (1983), no. 1, pp. 1–42.

[26] M. Crandall, H. Ishii, P. L. Lions; *User's guide to viscosity solutions of second order partial differential equations*, Bull. Am. Math. Soc. 27 (1992), no.1, pp. 1–61.

[27] R. Di Perna; *Compensated compactness and general systems of conservations laws*, Trans. Amer. Math. Soc. 292 (1985), pp. 383–420.

[28] R. Di Perna, A. Majda; *The validity of nonlinear geometrical opitcs for weak solutions of conservations laws*, Comm. Math. Phys. 98 (1985), pp. 313–347.

[29] P. Donat, J. Rauch *Dispersive nonlinear geometrical optics*, J. Math. Phys. 38 (1997), pp. 1484–1523.

[30] J. J. Duistermaat; *Oscillatory integrals, Lagrangian immersions and unfoldings of singularities*, Comm. Pure Appl. Math. XXVII (1974), pp. 207–281.

[31] J. J. Duistermaat, L. Hörmander; *Fourier integral operators II*, Acta Math. 128 (1972), pp. 183–269.

[32] C. Emmerich, A. Weinstein; *Geometry of the transport equation in multicomponent WKB approximations*, Comm. Math. Phys. 176 (1996), pp. 701–711.

[33] L. C. Evans; *Periodic homogenisation of certain fully nonlinear partial differential equations*, Proc. Roy. Soc. Edinburgh Sect. A 120 (1992), no. 3–4, pp. 245–265.

[34] L. C. Evans; *Towards a quantum analogue of weak KAM theory*, preprint (2003), available at `http://math.berkeley.edu/~evans`.

[35] V. M. Fedoryuk; *Partial Differential Equations V*, Encyclopidia of Math. Sci. 34, Springer Verlag (1999).

[36] C. Fermanian-Kammerer; *Semi-classical analysis of a Dirac equation without adiabatic decoupling*, to appear in Monatsh. f. Math. (2003).

[37] C. Fermanian-Kammerer, P. Gérard; *Mesures semi-classiques et croisement de modes*, Bull. Soc. Math. France, 130 (2002), no. 1, pp. 123–168.

[38] C. Fermanian-Kammerer, P. Gérard; *A Landau–Zener formula for non-degenerated involutive codimension* 3 *crossings*, to appear in Ann. Inst. Henri Poincaré (2003).

[39] G. Folland; *Harmonic Analysis in Phase Space*, Annals of Math. Studies 122, Princeton Univ. Press (1989).

[40] G. A. Francfort, F. Murat; *Oscillations and energy densities in the wave equation*, Comm. Part. Diff. Equ. 17 (1992), no. 11–12, pp. 1785–1865.

[41] N. Fusco, G. Moscariello; *On the homogenization of quasilinear divergence structure operators*, Annali Mat. Pura Appl. 146 (1987), pp. 1–13.

[42] I. Gasser, P. A. Markowich; *Quantum Hydrodynamics, Wigner transform and the classical limit*, Asympt. Anal. 14 (1997), pp. 97–116.

[43] I. Gasser, C. Lin, P. A. Markowich; *A review of dispersive limits of (non)linear Schrödinger-type equations*, Taiwanese J. Math. 4 (2000), no. 4, pp. 501–529.

[44] P. Gérard; *Microlocal defect measures*, Comm. Part. Diff. Equ. 16 (1991), pp. 1761–1794.

[45] P. Gérard; *Remarques sur l'analyse semi-classique de l'equation Schrödinger non linéaire*, Séminaire EDP de L'École Polytechnique (1992), lecture no. XIII.

[46] P. Gérard; *Oscillations and concentration effects in semilinear dispersive wave equations*, J. Funct. Anal. 141 (1996), no. 1, pp. 60–98.

[47] P. Gérard, P. A. Markowich, N. Mauser, F. Poupaud, *Homogenisation Limits and Wigner transforms*, Comm. Pure Appl. Math. 50 (1997), pp. 323–379.

[48] L. Gosse; *Using K-branch entropy solutions for multivalued geometrical optics computations*, J. Comp. Physics 180 (2002), pp. 155–182.

[49] L. Gosse, S. Jin, X. Li; *Two moment systems for computing multiphase semi-classical limits of the Schrödinger equation*, preprint (2003).

[50] L. Gosse, P. A. Markowich; *Multiphase semi-classical approximation of an electron in a one-dimensional crystalline lattice*, preprint available at `http://mailbox.univie.ac.at/peter.markowich`.

[51] E. Grenier; *Semi-classical limit of the nonlinear Schrödinger equation in small time*, Proc. AMS 126 (1998), no. 2, pp. 523–530.

[52] J. C. Guillot, J. Ralston, E. Trubowitz; *Semi-classical asymptotics in solid-state physics*, Comm. Math. Phys. 116 (1988), pp. 401–415.

[53] V. Guillemin, S. Sternberg; *Geometric asymptotics*, Math. Surveys, No. 14, AMS (1977).

[54] L. Hörmander; *Analysis of linear partial differential operators I–IV*, Grundlehren der Mathematischen Wissenschaften, Springer Verlag (1985).

[55] F. Hövermann, H. Spohn, S. Teufel; *Semi-classical limit for the Schrödinger equation with a short scale periodic potential*, Comm. Math. Phys. 215 (2001), pp. 609–629.

[56] J. Hunter, J. Keller; *Weakly nonlinear high frequency waves*, Comm. Pure Appl. Math. 36 (1983), pp. 547–569.

[57] J. Hunter, J. Keller; *Caustics of nonlinear waves*, Wave motion 9 (1987), pp. 429–443.

[58] V. Jikov, S. Kozlov, O. Oleinik; *Homogenization of differential operators and integral functionals*, Springer (1994).

[59] S. Jin, S. Osher; *A level set method for the computation of multivalued solutions to quasilinear hyperbolic PDEs and Hamilton–Jacobi equations*, preprint (2003), available at: `http://www.math.wisc.edu/~jin/research.html`.

[60] J. L. Joly, G. Metivier, J. Rauch; *Transparent nonlinear geometric optics and the Maxwell–Bloch equations*, J. Diff. Equ. 166 (2000), pp. 175–250.

[61] J. L. Joly, G. Metivier, J. Rauch; *Nonlinear oscillations beyond caustics*, Comm. Pure Appl. Math. 49 (1996), no. 5, pp. 443–527.

[62] S. Keraani; *Semi-classical limit of a class of Schrödinger equations with potential*, Comm. Partial Diff. Equ. 27 (2002), no. 3–4, pp. 693–704.

[63] H. Kramers; *Wellenmechanik und Halbzahlige Quantisierung*, Zeit. Phys 39 (1926), pp. 828–840.

[64] L. Landau; *Collected papers of L. Landau*, Pergamon Press (1965).

[65] P. Lax; *Asymptotic solutions of oscillatory initial value problems*, Duke Math. J. 24 (1957), pp. 627–646.

[66] P. D. Lax, C. Levermore; *The small dispersion limit of the Korteweg–de Vries equation I*, Comm. Pure Appl. Math. 36 (1983), no. 3, pp. 253–290.

[67] P. D. Lax, C. Levermore; *The small dispersion limit of the Korteweg–de Vries equation II*, Comm. Pure Appl. Math. 36 (1983), no. 5, pp. 571–593.

[68] P. D. Lax, C. Levermore; *The small dispersion limit of the Korteweg–de Vries equation III*, Comm. Pure Appl. Math. 36 (1983), no. 6, pp. 809–829.

[69] P. D. Lax, C. Levermore; *The semi-classical limit of the defocusing NLS Hierarchy*, Comm. Pure Appl. Math. 52 (1999), pp. 613–654.

[70] P. L. Lions, T. Paul; *Sur les measures de Wigner*, Rev. Math. Iberoamericana 9 (1993), pp. 553–618.

[71] P. L. Lions, P. E. Souganidis; *Correctors for the homogenization of Hamilton–Jacobi equations in the stationary ergodic setting*, to appear in Comm. Pure Appl. Math. (2003).

[72] D. Ludwig; *Uniform asymptotics near caustics*, Comm. Pure Appl. Math. XIX (1966), pp. 215–250.

[73] P. A. Markowich, F. Poupaud; *The Maxwell Equations in a Periodic Medium: Homogenization of the Energy Density*, Ann. Scuola Norm. Sup. Pisa Cl. Sci. (4) 23 (1996), no. 2, pp. 301–324.

[74] P. A. Markowich, N. Mauser; *The classical limit of a self-consistent quantum-Vlasov equation in 3D*, Math. Models Methods Appl. Sci. 3 (1993), no. 1, pp. 109–124.

[75] P. A. Markowich, N. Mauser, F. Poupaud; *A Wigner-function approach to (semi)classical limits: electrons in a periodic potential*, J. Math. Phys. 35 (1994), no. 3, pp. 1066–1094.

[76] P. A. Markowich, P. Pietra, C. Pohl; *Numerical approximation of quadratic observables of Schrödinger-type equations in the semi-classical limit*, Numer. Math. 81 (1999), no. 4, pp. 595–630.

[77] P. A. Markowich, P. Pietra, C. Pohl H.P. Stimming; *A Wigner-measure analysis of the Dufort–Frankel scheme for the Schrödinger equation*, SIAM J. Numer. Anal. 40 (2002), no. 4, pp. 1281–1310.

[78] V. P. Maslov, M. V. Feydoriuk; *Semi-Classical Approximation in Quantum Mechanics*, Reidel Dordrecht (1981).

[79] L. Miller; *Refraction of high-frequency waves density by sharp interfaces and semi-classical measures at the boundary*, J. Math Pures Appl (9) 79 (2000), no. 3, pp. 227–269.

[80] G. Namah; *Asymptotic solution of a Hamilton–Jacobi equation*, Asympt. Anal. 12 (1996), no. 4, pp. 355–370.

[81] J. Panati, H. Sohn, S. Teufel; *Space-adiabatic perturbation theory*, preprint math–ph/0201055, available at: `http://xxx.lanl.gov`.

[82] F. Poupaud, C. Ringhofer; *Semi-classical limits in a crystal with external potentials and effective mass theorems*, Comm. Part. Diff. Equ. 21 (1996), pp. 1897–1918.

[83] J. Rauch; *Lectures on Nonlinear Geometrical Optics*, IAS/Park City Math. Series 5, AMS (1999).

[84] M. Reed, B. Simon; *Methods of modern mathematical physics IV*, Academic Press (1987).

[85] D. Robert; *Semi-classical approximation in quantum mechanics. A survey of old and recent mathematical results*, Helv. Phys. Acta 71 (1998), pp. 44–116.

[86] O. Runborg; *Some new results in multiphase geometrical optics*, Math. Modeling and Num. Analysis 34 (2000), pp. 1203–1231.

[87] P.E. Souganidis; *Stochastic homogenization of Hamilton–Jacobi equations and some applications*, Asympt. Anal. 20 (1999), pp. 1–11.

[88] C. Sparber, P. A. Markowich, N. Mauser; *Wigner functions vs WKB-methods in multivalued geometrical optics*, Asympt. Anal. 33 (2003), no. 2, pp. 153–187.

[89] C. Sparber, P. A. Markowich; *Semi-classical Asymptotics for the Dirac–Maxwell system*, to appear in J. Math. Phys. (2003).

[90] H. Spohn; *Semi-classical limit of the Dirac equation and spin precession*, Ann. Physics 282 (2000), no. 2, pp. 420–431.

[91] L. Tartar; *Oscillations in nonlinear partial differential equations: compensated compactness and homogenization*, Lectures in Appl. Math. 23, AMS (1986).

[92] L. Tartar; *H-measures: a new approach for studying homogenization, oscillations and concentration effects in partial differential equations*, Proc. Roy. Soc. Edinburgh Sect. A 115 (1990), pp. 193–230.

[93] C. Sulem, P. L. Sulem; *The nonlinear Schrödinger equation*, Applied Math. Sciences 139, Springer (1999).

[94] P. Wall; *Some homogenization and corrector results for nonlinear monotone operators*, J. Nonl. Math. Phys. 5 (1998), no. 3, pp. 331–348.

[95] G. Wentzel; *Eine Verallgemeinerung der Quantenbedingung für die Zwecke der Wellenmechanik*, Zeit. Phys. 38 (1926), pp. 38–45.

[96] G. B. Whitham; *Linear and nonlinear waves*, Wiley (1974).

[97] E. Wigner; *On the quantum correction for the thermodynamical equilibrium*, Phys. Rev. 40 (1932), pp. 742–759.

[98] C. Zener; *Non-adiabatic crossing of energy-levels*, Proc. Roy. Soc. Lond. 137 (1932), pp. 696–702.

[99] P. Zhang, Y. Zheng, N. Mauser; *The limit from the Schrödinger–Poisson equation to the Vlasov–Poisson equation with general data in one dimension*, Comm. Pure Appl. Math. 55 (2002), no. 5, pp. 582–632.

Stan Osher accepts the ICIAM Pioneer Prize for 2003 from Olavi Nevanlinna. photography by *Happy Medium Photo Co.*

Alexander Mielke is Full Professor for Applied Analysis at the Universität Stuttgart. His research interests range from dynamical systems to nonlinear PDEs, in particular their applications in fluid and solid mechanics. Over the last 10 years he made major contributions to the area of pattern formation for PDEs on unbounded domains. Recently, the main emphasis of his work lies in the modelling of hysteretic behavior in elastic materials such as elastoplasticity and the shape memory effect.

Mielke studied mathematics at Stuttgart where he completed his PhD in 1984. In 1986 he spent one year at Cornell University as a postdoctoral student. From 1992 to 1999 he was full professor at the Universität Hannover. Since September 2000 Alexander has been coordinator of the DFG Priority Program Analysis, Modeling and Simulation of Multiscale Problems. In 1989 Mielke was awarded the Richard-von-Mises-Preis by the GAMM (Gesellschaft für Angewandte Mathematik und Mechanik) and the Heinz-Maier-Leibnitz-Preis by the German Ministry for Research. He was invited to be a speaker at the ICM 2002 meeting in Beijing.

Chapter 15
A New Approach to Elasto-Plasticity using Energy and Dissipation Functionals*

Alexander Mielke†

Abstract: We consider models for elasto-plasticity which are based on fully nonlinear elastostatics and rate-independent evolution laws for the plastic deformation and suitable hardening parameters. Accounting for finite strains leads to the multiplicative decomposition of the strain tensor and to a flow rule formulated on a Lie group. Our analysis is based on a recently developed energetic approach to general rate-independent material models which only uses two energy functionals, namely the elastic stored energy and the dissipation distance which plays the role of a metric on the space of internal variables. The evolution law can be reformulated as a static stability condition combined with an energy inequality. This work surveys results on the existence of solutions of an intrinsically associated incremental problem which has the form of a minimisation problem. Existence of solutions for the time-continuous problem remains open except for the one dimensional case.

Contents

[1]Research partially supported by SFB 404 "Multifield Problems in Continuum Mechanics"
†Institut für Analysis, Dynamik und Modellierung, Universität Stuttgart, Germany

1 Introduction

The mathematical theory of linearised elasto-plasticity was developed in the 1970s by J. J. Moreau [Mor76] and further developed subsequently up to efficient numerical implementations; see, e.g., [Joh76, HaR95]. This theory relies on the additive decomposition

$$\varepsilon = \tfrac{1}{2}(\mathrm{D}u + \mathrm{D}u^{\mathsf{T}}) = \varepsilon_{\mathrm{elast}} + \varepsilon_{\mathrm{plast}}\,,$$

of the linearised strain tensor ε, where $u : \Omega \to \mathbb{R}^d$ denotes the displacement. Moreover, the energy is assumed to be a quadratic functional such that the problem takes the form of a quasi-variational inequality. More general approaches including nonlinear hardening laws and viscoplastic effects can be found in [BeF96, Alb98, ACZ99, Che01a, Che01b]. A first local existence result for smooth solutions in finite-strain elasto-plasticity is obtained in [Nef02].

This work surveys current mathematical developments in the theory of elasto-plasticity which allows for large strains and which is based on the multiplicative decomposition

$$F = \mathrm{D}y = F_{\mathrm{elast}} F_{\mathrm{plast}}\,. \tag{1.1}$$

The main feature here is that, as in finite strain elasticity also called geometrically nonlinear elasticity, the nonlinearities arise from the multiplicative group of invertible matrices. The main question is to understand the interaction of functional analytical tools, mainly based on linear function spaces, and these algebraic nonlinearities.

Here, $y : \Omega \to \mathbb{R}^d$ is the deformation of the body $\Omega \subset \mathbb{R}^d$. The energy $\mathcal{E}$ stored in a deformed body depends only on the elastic part F_{elast} of the deformation tensor and suitable hardening parameters $p \in \mathbb{R}^m$, but not on the plastic part F_{plast}, which is contained in $\mathrm{SL}(\mathbb{R}^d)$, or another Lie group $\mathfrak{G}$ contained in $\mathrm{GL}_+(\mathbb{R}^d) = \big\{\, G \in \mathbb{R}^{d\times d} \;\big|\; \det G > 0 \,\big\}$. The energy functional takes the form

$$\mathcal{E}\big(t, y, (F_{\mathrm{plast}}, p)\big) = \int_\Omega \widehat{W}\big(x, \mathrm{D}y(x)F_{\mathrm{plast}}(x)^{-1}, p(x)\big)\,\mathrm{d}x - \big\langle \ell(t)\,, y\big\rangle\,,$$

where the external loading $\ell(t)$ is given via

$$\big\langle \ell(t)\,, y\big\rangle = \int_\Omega f_{\mathrm{ext}}(t,x)\cdot y(x)\,\mathrm{d}x + \int_\Gamma g_{\mathrm{ext}}(t,x)\cdot y(x)\,\mathrm{d}a\,.$$

To model the plastic effects one prescribes either a plastic flow law or, equivalently, a dissipation potential $\Delta : \Omega \times \mathrm{T}(\mathfrak{G}\times\mathbb{R}^m) \to [0,\infty]$. We consider $\Delta(x,\cdot,\cdot)$ as an infinitesimal metric which defines the global dissipation distance $D(x,\cdot,\cdot)$ on $\mathfrak{G}\times\mathbb{R}^m$. Thus, the second ingredient of the material model is the dissipation distance between two internal states $P_j = (F^{(j)}_{\mathrm{plast}}, p_j) : \Omega \to \mathrm{SL}(d)\times\mathbb{R}^m$:

$$\mathcal{D}(P_1,P_2) = \int_\Omega D\big(x, (F^{(1)}_{\mathrm{plast}}(x), p_1(x))\,,\, (F^{(2)}_{\mathrm{plast}}(x), p_2(x))\big)\,\mathrm{d}x\,.$$

Allowing for finite strains we are forced to avoid convexity of the stored-energy density $\widehat{W}$. It rather has to be polyconvex or quasiconvex and frame indifferent, see

[Bal77]. These notions work well together with the philosophy that $F = \mathrm{D}y$ is an element of $\mathrm{GL}_+(\mathbb{R}^d)$, i.e., we set $W(F) = \infty$ for $\det F \leq 0$. The aim of this work is to show that these assumptions can be connected naturally with the multiplicative decomposition (1.1) and the Lie group structure for $G = F_{\mathrm{plast}}$.

We follow the work in [MiT99, MTL02, Mie02a, Mie03a, MiR03] which shows that rate-independent evolution for elastic materials with internal variables ("standard generalised materials") can be formulated by energy principles as follows: A pair $(y, P) : [0, T] \times \Omega \to \mathbb{R}^d \times \mathrm{SL}(\mathbb{R}^d) \times \mathbb{R}^m$ is called a solution of the elasto-plastic problem associated with $\mathcal{E}(t, \cdot, \cdot)$ and $\mathcal{D}$, if **stability (S)** and the **energy inequality (E)** holds:

(S) For all $t \in [0, T]$ we have $\mathcal{E}\big(t, y(t), P(t)\big) \leq \mathcal{E}(t, \widetilde{y}, \widetilde{P}) + \mathcal{D}\big(P(t), \widetilde{P}\big)$.

(E) For all $s, t \in [0, T]$ with $s < t$ we have
$\mathcal{E}\big(s, y(s), P(s)\big) + \mathrm{Diss}\big(P, [s, t]\big) \leq \mathcal{E}\big(t, y(t), P(t)\big) - \int_s^t \langle \dot{\ell}(\tau), y(\tau) \rangle \, \mathrm{d}\tau$.

So far, we are not able to provide existence results for (S) & (E) in the present elasto-plastic setting. However, analogous models in phase transformations [MTL02, MiR03], in delamination [KMR02], in micromagnetism [Kru02, RoK02], and in fracture [FrM98, DMT02] have been treated with mathematical success. In these works two major restrictions had to be made: (i) $\mathcal{E}$ has to be convex in the strains (leading to infinitesimal strains) and (ii) the internal variable P has to lie in a closed convex subset of a Banach space. In finite-strain elasto-plasticity these two assumptions are clearly violated.

Since most of the above-mentioned existence results are based on time-incremental approximations we devote this work to an existence theory for the following incremental problem (IP). We consider this as a first step for finding solutions of (S) & (E) and comment on the problem of treating the time-continuous problem.

(IP) Incremental problem. For given $t_0 = 0 < t_1 < \ldots < t_N = T$ and P_0 find $(y_k, P_k) \in \mathrm{Argmin}\big\{\, \mathcal{E}(t_k, y, P) + \mathcal{D}(P_{k-1}, P) \;\big|\; (y, P) \,\big\}$, $k = 1, \ldots, N$.

Here "Argmin" denotes the set of all minimisers. Hence, (IP) consists of k minimisation problems which are coupled via the dissipation distance. Similar incremental minimisation problems are also used in the engineering community, cf. [OrR99, OrS99, MSS99, ORS00, MiL03, MSL02, HaH03]. Hence, it is justified to study the mathematical properties of (IP) even though its relation to (S) & (E) is not clear. In fact, existence and nonexistence for (IP) relates to questions of formation of microstructure, localisation or failure, see the discussions in [CHM02, Mie03a].

In Section 6 we introduce the notions of finite-strain elasto-plasticity in detail and establish the relation between the classical flow rules of elasto-plasticity with our energetic formulation (S) & (E). For a more extensive and mechanical treatment we refer to [Mie03a]. In Section 4 we study the incremental problem (IP) in specific function spaces. First we establish a rather general result which says that any solution $(y_k, P_k)_{k=1,\ldots,N}$ is stable in the sense of (S) and satisfies a two-sided discretised energy inequality replacing (E).

The key feature of the analysis of (IP), with $\mathcal{E}$ and $\mathcal{D}$ as given above, is to realise that the internal variables $P = (F_{\mathrm{plast}}, p)$ occur under the integral over the body Ω only in a local fashion. Hence, it is possible to minimise in (IP) pointwise in $x \in \Omega$ with respect to $P(x)$. This leads to the **condensed energy density**

$$W^{\text{cond}}(P_{\text{old}};F) = \min\{\,\widehat{W}(FG^{-1},p) + D\big(P_{\text{old}},(G,p)\big) \mid (G,p) \in \mathrm{SL}(\mathbb{R}^d)\times\mathbb{R}^m\,\}.$$

The first major assumption for our existence theory is that $W^{\text{cond}}\big((\mathbf{1},p_*);\cdot\big) : \mathbb{R}^{d\times d} \to \mathbb{R}_\infty := \mathbb{R}\cup\{\infty\}$ is polyconvex. The second major assumption is that W^{cond} and the dissipation distance D are coercive:

$$W^{\text{cond}}\big((\mathbf{1},p_*);F\big) \geq c|F|^{q_F} - C \quad\text{and}\quad D\big((\mathbf{1},p_*),(G,p)\big) \geq c|G|^{q_G} - C\,.$$

If the growth exponents satisfy $\frac{1}{q_F} + \frac{1}{q_G} \leq \frac{1}{q} < \frac{1}{d}$, then existence of solutions $(y_k, F^{(k)}_{\text{plast}}, p_k)$ for (IP) is obtained with $y_k \in \mathrm{W}^{1,q}(\Omega,\mathbb{R}^d)$ and $F^{(k)}_{\text{plast}} \in \mathrm{L}^{q_G}(\Omega,\mathbb{R}^{d\times d})$.

In Section 5 we supply a specific two-dimensional example in which all assumptions can be checked explicitly and are fulfilled for suitable parameter values. Thus, we provide a first existence theory for a multi-dimensional elasto-plastic incremental problem in the geometric nonlinear case. Moreover, we discuss a one-dimensional example to highlight the difficulties in proving existence of solutions for the time-continuous problem (S) & (E) by letting the step-size of the time discretisation tend to 0. Only by exploiting the very specific properties of the one-dimensional case, we obtain a convergence result for the incremental solution which implies that the time-continuous problem (S) & (E) has a solution as well.

In Section 6 we speculate about the solvability of (IP) and (S) & (E) after regularising the elastic energy by adding a gradient term in the form

$$\int_\Omega \kappa\,\|\mathrm{D}G\|^r\,\mathrm{d}x \quad\text{or}\quad \int_\Omega \kappa\,\|(\operatorname{curl} G)G^{\mathsf{T}}\|^r\,\mathrm{d}x\,.$$

We expect existence results for (IP), but solvability of (S) & (E) remains widely open.

2 Elasto-plasticity at finite strain

Here we describe the framework of multiplicative elasto-plasticity as developed in [Mie02a, Mie03a], where more details on the mechanical modeling are given.

We consider an elastic body $\Omega \subset \mathbb{R}^d$ which is bounded and has a Lipschitz boundary $\partial\Omega$. A deformation is a mapping $y : \Omega \to \mathbb{R}^d$ such that the deformation gradient $F(x) = \mathrm{D}y(x)$ exists for a.e. $x \in \Omega$ and satisfies

$$F(x) \in \mathrm{GL}_+(d) = \{\,F \in \mathbb{R}^{d\times d} \mid \det F > 0\,\}.$$

The internal plastic state at a material point $x \in \Omega$ is described by the plastic tensor $G = F_{\text{plast}} \in \mathrm{GL}_+(d)$ and a possibly vector-valued hardening variable $p \in \mathbb{R}^m$. We shortly write $P = (G,p)$ to denote the set of all plastic variables. The major assumption in finite-strain elasto-plasticity is the multiplicative decomposition of the deformation gradient F into an elastic and a plastic part

$$F = F_{\text{elast}}F_{\text{plast}} = F_{\text{elast}}G\,. \tag{2.1}$$

The main feature of this decomposition is that the elastic properties will depend only on F_{elast}, whereas previous plastic transformations through G are completely forgotten. However, the hardening variable p will record changes in G and may influence the elastic properties.

The deformation process is governed by two principles. First we have energy storage which gives rise to the equilibrium equations and second we have dissipation due to plastic transformations which give rise to the plastic flow rule. Energy storage is described by the Gibbs energy

$$\mathcal{E}(t,y,P) = \int_\Omega W\big(x, \mathrm{D}y(x), P(x)\big)\,\mathrm{d}x - \big\langle \ell(t)\,, y \big\rangle\,,$$

where $\big\langle \ell(t)\,, y\big\rangle = \int_\Omega f_{\text{ext}}(t,x)\cdot y(x)\,\mathrm{d}x + \int_{\Gamma_{\text{Neu}}} g_{\text{ext}}(t,x)\cdot y(x)\,\mathrm{d}a(x)$ denotes the process-time dependent loading. The major constitutive assumption is the multiplicative decomposition

$$W\big(x, F, (G,p)\big) = \widehat{W}(x, FG^{-1}, p)\,. \tag{2.2}$$

From now on we drop the variable x for notational convenience. However, the whole theory and analysis works in the inhomogeneous case as well.

The dissipational effects are usually modeled by prescribing yield surfaces. For our purpose it is more convenient and mathematically clearer to start on the other side, namely the dissipation metric. In mechanics this metric is called dissipation potential, since the dissipational friction forces are obtained from it via differentiation with respect to the plastic rates. We emphasise that the natural setup for the plastic transformation $G \in \mathrm{GL}_+(d)$ is that of an element of a Lie group $\mathfrak{G} \subset \mathrm{GL}_+(d)$. A usual assumption is plastic incompressibility, which gives $\mathfrak{G} = \mathrm{SL}(d) = \big\{\, G \;\big|\; \det G = 1 \big\}$. However, $\mathfrak{G} = \mathrm{GL}_+(d)$ or a single-slip system $\mathfrak{G} = \big\{\, \mathbf{1} + \gamma e_1 \otimes e_2 \;\big|\; \gamma \in \mathbb{R} \big\}$ may also be possible. A dissipation potential is a mapping

$$\Delta : \Omega \times \mathrm{T}(\mathfrak{G} \times \mathbb{R}^m) \to [0,\infty]\,,$$

which is called a dissipation metric, if it is continuous and if $\Delta(x,(G,p),\,\cdot\,)$ is convex and positively homogeneous of degree 1, i.e.,

$$\Delta\big(x, (G,p), \alpha(\dot G, \dot p)\big) = \alpha \Delta\big(x, (G,p), (\dot G, \dot p)\big) \text{ for } \alpha \geq 0\,.$$

(Again we drop the variable x for notational convenience.) This condition leads to rate-independent material behavior. Together with the multiplicative decomposition (2.1) one assumes **plastic indifference**

$$\Delta\big((G\widehat{G}, p), (\dot G \widehat{G}, \dot p)\big) = \Delta\big((G,p),(\dot G, \dot p)\big) \text{ for all } \widehat{G} \in \mathfrak{G}\,.$$

This amounts to the existence of a function $\widehat{\Delta} : \mathbb{R}^m \times \mathbb{R}^m \times \mathfrak{g} \to [0,\infty]$ such that

$$\Delta\big((G,p),(\dot G, \dot p)\big) = \widehat{\Delta}(p, \dot p, \dot G G^{-1})\,.$$

Here $\mathfrak{g} = \mathrm{T}_{\mathbf{1}}\mathfrak{G}$ is the Lie algebra associated with the Lie group $\mathfrak{G}$, and strictly speaking $\dot G G^{-1}$ is the right translation of $\dot G(t) \in \mathrm{T}_{G(t)}\mathfrak{G}$ to $\mathfrak{g} = \mathrm{T}_{\mathbf{1}}\mathfrak{G}$.

An important feature of our theory is the induced dissipation distance D on $\mathfrak{G} \times \mathbb{R}^m$ defined via (recall $P = (G,p)$)

$$D(P_0, P_1) = \inf\{ \int_0^1 \Delta(P(s), \dot{P}(s))\, \mathrm{d}s \mid P \in \mathrm{C}^1([0,1], \mathfrak{G}\times\mathbb{R}^m), P(0) = P_0, P(1) = P_1 \}.$$

It is important to note that we don't assume symmetry (i.e., $\Delta(P, -\dot{P}) = \Delta(P, \dot{P})$) which would contradict hardening. Thus, $D(\cdot\,, \cdot)$ will not be symmetric either. However, we will often use the triangle inequality

$$D(P_1, P_3) \leq D(P_1, P_2) + D(P_2, P_3),$$

which follows immediately from the definition. Plastic difference implies that the dissipation distance satisfies

$$D\big((G_1, p_1), (G_2, p_2)\big) = D\big((\mathbf{1}, p_1), (G_2 G_1^{-1}, p_2)\big). \tag{2.3}$$

Integration gives the bulk dissipation distance between two internal states $P_j : \Omega \to \mathfrak{G}\times\mathbb{R}^m$ via

$$\mathcal{D}(P_0, P_1) = \int_\Omega D\big(P_0(x), P_1(x)\big)\, \mathrm{d}x.$$

To make the energetic formulation mathematically rigorous we define the set of kinematically admissible deformations via

$$\mathcal{F} = \{ y \in \mathrm{W}^{1,q}(\Omega; \mathbb{R}^d) \mid y|_{\Gamma_{\mathrm{Dir}}} = y_{\mathrm{Dir}} \}, \tag{2.4}$$

where $\Gamma_{\mathrm{Dir}} = \partial\Omega \backslash \Gamma_{\mathrm{Neu}}$ is a part of the boundary with positive surface measure. Moreover, $y_{\mathrm{Dir}} = Y|_{\Gamma_{\mathrm{Dir}}}$ where $Y \in \mathrm{C}^1(\overline{\Omega}; \mathbb{R}^d)$ with $\mathrm{D}Y(x) \in \mathrm{GL}_+(d)$ for all $x \in \overline{\Omega}$. The integrability power q in $\mathrm{W}^{1,q}$ will be chosen larger than d in order to apply the theory of polyconvexity. The loading can then be considered as a function $\ell : [0,T] \to \mathrm{W}^{1,q}(\Omega, \mathbb{R}^d)^*$, where * denotes the dual space (space of all continuous linear forms).

The set of admissible internal states is simply

$$\mathcal{P} = \{ P : \Omega \to \mathfrak{G} \times \mathbb{R}^m \mid P \text{ measurable} \}. \tag{2.5}$$

Because of the image space, which is a manifold, it is not clear whether it is reasonable to consider $\mathcal{P}$ as a subset of a Banach space like $\mathrm{L}^{q_G}(\Omega, \mathbb{R}^{d\times d}) \times \mathrm{L}^{q_p}(\Omega, \mathbb{R}^m)$ or as a general manifold equipped with a metric associated with $\mathcal{D}$. The abstract theory in Section 3 will address the interplay between the topology on $\mathcal{Z} = \mathcal{F} \times \mathcal{P}$ and the metric $\mathcal{D}$.

Definition 2.1. *A process $(y, P) : [0,T] \to \mathcal{F} \times \mathcal{P}$ is called a solution of the elasto-plastic problem defined via $\mathcal{E}(t, \cdot\,, \cdot)$ and $\mathcal{D}$, if the stability condition (S) and the energy inequality (E) holds:*

(S) *For all $t \in [0,T]$ we have*
$\mathcal{E}\big(t, y(t), P(t)\big) \leq \mathcal{E}(t, \widetilde{y}, \widetilde{P}) + \mathcal{D}\big(P(t), \widetilde{P}\big)$ *for all* $(\widetilde{y}, \widetilde{P}) \in \mathcal{F} \times \mathcal{P}$.

(E) *For all $s, t \in [0,T]$ with $s < t$ we have* (2.6)
$\mathcal{E}\big(t, y(t), P(t)\big) + \mathrm{Diss}\big(P, [s,t]\big) \leq \mathcal{E}\big(s, y(s), P(s)\big) - \int_s^t \langle \dot{\ell}(r), y(r) \rangle\, \mathrm{d}r.$

Here $-\int_s^t\langle\dot\ell\,,y\rangle\,\mathrm{d}r = \int_s^t\langle\ell\,,\dot y\rangle\,\mathrm{d}r - \langle\ell\,,y\rangle\big|_s^t$ is called the reduced work of the external forces, since $\mathcal{E}$ denotes the Gibbs energy instead of the Helmholtz energy. The dissipation reads

$$\mathrm{Diss}\big(P,[s,t]\big) = \sup\big\{\,\textstyle\sum_{j=1}^N \mathcal{D}\big(P(t_{j-1}),P(t_j)\big)\;\big|\;N\in\mathbb{N}, s\le t_0<\ldots<t_N\le t\,\big\}$$

for general processes, which equals $\mathrm{Diss}\big(P,[s,t]\big) = \int_s^t\int_\Omega \Delta\big(P(r,x),\dot P(r,x)\big)\,\mathrm{d}x\,\mathrm{d}t$ for differentiable processes.

The major advantage of the energetic formulation via (S) and (E) is that neither derivatives of the constitutive functions W and Δ nor of the solution $(\mathrm{D}y,P)$ are needed. Nevertheless, (S) and (E) are strong enough to determine the physically relevant solutions. We refer to [MiT01] for uniqueness results under additional convexity assumptions. Moreover, it is shown in [Mie03a] that sufficiently smooth solutions (y,P) of (S) and (E) satisfy the classical equations of elasto-plasticity, namely the **equilibrium equation**

$$\left.\begin{array}{rcll} -\operatorname{div}T(t,x) &=& f_{\mathrm{ext}}(t,x) & \text{in } \Omega\,,\\ y(t,x) &=& y_{\mathrm{Dir}}(x) & \text{on } \Gamma_{\mathrm{Dir}}\,,\\ T(t,x)\nu(x) &=& g_{\mathrm{ext}}(t,x) & \text{on } \Gamma_{\mathrm{Neu}}\,, \end{array}\right\}$$

and the **flow rule**

$$0\in\partial^{\mathrm{sub}}_{\dot P}\Delta\big(P(t,x),\dot P(t,x)\big) - Q(t,x)\,, \tag{2.7}$$

where $T = \dfrac{\partial}{\partial F}W\big(\mathrm{D}y,P)\big) = \dfrac{\partial}{\partial F_{\mathrm{elast}}}\widehat{W}(\mathrm{D}y\,G^{-1},p)\,G^{-\mathsf{T}}$ is the first Piola–Kirchhoff stress tensor, $\partial^{\mathrm{sub}}_{\dot P}\Delta(P,\dot P)$ denotes the subgradient of the convex function $\Delta(P,\,\cdot\,)$: $\mathrm{T}_P(\mathfrak{G}\times\mathbb{R}^m)\to[0,\infty]$ and $Q\in\mathrm{T}^*_P(\mathfrak{G}\times\mathbb{R}^m)$ is the thermodynamically conjugated driving force to $P=(G,p)$, i.e.,

$$Q = -\frac{\partial}{\partial(G,p)}W\big(F,(G,p)\big) = \big(G^{-\mathsf{T}}F^{\mathsf{T}}\frac{\partial}{\partial F_{\mathrm{elast}}}\widehat{W}(FG^{-1},p)G^{-\mathsf{T}}, -\frac{\partial}{\partial p}\widehat{W}(FG{-}1,p)\big)\,.$$

Using the elastic domain $\mathbb{Q}(P) = \partial^{\mathrm{sub}}_{\dot P}\Delta(P,0)\subset\mathrm{T}^*_P(\mathfrak{G}{\times}\mathbb{R}^m)$, the Legendre–Fenchel transform shows that (2.7) is equivalent to the differential inclusion

$$\dot P\in\partial\,\mathcal{X}_{\mathbb{Q}(P)}(Q) = \mathrm{N}_Q\,\mathbb{Q}(P)\,, \tag{2.8}$$

where $\mathcal{X}_{\mathbb{Q}(P)}$ is the indicator function and $\mathrm{N}_Q\,\mathbb{Q}(P)$ denotes the exterior normal cone. If $\mathbb{Q}(P)$ is given by a yield function ϕ in the form $\mathbb{Q}(P) = \big\{\,Q\;\big|\;\phi(P,Q)\le 0\,\big\}$ and $\dfrac{\partial}{\partial Q}\phi(P,Q)\ne 0$ at $\phi(P,Q)=0$, then the flow rule (2.7) or (2.8) can be reformulated via the Karush–Kuhn–Tucker conditions:

$$\dot P = \lambda\frac{\partial}{\partial Q}\phi(P,Q)\,,\quad \lambda\ge 0,\quad \phi(P,Q)\le 0\,,\quad \lambda\phi(P,Q)=0\,.$$

3 Abstract setup of rate-independent problems

We show here that the energetic formulation derived above is a special case of an abstract formulation for rate-independent processes. This theory was developed in [MiT99, MiT01, MTL02] and takes its most nonlinear form in [MaM03]. Other applications are in the theory of shape-memory alloys [GMH02, MiR03], in ferromagnetism [Kru02, RoK02] or in fracture or delamination [DMT02, FrM98, KMR02]. However, we will see in the subsequent sections that the present state of the theory is not fully applicable in the theory of multiplicative elasto-plasticity.

We start with a topological space $\mathcal{Z}$ and denote convergence in this space by $z_k \xrightarrow{\mathcal{Z}} z$. The rate-independent system consists of two ingredients which both are considered to be energetic quantities. The time-dependent *energy functional* $\mathcal{E} : [0,T] \times \mathcal{Z} \to \mathbb{R}_\infty$ describes the energy-storage mechanism of the system. The *dissipation distance* $\mathcal{D} : \mathcal{Z} \times \mathcal{Z} \to [0,\infty]$ describes how the system dissipates energy. The latter is taken to be the minimal amount of dissipated energy when the system changes from one state into another. Hence, $\mathcal{D}$ should satisfy the triangle inequality:

$$\mathcal{D}(z_1, z_3) \leq \mathcal{D}(z_1, z_2) + \mathcal{D}(z_2, z_3) \quad \text{for all } z_1, z_2, z_3 \in \mathcal{Z}\,,$$

but we allow for the value ∞ and do not enforce symmetry; i.e., we allow for $\mathcal{D}(z_0, z_1) \neq \mathcal{D}(z_1, z_0)$ which is important for elasto-plasticity. For any given curve $z : [0,T] \to \mathcal{Z}$ we define the *dissipation* on $[s,t]$ via

$$\mathrm{Diss}_\mathcal{D}\big(z; [s,t]\big) = \sup\big\{ \textstyle\sum_1^N \mathcal{D}\big(z(\tau_{j-1}), z(\tau_j)\big) \;\big|\; N{\in}\mathbb{N}\,, s{=}\tau_0{<}\tau_1{<}\cdots{<}\tau_N{=}t \big\}\,. \tag{3.1}$$

Definition 3.1. *A curve $z : [0,T] \to X$ is called a* ***solution*** *of the rate-independent model $(\mathcal{E}, \mathcal{D})$, if* ***global stability (S)*** *and* ***energy inequality (E)*** *holds:*
(S) *For all $t \in [0,T]$ and all $\widehat{z} \in \mathcal{Z}$ we have $\mathcal{E}\big(t, z(t)\big) \leq \mathcal{E}(t, \widehat{z}) + \mathcal{D}\big(z(t), \widehat{z}\big)$.*
(E) *For all t_0, t_1 with $0 \leq t_0 < t_1 \leq T$ we have*

$$\mathcal{E}\big(t_1, z(t_1)\big) + \mathrm{Diss}_\mathcal{D}\big(z; [t_0, t_1]\big) \leq \mathcal{E}\big(t_0, z(t_0)\big) + \textstyle\int_{t_0}^{t_1} \partial_t \mathcal{I}\big(t, z(t)\big)\, \mathrm{d}t\,.$$

The stability condition (S) represents the fact that, while holding the process time t fixed, the system is in a stable state, which means that changing $z(t)$ into $\widehat{z}$ does not release more stored energy than has to be paid by the dissipation mechanism: $\mathcal{E}\big(t, z(t)\big) - \mathcal{E}(t, \widehat{z}) \leq \mathcal{D}\big(z(t), \widehat{z}\big)$. The energy inequality (E) says that the present stored energy plus the dissipated energy has to be less than the initial energy plus the work of the external forces.

Rate-independency manifests itself by the fact that the problem has no intrinsic time scale. It is easy to show that z is a solution of $(\mathcal{E}, \mathcal{D})$, if and only if the reparametrised curve $\widetilde{z} : t \mapsto z\big(\alpha(t)\big)$, with $\dot{\alpha} > 0$, is a solution of $(\widetilde{\mathcal{E}}, \mathcal{D})$, where $\widetilde{\mathcal{E}}(t, z) = \mathcal{E}\big(\alpha(t), z\big)$. In particular, the stability (S) is a static concept and the energy estimate (E) is rate-independent, since the dissipation defined via (3.1) is scale invariant like the length of a curve.

The major importance of the energetic formulation is that neither the given functionals $\mathcal{D}$ and $\mathcal{E}(t, \cdot)$ nor the solutions $z : [0,T] \to \mathcal{Z}$ must be differentiable. In fact, to make sense of such derivatives we would need to impose that $\mathcal{Z}$ is a (Banach) manifold or a suitable subset of a Banach space. Although this will be the case in many applications, it is better to avoid these concepts as long as possible.

A traditional "linear setup" is obtained, if we assume that $\mathcal{Z}$ is a closed convex subset of a Banach space X, that $\mathcal{E}(t,\cdot)$ is strictly convex and that $\mathcal{D}$ has the from $\mathcal{D}(z_0,z_1) = \|z_1 - z_0\|_X$. Then the energetic formulation (S) & (E) is equivalent to the differential inclusion

$$0 \in \partial^{\text{sub}}\big[\|\cdot\|_X\big]\big(\dot z(t)\big) + \partial^{\text{sub}}\big[\mathcal{E}\big(t,z(t)\big) + \mathcal{X}_{\mathcal{Z}}(\cdot)\big]\big(z(t)\big) \quad \text{for a.a. } t \in [0,T],$$

where ∂^{sub} denotes the subdifferential for convex functions. We refer to [CoV90, Vis01] for such doubly nonlinear problems and to [MiT01] for exact proofs of the implications between the different formulations.

To develop an existence theory for the problem (S)&(E), one needs to specify conditions on $\mathcal{Z}$, $\mathcal{E}$ and $\mathcal{D}$. We will not discuss the complete existence theory here, mainly because of the fact that in the case of elasto-plasticity the theory is not yet finished. Nevertheless, we give the main flavor of the theory and show how certain major steps work together. The main approach to the energetic formulation is based on time discretisation and exploiting the fact that the backward Euler step can be formulated as a minimisation problem.

To this end we choose discrete times $0 = t_0 < t_1 < \ldots < t_N = T$. For a given initial datum $z_0 \in \mathcal{Z}$ we formulate the time-incremental problem.

(IP) *For* $z_0 \in \mathcal{Z}$ *with* $\mathcal{E}(0,z_0) < \infty$ *find* $z_1,\ldots,z_N \in X$ *such that*
$$z_k \in \operatorname{Argmin}\{\, \mathcal{E}(t_k,z) + \mathcal{D}(z_{k-1},z) \mid z \in \mathcal{Z} \,\} \quad \text{for } k = 1,\ldots,N.$$

Here "Argmin" denotes the set of all minimisers. Note that the size of the time step does not enter here, which is due to rate-independence. Of course, the existence of minimisers is nontrivial in general. However, the following result shows that any solution satisfies a discretised version of (S) & (E) and thus justifies (IP) as an approximation to (S) & (E). The stability condition (S) can be reformulated as $\big(t,z(t)\big) \in \mathcal{S}$ for all $t \in [0,T]$, where the *stable set* $\mathcal{S}$ is given via

$$\mathcal{S} := \big\{\, (t,z) \in [0,T] \times X \mid \mathcal{E}(t,z) \le \mathcal{E}(t,\widehat z) + \mathcal{D}(z,\widehat z) \text{ for all } \widehat z \in X \,\big\}.$$

Theorem 3.2. *Assume* $(0,z_0) \in \mathcal{S}$, *then each solution* $(z_k)_{k=0,\ldots,N}$ *of (IP) satisfies the following properties. For* $k = 1,\ldots,N$
the state z_k *is stable at time* t_k, *i.e.,* $(t_k,z_k) \in \mathcal{S}$, *and*
$$\int_{t_{k-1}}^{t_k} \partial_s\mathcal{E}(s,z_k)\,\mathrm{d}s \le \mathcal{E}(t_k,z_k) - \mathcal{E}(t_{k-1},z_{k-1}) + \mathcal{D}(z_{k-1},z_k) \le \int_{t_{k-1}}^{t_k} \partial_s\mathcal{E}(s,z_{k-1})\,\mathrm{d}s.$$

This result shows that (IP) is intrinsically linked with (S) & (E). Its proof is a simple application of the minimisation property of z_k together with the triangle inequality for $\mathcal{D}$, *cf.* [MiT99, MTL02, MaM03].

To obtain existence for (IP) one typically makes the following assumptions: First we assume that the time dependence through the loading is controlled:

$$\big|\partial_t\mathcal{E}(t,z)\big| \le C_1 \quad \text{for all } (t,z) \in [0,T] \times \mathcal{Z}. \tag{3.2}$$

With this we define the set of energetically reachable states:

$$\mathcal{R} := \big\{ (t,z) \in [0,T] \times \mathcal{Z} \;\big|\; \mathcal{E}(t,z) + \mathcal{D}(z_0,z) \leq \mathcal{E}(0,z_0) + C_1 t + 1 \big\}, \tag{3.3}$$

and

$$\mathcal{R}_{\mathcal{Z}} := \big\{ z \in \mathcal{Z} \;\big|\; \exists t \in [0,T] \text{ with } (t,z) \in \mathcal{R} \big\}. \tag{3.4}$$

Then, the major assumptions are read:

$$\begin{aligned} &\text{(a)} \quad \mathcal{R} \text{ is a compact subset of } [0,T] \times \mathcal{Z}\,; \\ &\text{(b)} \quad \mathcal{E} : \mathcal{R} \to \mathbb{R}_\infty \text{ is lower semicontinuous;} \\ &\text{(c)} \quad \mathcal{D} : \mathcal{R}_{\mathcal{Z}} \times \mathcal{R}_{\mathcal{Z}} \to [0,\infty] \text{ is lower semicontinuous}\,. \end{aligned} \tag{3.5}$$

If the assumptions (3.2) and (3.5) hold, (IP) has a solution $(z_k)_{k=0,\ldots,N}$ which satisfies $(t_k,z_k) \in \mathcal{R}$. To show this one just uses induction over k and employs Theorem 3.2 and the triangle inequality in each step.

Finally we address the question whether the solutions of (IP) can be used to construct a solution of (S) & (E) as a limit of incremental solutions when the temporal step size tends to 0. In [MaM03] the following result was obtained.

Theorem 3.3. *Let the assumptions (3.2) and (3.5) hold and assume that the topology on $\mathcal{Z}$ is compatible with the metric $\mathcal{D}$ in the following way:*

If $(t_k,z_k),(t,z) \in \mathcal{R} \cap \mathcal{S}$, $t_k \to t$ and $\min\{\mathcal{D}(z,z_k),\mathcal{D}(z_k,z)\} \to 0$, then $z_k \stackrel{\mathcal{Z}}{\longrightarrow} z$.

Assume further that $\partial_t \mathcal{E} : \mathcal{R} \to [-C_1,C_1]$ is continuous and that the stable set $\mathcal{S}$ is closed (both in the topology of $\mathcal{Z}$).

Then, the problem (S) & (E) has a solution $z : [0,T] \to \mathcal{Z}$ which is obtained as pointwise limit of a subsequence of piecewise constant interpolants $Z^l : [0,T] \to \mathcal{Z}$ of solutions to (IP) for nested partitions with step size tending to 0, that is, for all $t \in [0,T]$ we have $Z^{l_m}(t) \stackrel{\mathcal{Z}}{\longrightarrow} z(t)$ for $m \to \infty$.
Moreover, the energy inequality (E) is an equality and $\mathcal{E}\big(t,z(t)\big) = \lim_{m\to\infty} \mathcal{E}\big(t,Z^{l_m}(t)\big)$ and $\mathrm{Diss}\big(z,[s,t]\big) = \lim_{m\to\infty} \mathrm{Diss}\big(Z^{l_m},[s,t]\big)$ for all $s,t \in [0,T]$ with $s < t$.

The major difficulty is to establish the closedness of $\mathcal{S}$, since it is defined only implicitly. A typical positive result is obtained when $\mathcal{D}$ and $\mathcal{E}$ are continuous on $\mathcal{R}$. In many applications $\mathcal{Z}$ is a Banach space and it is easy to show that $\mathcal{S}$ is closed in the strong topology, but there $\mathcal{R}$ is not compact. In the weak topology $\mathcal{R}$ would be compact, but since $\mathcal{E}$ and $\mathcal{D}$ are not weakly continuous it is difficult to show that $\mathcal{S}$ is closed. In Section 5 we give an example for elasto-plasticity where this dilemma can be studied explicitly.

4 Incremental problems in elasto-plasticity

Until now, no existence theory for the time continuous problem (S) & (E) is available, except for the case $d = 1$ given in the second part of Section 5. However, in computational plasticity [OrR99, MSS99, ORS00, MSL02, HaH03, MiL03] incremental problems are a fundamental tool and hence deserve a mathematical treatment in their right. It was realised in [OrR99, ORS00, CHM02, Mie03a] that existence of solutions for (IP) is not to be expected in general situations. In fact, nonexistence can be connected either with failure of the material due to localisation (e.g. in shear bands) or fracture or with formation of microstructure in material domains of positive measure. Here we present constitutive assumptions which allow us to prove existence of solutions for each incremental step, even though the abstract theory of Section 3 is not applicable.

We let $\mathcal{Z} = \mathcal{F} \times \mathcal{P}$, according to (2.4) and (2.5), and choose a time discretisation $0 = t_0 < \ldots < t_N = T$. The incremental problem assumes are more specific form as the states $z = (y, P)$ have two components and only one of them appears in the dissipation. For a stable initial state $(y_0, P_0) \in \mathcal{F} \times \mathcal{P}$ we consider the incremental problem

$$\textbf{(IP)}\quad \begin{array}{l} \text{For } k = 1, \ldots, N \text{ find } (y_k, P_k) \in \mathcal{F} \times \mathcal{P} \text{ such that} \\ (y_k, P_k) \in \operatorname{Argmin}\{\, \mathcal{E}(t_k, y, P) + \mathcal{D}(P_{k-1}, P) \mid (y, P) \in \mathcal{F} \times \mathcal{P} \,\}. \end{array} \tag{4.1}$$

The main point is to show that the set of global minimisers (Argmin$\{\cdots\}$) is nonempty, i.e. we have to find $z_k = (y_k, P_k) \in \mathcal{F} \times \mathcal{P}$ such that $\mathcal{I}_k(z_k) \leq \mathcal{I}_k(z)$ for all $z \in \mathcal{Z}$, where

$$\mathcal{I}_k(y, P) := \int_\Omega \left[W\big(\mathrm{D}y(x), P(x)\big) + D\big(P_{k-1}(x), P(x)\big) \right] \mathrm{d}x - \big\langle \ell(t_k)\,, y \big\rangle . \tag{4.2}$$

In fact, there are two ways to attack the problem. The first approach uses that y does not appear in the dissipation. One can define a reduced functional

$$\mathcal{E}^{\mathrm{red}} : [0, T] \times \mathcal{P} \to \mathbb{R}_\infty;\ (t, P) \mapsto \min\{\, \mathcal{E}(t, y, P) \mid y \in \mathcal{F} \,\},$$

and then apply the abstract theory with $\mathcal{Z} = \mathcal{P}$ to this new functional. This approach leads to the general problem that $\mathcal{E}^{\mathrm{red}}$ is defined only implicitly, that the derivative $\partial_t \mathcal{E}^{\mathrm{red}}$ may not exist, or may not be continuous.

We use a second approach which relies on the special structure that $P \in \mathcal{P}$ occurs under the integral only with its point values and no derivatives appear. Hence, we can minimise with respect to P for each point $x \in \Omega$ separately. We define the **condensed energy density**

$$W^{\mathrm{cond}}(P_{\mathrm{old}}; F) = \min\{\, W(F, P) + D(P_{\mathrm{old}}, P) \mid P \in \mathfrak{G} \times \mathbb{R}^m \,\},$$

the condensed functional

$$\mathcal{I}_k^{\mathrm{cond}}(y) = \int_\Omega W^{\mathrm{cond}}\big(P_{k-1}(x); \mathrm{D}y(x)\big)\, \mathrm{d}x - \big\langle \ell(t_k)\,, y \big\rangle .$$

and choose a measurable **update function**

$$\begin{array}{l} P^{\mathrm{upd}} : (\mathfrak{G} \times \mathbb{R}^m) \times \mathbb{R}^{d\times d} \to \mathfrak{G} \times \mathbb{R}^m \text{ with} \\ P^{\mathrm{upd}}(P_{\mathrm{old}}; F) \in \operatorname{Argmin}\{\, W(F, P) + D(P_{\mathrm{old}}, P) \mid P \in \mathfrak{G} \times \mathbb{R}^m \,\}. \end{array}$$

Lemma 4.1. *If $\widetilde{y} \in \mathcal{F}$ minimises $\mathcal{I}_k^{\rm cond}$ and if $\widetilde{P}(x) = P^{\rm upd}\big(P_{k-1}(x); {\rm D}\widetilde{y}(x)\big)$, then $(\widetilde{y},\widetilde{P})$ minimises $\mathcal{I}_k$. Moreover, $(y,P) \in \mathcal{F}\times\mathcal{P}$ minimises $\mathcal{I}_k$ if and only if y minimises $\mathcal{I}_k^{\rm cond}$ and $P(x) \in {\rm Argmin}\{\, W\big({\rm D}y(x),\cdot\big) + D\big(P_{k-1}(x),\cdot\big) \mid P \in \mathfrak{G}\times\mathbb{R}^m \,\}$.*

Hence, each step in (IP) reduces to a classical variational problem of nonlinear elasticity. We now state our main assumptions which are formulated in terms of $W^{\rm cond}$ and D. Thus, the assumptions are quite implicit, since in practice only the stored-energy density W and the dissipation potential Δ are given. In the next section we provide an example where all these conditions are satisfied. Note that the multiplicative decomposition (2.2) and the plastic indifference of the dissipation (2.3) implies $W^{\rm cond}\big((G_{\rm old},p_{\rm old});F\big) = W^{\rm cond}\big((\mathbf{1},p_{\rm old});FG_{\rm old}^{-1}\big)$.

(a) $W^{\rm cond}\big((1,\cdot);\cdot\big) : \mathbb{R}^m\times\mathbb{R}^{d\times d} \to [0,\infty]$ and $D(\cdot,\cdot) : (\mathfrak{G}\times\mathbb{R}^m)^2 \to [0,\infty]$ are lower semi-continuous.

(b) For each $p\in\mathbb{R}^m$ the function $W^{\rm cond}\big((1,p),\cdot\big) : \mathbb{R}^{d\times d}\to[0,\infty]$ is polyconvex.

(c) There exist $C,c>0, p_*\in\mathbb{R}^m$ and exponents $q_F,q_G,q_p\geq 1$ such that $D\big((\mathbf{1},p_*),(G,p)\big) \geq c|G|^{q_D}+c|p|^{q_p}-C$ and $W^{\rm cond}\big((\mathbf{1},p);F\big)\geq c|F|^{q_F}-C$ for all (F,G,p) with $D\big((1,p_*),(G,p)\big)<\infty$.

(d) $P^{\rm upd}\big((\mathbf{1},\cdot);\cdot\big) : \mathbb{R}^m\times\mathbb{R}^{d\times d}_+ \to \mathfrak{G}\times\mathbb{R}^m$ is Borel measurable.

(4.3)

The following result is established in [Mie03b].

Theorem 4.2. *Let the assumptions in (4.3) be satisfied such that $\frac{1}{q_F}+\frac{1}{q_G}\leq\frac{1}{q}<\frac{1}{d}$ holds, where q occurs in the definition of $\mathcal{F}\subset{\rm W}^{1,q}(\Omega,\mathbb{R}^d)$ in (2.4).*

Then, for each $P_0\in\mathcal{P}$ with $\mathcal{D}\big((1,p_*),P_0\big)<\infty$ and $\ell\in{\rm C}^0([0,T],{\rm W}^{1,q}(\Omega,\mathbb{R}^d)^*)$ the incremental problem (IP), see (4.1), has a solution $\big((y_k,P_k)\big)_{k=1,\ldots,N}$ with

$$y_k\in\mathcal{F} \text{ and } P_k = P^{\rm upd}\big(P_{k-1};{\rm D}y_k(\cdot)\big)\in\mathcal{P}\cap\big({\rm L}^{q_G}(\Omega,\mathbb{R}^{d\times d})\times{\rm L}^{q_p}(\Omega,\mathbb{R}^m)\big)\,.$$

Obviously, the result is proved by induction over $k=1,2,\ldots,N$. According to Lemma 4.1, the k-th minimisation problem for $\mathcal{I}_k$ (cf. (4.2)) reduces to minimisation of $\mathcal{I}_k^{\rm cond} : y\mapsto\int_\Omega W_k\big(x,{\rm D}y(x)\big)\,{\rm d}x - \big\langle \ell(t_k)\,,y\big\rangle$ with

$$W_k(x,F) = W^{\rm cond}\big(P_{k-1}(x);F\big) = W^{\rm cond}\big((\mathbf{1},p_{k-1}(x));FG_{k-1}(x)^{-1}\big)\,.$$

Clearly, $W_k : \Omega\times\mathbb{R}^{d\times d}\to[0,\infty]$ is measurable in x and lower semi-continuous in F, by (4.3c) it satisfies the lower bound

$$W_k(x,F)\geq c|FG_{k-1}(x)^{-1}|^{q_F} - C \geq c|F|^q - C\big|G_{k-1}(x)\big|^{q_*} - C\,.$$

Moreover, (4.3b) gives polyconvexity of $W_k(x,\cdot)$, since $W^{\rm cond}$ is polyconvex and the minors of the product FG_{k-1}^{-1} are linear combinations of products of the minors of F and G_{k-1}^{-1}. Hence, suitable existence results for polyconvex functionals apply. Induction works since $q_* \leq q_G$ holds.

We note that $\mathcal{I}_k : \mathcal{F} \times \mathcal{P} \to \mathbb{R}_\infty$ is not weakly lower semicontinuous because of the geometric nonlinearity coming from the multiplicative decomposition; i.e., $W\big(F,(G,p)\big) = \widehat{W}\big(FG^{-1},p\big)$. It is shown in [FKP94, LDR00] that weak lower semicontinuity of $\mathcal{I}_k$ implies that the map $(F,G,p) \mapsto W\big(F,(G,p)\big)+D\big(P_{k-1}(x),(G,p)\big)$ is cross-quasiconvex, which in turn implies convexity in $P=(G,p)$. However, this can only be achieved if $F_{\text{elast}} \mapsto \widehat{W}(F_{\text{elast}})$ is convex, but this contradicts the standard axioms of finite-strain elasto-plasticity, see [CHM02] and below. Of course, lower semi-continuity of $\mathcal{I}_k$ is not necessary and, as shown above, we may obtain minimisers without it.

5 Two examples

The crucial assumption of the above theory is the polyconvexity of W^{cond}. Since polyconvexity is in general hard to check, it is nontrivial to find a multi-dimensional example. Our example only works for the dimension $d=2$, since it depends on the fact that everything can be calculated explicitly. We consider the isotropic energy density

$$W : \begin{cases} \mathbb{R}^{2\times 2} \to \mathbb{R}_\infty\,, \\ F \mapsto \frac{1}{\alpha}(\nu_1^\alpha + \nu_2^\alpha) + V(\det F)\,, \end{cases} \tag{5.1}$$

where $\nu_1, \nu_2 \geq 0$ are the two singular values of F (i.e., the eigenvalues of $(F^\mathsf{T} F)^{1/2}$) and $V:\mathbb{R}\to[0,\infty]$ is convex, continuous and satisfies $V(\delta)=\infty$ for $\delta \leq 0$. For the plastic variables we take $P=(G,p)\in \mathrm{SL}(2)\times\mathbb{R}$ with the dissipation metric

$$\Delta(G,p,\dot{G},\dot{p}) = \begin{cases} A'(p)\dot{p} & \text{for } \dot{p} \geq \|\dot{G}G^{-1}\|\,, \\ \infty & \text{otherwise.} \end{cases} \tag{5.2}$$

Here, $A(p) = \mathrm{e}^{\alpha p/\sqrt{2}}$ and $\|\xi\|^2 = \sum \xi_{ij}^2$. The associated dissipation distance satisfies

$$D\big((\mathbf{1},p_0),E(s),p_1)\big) = \begin{cases} \mathrm{e}^{\alpha(p_0/\sqrt{2}+|s|)} - \mathrm{e}^{\alpha p_0/\sqrt{2}} & \text{for } p_1 \geq p_0 + \sqrt{2}|s|\,, \\ \infty & \text{otherwise,} \end{cases}$$

where $E(s) = \mathrm{diag}(\mathrm{e}^s, \mathrm{e}^{-s})$. Moreover, the condensed stored-energy density takes the explicit form

$$W^{\text{cond}}\big((1,p);F\big) = V(\nu_1\nu_2) - \mathrm{e}^{\alpha p/\sqrt{2}} + \begin{cases} \frac{2}{\alpha}\sqrt{\nu_1^\alpha(\nu_2^\alpha + b_p)} & \text{for } \nu_1^\alpha \geq \nu_2^\alpha + b_p\,, \\ \frac{1}{\alpha}\sqrt{\nu_1^\alpha + \nu_2^\alpha + b_p} & \text{for } |\nu_1^\alpha - \nu_2^\alpha| \leq b_p\,, \\ \frac{2}{\alpha}\sqrt{\nu_2^\alpha(\nu_1^\alpha + b_p)} & \text{for } \nu_2^\alpha \geq \nu_1^\alpha + b_p\,, \end{cases}$$

where $b_p = \alpha\mathrm{e}^{\alpha p/\sqrt{2}}$. The update functions can also be given explicitly. For details we refer to [Mie02a, HMM03, Mie03a, Mie03b].

By explicit calculations, it is shown in [Mie02b] that $W^{\rm cond}\big((1,p);\,\cdot\,\big)$ is polyconvex for $\alpha \geq 2$. Moreover, we have the lower bounds

$$D\big((\mathbf{1},p_*),(G,p)\big) \geq c\big(\|G\|^\alpha - 1\big) \quad \text{and} \quad W^{\rm cond}\big((\mathbf{1},p);F\big) \geq c\big(\|F\|^{\frac12\alpha} - b_p\big)$$

for all $F\in\mathbb{R}^{2\times2}, p_*, p\in\mathbb{R}$ and $G\in \mathrm{SL}(2)$ with $D\big((\mathbf{1},p_*),(G,p)\big) < \infty$. Thus, this two-dimensional example satisfies the assumptions (4.3) for $\alpha \geq 2$ with exponents $q_F = \frac12\alpha$ and $q_G = \alpha$. Hence, Theorem 4.2 is applies if $\dfrac12 = \dfrac1d > \dfrac1q \geq \dfrac{1}{q_W} + \dfrac{1}{q_D} = \dfrac{3}{\alpha}$ holds. Summarising, we obtain the following existence result.

Theorem 5.1. *Let $d=2$ and $\mathfrak{G} = \mathrm{SL}(2)$. With $\alpha > 6$, let $W:\mathbb{R}^{2\times2}\to[0,\infty]$ and $\Delta: \mathrm{T}(\mathfrak{G}\times\mathbb{R}) \to [0,\infty]$ be defined via (5.1) and (5.2), respectively. Assume that there exists a $p_*\in\mathbb{R}$, such that the initial condition $P_0\in\mathcal{P}$ satisfies $\mathcal{D}\big((\mathbf{1},p_*),P_0\big) < \infty$ and let $q = \frac13\alpha$.*

Then, for each $\ell : [0,T] \to \big(\mathrm{W}^{1,\frac13\alpha}(\Omega,\mathbb{R}^2)\big)^*$ the incremental problem (IP) (see (4.1)) has a solution $\big((y_k,P_k)\big)_{k=1,\dots,N} \in (\mathcal{F}\times\mathcal{P})^N$. Moreover, there exists a constant C which depends only on $\alpha, \ell,$ and P_0, but not on the partition $t_1,\dots,t_N$ nor on the solution, such that

$$\|y_k\|_{\mathrm{W}^{1,\frac13\alpha}} + \|G_k\|_{\mathrm{L}^\alpha} + \big\|\mathrm{e}^{\alpha p_k/\sqrt2}\big\|_{\mathrm{L}^1} \leq C \quad \textit{for } k=1,\dots,N\,.$$

As a second example we treat the one-dimensional situation, which seems to be trivial regarding the existence of solutions for the incremental problem. Here the major simplification is that polyconvexity is equal to convexity. However, the purpose of this example is to address the question of convergence of the incremental solutions towards a solution of the time-continuous problem (S) & (E). We will see, that general arguments, as given in Theorem 3.3, are not sufficient. Only in using the special one-dimensional structure, we are able to prove convergence (of a subsequence) and obtain finally an existence result for (S) & (E).

Again we treat a special case, but the analysis easily generalises to far more general constitutive laws W and Δ. We let

$$W(F) = \tfrac{1}{\alpha}(F^\alpha + F^{-\alpha}) \quad \text{for } F>0 \quad \text{and } \infty \text{ otherwise,}$$

$\mathfrak{G} = \mathrm{GL}_+(1) = (0,\infty)$, $P=(G,p)\in\mathfrak{G}\times\mathbb{R}$, and

$$\Delta\big((G,p),(\dot G,\dot p)\big) = \alpha \mathrm{e}^{\alpha p}\dot p \quad \text{for } \dot p \geq |\dot G/G| \quad \text{and } \infty \text{ otherwise.}$$

From this we find the condensed stored-energy density

$$W^{\rm cond}\big((1,p);F\big) = \frac1\alpha\begin{cases} 2\sqrt{1+b_pF^\alpha} - b_p & \text{for } F^\alpha \geq b_p + F^{-\alpha}\,,\\ F^\alpha + F^{-\alpha} & \text{for } |F^\alpha - F^{-\alpha}| \leq b_p\,,\\ 2\sqrt{1+b_pF^{-\alpha}} - b_p & \text{for } F^{-\alpha} \geq b_p + F^\alpha\,,\\ \infty & \text{for } F\leq 0\,,\end{cases}$$

with $b_p = \alpha e^{\alpha p}$. We see that $W^{\rm cond}\big((1,p);\,\cdot\,\big)$ is strictly convex for $\alpha \geq 2$ and that the theory of Section 4 applies for $\alpha > 3$, since the exponents in condition (4.3) are $q_F = \frac{1}{2}\alpha$ and $q_G = \alpha$. Thus, (IP) has a unique solution for each partition of $[0,T]$.

We now want to discuss the connection to the abstract theory of Section 3. Therefore, we restrict ourselves to the most simple nontrivial case. Let $\Omega = (0,1) \subset \mathbb{R}^1$, $\alpha = 6$ and $\mathcal{F} = \big\{\, y \in \mathrm{W}^{1,2}(\Omega) \mid y(0) = 0 \,\big\}$. The loading takes the form

$$\langle \ell(t)\,, y\rangle = \int_0^1 h_{\rm ext}(t,x)\, y(x)\,\mathrm{d}x + \sigma_1(t)\, y(1) = \int_0^1 H_{\rm ext}(t,x)\, y'(x)\,\mathrm{d}x\,,$$

where $H_{\rm ext}(t,x) = \sigma_1(t) + \int_x^1 h_{\rm ext}(t,\widetilde{x})\,\mathrm{d}\widetilde{x}$ with $H_{\rm ext} \in \mathrm{C}^1([0,T]\times\overline{\Omega})$. Moreover, we let $\mathcal{P} = \big\{\,(G,p) \in \mathrm{L}^6(\Omega)^2 \mid G \geq 0 \text{ a.e.}\,\big\}$ and $(G_0,p_0) \equiv (1,0)$.

With this definition it is clear that $\mathcal{R} \subset [0,T]\times\mathcal{R}_{\mathcal{Z}}$ for some bounded closed set $\mathcal{R}_{\mathcal{Z}}$ in $\mathrm{W}^{1,2}(\Omega)\times L^6(\Omega)^2$. It can be shown that $\mathcal{R}$ is also closed in the strong topology. However, on the one hand compactness of $\mathcal{R}$ in the strong topology fails since $\mathcal{R}$ contains "L^∞"-neighborhoods. On the other hand, compactness of $\mathcal{R}$ in the weak topology fails, since this is equivalent to the weak lower semi-continuity of $\mathcal{E}(t,\,\cdot\,,\,\cdot\,) + \mathcal{D}(P_{k-1},\,\cdot\,)$ on $\mathcal{F}\times\mathcal{P}$, which is not satisfied due to lacking cross-quasiconvexity.

As shown in [Mie03a, Mie03b], it is possible to characterise the stable set for this example explicitly:

$$\mathcal{S} = \big\{\,(t,y,G,p) \mid \big(y'(x), G(x), p(x)\big) \in M\big(H_{\rm ext}(t,x)\big) \text{ for a.a. } x \in \Omega\,\big\}\,,$$

where $M(H)=\big\{\,(F,G,p) \mid \Big|\Big(\frac{F}{G}\Big)^{\alpha}-\Big(\frac{F}{G}\Big)^{-\alpha}\Big| \leq \alpha e^{\alpha p},\ \Big(\frac{F}{G}\Big)^{\alpha-1}-\Big(\frac{F}{G}\Big)^{-\alpha-1}= G\,H\,\big\}$.
Again we see that $\mathcal{S}$ is closed in the strong topology but not in the weak topology, since the sets $M(H) \subset \mathbb{R}^3$ are not convex.

This shows that the abstract theory of Section 3 is not applicable. Nevertheless, it is possible to show that the incremental solutions converge to a solution of the time-continuous problem (S) & (E). Here we use that the compactness of $\mathcal{R}$ is solely used for the purpose of extracting converging subsequences. We can dispense with compactness if convergence can be shown by other means. In the present one-dimensional setting we can use the fact that the problem decouples into "zero-dimensional" plasticity problems for each $x \in \Omega$. Moreover, each of these problems is almost scalar such that monotonicity arguments can be used which imply convergence. As a conclusion the following existence and convergence result is obtained in [Mie03b].

Theorem 5.2. *For the above problem there exists a solution (y,G,p) of (S) & (E) (cf. (2.6)) with $(y,G,p) \in \mathrm{C}^0\big([0,T],\mathrm{W}^{1,\infty}(\Omega)\times\mathrm{L}^\infty(\Omega)^2\big)$. Moreover, there exists a constant $C > 0$ such that for each time discretisation $0 = t_0 < t_1 < \cdots < t_N = T$ the unique solution $(y_k,G_k,p_k)_{k=0,\dots,N}$ of the incremental problem (4.1) satisfies*

$$\big\|y(t_k){-}y_k\big\|_{\mathrm{W}^{1,\infty}}+\big\|G(t_k){-}G_k\big\|_{\mathrm{L}^\infty}+\big\|p(t_k){-}p_k\big\|_{\mathrm{L}^\infty} \leq C\max\{\,t_n{-}t_{n-1} \mid n{=}1,\dots,k\,\}$$

for $k = 1,\dots,N$.

The connection with the abstract theory is immediate if we equip $\mathcal{Z} = \mathcal{F} \times \mathcal{P}$ with the strong topology of $\mathrm{W}^{1,\infty}(\Omega) \times \mathrm{L}^\infty(\Omega)$. Then, essentially all the assumptions hold, except for the compactness of $\mathcal{R}$. In particular, $\mathcal{S}$ is closed, $\mathcal{E}$ and $\mathcal{D}$ are lower semicontinuous and $\partial_t \mathcal{E}$ is continuous.

6 Gradient theories

As we have seen in Sections 4 and 5, the above functionals $\mathcal{E}(t, \cdot)$ and $\mathcal{D}(\cdot, \cdot)$ are not weakly lower semi-continuous on spaces of the form $\mathrm{W}^{1,q} \times \mathrm{L}^r$. The major problem is the nonconvexity of the multiplicative term $\mathrm{D}y\, G^{-1}$ and of the Lie group structure underlying $\mathcal{D}$. One possible way out of these problems, which is also often used in engineering, cf. [Gur02, Sve02], consists in adding a regularising term involving the gradient terms of the plastic tensor $F_{\text{plast}} = G$. (Sometimes these gradient terms are also called "nonlocal terms".) Such terms introduce a length scale into the problem which prevent the formation of arbitrarily fine microstructures which are one of the main obstructions to existence of minimisers.

In this section we just want to indicate how terms involving the gradient $\mathrm{D}G$ (a tensor of third order) may help to establish existence for (IP) and for (S) & (E). Most of the results given here are derived in [MiM03].

We consider the same dissipation distance D and $\mathcal{D}$ as above. However, the energy functional now takes the form

$$\mathcal{E}(t, y, G) = \int_\Omega U(x, \mathrm{D}y, G, \mathrm{D}G)\,\mathrm{d}x - \langle \ell(t)\,, y \rangle\,,$$

where, for simplicity, we assume that the density U takes one of the two cases

$$U_{\text{grad}}(F, G, A) = \widehat{W}(FG^{-1}) + \kappa \|A\|^r,$$
$$U_{\text{curl}}(F, G, A) = \widehat{W}(FG^{-1}) + \kappa \|A_{\text{anti}} G^{\mathsf{T}}\|^r,$$

with $\kappa > 0$ and $r > d$. Note that A has the dimension of 1/length and hence κ must have a dimension like $(\text{length})^r$. Here A_{anti} denotes the anti-symmetric part of the tensor A, such that $(\mathrm{D}G)_{\text{anti}} = \operatorname{curl} G$. The density U_{curl} involves the tensor $(\operatorname{curl} G)\, G^{\mathsf{T}}$, which is known to be the physically most relevant tensor for measuring the density of geometrically necessary dislocations [Gur02, Sve02].

The major difference to the previous theory, is that the set $\mathcal{R}$ now provides a priori bounds for $\mathrm{D}G$ or $\operatorname{curl} G$ in $\mathrm{L}^r(\Omega, \mathbb{R}^{d\times d})$ as well. For $U = U_{\text{grad}}$ the set $\mathcal{R}$ is bounded in $[0, T] \times \mathcal{Z}$ with $\mathcal{Z} = \mathrm{W}^{1,q}(\Omega, \mathbb{R}^d) \times \mathrm{W}^{1,r}(\Omega, \mathbb{R}^{d\times d}) \times \mathrm{L}^{q_p}(\Omega, \mathbb{R}^m)$. In particular, weak lower semi-continuity of $\mathcal{E}$ on $\mathcal{Z}$ can now be established, since the dangerous product $\mathrm{D}yG^{-1}$ is under control. If

$$y_k \rightharpoonup y \;\text{ in }\; \mathrm{W}^{1,q}(\Omega, \mathbb{R}^d) \qquad \text{and} \qquad G_k \rightharpoonup G \;\text{ in }\; \mathrm{W}^{1,r}(\Omega, \mathbb{R}^{d\times d})\,,$$

then $\mathbb{M}(\mathrm{D}y_k\, G_k^{-1}) \rightharpoonup \mathbb{M}(\mathrm{D}yG^{-1})$ in $\mathrm{L}^1(\Omega)^{m_d}$, where $\mathbb{M}(F)$ denotes the vector of all minors of the matrix F. To show this, we use the product rule for minors and that G_k converges strongly to G. Moreover, the dissipation $\mathcal{D}$ which does not involve derivatives of G will be continuous with respect to this convergence.

More precisely, the incremental problem (IP) now consists in minimising the functional

$$\mathcal{I}_k : (y,G) \mapsto \int_\Omega W_k(x, \mathrm{D}y, G, \mathrm{D}g)\,\mathrm{d}x - \langle \ell(t_k)\,, y \rangle\,, \tag{6.1}$$

with the density $W_k(x,F,G,A) = U(x,F,G,A) + D_k^{\mathrm{cond}}(x,G)$ where the incrementally condensed distance D_k^{cond} is given by

$$D_k^{\mathrm{cond}}(x,G) = \min\{\, D\big(G_{k-1}(x), p_{k-1}(x)\big)\,, (G,p) \;\big|\; p \in \mathbb{R}^m \,\}\,.$$

The above arguments indicate that I_k for $U = U_{\mathrm{grad}}$ is weakly lower semi-continuous on $\mathrm{W}^{1,q}(\Omega,\mathbb{R}^d) \times \mathrm{W}^{1,r}(\Omega,\mathbb{R}^{d\times d})$. Coercivity can be obtained with arguments as in Section 4. Thus, the gradient regularisation provides an existence theory for (IP) under rather general and easily checkable conditions.

For the case $U = U_{\mathrm{curl}}$ the situation is more difficult. Using $\det G \equiv 1$ we first observe $G^{-1} = \mathrm{ad}\, G$ and find that the minors $\mathbb{M}_s(FG^{-1})$ of order s are exactly those minors in $\mathbb{M}_d\Big(\Big(\begin{smallmatrix} F \\ G \end{smallmatrix}\Big)\Big)$ which contain s rows from F and $d{-}s$ rows from G. Now, if

$$\begin{array}{rcll} y_k & \rightharpoonup & y & \text{in } \mathrm{W}^{1,q}(\Omega,\mathbb{R}^d)\,,\\ G_k & \rightharpoonup & G & \text{in } \mathrm{L}^{q_G}(\Omega,\mathbb{R}^{d\times d})\,,\\ \mathrm{curl}\, G_k & \rightharpoonup & \mathrm{curl}\, G & \text{in } \mathrm{L}^{q_c}(\Omega,\mathbb{R}^{d\times d\times d})\,, \end{array}$$

for suitable exponents q, q_G and $q_c > 1$, then $\mathbb{M}_s(\mathrm{D}_k\, G_k^{-1}) \rightharpoonup \mathbb{M}_s(\mathrm{D}\, G^{-1})$ and $\mathbb{M}_s(G_k) \rightharpoonup \mathbb{M}_s(G)$.

Thus, under suitable coercivity assumptions on U and D_k^{cond} the functional I_k in (6.1) can be shown to be weakly lower semi-continuous on a weakly closed subset $\mathcal{Z}$ of the Banach space

$$\mathrm{W}^{1,q}(\Omega,\mathbb{R}^d) \times \big\{\, G \in \mathrm{L}^{q_G}(\Omega,\mathbb{R}^{d\times d}) \;\big|\; \mathrm{curl}\, G \in \mathrm{L}^{q_c}(\Omega,\mathbb{R}^{d\times d\times d}) \,\big\}$$

for suitable exponents q, q_G and q_c. However, for this we additionally need that $G \mapsto D_k^{\mathrm{cond}}(x,G)$ is polyconvex, since only $\mathrm{curl}\, G$ is controlled.

Thus, the curl regularisation is also strong enough to handle the difficult multiplicative term $\mathrm{D}y\, G^{-1}$. But only under the nontrivial condition that $D_k^{\mathrm{cond}}(x,\cdot\,)$ is polyconvex we obtain existence for (IP).

Finally, we remark that both regularisations are not strong enough to do the limit from (IP) to (S) & (E). The major problem here is that the dissipation $\mathcal{D}$ only controls distances of the plastic variables $P = (G,p) \in \mathcal{P}$. However, in finite-strain elasto-plasticity there may be more than one global minimiser $y \in \mathcal{F}$ of $\mathcal{E}(t,\cdot\,,P)$. Thus, the compatibility condition between the topology on $\mathcal{Z} = \mathcal{F} \times \mathcal{P}$ and the dissipation distance $\mathcal{D}$, as stated in Theorem 3.3, does not hold. Of course, temporal oscillations between several global minimisers without any dissipation are unphysical and should be eliminated by adjusting the model in a suitable manner.

Acknowledgments The author is grateful for stimulating discussions with Andreas Mainik, Christian Miehe, Stefan Müller, Patrizio Neff, Tomaš Roubíček and Florian Theil. The research was partially supported by DFG through the SFB 404 (TP C11) *Multifield Problems in Continuum Mechanics.*

Bibliography

[ACZ99] J. Alberty, C. Carstensen, and D. Zarrabi; Adaptive numerical analysis in primal elastoplasticity with hardening. *Comput. Methods Appl. Mech. Engrg.*, 171(3–4), pp. 175–204, 1999.

[Alb98] H.-D. Alber; *Materials with Memory*, volume 1682 of *Lecture Notes in Mathematics.* Springer-Verlag, Berlin, 1998. Initial-boundary value problems for constitutive equations with internal variables.

[Bal77] J. Ball; Constitutive inequalities and existence theorems in nonlinear elastostatics. In *Nonlinear Analysis and Mechanics: Heriot-Watt Symposium (Edinburgh, 1976), Vol. I*, pp. 187–241. Res. Notes in Math., No. 17. Pitman, London, 1977.

[BeF96] A. Bensoussan and J. Frehse; Asymptotic behaviour of the time dependent Norton–Hoff law in plasticity theory and H^1 regularity. *Comment. Math. Univ. Carolin.*, 37, pp. 285–304, 1996.

[Che01a] K. Chełmiński; Coercive approximation of viscoplasticity and plasticity. *Asymptot. Anal.*, 26, pp. 105–133, 2001.

[Che01b] K. Chełmiński; Perfect plasticity as a zero relaxation limit of plasticity with isotropic hardening. *Math. Methods Appl. Sci.*, 24, pp. 117–136, 2001.

[CHM02] C. Carstensen, K. Hackl, and A. Mielke; Non-convex potentials and microstructures in finite-strain plasticity. *Proc. Royal Soc. London, Ser. A*, 458, pp. 299–317, 2002.

[CoV90] P. Colli and A. Visintin; On a class of doubly nonlinear evolution equations. *Comm. Partial Differential Equations*, 15(5), pp. 737–756, 1990.

[DMT02] G. Dal Maso and R. Toader; A model for quasi-static growth of brittle fractures: existence and approximation results. *Arch. Rat. Mech. Anal.*, 162, pp. 101–135, 2002.

[Efe03] M. Efendiev; On the compactness of the stable set for rate-independent processes. *Comm. Part. Diff. Eqns.*, 2003. Submitted.

[FKP94] I. Fonseca, D. Kinderlehrer, and P. Pedregal; Energy functionals depending on elastic strain and chemical composition. *Calc. Var. Partial Differential Equations*, 2(3), pp. 283–313, 1994.

[FrM98] G. Francfort and J.-J. Marigo; Revisiting brittle fracture as an energy minimization problem. *J. Mech. Phys. Solids*, 46, pp. 1319–1342, 1998.

[GMH02] S. Govindjee, A. Mielke, and G. Hall; The free-energy of mixing for n-variant martensitic phase transformations using quasi-convex analysis. *J. Mech. Physics Solids*, 50, pp. 1897–1922, 2002.

[Gur02] M. E. Gurtin; A gradient theory of single–crystal viscoplasticity that accounts for geometrically necessary dislocations. *J. Mech. Phys. Solids*, 50, pp. 5–32, 2002.

[HaH03] K. Hackl and U. Hoppe; On the calculation of microstructures for inelastic materials using relaxed energies. In C. Miehe, editor, *IUTAM Symposium on Computational Mechanics of Solids at Large Strains*, pp. 77–86. Kluwer, 2003.

[HaR95] W. Han and B. D. Reddy; Computational plasticity: the variational basis and numerical analysis. *Comput. Mech. Adv.*, 2(4), pp. 283–400, 1995.

[HMM03] K. Hackl, A. Mielke, and D. Mittenhuber; Dissipation distances in multiplicative elastoplasticity. In W. Wendland and M. Efendiev, editors, *Analysis and Simulation of Multifield Problems*. Springer-Verlag, 2003. In press.

[Joh76] C. Johnson; Existence theorems for plasticity problems. *J. Math. Pures Appl. (9)*, 55(4), pp. 431–444, 1976.

[KMR02] M. Kočvara, A. Mielke, and T. Roubíček; A rate-independent approach to the delamination problem. In preparation, 2002.

[Kru02] M. Kružík; Variational models for microstructure in shape memory alloys and in micromagnetics and their numerical treatment. In A. Ruffing and M. Robnik, editors, *Proceedings of the Bexbach Kolloquium on Science 2000 (Bexbach, 2000)*. Shaker Verlag, 2002.

[LDR00] H. Le Dret and A. Raoult; Variational convergence for nonlinear shell models with directors and related semicontinuity and relaxation results. *Arch. Ration. Mech. Anal.*, 154(2), pp. 101–134, 2000.

[MaM03] A. Mainik and A. Mielke; Existence results for energetic models for rate-independent materials. In preparation, 2003.

[Mie02a] A. Mielke; Finite elastoplasticity, Lie groups and geodesics on $\mathrm{SL}(d)$. In P. Newton, A. Weinstein, and P. Holmes, editors, *Geometry, Dynamics, and Mechanics*, pp. 61–90. Springer-Verlag, 2002.

[Mie02b] A. Mielke; Necessary and sufficient conditions for polyconvexity of isotropic functions. *J. Convex Analysis*, 2002. Submitted.

[Mie02c] A. Mielke; Relaxation via Young measures of material models for rate-independent inelasticity. *Preprint, February*, 2002.

[Mie03a] A. Mielke; Energetic formulation of multiplicative elasto-plasticity using dissipation distances. *Cont. Mech. Thermodynamics*, 2003. To appear.

[Mie03b] A. Mielke; Existence of minimizers in incremental elasto-plasticity with finite strains. *SIAM J. Math. Analysis*, 2003. Submitted.

[MiL03] C. Miehe and M. Lambrecht; Analysis of microstructure development in shearbands by energy relaxation of incremental stress potentials: large-strain theory for generalized standard solids. *Int. J. Numer. Meth. Eng.*, 2003. In press.

[MiM03] A. Mielke and S. Müller; Lower semi–continuity and existence of minimizers for a functional in elasto-plasticity. In preparation.

[MiR03] A. Mielke and T. Roubíček; A rate-independent model for inelastic behavior of shape-memory alloys. *Multiscale Model. Simul.*, 2003. Submitted.

[MiT99] A. Mielke and F. Theil; A mathematical model for rate-independent phase transformations with hysteresis. In H.-D. Alber, R. Balean, and R. Farwig, editors, *Proceedings of the Workshop on "Models of Continuum Mechanics in Analysis and Engineering"*, pp. 117–129. Shaker-Verlag, 1999.

[MiT01] A. Mielke and F. Theil; On rate-independent hysteresis models. *Nonl. Diff. Eqns. Appl. (NoDEA)*, 2001. To appear.

[Mor76] J.-J. Moreau; Application of convex analysis to the treatment of elastoplastic systems. In P. Germain and B. Nayroles, editors, *Applications of methods of functional analysis to problems in mechanics*, pp. 56–89. Springer-Verlag, 1976. Lecture Notes in Mathematics, 503.

[MSL02] C. Miehe, J. Schotte, and M. Lambrecht; Homogenization of inelastic solid materials at finite strain based on incremental minimization principles. Application to texture analysis of polycrystals. *J. Mech. Physics Solids*, 50, pp. 2123–2167, 2002.

[MSS99] C. Miehe, J. Schröder, and J. Schotte; Computational homogenization analysis in finite plasticity. Simulation of texture development in polycrystalline materials. *Comput. Methods Appl. Mech. Engrg.*, 171, pp. 387–418, 1999.

[MTL02] A. Mielke, F. Theil, and V. Levitas; A variational formulation of rate-independent phase transformations using an extremum principle. *Arch. Rational Mech. Anal.*, 162, pp. 137–177, 2002.

[Nef02] P. Neff; Finite multiplicative plasticity for small elastic strains with linear balance equations and grain boundary relaxation. *Cont. Mech. Thermodynamics*, 2002. To appear.

[OrR99] M. Ortiz and E. Repetto; Nonconvex energy minimization and dislocation structures in ductile single crystals. *J. Mech. Phys. Solids*, 47(2), pp. 397–462, 1999.

[ORS00] M. Ortiz, E. Repetto, and L. Stainier; A theory of subgrain dislocation structures. *J. Mech. Physics Solids*, 48, pp. 2077–2114, 2000.

[OrS99] M. Ortiz and L. Stainier; The variational formulation of viscoplastic constitutive updates. *Comput. Methods Appl. Mech. Engrg.*, 171(3–4), pp. 419–444, 1999.

[RoK02] T. Roubíček and M. Kružík; Mircrostructure evolution model in micromagnetics. *Zeits. angew. math. Physik*, 2002. In print.

[Rou02] T. Roubíček; Evolution model for martensitic phase transformation in shape-memory alloys. *Interfaces Free Bound.*, 4, pp. 111–136, 2002.

[Sve02] B. Svendsen; Continuum thermodynamic models for crystal plasticity including the effects of geometrically necessary dislocations. *J. Mech. Physics Solids*, 50, pp. 1297–1329, 2002.

[Vis01] A. Visintin; A new approach to evolution. *C.R.A.S. Paris*, 332, pp. 233–238, 2001.

Invited speaker Philippe Toint (Belgium). photography by *Happy Medium Photo Co.*

Harald Niederreiter is Professor of Mathematics and Computer Science at the National University of Singapore. He has done extensive research on quasi-Monte Carlo methods, random number generation, simulation methods, number theory, discrete mathematics, algebraic coding theory, and cryptology. He received his PhD in mathematics from the University of Vienna with the distinction *sub auspiciis praesidentis rei publicae.* Previous positions include Director of the Institute of Information Processing and Director of the Institute of Discrete Mathematics at the Austrian Academy of Sciences. Currently, he is an editor of the *Journal of Complexity* and is on the editorial boards of many other journals, among them *Mathematics of Computation* and *ACM Transactions on Modeling and Computer Simulation.*

Professor Niederreiter has been the principal speaker of an NSF–CBMS research conference in 1990 and an invited speaker at numerous other conferences, such as the 1977 Annual Meeting of the American Mathematical Society in St Louis and the International Congress of Mathematicians 1998 in Berlin. His awards include the Outstanding Simulation Publication Award 1995 and the Cardinal Innitzer Prize for Natural Sciences 1998. He is an elected member of the Austrian Academy of Sciences and of the German Academy of Natural Sciences and until recently was a member of the presidium of the latter academy.

Chapter 16

High-Dimensional Numerical Integration

Harald Niederreiter[*]

Abstract: Multiple integrals of high dimensions, often running into the hundreds or even thousands, occur in various applications, notably in computational finance. Classical methods for approximating multiple integrals based on Cartesian products of one-dimensional integration rules (trapezoidal rule, Simpson's rule, etc.) are efficient only in low-dimensional cases. For many years the method of choice for high-dimensional numerical integration was the statistical Monte Carlo method. However, the deterministic version of the Monte Carlo method, called the quasi-Monte Carlo method, offers a considerably faster convergence rate and is thus gradually replacing the statistical Monte Carlo method in many applications.

In this talk we discuss the above methods, with the emphasis being on the quasi-Monte Carlo method and its applications. A fascinating problem is that of the explicit construction of suitable node sets for quasi-Monte Carlo integration. Surprising connections with areas of discrete mathematics, such as coding theory, arise in this context.

Contents

[*]Department of Mathematics, National University of Singapore

1 Introduction

The task of multidimensional numerical integration arises in many problems of scientific computing. This task is, of course, of great interest *per se* as a fundamental issue in numerical analysis. There are, in addition, several important problems in numerical analysis that can be reduced to the approximate calculation of multiple integrals, such as the numerical solution of integral equations and integro-differential equations. Further afield, multidimensional numerical integration occurs over and over again as a standard numerical problem in computational physics, computational chemistry, engineering, and econometrics, to name just a few areas (see [11] for some pointers to the literature on these classical applications). Some lesser known applications have arisen in statistics (see [5], [26]) and stochastic differential equations (see [24]). Fairly recently, multidimensional numerical integration has become a big issue in computational finance (see Section 2 for a typical example).

In the traditional settings of multidimensional numerical integration in numerical analysis, computational physics, and computational chemistry, the dimension s of the integral is usually confined to a modest range such as $2 \leq s \leq 30$. However, in the modern applications to computational finance the dimension s can be of the order of magnitude 10^2 or even 10^3 (compare with Section 2), so this is the range we are thinking of when we speak of "high-dimensional" numerical integration. The approximate calculation of multiple integrals of such dimensions is an extremely challenging computational task. It should therefore not come as a surprise that rather few methods are available for this purpose. We will briefly discuss two methods of limited interest in this section. The most powerful methods for multidimensional numerical integration are the Monte Carlo method (see Section 3) and the quasi-Monte Carlo method (see Section 4).

In the case where the integration domain is a multidimensional interval, an obvious idea is to use Cartesian products of one-dimensional integration rules such as the trapezoidal rule and Simpson's rule. Without loss of generality we can assume that the integration domain is the s-dimensional unit cube $I^s := [0,1]^s$. For the sake of concreteness, let us take the trapezoidal rule

$$\int_0^1 f(u)\,\mathrm{d}u \approx \sum_{n=0}^{m} w_n\, f\left(\frac{n}{m}\right),$$

where m is a positive integer and the weights w_n are given by $w_0 = w_m = 1/(2m)$ and $w_n = 1/m$ for $1 \leq n \leq m-1$. The error involved in this approximation is $\mathrm{O}(m^{-2})$, provided that the integrand f has a continuous second derivative on $[0,1]$. The s-fold *Cartesian product* of the trapezoidal rule is then

$$\int_{I^s} f(\mathbf{u})\,\mathrm{d}\mathbf{u} \approx \sum_{n_1=0}^{m} \cdots \sum_{n_s=0}^{m} w_{n_1} \cdots w_{n_s}\, f\left(\frac{n_1}{m}, \ldots, \frac{n_s}{m}\right).$$

If the second partial derivatives of f with respect to each of its variables are continuous on I^s, then the integration error is $\mathrm{O}(m^{-2})$, which is $\mathrm{O}(N^{-2/s})$ in terms of the number $N = (m+1)^s$ of nodes. Note that in order to guarantee a prescribed level of accuracy, say an integration error that is in absolute value at most 10^{-2}, we must use roughly 10^s nodes, and so the required number of nodes increases exponentially with s. This phenomenon is often called the *curse of dimensionality*. In view of this phenomenon, Cartesian products of one-dimensional integration rules are usually employed in small dimensions only, say for $s \leq 6$.

Another classical approach to multidimensional numerical integration is based on rules that are exact for polynomials up to a certain degree or for trigonometric polynomials up to a certain degree. A detailed survey of integration rules that are exact for polynomials up to a certain degree is given in Cools [2]. A web-based project of listing all known integration rules of this type is described in Cools [3]. Because of the complexity of computing the nodes and weights for such integration rules, they are available in explicit form only for small dimensions and for a relatively small number of nodes. For integration rules that are exact for trigonometric polynomials up to a certain degree, we refer to Cools and Lyness [4] and the references therein.

2 An Example from Computational Finance

Modern computational finance abounds with tasks that lead to challenging problems of high-dimensional numerical integration. Prime examples of such problems arise, for instance, in the pricing of financial derivatives, that is, of financial instruments whose value is derived from that of a portfolio of underlying assets such as equities, bonds, and mortgages. In the following, we describe the case of MBS pricing in some detail.

In a *mortgage-based security (MBS)* the underlying asset is a portfolio of residential mortgages. A distinctive feature of this financial instrument is the prepayment privilege, that is, home owners can prepay their mortgages at any time. Let the mortgages run for M months, say $M = 360$ in typical cases. The present value v of the MBS is given by

$$v = \sum_{k=1}^{M} d_k c_k \,, \tag{2.1}$$

where d_k is the discount factor for month k and c_k is the cash flow for month k. If i_k is the interest rate for month k, with i_0 being the initial interest rate, then clearly

$$d_k = \prod_{j=0}^{k-1} \frac{1}{1+i_j} \,.$$

Note that c_k depends on the fraction p_k of remaining mortgages prepaying in month k and on the interest rates. From experience one knows that p_k also depends on the interest rates. Empirical data suggest that p_k can be modeled by

$$p_k = C_1 + C_2 \arctan(C_3 i_k + C_4)$$

with constants C_1, C_2, C_3, C_4.

Thus, in the final analysis, the variables determining v in (2.1) are the interest rates $i_1, \ldots, i_M$. These interest rates are typically modeled by

$$i_k = C_0 i_{k-1}\, e^{y_k} = C_0^k i_0\, e^{y_1 + \cdots + y_k},$$

where $y_1, \ldots, y_M$ are independent and normally distributed random variables and C_0 is a constant.

The expected present value $E(v)$ of the MBS is

$$E(v) = E\Big(\sum_{k=1}^{M} d_k c_k\Big),$$

and this expected value can be expressed as an integral over $\mathbb{R}^M$ with respect to the variables $y_1, \ldots, y_M$. After transforming normally distributed random variables into uniformly distributed random variables, we obtain an integral over the M-dimensional unit cube $[0,1]^M$. Since M is typically large (say $M = 360$ as mentioned above), this shows that pricing an MBS leads to a problem of high-dimensional numerical integration.

A breakthrough in the treatment of MBS pricing came with the work of Paskov and Traub [23] who applied the quasi-Monte Carlo method to this problem. This allowed for the first time the real-time pricing of mortgage-based securities. An excellent account of this work for nonspecialists is given in the book of Traub and Werschulz [33, Chapter 4]. Further results and interesting numerical data can be found in Tezuka [32].

Other types of financial derivatives which can be priced by multidimensional numerical integration are options. According to the famous Black–Scholes model (see [35, Chapter 3]), an option price can be expressed as an expected value, with the random variables being the underlying asset prices. This leads to the interpretation of the option price as an integral. High dimensionality of the integral can arise if the option payoff depends on many points along the path of a set of underlying assets (this happens, e.g., for Asian options) or if the number of underlying assets is large.

3 The Monte Carlo Method

The curse of dimensionality can be broken, in a certain sense, by the Monte Carlo method. This is a well-known numerical method based on random sampling. A comprehensive treatment of the Monte Carlo method can be found in the book of Fishman [6]. The more recent survey article of Caflisch [1] is also very useful.

The Monte Carlo method for numerical integration works for random variables on any probability space, but for the sake of consistency we give the description only for I^s with Lebesgue measure. We assume that the given integrand (or random variable) f is square integrable over I^s. Then the *Monte Carlo approximation* for the integral is

$$\int_{I^s} f(\mathbf{u})\, d\mathbf{u} \approx \frac{1}{N} \sum_{n=1}^{N} f(\mathbf{x}_n), \tag{3.1}$$

where $\mathbf{x}_1, \ldots, \mathbf{x}_N$ are independent random samples drawn from the uniform distribution on I^s. In the language of statistics, we thus estimate the expected value of a random variable by its sample means, which intuitively is a natural thing to do. The law of large numbers guarantees that with probability 1 we have

$$\lim_{N\to\infty} \frac{1}{N} \sum_{n=1}^{N} f(\mathbf{x}_n) = \int_{I^s} f(\mathbf{u})\,\mathrm{d}\mathbf{u}\,,$$

and so this numerical integration scheme converges almost surely.

We can, in fact, be more precise about the error committed in the approximation (3.1). It can be verified quite easily (see e.g. [13, Theorem 1.1]) that the square of the error in (3.1) is, on the average over all samples of size N, equal to $\sigma^2(f)N^{-1}$, where $\sigma^2(f)$ is the variance of f. Thus, with overwhelming probability we have

$$\int_{I^s} f(\mathbf{u})\,\mathrm{d}\mathbf{u} - \frac{1}{N} \sum_{n=1}^{N} f(\mathbf{x}_n) = \mathrm{O}\big(N^{-\frac{1}{2}}\big)\,. \tag{3.2}$$

A more precise form of this statement can be derived from the central limit theorem (see [13, p. 5]). It is an important observation that the convergence rate in (3.2) is independent of the dimension s, and this makes the Monte Carlo method attractive for high-dimensional problems.

There are many successful applications of the Monte Carlo method, but there are also some drawbacks of this method. In the first place, it is difficult to generate truly random samples. Furthermore, the Monte Carlo method for numerical integration provides only a probabilistic error bound. Thus, there is never any guarantee that the expected accuracy is achieved in a concrete computation. In some cases, the convergence rate in (3.2) is also considered too slow. The method introduced in the next section addresses these concerns.

4 The Quasi-Monte Carlo Method

The *quasi-Monte Carlo method* is a deterministic version of the Monte Carlo method. The random samples used in the implementation of the Monte Carlo method are replaced by *quasirandom points*, which are deterministic points with good distribution properties. The general idea is that the Monte Carlo error bound (3.2) describes the average performance of nodes $\mathbf{x}_1, \ldots, \mathbf{x}_N$, and there should exist nodes that perform better than average. These are the quasirandom points we are seeking.

We again consider this method in the context of numerical integration over an s-dimensional unit cube I^s. The approximation scheme is the same as for the Monte Carlo method, namely

$$\int_{I^s} f(\mathbf{u})\,\mathrm{d}\mathbf{u} \approx \frac{1}{N} \sum_{n=1}^{N} f(\mathbf{x}_n)\,,$$

but now $\mathbf{x}_1, \ldots, \mathbf{x}_N$ are deterministic points in I^s. The first important advantage of this approach is that instead of the probabilistic error bound (3.2), we get a deterministic error bound given in Theorem 2 below.

Before we can state this error bound, we need to introduce a quantity which measures the irregularity of distribution of the points $\mathbf{x}_1, \ldots, \mathbf{x}_N$ (see Definition 1 below). Let P be a point set consisting of the points $\mathbf{x}_1, \ldots, \mathbf{x}_N \in I^s$. For any subinterval J of I^s let $A(J;P)$ be the number of integers n with $1 \leq n \leq N$ such that $\mathbf{x}_n \in J$. We put

$$R(J;P) = \frac{A(J;P)}{N} - \mathrm{Vol}(J)\,.$$

Note that $R(J;P)$ is the difference between the relative frequency of the points of P in J and the volume of J, with the latter number being the ideal value of the relative frequency in the case of perfect uniform distribution of the points of P over I^s. Thus, if the points of P have a very uniform distribution over I^s, then the values of $R(J;P)$ should be close to 0 for all intervals J.

Definition 1. *The* star discrepancy D_N^* *of the point set* P *is given by*

$$D_N^* = D_N^*(P) = \sup_J \bigl|R(J;P)\bigr|\,,$$

where the supremum is extended over all intervals $J = \prod_{i=1}^{s} [0, u_i)$ *with* $0 < u_i \leq 1$ *for* $1 \leq i \leq s$.

We can now state the deterministic error bound for quasi-Monte Carlo integration which is known as the *Koksma–Hlawka inequality.* A detailed proof of this inequality can be found in [10, Section 2.5].

Theorem 2. *For any function* f *of bounded variation* $V(f)$ *on* I^s *in the sense of Hardy and Krause and any points* $\mathbf{x}_1, \ldots, \mathbf{x}_N \in [0,1)^s$ *we have*

$$\left|\int_{I^s} f(\mathbf{u})\,d\mathbf{u} - \frac{1}{N}\sum_{n=1}^{N} f(\mathbf{x}_n)\right| \leq V(f) D_N^*\,,$$

where D_N^* *is the star discrepancy of* $\mathbf{x}_1, \ldots, \mathbf{x}_N$.

Note that $V(f)$ is a measure for the oscillation of the function f. For the precise definition of the variation $V(f)$ in the sense of Hardy and Krause we refer to [13, p. 19].

The error bound in Theorem 2 leads to the conclusion that points $\mathbf{x}_1, \ldots, \mathbf{x}_N$ with small star discrepancy guarantee small errors in quasi-Monte Carlo integration over I^s. This raises the question of how small we can make the star discrepancy $D_N^*(P)$ of N points in I^s for fixed N and s. The least order of magnitude that can be achieved at present is

$$D_N^*(P) = \mathrm{O}\bigl(N^{-1}(\log N)^{s-1}\bigr)\,, \tag{4.1}$$

where the implied constant in the O-symbol is independent of $N \geq 2$. A point set P achieving (4.1) is called a *low-discrepancy point set.* The points in a low-discrepancy point set are an ideal form of quasirandom points. It is conjectured that the order of magnitude in (4.1) is best possible; i.e., that the star discrepancy of any $N \geq 2$ points in I^s is at least of the order of magnitude $N^{-1}(\log N)^{s-1}$. This conjecture is proved for $s = 1$ and $s = 2$ (see [10, Sections 2.1 and 2.2]).

A related concept is that of a *low-discrepancy sequence*, which is an infinite sequence S of points in I^s such that for all $N \geq 2$ the star discrepancy $D_N^*(S)$ of the first N terms of S satisfies

$$D_N^*(S) = \mathrm{O}\big(N^{-1}(\log N)^s\big)\,, \tag{4.2}$$

where the implied constant in the O-symbol is independent of N. It is again conjectured that the order of magnitude in (4.2) is best possible, but in this case the conjecture has been verified only for $s = 1$ (see [10, Section 2.2]).

It is clear from the Koksma–Hlawka inequality and (4.1) that if we apply the quasi-Monte Carlo method with an integrand f of bounded variation and with a low-discrepancy point set $\mathbf{x}_1, \ldots, \mathbf{x}_N \in [0,1)^s$, then

$$\int_{I^s} f(\mathbf{u})\,d\mathbf{u} - \frac{1}{N}\sum_{n=1}^{N} f(\mathbf{x}_n) = \mathrm{O}\big(N^{-1}(\log N)^{s-1}\big)\,. \tag{4.3}$$

This yields a significantly faster convergence rate than the convergence rate $\mathrm{O}\big(N^{-\frac{1}{2}}\big)$ for the Monte Carlo method in (3.2). Therefore, for many types of integrals the quasi-Monte Carlo method will outperform the Monte Carlo method.

In this paper we are able to discuss only a few important facts about the quasi-Monte Carlo method. A full treatment of the quasi-Monte Carlo method and many references can be found in the book [13]. Developments up to 1978 are covered in detail in the survey article [11]. A collection of more recent survey articles is available in the volume edited by Hellekalek and Larcher [7].

Besides the classical error bound in Theorem 2, there are also error bounds for quasi-Monte Carlo integration that apply to other regularity classes of integrands. The weakest regularity condition under which the quasi-Monte Carlo method converges is that the integrand be Riemann-integrable on I^s. For continuous functions f on I^s there is a bound due to Proinov [25] which says that

$$\left|\int_{I^s} f(\mathbf{u})\,d\mathbf{u} - \frac{1}{N}\sum_{n=1}^{N} f(\mathbf{x}_n)\right| \leq 4\omega\big(f\,;(D_N^*)^{\frac{1}{s}}\big)\,,$$

where $\omega(f\,;\,\cdot\,)$ is the modulus of continuity of f with respect to the maximum norm. An improvement on this bound for the case where the point set $\mathbf{x}_1, \ldots, \mathbf{x}_N$ satisfies special uniformity properties, which includes the case of (t, m, s)-nets to be discussed in Section 5, was recently obtained by Niederreiter [16].

For the actual implementation of the quasi-Monte Carlo method we need concrete constructions of low-discrepancy point sets and sequences. Except for some isolated examples, there are basically two families of quasirandom points that have been proposed in the literature, namely those forming digital nets and (t, s)-sequences (see Section 5) and those forming lattices.

Point sets with lattice structure lead to *lattice rules* for multidimensional numerical integration. The node set of a lattice rule is obtained by starting from an s-dimensional lattice L. That is, L is the set of all $\mathbb{Z}$-linear combinations of s linearly independent vectors in $\mathbb{R}^s$. If we request that $L \supseteq \mathbb{Z}^s$, then the intersection $L \cap [0,1)^s$ yields the node set of a lattice rule. Lattice rules are very well suited for the numerical integration of periodic functions with period interval I^s, but can also be adapted to work well for nonperiodic integrands. An in-depth treatment of lattice rules is given in the book of Sloan and Joe [28]. A shorter account was presented earlier in Niederreiter [13, Chapter 5]. Recent developments in the theory of lattice rules allow construction of such rules with node sets that are much larger than was previously thought possible. This is due to a technique of component-by-component construction of lattice rules, which was introduced by Sloan and Reztsov [29], and to the parallel theory of extensible lattice rules (see [8], [9]).

An intriguing question is that of the tractability (and of the strong tractability) of the quasi-Monte Carlo method, and more generally of multidimensional numerical integration, in the sense of complexity theory. For this aspect we refer to the recent papers of Sloan [27] and Sloan and Woźniakowski [30], [31] as well as to the references therein. One aim of this research is to explain the "unreasonable effectiveness" of the quasi-Monte Carlo method in high-dimensional numerical integration. After all, for N of practicable size and for dimensions s of the order of magnitude at least 10^2, the quantity $N^{-1}(\log N)^{s-1}$ in the quasi-Monte Carlo error bound (4.3) is much larger than the quantity $N^{-\frac{1}{2}}$ in the Monte Carlo error bound (3.2), but nevertheless the quasi-Monte Carlo method outperforms the Monte Carlo method in most numerical experiments for these dimensions.

5 Digital Nets and (t,s)-Sequences

Powerful constructions of low-discrepancy point sets and sequences are based on the theory of (t,m,s)-nets and (t,s)-sequences. The idea of a (t,m,s)-net is to consider point sets P which have a perfect uniform distribution; i.e., which satisfy $R(J;P) = 0$ for a large family of subintervals J of I^s. In the following, we fix the dimension $s \geq 1$ and an integer base $b \geq 2$.

Definition 1. *Let $m \geq 1$ and t be integers with $0 \leq t \leq m$. A (t,m,s)-*net in base *b is a point set of b^m points in I^s such that $R(J;P) = 0$ for all intervals*

$$J = \prod_{i=1}^{s} \left[a_i b^{-e_i}, (a_i+1) b^{-e_i}\right)$$

with integers a_i and e_i satisfying $e_i \geq 0$ and $0 \leq a_i < b^{e_i}$ for $1 \leq i \leq s$ and with $\mathrm{Vol}(J) = b^{t-m}$.

It is important that the parameter t (often called the *quality parameter* of the net) be as small as possible, because then the family of intervals J in Definition 1 becomes larger and so the condition on a (t,m,s)-net stricter. If t is small compared to m, then we get a low-discrepancy point set according to the following result (see [12], [13, Chapter 4]).

Theorem 2. *The star discrepancy $D_N^*(P)$ of a (t, m, s)-net P in base b satisfies*

$$D_N^*(P) = \mathrm{O}\big(b^t N^{-1}(\log N)^{s-1}\big),$$

where $N = b^m$ and where the implied constant depends only on b and s.

A (t, s)-sequence in base b is an infinite sequence for which certain b-adic segments in the sequence form (t, m, s)-nets with t independent of m. The precise definition is as follows.

Definition 3. *Let $t \geq 0$ be an integer. A sequence $\mathbf{x}_1, \mathbf{x}_2, \ldots$ of points in I^s is a* (t, s)-sequence in base b *if for all integers $k \geq 0$ and $m > t$ the points $[\mathbf{x}_n]_{b,m}$ with $kb^m < n \leq (k+1)b^m$ form a (t, m, s)-net in base b. Here $[\cdot]_{b,m}$ denotes truncation to m digits after the decimal point in the b-adic expansion (with the convention $1 = 0.\overline{b-1}$).*

Again, the quality parameter t of a (t, s)-sequence should be as small as possible. All (t, s)-sequences are low-discrepancy sequences according to the following theorem (see [12], [13, Chapter 4]).

Theorem 4. *The star discrepancy $D_N^*(S)$ of a (t, s)-sequence S in base b satisfies*

$$D_N^*(S) = \mathrm{O}\big(b^t N^{-1}(\log N)^s\big) \qquad \text{for all } N \geq 2,$$

where the implied constant depends only on b and s.

The *digital method* was introduced in [12] and provides a general framework for the construction of (t, m, s)-nets and (t, s)-sequences. We describe this method in the context of (t, m, s)-nets in a prime-power base q. Suppose we are given $m \geq 1$ and the dimension $s \geq 1$. Then we have to define q^m points in I^s for a (t, m, s)-net in base q. The first step is to choose a finite field $\mathbb{F}_q$ with q elements. If q is a prime p, then we can take for $\mathbb{F}_p$ the set $\{0, 1, \ldots, p-1\}$ of integers with arithmetic modulo p. If $q = p^e$ with an integer $e \geq 2$, then $\mathbb{F}_q$ can be represented as the residue class field of polynomials over $\mathbb{F}_p$ modulo an irreducible polynomial over $\mathbb{F}_p$ of degree e.

The next step is to choose $m \times m$ matrices $C^{(1)}, \ldots, C^{(s)}$ over $\mathbb{F}_q$; that is, one matrix for each coordinate of the points in I^s we want to construct. For a fixed column vector $\mathbf{a} \in \mathbb{F}_q^m$ of length m we compute the matrix-vector products $C^{(i)}\mathbf{a} \in \mathbb{F}_q^m$ for $1 \leq i \leq s$. On $\mathbb{F}_q^m$ we define the map

$$T(a_1, \ldots, a_m) = \sum_{j=1}^{m} \psi(a_j) q^{-j} \in [0, 1)$$

with a bijection $\psi : \mathbb{F}_q \to \{0, 1, \ldots, q-1\}$. Furthermore, we put

$$p^{(i)}(\mathbf{a}) = T(C^{(i)}\mathbf{a}) \in [0, 1) \qquad \text{for } 1 \leq i \leq s.$$

In this way we get the point

$$P(\mathbf{a}) = \left(p^{(1)}(\mathbf{a}), \ldots, p^{(s)}(\mathbf{a})\right) \in I^s.$$

By letting $\mathbf{a}$ range over all q^m possibilities in $\mathbb{F}_q^m$, we arrive at the desired q^m points in I^s.

The quality of this point set depends on the choice of the matrices $C^{(1)}, \ldots, C^{(s)}$. For $1 \leq i \leq s$ let $\mathbf{c}_j^{(i)} \in \mathbb{F}_q^m$, $1 \leq j \leq m$, be the row vectors of $C^{(i)}$. Then we have the following result (see [13, Theorem 4.28], [20, Theorem 8.2.4]).

Theorem 5. *Let $d \leq m$ be such that for any nonnegative integers $d_1, \ldots, d_s$ with $\sum_{i=1}^{s} d_i = d$ the vectors $\mathbf{c}_j^{(i)}$, for $1 \leq j \leq d_i$ and $1 \leq i \leq s$, are linearly independent over $\mathbb{F}_q$. Then the point set constructed above is an $(m-d, m, s)$-net in base q.*

By maximizing d, we get the minimal value of the quality parameter t for this point set. Similar construction principles are available for arbitrary bases and also for (t, s)-sequences (see [13, Chapter 4], [20, Section 8.2]). It is customary to refer to (t, m, s)-nets and (t, s)-sequences obtained by the digital method as *digital (t, m, s)-nets* and *digital (t, s)-sequences.*

In order to make the digital method work concretely, we have to solve the problem of combinatorial linear algebra arising from Theorem 5. In other words, we have to find systems $\{\mathbf{c}_j^{(i)} \in \mathbb{F}_q^m : 1 \leq i \leq s, 1 \leq j \leq m\}$ of vectors for which the number d in Theorem 5 is as large as possible. Various ways of designing such systems of vectors have been found, for instance by number theory, the theory of finite fields, coding theory, and algebraic geometry. We refer to the book of Niederreiter [13, Chapter 4] for developments up to 1992 and to Niederreiter [14] for an update on constructions of digital nets and (t, s)-sequences.

Let $t_q(s)$ denote the least value of t such that there exists a (t, s)-sequence in base q. Then $t_q(s) = 0$ for $1 \leq s \leq q$ and $t_q(s) \geq 1$ for $s \geq q+1$. Explicit constructions of digital $(0, s)$-sequences in base q for $1 \leq s \leq q$ can be found in [13, Section 4.5]. For fixed q, the order of magnitude of $t_q(s)$ as $s \to \infty$ is known, since we have

$$c_1(q)s \leq t_q(s) \leq c_2(q)s \qquad \text{for } s \geq q+1\,,$$

where $c_1(q)$ and $c_2(q)$ are positive constants depending only on q (see [19], [20, Chapter 8]). The upper bound is obtained by a construction using the digital method and global function fields over $\mathbb{F}_q$ with many rational places (compare with Section 6 for a construction of digital nets using global function fields).

6 Duality Theory and Codes

Several recent constructions of digital nets are based on a duality theory for digital nets which establishes an analogy with the problem of constructing good linear codes (see [34] for background on coding theory). The viewpoint of duality was introduced into the theory of digital nets by Niederreiter and Pirsic [18]. In this viewpoint, the problem of constructing digital (t, m, s)-nets in base q is reduced to that of constructing certain $\mathbb{F}_q$-linear subspaces of $\mathbb{F}_q^{ms}$. The vector space $\mathbb{F}_q^{ms}$ is endowed with a weight function which generalizes the classical Hamming weight in coding theory. There is then a known relationship between the quality parameter t of the digital net and the minimum distance (or minimum weight) of the corresponding $\mathbb{F}_q$-linear subspace. Small values of t correspond to large values of the minimum distance in this relationship.

The appropriate weight function V_m on $\mathbb{F}_q^{ms}$ is defined as follows. Let $m \geq 1$ and $s \geq 1$ be integers. First, we define a weight function v on $\mathbb{F}_q^m$ by putting $v(\mathbf{a}) = 0$ if $\mathbf{a} = \mathbf{0} \in \mathbb{F}_q^m$, and for $\mathbf{a} = (a_1, \ldots, a_m) \in \mathbb{F}_q^m$ with $\mathbf{a} \neq \mathbf{0}$ we set

$$v(\mathbf{a}) = \max\{j : a_j \neq 0\}.$$

Then we extend this definition to $\mathbb{F}_q^{ms}$ by writing a vector $\mathbf{A} \in \mathbb{F}_q^{ms}$ as the concatenation of s vectors of length m; i.e., $\mathbf{A} = \left(\mathbf{a}^{(1)}, \ldots, \mathbf{a}^{(s)}\right) \in \mathbb{F}_q^{ms}$ with $\mathbf{a}^{(i)} \in \mathbb{F}_q^m$ for $1 \leq i \leq s$, and putting

$$V_m(\mathbf{A}) = \sum_{i=1}^{s} v\left(\mathbf{a}^{(i)}\right).$$

Note that in the case $m = 1$, the weight $V_m(\mathbf{A})$ reduces to the Hamming weight of the vector $\mathbf{A}$. If for any $m \geq 1$ we define the distance $d_m(\mathbf{A}, \mathbf{B})$ of $\mathbf{A}, \mathbf{B} \in \mathbb{F}_q^{ms}$ by $d_m(\mathbf{A}, \mathbf{B}) = V_m(\mathbf{A} - \mathbf{B})$, then $\mathbb{F}_q^{ms}$ turns into a metric space, which for $m = 1$ is the well-known Hamming space in coding theory. As in coding theory, the concept of minimum distance relative to d_m, or equivalently V_m, plays a crucial role.

Definition 1. *For any nonzero $\mathbb{F}_q$-linear subspace $\mathcal{N}$ of $\mathbb{F}_q^{ms}$ we define the* minimum distance

$$\delta_m(\mathcal{N}) = \min_{\mathbf{A} \in \mathcal{N} \setminus \{\mathbf{0}\}} V_m(\mathbf{A}).$$

The basic step in [18] is to set up a duality between digital (t, m, s)-nets in base q and certain $\mathbb{F}_q$-linear subspaces of $\mathbb{F}_q^{ms}$. Rather than going into the technical details here, we state the one result from this duality theory that is of relevance for the construction of digital nets (compare also with [21, Lemma 1]). We assume $s \geq 2$ to avoid the trivial one-dimensional case.

Theorem 2. *Let q be a prime power and let $m \geq 1$ and $s \geq 2$ be integers. Then from any $\mathbb{F}_q$-linear subspace $\mathcal{N}$ of $\mathbb{F}_q^{ms}$ with $\dim(\mathcal{N}) \geq ms - m$ we can construct a digital (t, m, s)-net in base q with $t = m - \delta_m(\mathcal{N}) + 1$.*

As an example of a powerful construction of digital nets based on duality theory, we present a recent construction due to Niederreiter and Özbudak [17] which works with global function fields; i.e., algebraic function fields over finite fields. We refer to [15] for an introduction to global function fields. For the sake of simplicity, we consider only a special case of the construction of digital nets in [17]. This case can be seen as an analog of the well-known construction of algebraic-geometry codes (see [20, Chapter 6] for the latter).

Let $F/\mathbb{F}_q$ be a global function field with full constant field $\mathbb{F}_q$; i.e., $\mathbb{F}_q$ is algebraically closed in the global function field F. For a place P of F, we denote by ν_P the normalized discrete valuation of F corresponding to P. For a given dimension $s \geq 2$, we assume that F has at least s rational places, say $P_1, \ldots, P_s$ are s distinct rational places of F. For each $i = 1, \ldots, s$ let $t_i \in F$ be a local parameter at P_i, that is, $\nu_{P_i}(t_i) = 1$.

Next we choose an arbitrary divisor G of F and put $n_i = \nu_{P_i}(G)$ for $1 \leq i \leq s$. Consider the Riemann–Roch space

$$\mathcal{L}(G) = \{f \in F^* : \operatorname{div}(f) + G \geq 0\} \cup \{0\},$$

where $\operatorname{div}(f)$ is the principal divisor of $f \in F^*$. Then $\nu_{P_i}(f) \geq -n_i$ for $1 \leq i \leq s$ and any $f \in \mathcal{L}(G)$. Thus, there exists a uniquely determined $\mathbf{c}_f^{(i)} \in \mathbb{F}_q^m$ with

$$f \equiv \mathbf{c}_f^{(i)} \cdot \left(t_i^{-n_i+m-1}, t_i^{-n_i+m-2}, \ldots, t_i^{-n_i}\right) \bmod P_i^{-n_i+m},$$

where $\cdot$ denotes the standard inner product. Finally, we put

$$\mathbf{C}_f = \left(\mathbf{c}_f^{(1)}, \ldots, \mathbf{c}_f^{(s)}\right) \in \mathbb{F}_q^{ms}.$$

Proposition 3. *If $g \leq \deg(G) < ms$, then the $\mathbb{F}_q$-linear subspace $\mathcal{N} := \{\mathbf{C}_f : f \in \mathcal{L}(G)\}$ of $\mathbb{F}_q^{ms}$ satisfies $\dim(\mathcal{N}) \geq \deg(G) - g + 1$, and $\delta_m(\mathcal{N}) \geq ms - \deg(G)$, where g is the genus of F.*

Proof. Take any nonzero $f \in \mathcal{L}(G)$ and put $w_i(f) = \min(m, \nu_{P_i}(f) + n_i)$ for $1 \leq i \leq s$. The definition of $\mathbf{c}_f^{(i)}$ implies that $v(\mathbf{c}_f^{(i)}) = m - w_i(f)$ for $1 \leq i \leq s$. Therefore $V_m(\mathbf{C}_f) = \sum_{i=1}^s v(\mathbf{c}_f^{(i)}) = ms - \sum_{i=1}^s w_i(f)$. For each $i = 1, \ldots, s$ we have $\nu_{P_i}(f) \geq w_i(f) - n_i$, and so $f \in \mathcal{L}\left(G - \sum_{i=1}^s w_i(f)\, P_i\right)$. Since $f \neq 0$, we get $0 \leq \deg\left(G - \sum_{i=1}^s w_i(f)\, P_i\right) = \deg(G) - \sum_{i=1}^s w_i(f) = \deg(G) + V_m(\mathbf{C}_f) - ms$, and so $V_m(\mathbf{C}_f) \geq ms - \deg(G) > 0$. This shows that the $\mathbb{F}_q$-linear map $f \in \mathcal{L}(G) \mapsto \mathbf{C}_f$ is injective. Hence $\dim(\mathcal{N}) = \dim(\mathcal{L}(G)) \geq \deg(G) - g + 1$ by the Riemann–Roch theorem. It is clear from the above that $\delta_m(\mathcal{N}) \geq ms - \deg(G)$. □

Theorem 4. *If $m \geq \max(1, g)$ and $s \geq 2$, then we can construct a digital (g, m, s)-net in base q whenever there is a global function field $F/\mathbb{F}_q$ of genus g with at least s rational places.*

Proof. In the construction leading to Proposition 3 we choose a divisor G of F with $\deg(G) = ms - m + g - 1$. Hence $g \leq \deg(G) < ms$. Then the $\mathbb{F}_q$-linear subspace $\mathcal{N}$ of $\mathbb{F}_q^{ms}$ in Proposition 3 satisfies $\dim(\mathcal{N}) \geq ms - m$, with $\delta_m(\mathcal{N}) \geq m - g + 1$. The rest follows from Theorem 2. □

In the simple special case where $F = \mathbb{F}_q(x)$ is the rational function field over $\mathbb{F}_q$ (so that $g = 0$), we can choose s distinct rational places of F whenever $2 \leq s \leq q+1$.

The reason is that $\mathbb{F}_q(x)$ has altogether $q+1$ rational places, given by the infinite place and the linear polynomials $x-a$ with $a \in \mathbb{F}_q$. Theorem 4 yields then a digital $(0,m,s)$-net in base q for any $m \geq 1$ and $2 \leq s \leq q+1$.

The approach to digital nets via duality theory opens up the possibility of extending known constructions of good linear codes to obtain new constructions of digital nets. The above method of transferring the construction of algebraic-geometry codes to digital nets is just one case in point, with a few other such constructions already known and probably more to be discovered. Exciting research opportunities are beckoning here!

7 Conclusions

Of the basic methods for multidimensional numerical integration discussed in this paper, only two – the Monte Carlo method and the quasi-Monte Carlo method – appear to be suitable for high-dimensional numerical integration. Between these two, the quasi-Monte Carlo method seems to have the competitive edge for most problems arising in practice, although this may look "unreasonable" in the light of the comment at the end of Section 4. Thus, a subtitle of this paper could have been "The Unreasonable Effectiveness of the Quasi-Monte Carlo Method".

One possible explanation for this phenomenon may be that the Monte Carlo error bound (3.2) is pretty precise for random node sets (it represents, after all, the average performance over all possible node sets), whereas the quasi-Monte Carlo error bound (4.3) tends to be overly pessimistic for standard low-discrepancy point sets in most problem instances. It appears that the error in quasi-Monte Carlo integration behaves more like $O(N^{-1})$ in many practical problems. However, this explanation is more empirical than mathematical, and so the challenge remains to find an explanation based on solid theory. The papers quoted in the last paragraph of Section 4 take some steps towards a mathematical explanation. An interesting approach to the choice between the Monte Carlo method and the quasi-Monte Carlo method is to try combining the advantages of both methods. The survey article of Owen [22] provides access to the relevant literature.

Bibliography

[1] R. E. Caflisch; *Monte Carlo and quasi-Monte Carlo methods*, Acta Numerica, 7 (1998), pp. 1–49.

[2] R. Cools; *Constructing cubature formulae: the science behind the art*, Acta Numerica, 6 (1997), pp. 1–54.

[3] R. Cools; *An encyclopaedia of cubature formulas*, J. Complexity, 19 (2003), pp. 445–453.

[4] R. Cools and J. N. Lyness; *Three- and four-dimensional K-optimal lattice rules of moderate trigonometric degree*, Math. Comp., 70 (2001), pp. 1549–1567.

[5] K.-T. Fang; *Some applications of quasi-Monte Carlo methods in statistics*, in Monte Carlo and Quasi-Monte Carlo Methods 2000, K.-T. Fang, F. J. Hickernell, and H. Niederreiter, eds., Springer, Berlin, 2002, pp. 10–26.

[6] G. S. Fishman; *Monte Carlo: Concepts, Algorithms, and Applications*, Springer, New York, NY, 1996.

[7] P. Hellekalek and G. Larcher, eds.; *Random and Quasi-Random Point Sets*, Lecture Notes in Statistics, Vol. 138, Springer, New York, NY, 1998.

[8] F. J. Hickernell, H. S. Hong, P. L'Ecuyer, and C. Lemieux; *Extensible lattice sequences for quasi-Monte Carlo quadrature*, SIAM J. Sci. Comput., 22 (2000), pp. 1117–1138.

[9] F. J. Hickernell and H. Niederreiter; *The existence of good extensible rank-1 lattices*, J. Complexity, 19 (2003), pp. 286–300.

[10] L. Kuipers and H. Niederreiter; *Uniform Distribution of Sequences*, Wiley, New York, NY, 1974.

[11] H. Niederreiter; *Quasi-Monte Carlo methods and pseudo-random numbers*, Bull. Amer. Math. Soc., 84 (1978), pp. 957–1041.

[12] H. Niederreiter; *Point sets and sequences with small discrepancy*, Monatsh. Math., 104 (1987), pp. 273–337.

[13] H. Niederreiter; *Random Number Generation and Quasi-Monte Carlo Methods*, SIAM, Philadelphia, PA, 1992.

[14] H. Niederreiter; *Constructions of (t, m, s)-nets*, in Monte Carlo and Quasi-Monte Carlo Methods 1998, H. Niederreiter and J. Spanier, eds., Springer, Berlin, 2000, pp. 70–85.

[15] H. Niederreiter; *Algebraic function fields over finite fields*, in Coding Theory and Cryptology, H. Niederreiter, ed., World Scientific, Singapore, 2002, pp. 259–282.

[16] H. Niederreiter; *Error bounds for quasi-Monte Carlo integration with uniform point sets*, J. Comput. Appl. Math., 150 (2003), pp. 283–292.

[17] H. Niederreiter and F. Özbudak; *Constructions of digital nets using global function fields*, Acta Arith., 105 (2002), pp. 279–302.

[18] H. Niederreiter and G. Pirsic; *Duality for digital nets and its applications*, Acta Arith., 97 (2001), pp. 173–182.

[19] H. Niederreiter and C. P. Xing; *Low-discrepancy sequences and global function fields with many rational places*, Finite Fields Appl., 2 (1996), pp. 241–273.

[20] H. Niederreiter and C. P. Xing; *Rational Points on Curves over Finite Fields: Theory and Applications*, London Math. Soc. Lecture Note Series, Vol. 285, Cambridge University Press, Cambridge, 2001.

[21] H. Niederreiter and C. P. Xing; *Constructions of digital nets*, Acta Arith., 102 (2002), pp. 189–197.

[22] A. B. Owen; *Monte Carlo, quasi-Monte Carlo, and randomized quasi-Monte Carlo*, in Monte Carlo and Quasi-Monte Carlo Methods 1998, H. Niederreiter and J. Spanier, eds., Springer, Berlin, 2000, pp. 86–97.

[23] S. H. Paskov and J. F. Traub; *Faster valuation of financial derivatives*, J. Portfolio Management, 22 (1995), no. 1, pp. 113–120.

[24] E. Platen; *An introduction to numerical methods for stochastic differential equations*, Acta Numerica, 8 (1999), pp. 197–246.

[25] P. D. Proinov; *Discrepancy and integration of continuous functions*, J. Approx. Theory, 52 (1988), pp. 121–131.

[26] J. E. H. Shaw; *A quasirandom approach to integration in Bayesian statistics*, Ann. Statistics, 16 (1988), pp. 895–914.

[27] I. H. Sloan; *QMC integration – beating intractability by weighting the coordinate directions*, in Monte Carlo and Quasi-Monte Carlo Methods 2000, K.-T. Fang, F. J. Hickernell, and H. Niederreiter, eds., Springer, Berlin, 2002, pp. 103–123.

[28] I. H. Sloan and S. Joe; *Lattice Methods for Multiple Integration*, Oxford University Press, Oxford, 1994.

[29] I. H. Sloan and A. V. Reztsov; *Component-by-component construction of good lattice rules*, Math. Comp., 71 (2002), pp. 263–273.

[30] I. H. Sloan and H. Woźniakowski; *Tractability of multivariate integration for weighted Korobov classes*, J. Complexity, 17 (2001), pp. 697–721.

[31] I. H. Sloan and H. Woźniakowski; *Tractability of integration in non-periodic and periodic weighted tensor product Hilbert spaces*, J. Complexity, 18 (2002), pp. 479–499.

[32] S. Tezuka; *Financial applications of Monte Carlo and quasi-Monte Carlo methods*, in [7], pp. 303–332.

[33] J. F. Traub and A. G. Werschulz; *Complexity and Information*, Cambridge University Press, Cambridge, 1998.

[34] J. H. van Lint; *Introduction to Coding Theory*, 3rd ed., Springer, Berlin, 2000.

[35] P. Wilmott, S. Howison, and J. Dewynne; *The Mathematics of Financial Derivatives*, Cambridge University Press, Cambridge, 1995.

Ernie Tuck recently retired from the University of Adelaide as the Elder Professor of Applied Mathematics. His interests are in applied and industrial mathematics, maritime engineering, aeronautical engineering, defence science and technology, and oceanography. He has published over 150 papers covering a wide range of topics in ship hydrodynamics, aerodynamics, hydraulics, biofluid dynamics, and games theory. His work is distinguished by its originality and flair, at times its basic simplicity, its computational excellence, and its successful application to real problems.

After first degrees from the University of Adelaide, Tuck obtained his PhD degree at the University of Cambridge. He has held academic positions at Manchester University, the David Taylor Model Basin, and CalTech. At Adelaide he was Chair of the Department of Applied Mathematics for several periods, and has served as Dean of the Faculty of Mathematical and Computer Sciences. He has also worked with the US Navy and Australian Navy Departments, the Shell Oil Company, and the Australian Defence Science Technology Organisation.

Chapter 17

Computation and Minimisation of Ship Waves

*Ernest O. Tuck**

Abstract: Bodies moving steadily forward at or near a water surface make a beautiful and complex V-shaped pattern of ship waves on that surface. This pattern is essentially of the same form whether the body is an actual ship, or a duck, dolphin, hovercraft, submarine, ekranoplan (ground-effect airplane), catamaran, trimaran, rowing shell, yacht, planing boat, etc. Ship waves come at a price, and are almost always undesirable, for reasons such as propulsive inefficiency, environmental damage, and ease of detection. The above list includes bodies whose aim is generally to avoid making waves, by keeping away from the water surface as far as possible, but this is not always feasible or desirable.

Computation of ship-wave patterns is a challenging task that was begun by the Australian mathematician J. H. Michell in 1898, but only really became possible with sufficient fine detail and accuracy late in the 20th century. In some cases, the task can be simplified to 'merely' that of quadruple numerical integration at hundreds of thousands of points, but with unpleasant characteristics such as rapidly oscillating integrands. In other cases, such as for planing surfaces, integral equations must be solved over a two-dimensional wetted domain that is not fully known in advance. Examples of recent computations which retain the spirit of Michell will be given. These computations allow near-photographic detail to be captured in minutes of inexpensive PC time.

A separate analytical and computational challenge is that of design optimisation, to minimise a measure of the ship-wave magnitude, such as the net energy lost to infinity, or wave resistance. For conventional ships, such hull-design considerations have always been an important part of the ancient discipline of naval architecture. For unconventional vessels, there is less guidance from experience or from nature, so mathematical and computational optimisation can play a significant role. Examples to be discussed include optimal hull-spacing choices for multihull vessels, and optimised pressure patches modelling hovercraft or planing surfaces.

*The University of Adelaide

Contents

1 Introduction

Ships make waves. Sometimes these waves are like Figure 1. Sometimes they are like Figure 2. In fact, these pictures represent the two extremes: the ambient sea is seldom quite as calm as in Figure 1, and most ships avoid seas as rough as in Figure 2. Most of the life of most ships is spent in seas nearer to the calm state than the rough state. This paper will therefore concentrate on ships moving steadily forward at constant speed into a totally calm sea. The effect of (small to moderate) ambient waves can be added later if required.

Ships and other objects (including animals) moving on or near a water surface are influenced by three dominant forces, namely inertia, gravity and viscosity. Waves are essentially a balance between the first two of these. Viscosity (i.e. friction) in fact plays only a very small role in most of the fluid domain surrounding the body, as is the case for most flows at human scales (metres or more), whether or not a free surface is present. However, viscosity dominates very close to the surface of the body, in a thin (centimetres or less) boundary layer, where it provides a significant fraction of the total drag. In the present paper we shall for the most part neglect viscosity. Again, the effect of viscosity can be added later if required.

Indeed this additive characteristic of viscous and wave effects on ships was the main thrust of the mid-19th century work of the pioneering naval architect William Froude, and is the basis for use of model towing tanks for ship design. Essentially a small scaled model is used to determine wave drag, which is then scaled up to full scale by what is now called "Froude scaling" (speeds proportional to the square root of length), and then viscous effects are added via an empirical skin friction formula dependent on wetted surface area. Froude's experimental work on this scaling principle was done long before the concept of the boundary layer; e.g., as formalised by Ludwig Prandtl in 1912 giving a rational justification for it.

Once viscosity is neglected, the mathematical and computational task is to solve a boundary-value problem for Laplace's equation, subject to a Neumann boundary condition on both the body surface and the free water surface. However, the latter surface is unknown in advance, so we must add a further boundary condition that the pressure (obtained from the flow via Bernoulli's equation) be

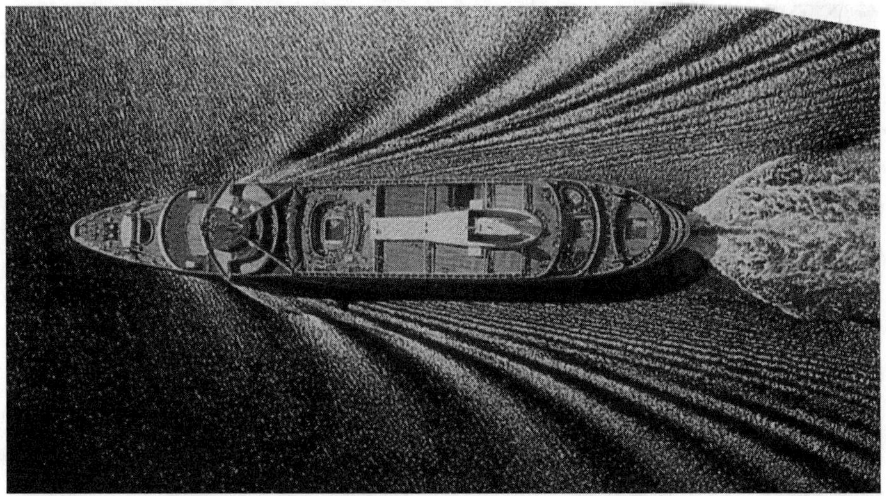

Figure 1. *A cruise ship in a calm sea*

constant on the free surface, in order to simultaneously determine that free surface. The combination of two (nonlinear) boundary conditions on the (unknown) free surface may be called the Stokes conditions, and the resulting boundary-value problem a "Neumann–Stokes" problem. Although at first sight one might consider this to be a straightforward task, the parenthetic qualifiers "nonlinear" and "unknown" in the previous sentence give indications of trouble ahead, and it was late in the 20th century before anything close to a complete or useful solution of the Neumann-Stokes problem was possible.

2 Thin Ships

Meanwhile, however, even in the 19th century, a remarkable approximate solution to the Neumann-Stokes problem was presented by the Australian mathematician John Henry Michell [11]. Michell assumed that the ship was thin, lying close to a vertical plane, and therefore was able to approximate the Neumann boundary condition on the hull by a simpler "Michell" boundary condition on that plane. Not only that, but thinness of the ship guaranteed that it made small waves, and hence the nonlinear Stokes conditions on the unknown free surface could be replaced by a linear "Kelvin" condition on the known undisturbed free surface, which is a horizontal plane. The resulting "Michell-Kelvin" problem, having only plane coordinate boundaries, was immediately solvable analytically by Fourier methods.

Michell's main concern was for the wave resistance, or inviscid drag D due to the energy left behind in the wave system. His final result can be written

$$D = \frac{2}{\pi}\,\rho U^2 k_0^4 \int_{-\pi/2}^{\pi/2} \mathrm{d}\theta \sec^5\theta \left| \iint_W \mathrm{d}x\ \mathrm{d}z\ Y(x,z) e^{ik_0 x \sec\theta + k_0 z \sec^2\theta} \right|^2. \tag{2.1}$$

Figure 2. *A tanker in heavy seas*

Here $y = \pm Y(x, y)$ is the hull surface, in a frame of reference with x from bow to stern, y to starboard, and z upward from the free-surface (so $z < 0$ here). The (x, z) integral in (2.1) is over the centreplane W of the ship, and ρ is the water density, U the ship speed and $k_0 = g/U^2$ where g is gravity. The form of Michell's integral quoted here is for pointed hulls with $Y = 0$ at the ends; if this condition is not met, e.g. for transom-hulled vessels, some extra terms must be added. The water is assumed to be of infinite depth.

Numerical evaluation of Michell's integral is a triple quadrature, a daunting task in 1898, and not entirely a simple one in 2003. Nevertheless Michell himself did present one numerical result, for an idealised choice of $Y(x, z)$ such that the (x, z) integral could be evaluated analytically, so he only had the θ integration to compute numerically by Simpson's rule. More than 100 years later, we can evaluate the full Michell triple integral in less than a twentieth of a second with high precision, for real ships whose offsets $Y(x, z)$ are prescribed as data.

The 20th century history of Michell's integral is a somewhat sad one, reviewed by Wehausen [20] and Tuck [15]. Though published in a very well-known journal, it lay almost unknown for a quarter of a century, until Wigley [23] and Havelock [7] eventually gave it credit in the 1920s, and then in long series of papers used Michell's integral to explore the way in which various changes in simplified ship forms affected wave resistance. Later, Weinblum [21] and others [2] developed pre-computer numerical tools for its evaluation for actual ships, and used it in drag

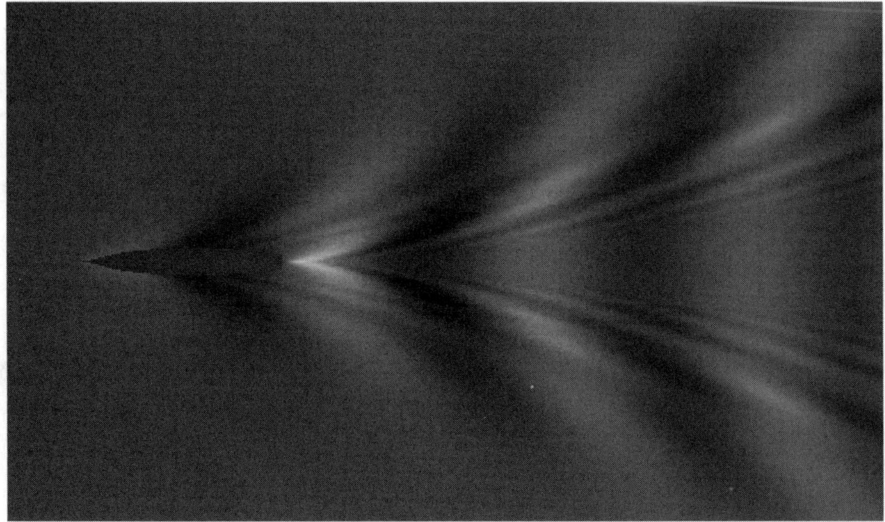

Figure 3. *Free-surface elevation due to destroyer at 30 knots*

minimisation studies [22]. With the dawn of the computer age, this sort of work was carried a little further [9], but by then the approximate nature of the original formula was a cause for (misplaced) concern, and the power of the computer was increasingly turned instead toward the much more difficult task of solving more exact models, such as the Neumann-Stokes problem, or even models including viscosity. However, Michell's integral remains a useful and often remarkably accurate estimator of wave resistance; see [20, 15, 19].

Michell also provided formulae enabling determination of flow quantities other than wave resistance. In particular, the actual wave elevation $z = Z(x, y)$ made by the thin ship $y = \pm Y(x, z)$ (with pointed ends) can be written as the quadruple integral

$$Z(x,y) = \frac{1}{\pi^2} \iint_W \mathrm{d}\xi \; \mathrm{d}\zeta \; Y(\xi,\zeta) \, \Re \int_{-\pi/2}^{\pi/2} \mathrm{d}\theta \int_0^\infty \mathrm{d}k \qquad \frac{k^2}{k - k_0 \sec^2\theta} \exp\bigl[-ik(x-\xi)\cos\theta - iky\sin\theta + k\zeta\bigr] \,. \tag{2.2}$$

The path of k-integration in (2.2) passes above the pole at $k = k_0 \sec^2\theta$, so guaranteeing that waves occur only for $x > 0$.

Clearly (2.2) is much harder to compute from than (2.1), not least because of the extra k integral, and Michell himself made no attempt to compute ship waves in 1898. Indeed until computers arrived, this was an almost impossible task, and even then little progress was made until late in the 20th century.

An important advance was made by Newman [12] who in effect reduced a significant part of the k and θ integration task to evaluation of a Chebyshev polynomial approximation. The speed at which this can be done is such that computation of

$Z(x, y)$ averages, for each (x, y) point, a time not more than that taken for one evaluation of the wave resistance D. Such speed is necessary, since to capture a realistic representation of a complete ship wave field with its complex mix of superposed waves, short and long, near and far, transverse and diverging, requires about 100,000 points. This takes about 5 minutes to compute on a current 2 GHz PC. There are many other factors influencing total computer times for such detailed wave fields, such as the balance between near-field and far-field contributions, the efficiency of handling rapid oscillations near $|\theta| = \pi/2$, and the extent to which parts of the work done at one point can be saved and used again at other points. For further discussion of these matters, see [18, 13, 19].

Figure 3 is an example of a computation performed in this way for a destroyer hull at 30 knots full-scale speed. This figure is simply produced by colouring about 100,000 (x, y) pixels a shade of blue whose lightness is proportional to the computed value of $Z(x, y)$, with the deepest troughs nearly black and the highest crests nearly white. The red outline of the ship itself is less well represented pictorially, but is modelled quite smoothly in the computation.

3 Pressure Distributions

Another type of travelling disturbance that potentially makes small waves, allowing linearisation of the free-surface boundary conditions, is a surface pressure distribution with a prescribed small excess $p = P(x, y)$ over atmospheric pressure. This could directly be a moving meteorological disturbance or the pressure exerted by a hovercraft, or indirectly be a model for a vessel of small draft, on the hull of which the hydrodynamic pressure is $P(x, y)$.

The wave resistance of such a pressure field is

$$D = \frac{k_0^3}{2\pi\rho g} \int_{-\pi/2}^{\pi/2} \mathrm{d}\theta \sec^5\theta \left| \iint_W \mathrm{d}x \; \mathrm{d}y \; P(x, y) e^{ik_0 \sec^2\theta (x\cos\theta + y\sin\theta)} \right|^2, \tag{3.1}$$

and the wave field $z = Z(x, y)$ is

$$Z(x, y) = \frac{1}{2\pi^2\rho g} \iint_W \mathrm{d}\xi \; \mathrm{d}\eta \; P(\xi, \eta) \; \Re \int_{-\pi/2}^{\pi/2} \mathrm{d}\theta \int_0^\infty \mathrm{d}k$$
$$\frac{k^2}{k - k_0 \sec^2\theta} \exp\left[-ik(x - \xi)\cos\theta - ik(y - \eta)\sin\theta\right]. \tag{3.2}$$

The formulae (3.1) and (3.2) are similar to the corresponding Michell formulae (2.1) and (2.2), a double integral over a portion W of the plane $z = 0$ replacing a double integral over a portion of the plane $y = 0$, and similar numerical procedures enable efficient computation of 100,000-point wave fields, for pressures $P(x, y)$ inputted as data. For examples, see [19].

4 Flat Ships

Michell used "thinness" (i.e. small beam) of the hull to linearise the Neumann-Stokes problem for a ship. This is not the only way that we can restrict the hull shape to ensure small waves. In particular, another option is "flatness" (i.e. small draft). If the equation of the ship is now written $z = Z_0(x, y)$, we therefore assume Z_0 is small.

In one mathematical sense, flatness is more attractive as an approximation than thinness, since in the formal limit as $Z_0 \to 0$, the only boundary is where $z = 0$. Physically, there is no longer any disturbance "in" the water. Again, Fourier methods allow an immediate (partial) solution, and the disturbed flow field can be seen to be caused by an apparent travelling pressure distribution $p = P(x, y)$ on the free surface. Since we now have efficient tools as above to compute the wave resistance D and wave field $z = Z(x, y)$ for such pressure distributions, this solves the flat-ship problem once the pressure $P(x, y)$ is known.

But unfortunately, given the hull shape $Z_0(x, y)$, the corresponding pressure distribution $P(x, y)$ is not immediately known. An early attempt at a flat-ship theory was that of Hogner [8] (see also [20], pp.170–171) who assumed in effect that $P(x, y) = -\rho g Z_0(x, y)$. This is just a "hydrostatic" approximation, since it is the fluid pressure acting on the hull of the vessel when it is at rest, noting that in general $Z_0 < 0$ so $P > 0$. One might expect that the Hogner theory would be valid (if at all) for relatively low speeds, where the hydrostatic approximation retained some validity.

In fact, the correct condition ultimately determining the pressure $P(x, y)$ is just $Z(x, y) = Z_0(x, y)$. That is, over the limiting portion W of the plane $z = 0$ that is occupied by the ship, we require that the apparent free-surface elevation produced by the pressure $P(x, y)$ is identical to the elevation (usually negative) of the corresponding point of the ship's surface. Equating $Z(x, y)$ as computed from (3.2) to the input $Z_0(x, y)$ gives an integral equation for the unknown $P(x, y)$ on the portion W of the (x, y) plane, and hence the basic task in solving the flat-ship problem is to solve such an integral equation. This makes the flat-ship problem very much more difficult computationally than the thin-ship problem.

However, in principle we have all the tools required. It is convenient to envisage the problem in discrete form on some (x, y) grid, with a vector $\mathbf{p} = \{P(x, y)\}$ of pressures corresponding to a vector $\mathbf{z} = \{Z(x, y)\}$ of elevations. Then (3.2) tells us that $\mathbf{z} = \mathbf{A}\mathbf{p}$, for some matrix $\mathbf{A}$ obtained by discretising the kernel function. The most straightforward way to find the elements of the matrix $\mathbf{A}$ is just to run the code that determines $Z(x, y)$ many times, using some basis set of $\mathbf{p}$-vectors. Now given $\mathbf{z} = \{Z_0(x, y)\}$, we simply have to invert that system of linear equations to find $\mathbf{p} = \mathbf{A}^{-1}\mathbf{z}$.

There are a number of difficulties with this approach, encountered and appreciated by many who have attempted to solve this problem in the past; e.g., [14, 5, 4]. From the computational point of view, the matrix $\mathbf{A}$ appears to be ill-conditioned, and most authors have seen grid-scale oscillations in the output pressures. Physically, this is associated with short diverging waves with crests nearly parallel to the ship's track that spring almost uncontrollably from each discontinuity induced in

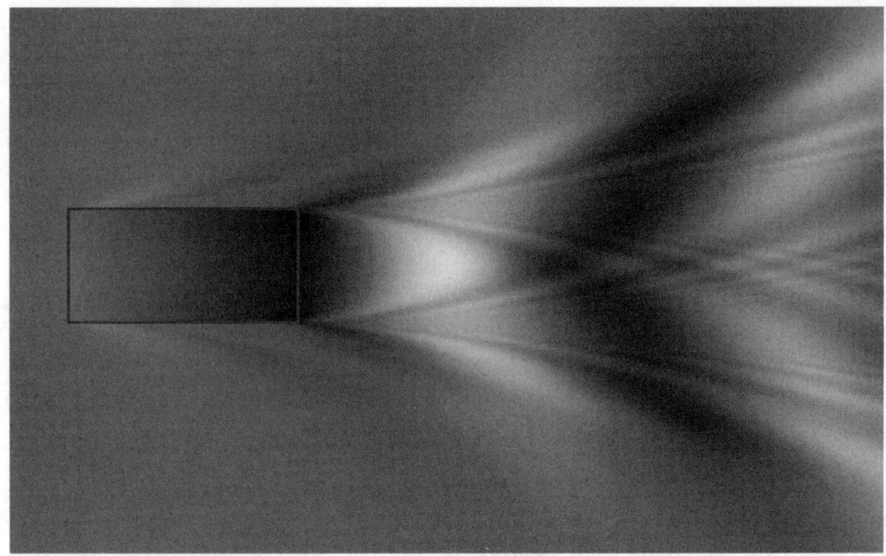

Figure 4. *Free-surface elevation due to "flat" plate at $F = 0.5$*

the pressure distribution by numerical approximation of the integrals. Partial cures are possible by various averaging techniques, but it remains difficult to produce smooth pressure outputs.

Interestingly, though, perhaps it is not always necessary to do so. If the pressure $P(x, y)$ depends over-sensitively on the hull shape $Z_0(x, y)$, it follows that the hull shape $Z_0(x, y)$ is not very sensitive to the pressure $P(x, y)$. The present author's experience is that even when the inversion code produced somewhat "wobbly" pressures $P(x, y)$, a back-substitution showed that this pressure nevertheless yielded quite smooth and believable wave fields $Z(x, y)$ agreeing with $Z_0(x, y)$, and these results were essentially unchanged when the wobbles in $P(x, y)$ were smoothed away, either by eye or systematically by averaging or least square techniques.

5 Planing Surfaces

A "planing surface" is in principle just another name for a flat ship. However, it is appropriate in introducing this alternate terminology to highlight special considerations which are of particular importance in the application to actual planing boats, an important feature of which is their high speed. Indeed, it is generally felt that planing is a state that is achieved for any vessel only, if at all, when the speed is sufficiently high.

At low speed the vessel moves forward with its hull wetted to an extent that is little changed from its static state. As the speed increases, for some hulls there may come a point where (often quite suddenly and simultaneously): significant hydrodynamic lifting forces are induced, the trim and wetted domain change dramatically,

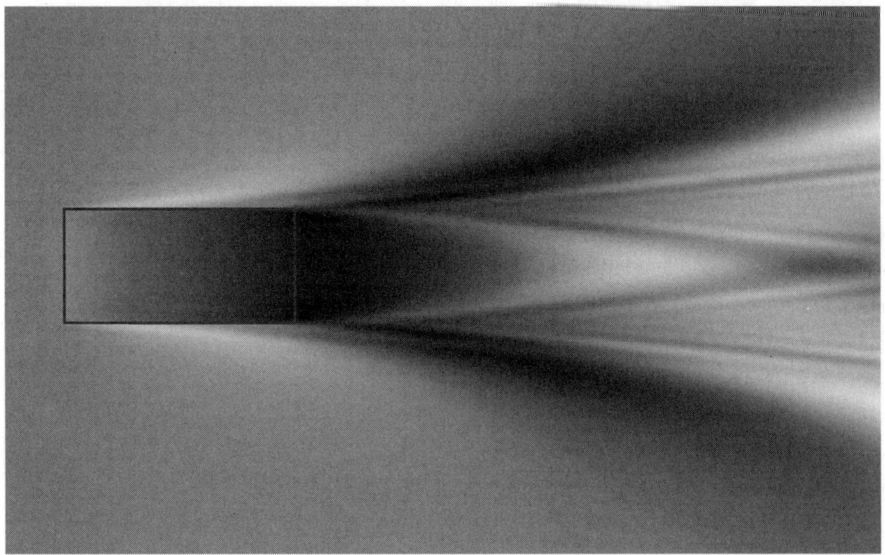

Figure 5. *Free-surface elevation due to "flat" plate at* $F = 1.0$

the boat lifts out of the water, with a decrease in drag or increase in speed at fixed propulsive force, and the flow near the stern changes its character. If there is a transom (a suddenly cut off stern end) the flow will detach smoothly from the edges of the transom rather than wetting it, but even if there is a pointed stern, smooth detachment and de-wetting may occur prior to the actual stern ending. In any case, once this happens, there is a "hole in the water" at atmospheric pressure immediately behind the stern, and the smoothness of the detachment of the water from the stern to form this hole indicates that the pressure must take the atmospheric value not just on the surface of the water, but also at that detachment location on the hull.

Mathematically, this means that any pressure $P(x, y)$ that is an acceptable output from the code described above for a flat ship must, at least at speeds above that where planing commences, vanish at the trailing end of the wetted domain. This is an extra condition, equivalent to the Kutta condition of aerodynamics, which is not in general satisfied by the unique inversions of (3.2) for a given $Z = Z_0(x, y)$ defined on a given wetted portion W of the (x, y) plane.

Nor indeed should we expect it to be satisfied! A given hull shape $Z_0(x, y)$ will be wetted when planing to an extent that must be determined by the flow, and hence the domain W over which the integral equation is solved must be determined as part of the solution, not fixed. Although this dependency of wetted domain on hydrodynamics holds in principle for any surface-piercing body at any speed, the change in wettedness accompanied by changes in trim and draft (or "squat") is relatively small for conventional ships at low speeds. However, it is an essential feature of planing that *large* changes in wettedness and trim occur as the speed

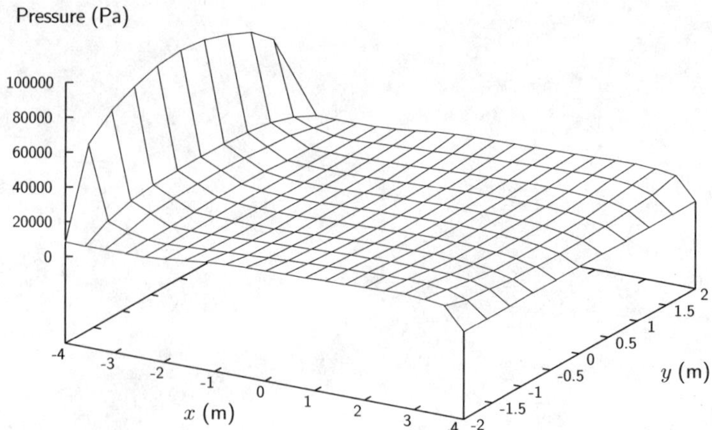

Figure 6. *Pressure on "flat" plate at* $F = 1$

increases, and for example the wetted length of a planing boat at high speed can be only a small fraction of that at rest. This requirement to determine the domain W of integration in the (x, y) plane, simultaneously with solution of the integral equation for $P(x, y)$ on that domain, presents a real challenge to any computational procedure for solving the planing problem.

Alternatively, if we choose to fix the wetted domain W, we must allow an equivalent degree of freedom in the hull shape $Z_0(x, y)$. This can be computationally more convenient, though less relevant to real planing boat design. For example, if we set $Z_1(x, y) = Z_0(x, y) + C(y)$, then the program determines $C(y)$, thus distorting the input hull $Z_0(x, y)$ (by bending it about a longitudinal axis) to a new shape $Z_1(x, y)$ on the same wetted domain W, such that the Kutta condition $P(x, y) = 0$ is satisfied at the trailing end. Whether we like it or not, the program then solves for the flow about the boat $z = Z_1(x, y)$, not the input boat $z = Z_0(x, y)$!

Although much more work is needed to bring this sort of planing surface program to a useful state, Figures 4 and 5 show examples of free-surface elevations $Z(x, y)$ computed in this way for a flat plate $Z_0(x, y) = -\alpha x$ at constant angle of attack α on a fixed 2:1 rectangular domain W. These are for two different speeds, namely for Froude numbers $F = 0.5$ and $F = 1.0$ respectively, where $F = U/\sqrt{gL}$, with L the length. In fact, at both speeds the function $C(y) \approx -0.7\alpha\, L/2$ is almost constant, so to a large extent the true body $z = Z_1(x, y)$ is still a flat plate, but simply shifted down by the flow without distortion, with the forward 30% of the plate above the undisturbed water level (but still wetted). The scale of the highest crests and deepest troughs shown is about $\pm 1.5\alpha\, L/2$; i.e. about twice the maximum plate submergence at the stern. Note the transverse waves (e.g., as marked by the dark trough for $F = 0.5$) at low speeds, disappearing at the higher speed $F = 1$ to leave a mainly diverging pattern. Note also the familiar "rooster tail" of white (highest crest) water behind the stern, which moves further downstream as the speed increases.

Figure 6 shows an example at $F = 1$ of the corresponding pressure distribution $P(x, y)$, for a plate with $L = 8$ metres and $\alpha L/2 = 1$ metre. The basic "bedstead" shape of the pressure function $P(x, y)$ seems to be smooth in both directions, with a leading-edge singularity (inverse square root and only crudely approximated here by a large finite pressure) modelling a splash, finite non-zero values at the sides, and a square-root approach to zero at the trailing edge. However, the rapidly-varying character of the diverging waves tends to compromise this smoothness computationally, and requires averaging techniques in the solution of the integral equation to avoid spurious rapid oscillations in the output pressure.

In fact, the speeds in these examples are still relatively low for application to actual planing boats (though high relative to conventional ship speeds), and the large waves made at these speeds are undesirable. There is some interest in design of large vessels which must operate at such speeds, and hence interest in minimising these waves. On the other hand, at even higher true planing speeds, wave effects are essentially negligible irrespective of hull shape, and the planing surface behaves like a lifting surface in aerodynamics. There are then computational methods [3] using aerodynamic lifting-surface theory for analysing the flow, but even these have seen little application so far to real planing-boat design, which remains largely empirical.

6 Design, Detection and Optimisation

The above describes flow *analysis* tools for a given body, and in particular for given planing surfaces reveals computational challenges that are far from resolved. Somewhat different challenges are presented by *design* tasks, where the body shape is (fully or partially) output rather than input. In some cases, the design task can be simpler than the analysis task.

If the body shape is to be determined, then something else must be given. One possibility is to design a body to produce a given wave field. Although this is a somewhat unlikely *design* task, it is essentially the same as a *detection* task. That is, we may wish to determine the shape or character of a vessel that is not readily observable directly, by examining the wave field that it produces. Although this is a very interesting and potentially important application, it will not be further discussed here.

Another design objective is *optimisation*. In particular, it has always been one of the main tasks of the naval architect to minimise the drag on a vessel, which is the sum of a viscous component and a wave component. Viscous drag is largely insensitive to hull shape, being mostly just proportional to the total wetted surface area. Hence so long as this wetted area is kept as small as possible, the residual design effort goes into minimising the wave resistance D.

For conventional thin ships, Michell's integral (2.1) is very well adapted to this task, being a positive definite quadratic functional in the hull shape function $Y(x, z)$, and there have been a number of optimisation studies [9], [20] pp. 205–214, of its minimisation subject to various constaints, the most important of which is fixed displacement volume $2 \iint Y \, dx \, dz$. The results have not been too satisfactory, with unacceptable features like negative offsets $Y(x, z)$, corrugations in the designed hull at low speeds, and end singularities at higher speeds. Some of these features are perhaps inevitable: corrugations are a fruitless attempt to cancel waves that are much shorter than the ship, while the end singularities are a fruitless attempt to place as much of the displacement as far apart as possible in order to cancel waves that are much longer than the ship. Features such as negative offsets can be eliminated by including non-negativity constraints in the optimising code. This area is one that perhaps could benefit from further research, in an era of cheap high-speed computing capability, and sophisticated optimisation algorithms.

Unconventional vessels allow more scope for optimisation, and we discuss here some drag minimisation issues for multihulls (catamarans, trimarans, etc) and for pressure distributions modelling hovercraft or planing surfaces.

7 Optimal Spacing for Multihulls

Michell's thin-ship theory provides a linear relationship (2.2) between hull shape Y and waves Z, which can (with some reservations) be extended by simple linear superposition to collections of separate thin hulls. [The reservations are associated with vertical-axis vortices and side forces induced by each hull on every other hull, but these are generally felt to be negligible for vessels of sufficiently small draft.] The wave resistance D depends quadratically rather than linearly on the hull shape function Y, but the Michell integral (2.1) can be written

$$D = \tfrac{2}{\pi}\,\rho U^2 k_0^4 \int_{-\pi/2}^{\pi/2} \mathrm{d}\theta \sec^5\theta \left|\Omega(\theta)\right|^2, \tag{7.1}$$

involving the modulus squared of a complex spectrum function $\Omega(\theta)$, which for a single hull is given by

$$\Omega(\theta) = \iint_W \mathrm{d}x\ \mathrm{d}z\ Y(x,z) e^{ik_0 x \sec\theta + k_0 z \sec^2\theta}\,, \tag{7.2}$$

and thus depends linearly on the hull shape Y.

Suppose now that we have N hulls, with the jth hull having equation

$$y = b_j + Y_j(x - a_j, z)\,, \tag{7.3}$$

on its centreplane W_j, so that its centre point is located at $(x,y) = (a_j, b_j)$. Then the linearly superposed spectrum is

$$\Omega(\theta) = \sum_{j=0}^{N} e^{-ik_0\sec^2\theta(a_j\cos\theta + b_j\sin\theta)} \iint_{W_j} \mathrm{d}x\ \mathrm{d}z\ Y_j(x,z) e^{ik_0 x\sec\theta + k_0 z\sec^2\theta}\,. \tag{7.4}$$

A simple but instructive special case is where all Y_j functions are proportional to each other; i.e.,

$$Y_j(x,z) = \beta_j Y_0(x,z)\,, \tag{7.5}$$

for constants β_j. Physically, each hull has the same shape and centreplane W, but a varying beam proportional to β_j. Then

$$\Omega(\theta) = \Omega_0(\theta)\,F(\theta)\,, \tag{7.6}$$

where

$$\Omega_0(\theta) = \iint_W \mathrm{d}x\ \mathrm{d}z\ Y_0(x,z)e^{ik_0 x \sec\theta + k_0 z \sec^2\theta}\,, \tag{7.7}$$

and

$$F(\theta) = \sum_{j=0}^{N} \beta_j e^{-ik_0 \sec^2\theta (a_j \cos\theta + b_j \sin\theta)}\,. \tag{7.8}$$

Now since D depends only on the modulus squared of Ω, for a fixed basic hull shape Y_0 and hence a fixed basic spectrum Ω_0, it depends only on the modulus squared of $F(\theta)$.

It is therefore possible to perform quite simple optimisation studies by minimising $|F(\theta)|^2$ as a function of the arrangement pattern (a_j, b_j) and beams β_j, divorcing this from the separate question of the optimum choice of the shape $Y_0(x,z)$ of the hulls. Such a minimisation is potentially confused by the dependence of $F(\theta)$ on the wave angle θ, but any choice which reduces $|F(\theta)|^2$ over a significant range of angles θ is likely to be desirable when this quantity is integrated to give the net wave resistance.

A systematic set of such studies was reported in [16]. The simplest example is a conventional (side-by-side) catamaran $N = 2$, where we can set $\beta_1 = \beta_2 = 1/2$, $a_1 = a_2 = 0$ and $b_1 = -b_2 = b$, where $2b$ is the hull centreplane separation. Then

$$F(\theta) = \cos(k_0 b \sec^2\theta \sin\theta) = C(\theta)\,. \tag{7.9}$$

Good hull-separation choices then are such that $C(\theta) = 0$ for some $\theta = \theta_0$ of importance; e.g., where the basic spectrum $\Omega_0(\theta)$ has a maximum. The smallest such separation is therefore $2b = (\pi/k_0)\cos^2\theta_0/\sin\theta_0$. Importantly, however, nothing can be done by adjusting the hull separation of conventional catamarans to eliminate or even reduce *transverse* waves having $\theta = 0$, since $F(0) = 1$ irrespective of the separation $2b$.

A somewhat more interesting example is a "staggered" trimaran $N = 3$, which is essentially the above catamaran, but with a third centrally placed hull having $a_3 = -s$, $b_3 = 0$ and $\beta_3 = 1$, located a distance s ahead of the side hulls. We have chosen the central hull to have exactly half of the total displacement of the vessel, so its beam is twice that of each of the side hulls, which is optimal for cancellation of transverse waves. Then

$$F(\theta) = C(\theta) + e^{ik_0 s \sec\theta}\,, \tag{7.10}$$

where $C(\theta)$ is again given by (7.9). Indeed, transverse waves are easily seen to be totally cancelled, with $F(0) = 0$, if the (smallest) longitudinal stagger is chosen to be $s = \pi/k_0$. Physically, this simply means that the stagger s is half of the transverse wavelength $2\pi/k_0$, so the transverse waves made by the central hull are exactly out of phase (and equal in amplitude because of the choice of displacement) to the transverse waves made by the side hulls. With that choice of s, we have finally

$$\left|F(\theta)\right|^2 = 1 + C(\theta)^2 + 2C(\theta)\cos(\pi\sec\theta)\,, \tag{7.11}$$

and our remaining task is to choose the best separation $2b$, in order to make $\left|F(\theta)\right|^2$ as small as possible over a large range of θ values.

It is not hard to see (e.g., graphically) that it is optimal to make both $C(\theta)$ and $\cos(\pi \sec\theta)$ vanish simultaneously at some angle θ_0. A good choice appears to be $\sec\theta_0 = \frac{5}{2}$ ($\theta_0 \approx 66°$) and $k_0 b \sec^2\theta_0 \sin\theta_0 = \frac{1}{2}\pi$, so $k_0 b/\pi = b/s = \cos^2\theta_0/2\sin\theta_0 \approx 0.262$. With that choice, the spectral energy $|F(\theta)|^2$ remains below about a quarter of that for a monohull, for a large range of angles, about $|\theta| < 71°$. Notably the condition $b/s \approx 0.262$ indicates that the trimaran forms an "arrow" configuration with angle about 29°, so lying inside the Kelvin angle of 39°. Actual computations from Michell's integral [16] confirm this to be a configuration with very low wave resistance near the design speed $U = \sqrt{gs/\pi}$. Other examples with especially low wave resistance are studied in [16], including a laterally asymmmetric (staggered) catamaran ($N = 2$), and a tetra-hull ($N = 4$) in a "diamond" arrangement.

8 Optimal Pressure Distributions

Design of pressure distributions $P(x, y)$ to minimise an objective such as wave resistance may be a worthwhile task for vessels such as hovercraft or surface-effect ships, where there is some element of direct control of the pressure distribution. However, it is potentially of even greater significance for planing surfaces, where the pressure $P(x, y)$ is that felt under the hull due to the hydrodynamics of the flow. If we can make such an optimal choice of $P(x, y)$, then this leads immediately to a hull shape $Z_0(x, y)$, without the need to solve the integral equation (3.2).

The formula (3.1) allows wave resistance minimisation with respect to choice of pressure $P(x, y)$, subject to a fixed weight constraint of given $\iint P\,dx\,dy$, and there were early studies on such optimisations; e.g. [10, 1], and more recently [6]. Typically the results are oscillatory, and in particular possess unacceptable large swings between positive and negative pressures.

An additional constraint of importance in practice, both directly for hovercraft and indirectly for planing boats, is non-negativity of the pressure; i.e., the inequality constraint $P(x, y) \geq 0$. When this constraint is implemented [17], perhaps not surprisingly, we find zones of zero pressure, replacing but not necessarily coinciding with the negative-pressure zones of the unconstrained solutions, interspersed with zones of positive pressures. In effect, this implies that the optimal pressure distribution is not a single patch, but rather a "multi-patch", analogous to a multi-hull ship as above.

At relatively high speeds the optimal positive pressure is confined to the extreme bow and stern ends. The end pressure varies in magnitude smoothly across the width, decreasing toward zero at the sides. In fact, an aerodynamic analogy (with exact equivalence between wave resistance and induced drag in the limit as $F \to \infty$) shows that at sufficiently high speed this lateral variation must be elliptic.

In the case of a 2:1 rectangle, so long as $F > 0.96$ the optimum consists *only* of these two bow and stern pressure patches, and the best result is achieved in the apparently unrealistic limit where their longitudinal extent is zero and their magnitude is large. Such Froude numbers are high enough that transverse wave cancellation by bow-stern interaction is ineffective, and the best that can be achieved is to place the disturbing pressures as far apart longitudinally as possible. At lower speeds, it is desirable to include a third patch of positive pressure, located amidships, and this patch grows in importance and length as the speed is reduced, with the end lines withering away, until the optimal configuration is a single patch.

However, the simple bow-stern pressure-line optimum holds in a speed range near $F = 1$ of considerable interest, and has a wave resistance typically less than a quarter of that for a constant-pressure patch bearing the same total weight. For a hovercraft modelled by a constant-pressure patch, this occurs at "hump" speed, which is a barrier that has to be overcome before the vessel moves to a higher operating speed where wave-making is not a factor, so there is only a relatively minor operational importance attached to reducing the size of this hump. In the indirect application to a planing surface, on the other hand, the reduced wave resistance may be a real benefit for large boats incapable of reaching these higher speeds, whose operating speed may of necesssity be near $F = 1$.

At first sight, the conclusion that the optimal pressure distribution is highly concentrated at the bow and stern of the vessel is disappointing, and perhaps not achievable. However, at least for the planing application, it suggests that a "tandem" configuration consisting of two distinct planing surfaces, each of high aspect-ratio (small length relative to width), may be optimal.

Throughout the present paper we have until now discussed only fully three-dimensional bodies and flows. However, it is interesting that this conjectured optimum points toward locally two-dimensional flows near each of these two high-aspect-ratio planing surfaces, and (especially since on the small local lengthscale the Froude number is high and wave effects locally negligible) there is scope for simplified two-dimensional flow analysis. It should however be noted that although the *far-field* waves made by the tandem configuration are small, so yielding low wave resistance, this does not necessarily imply a small free-surface elevation everywhere, and indeed, there are indications (see e.g. [19]) of a large trough between the bow and stern patches.

9 Acknowledgement

This work was supported by the Australian Research Council, and also by the Surveillance Systems Division of DSTO Australia. Contributions from and discussions with David Scullen and Leo Lazauskas are gratefully acknowledged.

Bibliography

[1] Bessho M.; "On the problem of the minimum wave-making resistance of ships," *Memoirs of the Defence Academy of Japan*, Vol. 2, 1962, pp. 1–30.

[2] Birkhoff G., Korvin-Kroukovsky B.V. and Kotik J.; "Theory of the wave resistance of ships: The significance of Michell's integral", *Trans. Soc. Nav. Archs. Mar. Engrs.*, Vol. 62, 1954, pp. 359–371.

[3] Casling E. M. and King G. W.; "Calculation of the wetted area of a planing hull with a chine", *Journal of Engineering Mathematics*, Vol. 14, 1979, pp. 191–205.

[4] Cheng X. and Wellicome J. F.; "Study of planing hydrodynamics using strips of transversely variable pressure", *Journal of Ship Research*, Vol. 38, 1994, pp. 30–41.

[5] Doctors L. J.; "Representation of three-dimensional planing surfaces by finite elements", *Proceedings of the 1st Conference on Numerical Ship Hydrodynamics*, Office of Naval Research, 1975, pp. 517–537.

[6] Doctors L. J.; "Optimal pressure distributions for river-based air-cushion vehicles", *Schiffstechnik*, Vol. 44, 1997, pp. 32–36.

[7] Havelock T. H.; "Studies in wave resistance: influence of the form of the waterplane of the ship", *Proceedings of the Royal Society of London*, Series A, Vol. 103, 1923, pp. 571–585.

[8] Hogner E.; "Schiffsform und Wellenwiderstand", *Hydromechanische Probleme des Schiffsantriebs*, Hamburg, 1932, pp. 99–114.

[9] Lin W.-C., Webster W. C. and Wehausen J. V.; "Ships of minimum wave resistance", *Int. Sem. Theor. Wave Resist.*, Ann Arbor, 1963. Proc. University of Michigan.

[10] Maruo H., (in Japanese), *Journal of the Society of Naval Architects of Japan*, Vol. 81, (1949).

[11] Michell J. H.; "The wave resistance of a ship." *Philosophical Magazine*, Series 5, Vol. 45, 1898, pp. 106–123.

[12] Newman J. N.; "Evaluation of the wave-resistance Green function: Part 1 – The double integral", *Journal of Ship Research*, Vol. 31, 1987, pp. 79–90.

[13] Noblesse F., "Analytical representation of ship waves" (23rd Weinblum Memorial Lecture), *Schiffstechnik*, Vol. 48, 2001, pp. 23–48.

[14] Oertel R. P.; "The steady motion of a flat ship, with an investigation of the flow near the bow and stern", Ph.D. Thesis, The University of Adelaide, 1975.

[15] Tuck E. O.; "The wave resistance formula of J.H. Michell (1898) and its significance to recent research in ship hydrodynamics", *ANZIAM Journal*, Vol. 30, 1989, pp. 365–377.

[16] Tuck E. O. and Lazauskas L.; "Optimum spacing of a family of multihulls", *Schiffstechnik*, Vol. 45, 1998, pp. 180–195.

[17] Tuck E. O. and Lazauskas L., "Free-surface pressure distributions with minimum wave resistance", *ANZIAM Journal*, Vol. 43, 2001, pp. E75–E101.

[18] Tuck E. O., Scullen D. C. and Lazauskas L.; "Ship-wave patterns in the spirit of Michell", *Proceedings of the IUTAM Symposium on Free-Surface Flows*, Birmingham, July 2000, ed. A. C. King and Y. D. Shikhmurzaev, Kluwer Academic Publishers, Dordrecht, 2001, pp. 311–318.

[19] Tuck E. O., Scullen D. C. and Lazauskas L.; "Wave patterns and minimum wave resistance for high-speed vessels", *24th Symposium on Naval Hydrodynamics*, Fukuoka, July 2002. Proc. ONR, Washington DC, 2003.

[20] Wehausen J. H.; "The wave resistance of ships", *Advances in Applied Mechanics*, Vol. 13, 1973, pp. 93–245.

[21] Weinblum G.; "Anwendungen der Michellschen Widerstandstheorie", *Jahrb. Schiffbautech. Ges.*, Vol. 31, 1930, pp. 389–436.

[22] Weinblum G.; "A systematic evaluation of Michell's integral", US Department of the Navy, *David Taylor Model Basin*, Report 886, 1955.

[23] Wigley W. C. S.; "Ship wave resistance. A comparison of mathematical theory with experimental results", *Trans. Inst. Nav. Archs.*, Vol. 68, 1926, pp. 124–137.

Henk van der Vorst has been a full professor in numerical analysis since 1990 with the Department of Mathematics, Utrecht University in the Netherlands. His current research interests include iterative solvers for linear systems, for large sparse eigensystems, and the design of algorithms for parallel computers. He did his PhD on preconditioning techniques at Utrecht University in 1982 after a 14-year period outside academia. Together with Meijerink, he suggested the use and construction of general incomplete LU-Decompositions as preconditioners (1977). This is still an often-cited paper.

From 1984 until 1990 Vorst was full professor at the Delft University of Technology. Early in 1991 he co-authored "Solving Linear Systems on Shared Memory Computers" with Dongarra, Duff and Sorensen. This book was reprinted in 1993, and has been translated into Japanese. A follow-up was published in 1998. More recently he proposed the now popular Bi-CGSTAB method (1992), and, together with Vuik, the class of GMRES methods (1994), that admit variable preconditioning in a robust way. The paper on Bi-CGSTAB was the most cited paper in mathematics written in the 1990s (ISI). For a joint paper with Sleijpen on Jacobi–Davidson methods for eigenvalue problems (1996), he received the SIAG/LA Prize in 1997. In 1994 he co-authored Templates for Linear Systems, including Preconditioning. This book has been rather influential in standardising algorithms; most of the templates have been included in MATLAB. A similar book on eigenproblems appeared in 2000. All together he has (co-)authored over 100 publications. He is (associate) editor of seven journals related to scientific computing, including *SISC*, *JCAM*, *APNUM*, and *Parallel Computing.*

Chapter 18

Iterative Solution Methods: Aims, Tools, Craftmanship

Henk A. van der Vorst[*]

Abstract: The cooling of an infant's brain, pollution of groundwater, the forces over a space-rocket, the flows of the ocean, and design of electronic devices have in common that, when the modelling is complete, very large linear systems have to be solved. Here 'very large' may mean the order of more than a billion unknowns. This may seem an impossible task, yet many of these problems have been solved successfully by iterative solution methods. Modern methods include popular algorithms such as Conjugate Gradients, GMRES, and Bi-CGSTAB. These are examples of the large family of Krylov subspace methods.

However, besides many reports of great success we see also many reports of failure, so that the natural question is how and when they can be used effectively. The answer to this question requires some basic insight into how they work, and into the circumstances where they may fail. This depends strongly on spectral properties of the matrix of the system, and the way to improve them is preconditioning. We discuss these aspects, and see how we can monitor the effects of preconditioning with the help of the readily available iteration parameters.

Contents

[*]Mathematical Institute, Utrecht, The Netherlands

1 Introduction

In this paper, we will present an overview of currently available numerical approaches for the iterative solution of large linear systems of equations $Ax = b$, with A a nonsingular $n \times n$ matrix. The aim is to solve such systems in reasonable time. The iterative methods are so-called Krylov projection type methods and they include popular methods such as Conjugate Gradients, MINRES, Bi-Conjugate Gradients, QMR, Bi-CGSTAB, and GMRES. They represent the tools that we have available.

Iterative methods are often used in combination with so-called preconditioning operators (easily invertible approximations for the operator of the system to be solved). We will give a brief overview of the various preconditioners that exist and we will sketch the ideas for tailor-made preconditioners for specific problems. This is the point where craftmanship comes in.

1.1 The aims of iterative methods

For the solution of a linear system $Ax = b$ we have the choice between direct and iterative methods. The usual *pro* arguments for iterative methods are based on economy of computer storage and CPU time. On the *con* side, it should be noted that the usage of iterative methods requires some expertise. If CPU-time and computer storage are not really at stake, then it would be unwise to consider iterative methods for the solution of a given linear system. The question remains whether there are situations where iterative solution methods are really preferable and here we will provide an argument.

Dense linear systems, and sparse systems with a suitable non-zero structure, are most often solved by a so-called direct method, such as Gaussian elimination. A direct method leads, in the absence of rounding errors, to the exact solution of the given linear system in a finite and fixed amount of work. Rounding errors can be handled fairly well by pivoting strategies. Problems arise when the direct solution scheme becomes too expensive for the task. For instance, the elimination steps in Gaussian elimination may cause some zero entries of a sparse matrix to become nonzero entries, and nonzero entries require storage as well as CPU time. This is what may make Gaussian elimination, even with strategies for the reduction of the so-called fill-in, expensive.

In order to get a more quantitative impression of this, we consider a sparse system related to the discretisation of a second order PDE over a (not necessarily regular) grid, with about m unknowns per dimension. Think, for instance, of a finite element discretisation over an irregular grid. In a 3D situation this leads typically to a bandwidth $\sim n^{\frac{2}{3}}$ ($\approx m^2$ and $m^3 \approx n$, where $1/m$ is the (average) gridsize). When taking proper account of the band structure, the number of flops is then usually $\mathrm{O}(nm^4) \sim n^{2\frac{1}{3}}$ [16, 10]. We make the caveat 'usually', because it may happen that fill-in is very limited when the sparsity pattern of the matrix is special.

In order to be able to quantify the amount of work for iterative methods, we have to be a little more specific. Let us assume that the given matrix is symmetric positive definite, in which case we may use the Conjugate Gradient (CG) method. The error reduction per iteration step of CG is $\sim \frac{\sqrt{\kappa}-1}{\sqrt{\kappa}+1}$, with $\kappa = \|A\|_2 \, \|A^{-1}\|_2$ [16].

For discretised second order PDEs over grids with gridsize $\frac{1}{m}$, it can be shown that $\kappa \sim m^2$ (see, for instance, [28]). Hence, for 3D problems we have that $\kappa \sim n^{\frac{2}{3}}$. In order to have an error reduction by a factor of ϵ, the number j of iteration steps must satisfy

$$\left(\frac{1-\frac{1}{\sqrt{\kappa}}}{1+\frac{1}{\sqrt{\kappa}}}\right)^j \approx \left(1-\frac{2}{\sqrt{\kappa}}\right)^j \approx \mathrm{e}^{-\frac{2j}{\sqrt{\kappa}}} < \epsilon\,,$$

and hence

$$j \sim -\tfrac{1}{2}\log \epsilon\sqrt{\kappa} \approx -\tfrac{1}{2}\log \epsilon n^{\frac{1}{3}}\,.$$

If we assume the number of flops per iteration to be $\sim fn$ (f stands for the average number of nonzeros per row of the matrix and the overhead per unknown introduced by the iterative scheme, f is typically a modest number, say of order 10–15), then the required number of flops for a reduction of the initial error with ϵ is $\sim -fn^{\frac{4}{3}}\log\epsilon$.

From comparing the flops counts for the direct scheme with those for the iterative CG method we conclude that the CG method may be preferable if we have to solve one system at a time, and if n is large, or f is small, or ϵ is modest.

The above arguments are quite nicely illustrated by observations made by Horst Simon [32]. He presented examples of systems with some 5×10^9 unknowns, related to the modeling of forces acting on a space rocket. From extrapolation of the CPU times observed for a representative small model problem, he estimated the CPU time for the most efficient direct method as 520 040 years, provided that the computation can be carried out at a speed of 1 TFLOPS.

On the other hand, Simon's extrapolated guess for the CPU time with preconditioned conjugate gradients, still assuming a processing speed of 1 TFLOPS, was 575 seconds. The actual processing speed for iterative methods is usually a factor lower than for direct methods; but, nonetheless, it is obvious that the differences in CPU time requirements are gigantic. The ratio of the two times is of order n, just as we might have expected from our previous arguments.

Also, the requirements for memory space for the iterative methods are typically smaller by orders of magnitude. This is often the argument for the use of iterative methods in 2D situations, when flops-count for both classes of method are more or less comparable.

For general background on linear algebra for numerical applications, see [16, 34]. Modern iterative methods for linear systems are discussed in [39]. Throughout this paper we will give other useful pointers to literature.

Some useful state of the art papers have appeared; we mention a paper on the history of iterative methods by [31]. An overview on parallelisable aspects of sparse matrix techniques is presented in [11]. A state-of-the-art overview for preconditioners is given in [6]. For implementation aspects of iterative methods see [5, 9].

1.2 The basic ideas

The basic idea behind iterative methods is to replace the given system $Ax = b$ by some nearby system $Kx_0 = b$ that can be more easily solved and take x_0 as an approximation for x. Obviously, we want the correction z that satisfies

$$A(x_0 + z) = b\,,$$

and this leads to a new linear system $Az = b - Ax_0$. Again, we solve this system by a nearby system, and most often one takes K again:

$$Kz_0 = b - Ax_0\,.$$

The correction procedure can now be repeated for $x_1 = x_0 + z_0$, and so on, which gives us an iterative method.

For the basic or Richardson iteration, introduced here, it follows that

$$x_{k+1} = x_k + z_k = x_k + K^{-1}r_k\,, \tag{1.1}$$

with $r_k = b - Ax_k$. We use K^{-1} only for notational purposes; we (almost) never compute inverses of matrices explicitly. With $K^{-1}b$, we refer to the vector $\widetilde{b}$ that is solved from $K\widetilde{b} = b$. The matrix K is called the *preconditioner*. In fact, the operator K^{-1} applied to a vector w formally represents the process of constructing an approximation for the solution v of $Av = w$. The operator K, and likewise K^{-1}, does not need to be represented as an array of coefficients. For instance, K^{-1} may represent the effect of a few steps of multigrid or a few steps of some iterative method. In the context of most methods to be discussed here, one should take care that K represents formally the same fixed operator for all iteration steps.

In order to simplify our formulas we will take $K = I$ and assume that the presented iteration schemes are applied to the preconditioned system $K^{-1}Ax = K^{-1}b$.

From now on we will also assume that $x_0 = 0$ to simplify future formulas. This does not mean a loss of generality, because the situation $x_0 \neq 0$ can be transformed with a simple shift to the system $Ay = b - Ax_0 = \bar{b}$, for which obviously $y_0 = 0$.

For the simple Richardson iteration, it is easily shown that

$$x_k \in \text{span}\{r_0, Ar_0, \ldots, A^{k-1}r_0\} \equiv \mathcal{K}^k(A; r_0)\,. \tag{1.2}$$

The k-dimensional space spanned by a given vector v, and increasing powers of A applied to v, up to the $(k-1)$-th power, is called the k dimensional Krylov subspace, generated with A and v, denoted by $\mathcal{K}^k(A;v)$.

It turns out that, at the expense of relatively little additional work compared to the Richardson method, we can identify much better approximations for the solution from the Krylov subspaces. This has led to the class of Krylov subspace methods, which contains very effective iterative methods.

2 Tools: Krylov subspace methods

2.1 The Krylov subspace approach

Methods that attempt to generate better approximations from the Krylov subspace are often referred to as Krylov subspace methods. Because optimality usually refers to some sort of projection, they are also called Krylov projection methods. The Krylov subspace methods, for identifying suitable $x \in \mathcal{K}^k(A;r_0)$, can be distinguished in four different classes:

1. The *Ritz–Galerkin approach*: Construct the x_k for which the residual is orthogonal to the current subspace: $b - Ax_k \perp \mathcal{K}^k(A;r_0)$.

2. The *minimum norm residual approach*: Identify the x_k for which the Euclidean norm $\|b - Ax_k\|_2$ is minimal over $\mathcal{K}^k(A;r_0)$.

3. The *Petrov–Galerkin approach*: Find an x_k so that the residual $b - Ax_k$ is orthogonal to some other suitable k-dimensional subspace.

4. The *minimum norm error approach*: Determine x_k in $A^{\mathsf{T}}\mathcal{K}^k(A^{\mathsf{T}};r_0)$ for which the Euclidean norm $\|x_k - x\|_2$ is minimal.

The Ritz–Galerkin approach leads to well-known methods such as Conjugate Gradients, the Lanczos method, the Full Orthogonalisation Method (FOM), and Generalised Conjugate Gradients (GENCG). The minimum norm residual approach leads to methods like GMRES, MINRES, and ORTHODIR. The main disadvantage of these two approaches is that, for most unsymmetric systems, they lead to long and therefore expensive recurrence relations for the approximate solutions. This can be relieved by selecting other subspaces for the orthogonality condition (the Galerkin condition). If we select the k-dimensional subspace in the third approach as $\mathcal{K}^k(A^{\mathsf{T}};s_0)$, then we obtain the Bi-CG and QMR methods, and these methods work with short recurrences. The fourth approach is not so obvious, but for $A = A^{\mathsf{T}}$ it leads to the SYMMLQ method of [26]. Hybrids of these approaches have been proposed, like CGS, Bi-CGSTAB, BiCGSTAB(ℓ), TFQMR, FGMRES, and GMRESR. For references to all these methods and further details, see [17, 29, 39]

In order to facilitate the construction of approximations corresponding to the four different approaches, we need a suitable basis for the Krylov subspace and the usual way to this is to construct an orthonormal basis [1], as is schematically shown in Figure 1.

```
v_1 = r_0/||r_0||_2;
for j = 1, ..., m - 1
    t = Av_j;
    for i = 1, ..., j
        h_{i,j} = v_i^T t;
        t = t - h_{i,j} v_i;
    end;
    h_{j+1,j} = ||t||_2;
    v_{j+1} = t/h_{j+1,j};
end
```

Figure 1. *Arnoldi's method with modified Gram–Schmidt orthogonalisation*

The orthogonalisation can be conveniently expressed in matrix notation. Let V_j denote the matrix with columns v_1 up to v_j then it follows that

$$A\,V_{m-1} = V_m\,H_{m,m-1}\,. \tag{2.1}$$

The $m \times (m-1)$ matrix $H_{m,m-1}$ is upper Hessenberg, and its elements $h_{i,j}$ are defined by the Arnoldi algorithm (cf. Fig. 1).

We see that this orthogonalisation becomes increasingly expensive for increasing dimension of the subspace, since the computation of each $h_{i,j}$ requires an inner product and a vector update.

Note that if A is symmetric, then so is $H_{m-1,m-1} = V_{m-1}^{\mathsf{T}} A\,V_{m-1}$, so that in this situation $H_{m-1,m-1}$ is tridiagonal. This means that in the orthogonalisation process, each new vector has to be orthogonalised with respect to the previous two vectors only, since all other inner products vanish. The resulting three-term recurrence relation for the basis vectors of $\mathcal{K}^m(A;r_0)$ is known as the *Lanczos method* ([20]) and some very elegant methods are derived from it. In this symmetric case the orthogonalisation process involves constant arithmetical costs per iteration step: one matrix vector product, two inner products, and two vector updates.

2.2 The Ritz–Galerkin approach

The Ritz–Galerkin conditions imply that $r_k \perp \mathcal{K}^k(A;r_0)$, and this is equivalent to

$$V_k^{\mathsf{T}}(b - Ax_k) = 0\,.$$

Since $b = r_0 = \|r_0\|_2\,v_1$, it follows that $V_k^{\mathsf{T}} b = \|r_0\|_2\,e_1$ with e_1 the first canonical unit vector in $\mathbb{R}^k$. With $x_k = V_k\,y$ we obtain

$$V_k^{\mathsf{T}} A\,V_k\,y = \|r_0\|_2\,e_1\,.$$

This system can be interpreted as the system $Ax = b$ projected onto the subspace $\mathcal{K}^k(A;r_0)$.

Obviously we have to construct the $k \times k$ matrix $V_k^{\mathsf{T}} A V_k$, but this is immediately available from the orthogonalisation process:

$$V_k^{\mathsf{T}} A V_k = H_{k,k} ,$$

so that the x_k for which $r_k \perp \mathcal{K}^k(A; r_0)$ can be easily computed by first solving $H_{k,k}\, y = \|r_0\|_2 e_1$, and then forming $x_k = V_k\, y$. This algorithm is known as FOM or GENCG, see [30].

When A is symmetric, then $H_{k,k}$ reduces to a tridiagonal matrix $T_{k,k}$, and the resulting method is known as the *Lanczos* method ([21]). When A is in addition positive definite then we obtain, at least formally, the *Conjugate Gradient* method. In commonly used implementations of this method, one implicitly forms an *LU* factorisation for $T_{k,k}$, without generating $T_{k,k}$ itself, and this leads to very elegant short recurrences for the x_j and the corresponding r_j; see the algorithm presented in Figure 2. This algorithm includes preconditioning with an operator K, which should be a fixed approximation for A throughout the entire iteration process.

x_0 is an initial guess, $r_0 = b - Ax_0$
for $i = 1, 2, ...$
  Solve $Kw_{i-1} = r_{i-1}$
  $\rho_{i-1} = r_{i-1}^H w_{i-1}$
  if $i = 1$
    $p_i = w_{i-1}$
  else
    $\beta_{i-1} = \rho_{i-1}/\rho_{i-2}$
    $p_i = w_{i-1} + \beta_{i-1} p_{i-1}$
  endif
  $q_i = Ap_i$
  $\alpha_i = \rho_{i-1}/p_i^H q_i$
  $x_i = x_{i-1} + \alpha_i p_i$
  $r_i = r_{i-1} - \alpha_i q_i$
  if x_i accurate enough **then quit**
end

Figure 2. *Conjugate Gradients with preconditioning K*

The positive definiteness is necessary to guarantee the existence of the *LU* factorisation, but it also guarantees that $\|x_k - x\|_A$ is minimal[1] over all possible x_k from the Krylov subspace of dimension k.

2.3 The minimum norm residual approach

We look for an $x_k \in \mathcal{K}^k(A; r_0)$, that is $x_k = V_k\, y$, for which $\|b - Ax_k\|_2$ is minimal. This norm can be rewritten, with $\rho \equiv \|r_0\|_2$, as

$$\|b - Ax_k\|_2 = \|b - A\, V_k\, y\|_2 = \|\, \rho V_{k+1} e_1 - V_{k+1} H_{k+1,k}\, y\|_2 ,$$

[1]The A-norm is defined by $\|y\|_A^2 \equiv (y, y)_A \equiv (y, Ay)$.

using the orthogonality relation (2.1). Now we exploit the fact that V_{k+1} is an orthonormal transformation with respect to the Krylov subspace $K^{k+1}(A; r_0)$:

$$\|b - Ax_k\|_2 = \|\rho e_1 - H_{k+1,k}\, y\|_2 ,$$

and the last norm can simply be minimised by solving the minimum norm least squares problem for the $k+1$ by k matrix $H_{k+1,k}$ and right-hand side $\|r_0\|_2\, e_1$. The least squares problem is solved by constructing a QR factorisation of $H_{k+1,k}$ and, because of the upper Hessenberg structure, this can conveniently be done with Givens transformations.

The GMRES method [30] is based upon this approach. In order to avoid excessive storage requirements and computational costs for the orthogonalisation, GMRES is usually restarted after each cycle of m iteration steps. This algorithm is referred to as GMRES(m); the not-restarted version is often called 'full' GMRES. There is no simple rule to determine a suitable value for m; the speed of convergence over cycles of GMRES(m) may drastically vary for nearby values of m.

For an excellent overview of GMRES and related variants, such as FGMRES, see [29].

2.4 The Petrov–Galerkin approach

For unsymmetric systems we cannot, in general, reduce the matrix A to a tridiagonal system in a lower-dimensional subspace, by orthogonal projections. The reason is that we cannot create an orthogonal basis for the Krylov subspace by a three-term recurrence relation [14]. We can, however, obtain a suitable non-orthogonal basis with a three-term recurrence, by requiring that this basis is orthogonal with respect to some other basis.

For this other basis, we select a convenient basis for the Krylov subspace generated with A^{T} and starting vector w_1. It can be shown that a basis $v_1, \ldots, v_i$ for $\mathcal{K}^i(A; v_1)$ can be created with a three-term recurrence relation, so that the v_j are orthogonal with respect to the w_k, for $k \neq j$. The w_i are generated with the same recurrence relation as for the v_j but with A replaced by A^{T}.

In matrix notation this leads to $W_i^{\mathsf{T}} A V_i = D_i T_{i,i}$, and also that $V_i^{\mathsf{T}} A^{\mathsf{T}} W_i = D_i T_{i,i}$, with $D_i = W_i^{\mathsf{T}} V_i$ a diagonal matrix and $T_{i,i}$ a tridiagonal matrix. These bi-orthogonal sets of vectors form the basis for methods as Bi-CG and QMR.

Bi-CG is not as robust as CG. It may happen, for instance, that $w_i^{\mathsf{T}} v_i = 0$ and then the method breaks down. Bi-CG is based on an LU decomposition of $T_{i,i}$, but since $T_{i,i}$ is not necessarily positive definite or so, an LU decomposition in bidiagonal L and U may not exist, which gives another breakdown of the method. Fortunately, these circumstances do not occur frequently, but one has to be aware of them and carry out the required checks. There exist techniques to repair these breakdowns, but they require complicated coding. It may be convenient just to restart when a breakdown occurs, giving up some of the efficiency of the method and, of course, with a chance that breakdown occurs again. For a full treatment of Bi-CG, see [39].

2.5 Bi-CGSTAB

Sonneveld [33] showed that the two operations with A and A^{T} per iteration of Bi-CG can be replaced by two operations with A and with the effect that i iterations with Bi-CG are applied twice: once with the starting residual r_0 and then again with the same iteration constants on r_i. Surprisingly, this can be done for virtually the same computational costs as for Bi-CG, but the result is a method that often converges about twice as fast as Bi-CG and known as CGS.

Sonneveld's principle was further perfected in [37] for the construction of Bi-CGSTAB in which the Bi-CG operations with A^{T} are replaced by operations with A in order to carry out a GMRES(1) reduction on top of each Bi-CG iteration.

The computational costs for Bi-CGSTAB are, per iteration, about the same as for Bi-CG. However, because of the additional GMRES(1) steps after each Bi-CG step, Bi-CGSTAB converges often considerably faster. Of course, Bi-CGSTAB may suffer from the same breakdown problems as Bi-CG. In an actual code one should test for such situations and take appropriate measures; e.g., restart with a different $\widetilde{r}$ $(= w_1)$ or switch to another method (for example GMRES).

The method has been further generalised to Bi-CGSTAB(ℓ), which generates iterates that can be interpreted as the product of Bi-CG and repeated GMRES(ℓ); for more details and further references see [39]. Software for this method is available from NAG.

2.6 General remarks on Krylov methods

The choice for a method is a delicate problem. If the matrix A is symmetric positive definite, then the choice is easy: Conjugate Gradients. For other types of matrices the situation is very diffuse. GMRES, is the most robust method, but in terms of work per iteration step it is also relatively expensive. Bi-CG, which was suggested by [15], is a relatively inexpensive alternative. The main disadvantage of Bi-CG is that it involves per iteration an operation with A and one with A^{T}. Bi-CGSTAB, is an efficient combination of Bi-CG and repeated 1-step GMRES, avoiding operations with A^{T}. Bi-CGSTAB requires about as many operations per iteration as Bi-CG. A more thorough discussion on Krylov methods is given in [39]. Other useful sources of information on iterative Krylov subspace methods include [3, 17, 19, 24], and [29].

2.7 Understanding the convergence behaviour

The approximation x_i for the solution of $Ax = b$ is taken from the shifted i-dimensional Krylov subspace $x_0 + \mathcal{K}^i(A; r_0)$ and hence it can be expressed as

$$x_i = x_0 + Q_{i-1}(A)\, r_0 \,,$$

in which Q_{i-1} denotes a polynomial of degree $i-1$. From this it follows that

$$r_i = r_0 - A\, Q_{i-1}(A)\, r_0 = P_i(A)\, r_0 \,,$$

in which P_i is a polynomial of degree i with $P_i(0) = 1$, and (if A is nonsingular) that

$$x_i - x = P_i(A)\,(x - x_0)\,. \tag{2.2}$$

Note that these observations hold for general matrices and for all approaches that select x_i from the shifted Krylov subspace. As we will see, the observation that $P_i(0) = 1$ is a very important one that will help us to understand differences in convergence behaviour.

For the Ritz–Galerkin approach it can be proved that the zeros of P_i are the eigenvalues of the reduced matrix $H_{i,i}$ and, together with the normalisation constraint $P_i(0) = 1$ this fixes the solution completely (see, for instance, [39, Chapter 5.3.1]).

As an example of the convergence behaviour of Krylov methods, we consider the Ritz–Galerkin approach for symmetric matrices. In that case we have that the eigenvalues of the matrix are real and the eigenvectors form an orthonormal system. Also, the eigenvalues of the now tridiagonal matrix $H_{i,i}$ are real and this makes it possible to represent and interpret the reduction of the approximation errors $\|x_i - x\|_2$ in an easy way. Much, however, carries over to the asymmetric case.

We denote the eigenvalues and the orthonormalised eigenvectors of A by λ_j, z_j, and write $r_0 = \sum_j \gamma_j z_j$. It follows that

$$r_i = P_i(A)\,r_0 = \sum_j \gamma_j P_i(\lambda_j)\,z_j\,. \tag{2.3}$$

Clearly, the value of $P_i(\lambda_j)$ represents the reduction of the error in the direction of z_j. Of course, for convergence, the reduction factors have to become small for increasing i and for all λ_j. We illustrate for a simple example what we may expect. Let A be a 100×100 symmetric matrix with eigenvalues $2, 4, 6, 8, \ldots, 200$. For the right-hand side we select a vector b such that the solution x of $Ax = b$ has equal components γ_j in all eigenvector directions, and we take $x_0 = 0$.

The polynomial P_i has been plotted for $i = 10$ in Figure 3. The value $P_{10}(0) = 1$ has been indicated with an encircled $*$.

From the theory for the eigenvalues of $H_{i,i}$, the so-called Ritz values, it is known that these Ritz values are in the interval spanned by the minimum and maximum eigenvalues of A and that they move from inside to the eigenvalues of A when i is increased. In particular, if we assume for simplicity that all eigenvalues λ_j are different and that they are ordered as $\lambda_1 < \lambda_2 < \cdots < \lambda_n$, and that the Ritz values $\theta_j^{(i)}$ of $H_{i,i}$ are ordered likewise, then it can be shown that $\theta_j^{(i)}$ moves, from the right, towards λ_j for increasing i. Similarly, θ_{i-k} moves, from the left, towards λ_{n-k} for increasing i and $k = 0, \ldots, i$. This means that all zeros of P_i are inside the eigenvalue interval $[\lambda_1, \lambda_n]$ and outside this interval the polynomial increases rapidly. This means that the normalisation $P_i(0)$ to 1 forces the polynomial to be small over the eigenvalue interval of A, and this is what we see in Figure 3.

When we replace the eigenvalues of A by the shifted eigenvalues $\lambda + \sigma$ (and keep the same b), then the Ritz values shift accordingly with σ. This implies that,

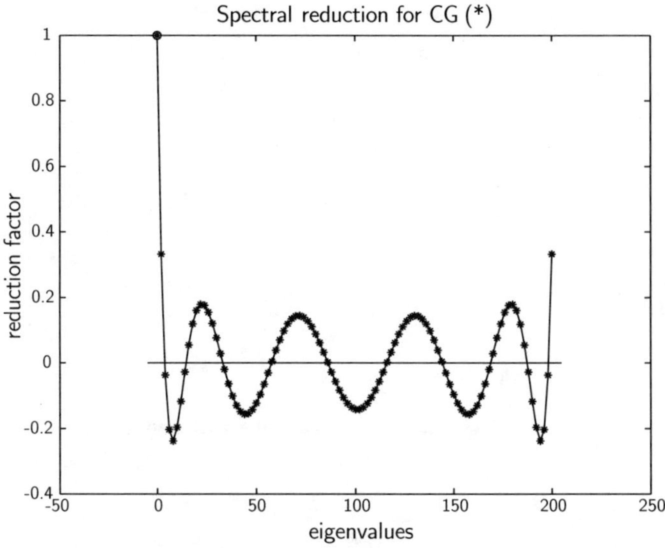

Figure 3. *The iteration polynomial P_{10} and the reduction factors for all eigendirections indicated by a* $*$

because of the normalisation $P_i(0) = 1$, the reduction factors after i iterations should now be much smaller than for the unshifted case, and this is, for $\sigma = 10$, displayed in Figure 4. There we see the same polynomial as in Figure 3, but now shifted over $\sigma = 10$. This indicates that, for symmetric positive definite A, the condition number λ_n/λ_1, plays an important role in the convergence of the conjugate gradients method (which represents the implementation of the Ritz–Galerkin approach for the symmetric positive definite case). We will see this later in more detail.

It is now obvious that we may expect all sorts of problems with the Ritz–Galerkin approach if A is symmetric indefinite, because in that case the eigenvalue interval contains the origin. Since the Ritz values move from inside to the exterior eigenvalues, they may occasionally be close to the origin, even when A itself has no eigenvalue that is close to the origin. The normalisation constraint $P_i(0) = 1$ may then lead to some very large values of $P_i(\lambda_j)$. In fact, the method may even break down at iteration i when a Ritz value $\theta_j^{(i)}$ happens to be zero. There is no way to circumvent these problems in the standard Ritz–Galerkin approach. In Figure 5 we see the above illustrated for a shift $\sigma = -15$.

For the minimum residual approach, the situation is different. In that case, the iteration polynomial P_i has the harmonic Ritz values of $H_{i+1,i}$ as its zeros; see, for instance, [25]. The harmonic Ritz values are the eigenvalues of the matrix $H_{i,i}^{-1}H_{i+1,i}^{\mathsf{T}}H_{i+1,i}$ and these harmonic Ritz values are shift dependent. The inverses of these harmonic Ritz values can be interpreted as Ritz values for the matrix A^{-1}. In particular, for the symmetric indefinite case this implies that the negative harmonic Ritz values are smaller than the largest negative eigenvalue λ_- (i.e. the

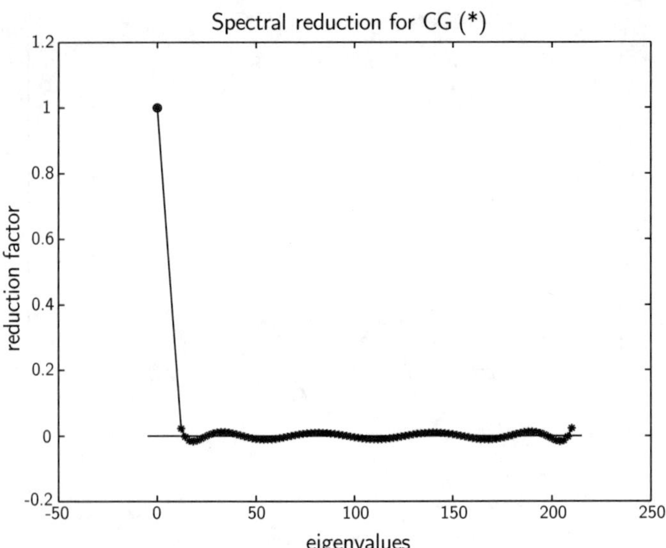

Figure 4. *The iteration polynomial P_{10} and the reduction factors for all eigen-directions indicated by a $*$, for shift $\sigma = 10$*

one closest to the origin) and the positive harmonic Ritz values are always larger than the smallest positive eigenvalue λ_+. This means that P_i has no zero in the interval $[\lambda_-, \lambda_+]$ and hence, if these eigenvalues are relatively well away from the origin, then $P_i(0)$ can potentially be relatively large with respect to the values $P_i(\lambda_j)$. This means that the minimum residual approach can be useful for indefinite systems, much more than the Ritz–Galerkin approach; the latter may even break down. However, the constraint $P_i(0) = 1$ implies that we may expect at least phases of slow convergence if A has eigenvalues close to the origin. As we will see later, this slow convergence may be followed by more rapid convergence whence these eigenvalues close to zero have been approximated well enough by harmonic Ritz values. For more information on harmonic Ritz values, see [38, section 25].

Note that only those λ_j play a role in the process for which $\gamma_j \neq 0$. In particular, if A happens to be semidefinite, i.e., there is a $\lambda = 0$, then this is no problem for the minimisation process as long as the corresponding coefficient γ is zero as well.

We now return to the positive definite symmetric case, because that situation is easier to analyse. For the conjugate gradient method, it can be shown that $(x_i - x)^\top A(x_i - x)$ is minimal over all possible $x_i \in \mathcal{K}^i(A; r_0)$. Upper bounds on the error (in A-norm) are obtained by observing that

$$\|x_i - x\|_A^2 = \sum_j \frac{\gamma_j^2}{\lambda_j} P_i^2(\lambda_j) \;\leq\; \sum_j \frac{\gamma_j^2}{\lambda_j} Q_i^2(\lambda_j) \;\leq\; \max_{\lambda_j} Q_i^2(\lambda_j) \sum_j \frac{\gamma_j^2}{\lambda_j},$$

for any arbitrary polynomial Q_i of degree i with $Q_i(0) = 1$, where the maximum is

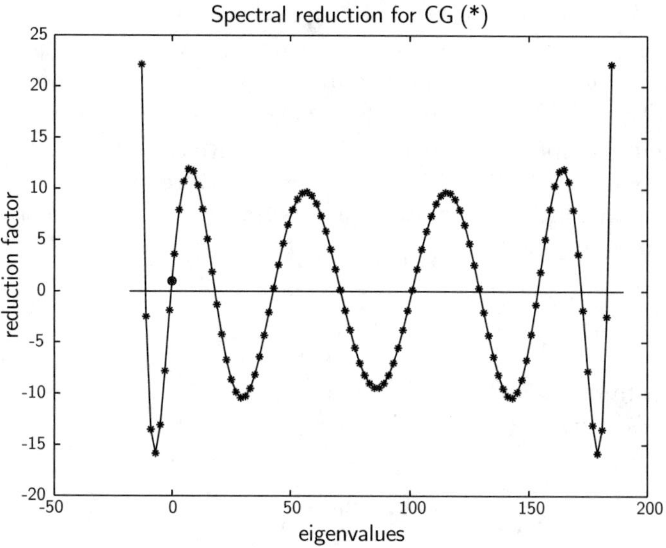

Figure 5. *The iteration polynomial P_{10} and the reduction factors for all eigendirections indicated by a $*$, for shift $\sigma = -15$*

taken, of course, only over those λ for which the corresponding $\gamma \neq 0$.

We get descriptive upper bounds by selecting appropriate polynomials for Q_i. A very well-known upper bound is obtained by taking for Q_i the ith degree Chebyshev polynomial C_i transformed to the interval $[\lambda_{\min}, \lambda_{\max}]$ and scaled such that its value at 0 is equal to 1. With $\kappa \equiv b/a = \lambda_{\max}/\lambda_{\min}$, this leads to the following upper bound for the A-norm of the error [8, 16, 2]:

$$\|x_i - x\|_A \leq 2\Big(\frac{\sqrt{\kappa}-1}{\sqrt{\kappa}+1}\Big)^i \|x_0 - x\|_A \,. \tag{2.4}$$

This upper bound shows that we have fast convergence for small condition numbers.

Upper bounds as in (2.4) show that we have global convergence, but they do not help us to explain all sorts of local effects in the convergence behaviour of CG. A very well known effect is the so-called 'super linear' convergence: in many situations one observes that the average speed of convergence seems to increase as the iteration proceeds. It can be shown that as soon as one of the extreme eigenvalues is modestly well approximated by a Ritz value, the procedure converges from then on as a process in which this eigenvalue is absent; i.e., a process with a reduced condition number. Note that super linear convergence behaviour in this connection is used to indicate linear convergence with a factor that is gradually decreased during the process as more and more of the extreme eigenvalues are sufficiently-well approximated. Since the convergence of Ritz values towards a particular eigenvalue depends on how well that eigenvalue is separated from the remaining spectrum [27], this underlines the importance of achieving at least local clustering of eigenvalues when preconditioning is employed.

The local convergence behavior of CG, and especially the occurrence of super linear convergence, was first explained in a qualitative sense in [8], and later in a quantitative sense in [35]. In both papers it was linked to the convergence of eigenvalues (Ritz values) of $T_{i,i}$ towards eigenvalues of A, for increasing i.

For instance, suppose that at iteration i we have that $\theta_1^{(i)} \approx \lambda_1$, then after the i-th step we may expect for iterations $i+k$, $k = 1, \ldots$, reductions that are bounded by an expression in which the condition number $\kappa_2 = \lambda_n/\lambda_2$ for the remaining part of the spectrum is active:

$$\|x - x_{i+k}\|_A \;\leq\; 2F_i^2\left(\frac{\sqrt{\kappa_2}-1}{\sqrt{\kappa_2}+1}\right)^k \|x - x_i\|_A \,.$$

Here F_i is a constant larger than 1. We do not need to have accurate approximations for λ_1 by the first Ritz values. For instance, if

$$\frac{\theta_1^{(i)} - \lambda_1}{\lambda_1} < 0.1 \quad \text{and} \quad \frac{\theta_1^{(i)} - \lambda_1}{\lambda_2 - \lambda_1} < 0.1\,,$$

then we already have that $1.0 < F_i < 1.25$. For more details we refer to [35].

3 Craftmanship: how to make things work?

3.1 Preconditioning

As we have argued, the various Krylov subspace methods are not robust in the sense that they can be guaranteed to lead to acceptable approximate solutions within modest computing time and storage (modest with respect to alternative solution methods). For some methods (for instance, full GMRES) it is obvious that they lead, in exact arithmetic to the exact solution in maximal n iterations, but that may not be very practical. Other methods are restricted to specific classes of problems (CG, MINRES) or occasionally suffer from such nasty side-effects as stagnation or break down (Bi-CG, Bi-CGSTAB). Such poor convergence depends in a very complicated way on spectral properties (eigenvalue distribution, field of values, condition of the eigensystem, etc.) and this information is not available in practical situations.

The trick is then to try to find some nearby operator K such that $K^{-1}A$ has better (but still unknown) spectral properties. This is based on the observation that for $K = A$, we would have the ideal system $K^{-1}Ax = Ix = K^{-1}b$ and all subspace methods would deliver the true solution in one single step. The hope is that for K in some sense close to A a properly selected Krylov method applied to, for instance, $K^{-1}Ax = K^{-1}b$, would need substantially fewer iterations to yield a good enough approximation for the solution of the given system $Ax = b$. An operator that is used with this purpose is called a *preconditioner* for the matrix A.

The general problem of finding an efficient preconditioner is to identify a linear operator K (the *preconditioner*) with the properties that[2]:

[2] The presentation in this chapter has partial overlap with [9, Chapt. 9]

1. K is a good approximation to A in some sense,

2. the cost of the construction of K is not prohibitive,

3. the system $Ky = z$ is much easier to solve than the original system.

There is a great freedom in the definition and construction of preconditioners for Krylov subspace methods. Note that in all the Krylov methods, one never needs to know individual elements of A, and one never has to modify parts of the given matrix. It is always sufficient to have a rule (subroutine) that generates, for given input vector y, the output vector z, that can *mathematically* be described as $z = Ay$. This holds also for the nearby operator: it does not have to be an explicitly given matrix. For some methods, in particular Flexible GMRES and GMRESR, it is permitted that the operator K is (slightly) different for different input vectors (*variable preconditioning*). This plays an important role in the solution of nonlinear systems, if the Jacobian of the system is approximated by a Frechet derivative, and it is also attractive in some domain decomposition approaches (in particular, if the solution per domain itself is obtained by some iterative method again).

We note that symmetry of A and K does not imply symmetry of $K^{-1}A$, which would be necessary for some methods. However, if K is symmetric positive definite then $[x, y] \equiv (x, Ky)$ defines a proper inner product. It is easy to verify that $K^{-1}A$ is symmetric with respect to the new inner product $[\cdot, \cdot]$, so that we can use methods like MINRES, SYMMLQ, and CG (when A is positive definite as well) in this case. Popular formulations of preconditioned CG are based on this observation.

The choice of K varies from purely "black box" algebraic techniques which can be applied to general matrices to "problem dependent" preconditioners which exploit special features of a particular problem class. Examples of the last class are discretised PDE's, where the preconditioner is constructed as the discretisation of a nearby (easier to solve) PDE. Although problem dependent preconditioners can be very powerful, there is still a practical need for efficient preconditioning techniques for large classes of problems.

We will give an overview of some of the more popular preconditioning techniques. The reader is referred to [3, 24, 29] for more complete overviews of (classes of) preconditioners. See [6] for a very readable introduction to various concepts of preconditioning and for many references to specialised literature.

3.2 Incomplete LU factorisations

Originally, preconditioners were based on direct solution methods in which part of the computation is skipped. This leads to the notion of *Incomplete LU* (or *ILU*) *factorisation* [22]. Standard Gaussian elimination is equivalent to factoring the matrix A as $A = LU$, where L is lower triangular and U is upper triangular. In actual computations these factors are explicitly constructed. The main problem in sparse matrix computations is that the factors of A are often a good deal less sparse than A, which makes solving expensive. The basic idea behind the ILU preconditioner is to modify Gaussian elimination to allow fill-in at only a restricted set of positions in the LU factors.

The following theorem collects results for the situation where we determine *a priori* the positions of the elements that we wish to ignore during the Gaussian elimination process. Note that this is not a serious restriction because we may also neglect elements during the process according to certain criteria and this defines the positions implicitly. The indices of the elements to be ignored are collected in a set $\mathcal{S}$:

$$S \subset S_n \equiv \{(i,j) \mid i \neq j, 1 \leq i \leq n, 1 \leq j \leq n\}\,. \tag{3.1}$$

We can now formulate the theorem that guarantees the existence of incomplete decompositions for the M-matrix A (cf. [22, Th.2.3]).

Theorem 18.1. *Let $A = (a_{i,j})$ be an $n \times n$ M-matrix*[3]*, then there exists for every $S \subset S_n$ a lower triangular matrix $\widetilde{L} = (\ell_{i,j})$, with $\ell_{i,i} = 1$, an upper triangular matrix $\widetilde{U} = (u_{i,j})$, and a matrix $N = (n_{i,j})$ with $\ell_{i,j} = 0$ and $u_{i,j} = 0$ if $(i,j) \in S$, and $n_{i,j} = 0$ if $(i,j) \notin S$, such that the splitting $A = \widetilde{L}\widetilde{U} - N$ leads to a convergent iteration (1.1). The factors $\widetilde{L}$ and $\widetilde{U}$ are uniquely defined by S.*

When A is symmetric and positive definite, then it is obvious to select S so that it defines a symmetric sparsity pattern and then one can rewrite the factorisation so that the diagonals of $\widetilde{L}$ and $\widetilde{U}$ are equal. This is known as an incomplete Cholesky decomposition.

A commonly used strategy is to define S by:

$$S = \{(i,j) \mid a_{i,j} \neq 0\}\,. \tag{3.2}$$

That is, the only non-zeros allowed in the LU factors are those for which the corresponding entries in A are non-zero. It is easy to show that the elements $k_{i,j}$ of K match those of A on the set S:

$$k_{i,j} = a_{i,j} \quad \text{if} \quad (i,j) \in S\,, \tag{3.3}$$

and this can be exploited for implementation purposes.

A well-known variant on ILU is the so-called *Modified ILU* (MILU) factorisation [12, 18]. For this variant the condition (3.3) is replaced by

$$\sum_{j=1}^{n} k_{i,j} = \sum_{j=1}^{n} a_{i,j} + ch^2 \quad \text{for} \quad i = 1, 2, \ldots, n\,. \tag{3.4}$$

The term ch^2 is for grid-oriented problems with mesh-size h. In our context, the row-sum requirement in (3.4) amounts to an additional correction to the diagonal entries. The correction leads to the observation that $Kz \approx Az$ for almost constant z (in fact this was the motivation for the construction of these preconditioners). This

[3]The nonsingular matrix A is called an M-matrix if all its off-diagonal elements are nonpositive and if all elements of A^{-1} are nonnegative

results in very fast convergence for problems where the solution is locally smooth. However, quite the opposite may be observed for problems where the solution is far from smooth.

3.3 Variants of ILU Preconditioners

Many variants on the idea of incomplete or modified incomplete decomposition have been proposed in the literature. These variants are designed to either reduce the total computational work or to improve the performance on vector or parallel computers or to handle special problems.

A natural approach is to allow more fill-in in the LU factor (that is a larger set S). Several possibilities have been proposed. The most obvious variant is to allow more fill-ins in specific locations in the LU factors; for example allowing more nonzero bands in the $\widetilde{L}$ and $\widetilde{U}$ matrices (that is larger stencils), see [4, 18, 23]. The most common location-based criterion is to allow a set number of levels of fill-in, where original entries have level zero, original zeros have level ∞ and a fill-in in position (i, j) has level determined by

$$\text{Level}_{ij} = \min_{1 \leq k \leq \min(i,j)} \{\text{Level}_{ik} + \text{Level}_{kj} + 1\}.$$

In the case of simple discretisations of partial differential equations, this gives a simple pattern for incomplete factorisations with different levels of fill-in. For example, if the matrix is from a 5-point discretisation of the Laplacian in two dimensions, level 1 fill-in will give the original pattern plus a diagonal inside the outermost band.

The other main criterion for deciding which entries to omit is to replace the *drop-by-position* strategy by a *drop-by-size* one. That is, a fill-in entry is discarded if its absolute value is below a certain threshold value. For regular problems with M-matrices, it is interesting that the level fill-in and drop strategies give a somewhat similar incomplete factorisation, because the numerical value of successive fill-in levels decreases markedly, reflecting the characteristic decay in the entries in the factors of the LU decomposition of A. For general problems, however, the two strategies can be significantly different. Since it is usually not known *a priori* how many entries will be above a selected threshold, the dropping strategy is normally combined with restricting the number of fill-ins allowed in each column. When using a threshold criterion, it is possible to change it dynamically during the factorisation to attempt to achieve a target density of the factors. Useful introductions to these techniques are given in [3, 24, 29].

A point of concern is that for non-M-matrices the incomplete factors of A may be very ill-conditioned. For instance, it has been demonstrated in [36] that, if A comes from a 5-point finite-difference discretisation of $\Delta u + \beta(u_x + u_y) = f$, then for β sufficiently large, the incomplete LU factors may be very ill-conditioned even though A has a very modest condition number. Remedies for reducing the condition numbers of $\widetilde{L}$ and $\widetilde{U}$ have been discussed in [13, 36].

3.4 Hybrid Techniques

In the classical incomplete decompositions one ignores fill-in right from the start of the decomposition process. However, it might be a good idea to delay this until the matrix becomes too dense. This leads to a hybrid combination of direct and iterative techniques. One such approach has been described in [7]; we will treat it in some detail here.

We first permute the given matrix of the linear system $Ax = b$ to a doubly bordered block diagonal form:

$$\widetilde{A} = P^{\mathsf{T}} AP = \begin{bmatrix} A_{00} & 0 & \cdots & 0 & A_{0m} \\ 0 & A_{11} & \ddots & \vdots & A_{1m} \\ \vdots & \ddots & \ddots & 0 & \vdots \\ 0 & \cdots & 0 & A_{m-1m-1} & \vdots \\ A_{m0} & A_{m1} & \cdots & \cdots & A_{mm} \end{bmatrix}. \tag{3.5}$$

We permute the right-hand side b as well to $\widetilde{b} = P^{\mathsf{T}} b$, which leads to the system

$$\widetilde{A}\widetilde{x} = \widetilde{b}, \tag{3.6}$$

with $x = Px$.

The parts of $\widetilde{b}$ and $\widetilde{x}$ that correspond to the block ordering, will be denoted by $\widetilde{b}_i$ and $\widetilde{x}_i$. The first step in the (parallelisable) algorithm will be to eliminate the unknown parts $\widetilde{x}_0, \cdots, \widetilde{x}_{m-1}$, which is done by the Algorithm in Fig. 6.

Parallel_for $i = 0, 1, ..., m-1$
  Decompose A_{ii}: $A_{ii} = L_{ii}U_{ii}$
  $L_{mi} = A_{mi}U_{ii}^{-1}$
  $U_{im} = L_{ii}^{-1}A_{im}$
  $y_i = L_{ii}^{-1}\widetilde{b}_i$
  $S_i = L_{mi}U_{im}$
  $z_i = L_{mi}\, y_i$
end
$S = A_{mm} - \sum_{i=0}^{m-1} S_i$
$y_m = \widetilde{b}_m - \sum_{i=0}^{m-1} z_i$
Solve $Sx_m = y_m$
Parallel_for $i = 0, 1, ..., m-1$
  $x_i = U_{ii}^{-1}(y_i - U_{im}x_m)$
end

Figure 6. *Parallel elimination*

Note that S in Fig. 6 denotes the Schur complement after the elimination of the blocks 0, 1, ..., $m-1$. In many relevant situations, direct solution of the reduced system $Sx_m = y_m$ requires the dominating part of the total computational costs, and this is where we bring in the iterative component of the algorithm.

The next step is to construct a preconditioner for the reduced system. This is based on discarding small elements in S. The elements larger than some threshold value define the preconditioner C:

$$c_{ij} = \begin{cases} s_{ij} & \text{if } |s_{ij}| > t|s_{ii}| \quad \text{or} \quad |s_{ij}| > t|s_{jj}| \\ 0 & \text{elsewhere} \end{cases} \tag{3.7}$$

with a parameter $0 \leq t < 1$. In the experiments, reported in [7] the value $t = 0.02$ turned out to be satisfactory, but this may need some experimentation for specific problems.

When we take C as the preconditioner, then we have to solve systems like $Cv = w$, and this requires decomposition of C. In order to prevent too much fill-in, it is suggested to reorder C with a minimum degree ordering. The system $Sx_m = y_m$ is then solved with, for instance, GMRES with preconditioner C. For the examples described in [7] it turns out that the convergence of GMRES was not very sensitive to the choice of t. The preconditioned iterative solution approach for the reduced system offers also opportunities for parallelism, although in [7] it is shown that even in serial mode the iterative solution (to sufficiently high precision) is often more efficient than direct solution of the reduced system.

In [7] heuristics are described for the decision on when the switch from direct to iterative should take place. These heuristics are based on mild assumptions on the speed of convergence of GMRES.

3.5 Monitoring the effect of preconditioning

There are only very few specialised cases where it is known *a priori* how to construct a good preconditioner and there are few proofs of convergence except in very idealised cases. For a general system, however, the following approach may help to build up one's insight into what is happening. For a representative linear system, one starts with unpreconditioned GMRES(m), with m as high as possible. In one cycle of GMRES(m), the method explicitly constructs an upper Hessenberg matrix of order m, denoted by $H_{m,m}$. This matrix is reduced to upper triangular form but, before this takes place, one should compute the eigenvalues of $H_{m,m}$, called the Ritz values. These Ritz values usually give a fairly good impression of the most relevant parts of the spectrum of A. Then one does the same with the preconditioned system and inspects the effect on the spectrum. If there is no specific trend of improvement in the behaviour of the Ritz values (Ideally, the eigenvalues of $K^{-1}A$ should cluster around 1.), when we try to improve the preconditioner, then obviously we have to look for another class of preconditioner. If there is a positive effect on the Ritz values, then this may give us some insight into how much more the preconditioner has to be improved in order to be effective. At all times, we have to check whether the construction of the preconditioner and its costs per iteration are still inexpensive enough to be balanced by an appropriate reduction in the number of iterations.

With preconditioning, either explicit or implicit (through a redefinition of the inner product) we are generating a Krylov subspace for the preconditioned operator.

This implies that the reduced matrix $H_{k,k}$ (cf. 2.1) gives information about the preconditioned matrix: in particular, the Ritz values approximate eigenvalues of the preconditioned matrix. The generated Krylov subspace cannot be used in order to obtain information as well for the *unpreconditioned* matrix.

Also for other methods, including CG, Bi-CG, MINRES, and Bi-CGSTAB, approximations for the eigenvalues of the (preconditioned) matrix can be computed from the (implicitly) generated matrix $T_{i,i}$. This matrix can easily be reconstructed from the iteration parameters. For more details on this, see [39].

Bibliography

[1] W. E. Arnoldi; The principle of minimized iteration in the solution of the matrix eigenproblem. *Quart. Appl. Math.*, 9, pp. 17–29, 1951.

[2] O. Axelsson; Solution of linear systems of equations: iterative methods. In V. A. Barker, editor, *Sparse Matrix Techniques*, pages 1–51, Berlin, 1977. Copenhagen 1976, Springer-Verlag.

[3] O. Axelsson; *Iterative Solution Methods.* Cambridge University Press, Cambridge, 1994.

[4] O. Axelsson and V. A. Barker; *Finite Element Solution of Boundary Value Problems. Theory and Computation.* Academic Press, New York, NY, 1984.

[5] R. Barrett, M. Berry, T. Chan, J. Demmel, J. Donato, J. Dongarra, V. Eijkhout, R. Pozo, C. Romine, and H. van der Vorst; *Templates for the Solution of Linear Systems: Building Blocks for Iterative Methods.* SIAM, Philadelphia, PA, 1994.

[6] M. Benzi; Preconditioning techniques for large linear systems: A survey. *J. Comput. Phys.*, 182, pp. 418–477, 2002.

[7] C. W. Bomhof and H. A. van der Vorst; A parallel linear system solver for circuit-simulation problems. *Num. Lin. Alg. Appl.*, 7, pp. 649–665, 2000.

[8] P. Concus, G. H. Golub, and D. P. O'Leary; A generalized conjugate gradient method for the numerical solution of elliptic partial differential equations. In J. R. Bunch and D. J. Rose, editors, *Sparse Matrix Computations.* Academic Press, New York, 1976.

[9] J. J. Dongarra, I. S. Duff, D. C. Sorensen, and H. A. van der Vorst; *Numerical Linear Algebra for High-Performance Computers.* SIAM, Philadelphia, PA, 1998.

[10] I. S. Duff, A. M. Erisman, and J. K. Reid; *Direct methods for sparse matrices.* Oxford University Press, London, 1986.

[11] I. S. Duff and H. A. van der Vorst; Developments and trends in the parallel solution of linear systems. *Parallel Computing*, 25, pp. 1931–1970, 1999.

[12] T. Dupont, R. P. Kendall, and H. H. Rachford Jr; An approximate factorization procedure for solving self-adjoint elliptic difference equations. *SIAM J. Numer. Anal.*, 5(3), pp. 559–573, 1968.

[13] H. C. Elman; Relaxed and stabilized incomplete factorizations for non-self-adjoint linear systems. *BIT*, 29, pp. 890–915, 1989.

[14] V. Faber and T. A. Manteuffel; Necessary and sufficient conditions for the existence of a conjugate gradient method. *SIAM J. Numer. Anal.*, 21(2), pp. 352–362, 1984.

[15] R. Fletcher; *Conjugate gradient methods for indefinite systems*, volume 506 of *Lecture Notes Math.*, pages 73–89. Springer-Verlag, Berlin–Heidelberg–New York, 1976.

[16] G. H. Golub and C. F. Van Loan; *Matrix Computations.* The Johns Hopkins University Press, Baltimore, 1996.

[17] A. Greenbaum; *Iterative Methods for Solving Linear Systems.* SIAM, Philadelphia, 1997.

[18] I. Gustafsson; A class of first order factorization methods. *BIT*, 18, pp. 142–156, 1978.

[19] W. Hackbusch; *Iterative Solution of Large Sparse Systems of Equations.* Springer-Verlag, Berlin, 1994.

[20] C. Lanczos; An iteration method for the solution of the eigenvalue problem of linear differential and integral operators. *J. Res. Natl. Bur. Stand*, 45, pp. 225–280, 1950.

[21] C. Lanczos; Solution of systems of linear equations by minimized iterations. *J. Res. Natl. Bur. Stand*, 49, pp. 33–53, 1952.

[22] J. A. Meijerink and H. A. van der Vorst; An iterative solution method for linear systems of which the coefficient matrix is a symmetric M-matrix. *Math. Comp.*, 31, pp. 148–162, 1977.

[23] J. A. Meijerink and H. A. van der Vorst; Guidelines for the usage of incomplete decompositions in solving sets of linear equations as they occur in practical problems. *J.of Comp. Physics*, 44, pp. 134–155, 1981.

[24] G. Meurant; *Computer Solution of Large Linear Systems.* North-Holland, Amsterdam, 1999.

[25] C. C. Paige, B. N. Parlett, and H. A. van der Vorst; Approximate solutions and eigenvalue bounds from Krylov subspaces. *Num. Lin. Alg. Appl.*, 2(2), pp. 115–134, 1995.

[26] C. C. Paige and M. A. Saunders; Solution of sparse indefinite systems of linear equations. *SIAM J. Numer. Anal.*, 12, pp. 617–629, 1975.

[27] B. N. Parlett; *The Symmetric Eigenvalue Problem.* Prentice-Hall, Englewood Cliffs, N.J., 1980.

[28] A. Quarteroni and A. Valli; *Numerical Approximation of Partial Differential Equations.* Springer Verlag, Berlin, 1994.

[29] Y. Saad; *Iterative Methods for Sparse Linear Systems, Second Edition.* SIAM, Philadelphia, 2003.

[30] Y. Saad and M. H. Schultz; GMRES: a generalized minimal residual algorithm for solving nonsymmetric linear systems. *SIAM J. Sci. Statist. Comput.*, 7, pp. 856–869, 1986.

[31] Y. Saad and H. A. van der Vorst; Iterative solution of linear systems in the 20-th century. *J. Comp. and Appl. Math.*, 123 (1-2), pp. 1–33, 2000.

[32] H. D. Simon; Direct sparse matrix methods. In James C. Almond and David M. Young, editors, *Modern Numerical Algorithms for Supercomputers*, pages 325–444, Austin, 1989. The University of Texas at Austin, Center for High Performance Computing.

[33] P. Sonneveld; CGS: a fast Lanczos-type solver for nonsymmetric linear systems. *SIAM J. Sci. Statist. Comput.*, 10, pp. 36–52, 1989.

[34] G.W. Stewart; *Matrix Algorithms, Vol. I: Basic Decompositions.* SIAM, Philadelphia, 1998.

[35] A. van der Sluis and H. A. van der Vorst; The rate of convergence of conjugate gradients. *Numer. Math.*, 48, pp. 543–560, 1986.

[36] H. A. van der Vorst; Iterative solution methods for certain sparse linear systems with a non-symmetric matrix arising from PDE-problems. *J. Comp. Phys.*, 44, pp. 1–19, 1981.

[37] H. A. van der Vorst; Bi-CGSTAB: A fast and smoothly converging variant of Bi-CG for the solution of non-symmetric linear systems. *SIAM J. Sci. Statist. Comput.*, 13, pp. 631–644, 1992.

[38] H. A. van der Vorst; Computational methods for large eigenvalue problems. In P. G. Ciarlet and J. L. Lions, editors, *Handbook of Numerical Analysis*, volume VIII, pages 3–179. North-Holland, Amsterdam, 2002.

[39] H. A. van der Vorst; *Iterative Krylov Methods for Large Linear Systems.* Cambridge University Press, Cambridge, UK, 2003.

Invited speaker Ernie Tuck (Australia). photography by *Happy Medium Photo Co.*

Ying Lung-An is professor of the School of Mathematical Sciences of the Beijing University. In his 41-year research career, his interests have been in nonlinear partial differential equations and numerical methods for partial differential equations. He has published four books and 84 papers in this area, including work on finite element methods, hybrid finite element methods, vortex methods, inhomogeneous systems of hyperbolic conservation laws, combustion theory, numerical schemes for conservation laws, and interface problems.

Professor Ying was born in Beijing and graduated at the Beijing University in 1960. He was within the first group of 52 visiting scholars from the People's Republic of China to the United States of America in 1978. He has been awarded 14 prizes, including the Award of the Chinese National Conference of Science, the National Prize on Natural Science, the National Prize for the Accomplishments of Education, the State Education Commission's Prize for the Advancement of Science and Technology, and the Pei-yuan Chou Prize. He was the chairman of the Department of Mathematics, Beijing University, 1991–1995; president of the Beijing Computational Mathematics Society, 1990-1993; vice president of the Chinese Computational Mathematical Society, 1991–1998; and since 2000 the chief editor of *Advances in Mathematics* (China).

Chapter 19

Interface Problems and their Applications

Lung-an Ying[*]

Abstract: Many scientific problems involve the solutions to interface problems of partial differential equations; for example, interfaces often play a crucial role in material science, fracture mechanics, and fluid mechanics. At interfaces the solutions are generally singular, and the physical phenomena often extreme.

We present mathematical results on the asymptotic properties of the solutions in the neighborhood of singular points. One important particular case, that of re-entrant corners, has been studied extensively. Recently we have studied more general problems for elliptic partial differential equations. In this paper we present results for interface problems of both linear and nonlinear elliptic equations and systems.

We also present different approaches which have been introduced in numerical computations to deal with singularities. Some of the effective approaches are: locally refined grids, infinite elements, post-processing, adaptive approaches, extrapolation, and defect correction. We state the results of *a priori* and *a posteriori* error estimates, show some examples of applications, and compare methods.

Contents

[*]Key Laboratory of Pure and Applied Mathematics, School of Mathematical Sciences, Peking University, People's Republic of China.

1 Introduction

It is known that the solutions are regular for a large class of elliptic or parabolic partial differential equations, provided that the known functions in the equations are regular, and the boundary data are regular. On the contrary if we consider some equations with discontinuous coefficients, the solutions are no longer regular, and the same also true if the boundary of the physical domain is not smooth.

One example is the crack in materials. At the tip of a crack the stress is singular; this is called 'stress concentration'. Stress intensity factors are defined in fracture mechanics, describing the phenomena and measuring the strength of stress concentration. Interface problems are another example; the common boundaries of different kinds of materials neighboring each other are called 'interfaces'. Coefficients of the governing equations are discontinuous on interfaces. As a result, the solutions of these equations are singular in the neighborhoods of the interfaces.

From the general theory of elliptic partial differential equations, if the coefficients are bounded but not necessary continuous, the solutions satisfy the Hölder condition:

$$\big|u(x) - u(y)\big| \leq C|x - y|^{\gamma}, \tag{1.1}$$

where x and y are two points, C is a constant, and γ is also a constant, satisfying $0 < \gamma < 1$. By (1.1) one can see that the first-order derivatives of the solutions are not bounded in general. In the application of partial differential equations in science and technology the first-order derivatives frequently represent some important quantities; for instance velocity, strain, and stress. Equation (1.1) does not provide enough information to understand the effect of singularities in these quantities. However (1.1) is valid for very general equations with discontinuous coefficients, and usually the coefficients are piecewise smooth for the equations in the real world. For these kinds of problem it is possible to know more about the solutions; namely, the behavior of the derivatives of the solutions near singular points. We want to know the asymptotic expression of the derivatives, $\sigma \sim K|x - x_0|^{\gamma-1}$, where x_0 is a singular point. The coefficients K and the negative exponents $\gamma - 1$ are extremely important. The exponents may be real or complex, reflecting important phenomena.

One branch of fracture mechanics, the interface fracture mechanics, has been developed extensively. Williams [34] discovered the phenomenon of 'oscillatory singularity'; that is, if the materials are different on either side of a crack, then in general the exponent is a complex number. This problem has since been well-studied, by Suo–Huchinson [31] and Rice [29, 30].

On the other hand, singularities can be present not only at cracks, but also in any composite materials. The problem of interface singularities is more general. The case of two materials has been investigated by many authors. Bogy [7, 8, 9] developed a theory using the Dunders constants. Lu and Erdogan [27] employed the singular integral equation as a tool to study this problem. Experiments in laboratories have been carried out by Pitkethy, Favse, and Gaur [28].

From a mathematical point of view we are interested in the structure of solutions near singular points, and in developing efficient methods to solve for the coefficients K and exponents γ numerically. These problems have been extensively studied. In this survey we will present some results on these subjects, and some applications which we have made.

2 Mathematical theory of singularities

To begin with, we consider the problems of reentrant corners, which include the crack problem for a homogeneous material.

For simplicity we take the Laplace equation

$$\Delta u = 0\,,$$

as an example. Let o be a reentrant corner at the boundary with the angle $\alpha > \pi$, and assume that the solution u satisfies the Neumann boundary condition $\dfrac{\partial u}{\partial \nu} = 0$ near o, where ν is the exterior unit normal vector along the boundary. Then locally the solution can be expressed in terms of

$$u = \sum_{n=0}^{\infty} c_n r^{\frac{n\pi}{\alpha}} \cos \frac{n\pi}{\alpha}\vartheta\,,$$

where (r, ϑ) are the polar coordinates. As $r \to 0$, the gradient $\nabla u \sim r^{\frac{\pi}{\alpha}-1}$ goes to infinity, and c_1 is the stress intensity factor.

For interface problems the problems can be analyzed by means of the Mellin transform and the method of separation of variables. Kellogg [17] proved the following theorem:

Theorem 2.1. *For the equation*

$$\sum_{i,j=1}^{2} \frac{\partial}{\partial x_j}\Big(a_{ij}(x)p(x)\frac{\partial u}{\partial x_i}\Big) + \sum_{i=1}^{2} b_i(x)\frac{\partial u}{\partial x_i} + c(x)u = f(x)\,,$$

where $a_{ij}(x)$ are smooth and symmetric and $p(x)$ is a piecewise constant function, the solution u can be expressed as $u = v + w$ near a singular point, and v, w satisfy

$$v = \sum_{n=1}^{N} c_n r^{\alpha_n}\varphi_n(\vartheta)\,, \qquad w \in H^2,$$

and

$$\sum_{n=1}^{N} |c_n| + \|w\|_2 \le C(\|u\|_1 + \|f\|_0)\,,$$

where C is a positive constant independent of the solution u, φ_n are continuous and piecewise-smooth functions, also independent of u, and H^s, $\|\cdot\|_s$ are the standard notations for the Sobolev spaces and norms in the Sobolev spaces.

Blumenfield [6] proved the same result for another equation

$$\sum_{i,j=1}^{2} \frac{\partial}{\partial x_j}\Big(p(x)\frac{\partial u}{\partial x_i}\Big) + \sum_{i=1}^{2} b_i(x)\frac{\partial u}{\partial x_i} + c(x)\,u = f(x)\,,$$

where $p(x)$ is a discontinuous and piecewise-smooth function.

The infinite element method is used to solve the singular problem numerically, which will be explored in the next section. It also provides an approach for analytic study. Ying [40] proved the following theorem:

Theorem 2.2. *For the equation*

$$\sum_{i,j=1}^{2} \frac{\partial}{\partial x_j}\left(a_{ij}(x)\frac{\partial u}{\partial x_i}\right) + \sum_{i=1}^{2} b_i(x)\frac{\partial u}{\partial x_i} + c(x)\,u = f(x)\,,$$

where $a_{ij}(x)$ are discontinuous and piecewise-smooth functions, the solution u can be expressed as $u = v + w$ near a singular point, and v, w satisfy

$$v = \sum_{n=1}^{N} c_n\, r^{\alpha_n} \log^{m_n} r\, \varphi_n(\vartheta)\,,$$

and

$$\sum_{n=1}^{N} |c_n| + \left\| \frac{D^2 w}{(|\log r| + 1)^M} \right\|_0 \le C(\|u\|_1 + \|f\|_0)\,,$$

where C is a positive constant independent of the solution u, φ_n are continuous and piecewise-smooth functions, also independent of u.

Using the same approach Wu [35] studied systems of elliptic equations,

$$\sum_{k=1}^{m} \sum_{i,j=1}^{2} \frac{\partial}{\partial x_j}\left(a_{ijkl}(x)\frac{\partial u_k}{\partial x_i}\right) = f_l(x)\,, \quad l = 1, \ldots, m\,,$$

which satisfy the elliptic condition:

$$\sum_{i,j,k,l} \int_\Omega a_{ijkl}(x)\frac{\partial v_k}{\partial x_i}\frac{\partial v_l}{\partial x_j}\, \mathrm{d}x \ge \alpha \sum_k \int_\Omega |Dv_k|^2\, \mathrm{d}x\,, \quad \forall v_k \in H_0^1(\Omega)\,.$$

a_{ijkl} are discontinuous and piecewise smooth. The result was the same as Theorem 2.2. The elliptic condition is essential, since it is known that for general systems the Hölder continuity may fail to be true.

For nonlinear equations Wu [36] and Ying [42] have proved the parallel results. Let u be a solution to the equation

$$\sum_{i,j=1}^{2} \frac{\partial}{\partial x_j}\left(a_{ij}(x,u)\frac{\partial u}{\partial x_i}\right) = \sum_{i=1}^{2} \frac{\partial f_i(x)}{\partial x_i}\,,$$

where a_{ij} are discontinuous and piecewise smooth, then $u = v + w$ near a singular point, and v, w satisfy

$$v = \sum_{n=1}^{N} c_n r^{\alpha_n} \log^{m_n} r\, \varphi_n(\vartheta)\,,$$

and

$$\sum_{n=1}^{N} |c_n| + \left\| \frac{D^2 w}{(|\log r| + 1)^M} \right\|_0 \le C ,$$

where C depends on a_{ij}, f_i, and $\|u\|_1$ only.

For this equation Ying [43] obtained the Hölder norm estimate for first-order derivatives:

$$\|w\|_{C^{1,\alpha}} \le C .$$

The following is a brief outline of the proof for these theorems.

A. We consider a homogeneous equation with piecewise-constant coefficients:

$$\sum_{i,j=1}^{2} \frac{\partial}{\partial x_j}\left(a_{ij}\frac{\partial u}{\partial x_i}\right) = 0 .$$

Let o be the singular point. We define $y_0 = u|_{r=1} \in H^{\frac{1}{2}}$, $y_1 = u|_{r=\xi} \in H^{\frac{1}{2}}$, where $0 < \xi < 1$. Then $y_1 = Xy_0$, and X is a compact operator. y_0 can be decomposed in terms of the spectrum of X. It is easy to see that X^k is a semigroup of operators in $H^{\frac{1}{2}}$ and $u|_{r=\xi^k} = X^k y_0$. Therefore $u|_{r=\xi^k}$ are decomposed for all k. Then it can be proved that $u = v + w$, with $w \in H^2$ and

$$v = \sum_{n=1}^{N} c_n r^{\alpha_n} \log^{m_n} r\, \varphi_n(\vartheta) .$$

B. The inhomogeneous equation

$$\sum_{i,j=1}^{2} \frac{\partial}{\partial x_j}\left(a_{ij}\frac{\partial u}{\partial x_i}\right) = f(x)$$

is studied. Let $\Omega_k = \{(r,\vartheta) \mid \xi^{k+1} < r < \xi^k\}$. The argument proceeds via the following steps:

Step 1. Solving the equation in R^2 if $\operatorname{supp} f \subset \overline{\Omega_1}$.

Step 2. Decomposition: $f = \sum_k f_k$, where $f_k = f$ within Ω_k and is 0 elsewhere.

Step 3. Scaling: $y = \xi^{-k} x$.

Step 4. Use the result of the first step to estimate a particular solution $u = \sum_k u_k$, where u_k is the solution corresponding to f_k and u is regular.

C. A particular inhomogeneous equation is then studied. Let $\psi(x) = r^\alpha \log^m r\varphi(\vartheta)$, then we solve

$$\sum_{i,j=1}^{2} \frac{\partial}{\partial x_j}\left(a_{ij}\frac{\partial u}{\partial x_i}\right) = \sum_{i,j=1}^{2} \frac{\partial}{\partial x_j}\left(\psi_1(x)\frac{\partial \psi_2(x)}{\partial x_i}\right),$$

to get a particular solution.

D. Finally we study the general equation

$$\sum_{i,j=1}^{2} \frac{\partial}{\partial x_j}\Big(a_{ij}(x,u)\frac{\partial u}{\partial x_i}\Big) = f(x)\,,$$

which can be rewritten as

$$\sum_{i,j=1}^{2} \frac{\partial}{\partial x_j}\Big(a_{ij}(0,0)\frac{\partial u}{\partial x_i}\Big) = -\sum_{i,j=1}^{2} \frac{\partial}{\partial x_j}\Big(a_{ij}(x,u) - a_{ij}(0,0)\frac{\partial u}{\partial x_i}\Big) + f(x)\,.$$

The assertion is proved by induction. We take the solution of the homogeneous equation as the first approximation, then put it to the right-hand side; next use B and C to get a particular solution to this inhomogeneous equation, then repeat the procedure. After careful analysis, this sequence can be proved to be convergent.

Some problems are still open. One can get some expressions for three-dimensional equations with constant coefficients. However, to the author's knowledge, the structure of solutions for general linear and nonlinear equations are still unknown.

3 Numerical methods

The finite-element method is widely applied in structure computation. However without special approaches it converges slowly for interface problems. For second-order elliptic equations the solutions belong to $H^{1+\varepsilon}$, by the general theory, where ε is positive but can be arbitrarily small. The convergence rate is:

$$\|u - u_h\|_1 \le Ch^{\varepsilon}|u|_{1+\varepsilon}\,,$$

where u is the exact solution, and u_h is the finite-element approximation, $\|\cdot\|_s$ and $|\cdot|_s$ are the Sobolev norms and semi-norms in H^s, and h is the maximum diameter of the elements. On the other hand, to evaluate also the coefficients K and exponent γ needs a special approach.

One natural approach is that of local refinement. Babuska [2] introduced the 'adaptive algorithm' and proved that

$$\|u - u_h\|_1 \le C(h + h_j^{\varepsilon})\,,$$

for first-order elements, where h is the maximum diameter of elements, and h_j is the minimum diameter of elements. From this estimate we can see that the order of one for h is optimal, and the small exponent ε is applied on h_j, which by local refinement is small; therefore this technique improves the precision of finite-element approximate solutions.

Lin, C. Chen, and Huang introduced a kind of graded mesh. The diameter h_T of an element T is designed to satisfy the following condition, close to singular points:

$$Ch \max_T \phi(x) \le h_T \le h \min_T \phi(x)\,,$$

where $\phi(x) = \prod_{i=1}^{l} |x - Q_i|^{\alpha_i}$, Q_i are the singular points, $0 \le \alpha_i < 1$, and h is the maximum diameter. For first-order elements, optimal error estimates were obtained

$$\|u - u_h\|_1 \le Ch\,.$$

Adaptive algorithms are usually carried out by means of an *a posteriori* theory. A number of authors have studied this problem; for example, Babuska–Rheinboldt [4], Dörfler [14], Lin et al. [26], Z. Chen [11], Zienkiewicz and Zhu [44]. The following technique was introduced by Chen:

Consider the equation:

$$-\nabla\big(a(x,y)\nabla\big)u = f(x,y)\,,$$

where $a(x,y)$ is a piecewise constant function. For triangular elements and linear interpolation let T be elements and $S \subset \partial T$ be a side of one element. Denote by h_T the diameter of T and h_s the length of S. Let ω_T be a macro element: $T \subset \omega_T$. Define

$$\Lambda_T = \frac{a_T}{\min\{a_K \mid K \subset \omega_T\}}\,.$$

Then the *a posteriori* estimator η_h is defined as

$$\eta_h^2 = \sum_T \Big\{\Lambda_T \|a_T^{-\frac{1}{2}} h_T f\|_{0,T}^2 + \sum_{S\subset\partial\Omega} \Lambda_s \, \|a_T^{-\frac{1}{2}}\, h_S^{\frac{1}{2}} J_S\|_{0,S}^2\Big\}\,,$$

where $J_S = \big[a\nabla u_h\big]\cdot\nu$, $[\,\cdot\,]$ is the jump of a function along boundary and ν is the unit normal vector. It was proved that

$$\|u - u_h\| \le C\eta_h\,,$$

where $\|\cdot\|$ is the energy norm. A local *a posteriori* estimator is defined as:

$$\eta_S^2 = \sum_{i=1,2} \Lambda_{T_i} \Big\{\|h a_{T_i}^{-\frac{1}{2}} f\|_{0,T_i}^2 + \tfrac{1}{2}\|a_T^{-\frac{1}{2}} h_S^{\frac{1}{2}} J_S\|_{0,S}^2\Big\}\,,$$

where T_1, T_2 are two elements neighboring a given side S. Then it can be proved that

$$\eta_s^2 \le C\,\|u - u_h\|_{\Omega_S}^2 + C \sum_{i=1,2} \Lambda_{T_i}\, \|a_T^{-\frac{1}{2}} h(f - f_{T_i})\|_{1,T_i}^2\,,$$

where $\Omega_S = T_1 \cup T_2$, and $f_{T_i} = |T_i|^{-1}\int_{T_i} f\,\mathrm{d}x$. Using this *a posteriori* estimator one can design an adoptive algorithm.

The patch recovery technique was introduced by Zienkiewicz and Zhu, which can be applied to the interface problems. Let

$$G\,u_h(z) = \sum_j a_j \big(\nabla u_h\big)_{T_j}\,, \qquad \sum_j a_j = 1\,,$$

where z is a node, T_j are elements neighboring z. Let φ_z be basic functions, then an *a posteriori* estimator can be defined as

$$\eta_T = \|G\,u_h - \nabla u_h\|_{0,T}\,, \qquad G\,u_h = \sum G\,u_h(z)\,\varphi_z\,.$$

Then the following *a posteriori* estimate can be proved:

$$C\|u-u_h\|_{1,\Omega}-\varepsilon\le\eta\le C\|u-u_h\|_{1,\Omega}+C\varepsilon\,.$$

Also it implies an adoptive algorithm.

To improve the precision of the results, super-convergence theory and technique can also be applied. Lin and Yan [25] obtained the following result: For "good" local refinement and bilinear elements

$$\|u_h-u_I\|_1\le C\big(h^{\frac{3}{2}}+h_j^{\varepsilon}\big)\,,$$

where u_I is the interpolation of the true solution. Super-convergence results can be achieved after a post-processing algorithm, namely, defect correction or extrapolation.

C. Chen [10] considered a kind of strong regular mesh, which is defined such that if all of two neighboring triangular elements constitute an "approximate" parallelogram. Let $\overline{\nabla u_h}$ be the average of ∇u_h, and $I_h\overline{\nabla u_h}$ be the linear interpolation of $\overline{\nabla u_h}$, then it was proved that

$$\big|\nabla u-I_h\overline{\nabla u_h}\big|<Ch^2|\log h|^{\frac{1}{2}}\,.$$

Again, defect correction can be applied to improve the precision.

The infinite-element method was introduced by Thatcher [32] and Ying [37] to solve singular solutions and for the exterior problems. Han [16] applied it to interface problems. The algorithm is the following:

Let o be a singular point. Around o let Ω_0 be a polygonal domain with is star shape with respect to o. Let Γ_0 be the boundary of Ω_0. Take a constant $0<\xi<1$, then use $\xi,\xi^2,\dots,\xi^k,\dots$ as the constants of scaling to generate a sequence of geometrically shrinking images Γ_k of Γ_0. The domain Ω_0 is subdivided into annular regions by the straight lines starting from o. Then the domain Ω_0 is further subdivided such that the meshes between Γ_k and Γ_{k+1} are similar to each other geometrically. The associated stiffness matrix is then an $\infty\times\infty$ block tri-diagonal Toeplitz matrix

$$\begin{pmatrix} K_0 & -A^T & & \\ -A & K & -A^T & \\ & -A & K & \ddots \\ & & \ddots & \ddots \end{pmatrix}.$$

This leads to a system of infinitely-many equations for infinitely-many unknowns; but these can be reduced by analytical techniques to a finite system. For homogeneous equations, one inhomogeneous term in the zero-th component appears in this algebraic system. Let y_k be a vector consisting of the unknowns on Γ_k, then there is a transfer matrix X such that $y_{k+1}=Xy_k$. Substituting it and $y_k=Xy_{k-1}$ into the equation

$$-Ay_{k-1}+Ky_k-A^Ty_{k+1}=0$$

yields an equation for X,

$$A^T X^2 - KX + A = 0\,.$$

Let the eigen-pair for X be (λ, g), then it can be proved that $|\lambda| \leq 1$. From

$$(\lambda^2 A^T - \lambda K + A)g = 0\,,$$

one has

$$\begin{pmatrix} K & -A \\ I & 0 \end{pmatrix} \begin{pmatrix} \lambda g \\ g \end{pmatrix} = \lambda \begin{pmatrix} A^T & 0 \\ 0 & I \end{pmatrix} \begin{pmatrix} \lambda g \\ g \end{pmatrix}.$$

This eigenvalue problem can be solved. The set of eigenvalues of X is a subset of that for the above eigenvalue problem. Using the property of $|\lambda| \leq 1$, one gets the full set of eigenvalues of X: $\lambda_1, \lambda_2, \ldots, \lambda_n$, and the eigenvectors, $g_1, g_2, \ldots, g_n$ as well. The expression for X is:

$$X = T\Lambda T^{-1}\,, \quad T = (g_1, g_2, \ldots, g_n)\,, \quad \Lambda = \mathrm{diag}(\lambda_1, \lambda_2, \ldots, \lambda_n)\,.$$

The approximate solution can be decomposed as $u_h|_{\Gamma_k} = \sum_i c_i \lambda_i^k g_i$, which leads to $u_h = \sum_i d_{hi} r^{\alpha_{hi}} \varphi_{hi}(\vartheta)$, where $\alpha_{hi} = (\log \lambda_i)/(\log \xi)$, and d_{h1} is the stress-intensity factor for normalized φ_{hi}.

Optimal error estimates can be proved to be $\|u - u_h\|_1 \leq C \inf_{v_h \in V_h} \|u - v_h\|_1$, where V_h is the infinite-element space. The convergence of singularities is proved for linear elements:

$$|\alpha_i - \alpha_{hi}| < Ch\,, \qquad |d_i - d_{hi}| < ch\,.$$

Superconvergence is valid for the infinite-element solutions:

$$\|u_h - u_I\|_1 \leq Ch^2\, |\log h|\,,$$

then post-processing can be used to improve the precision.

4 Application

We show one application in dental science here. The problem of bond strengths in metal–ceramic systems has been of concern since the method of porcelain-fused-to-metal was used for prosthetics. The exact definition of bond-strength needs to be described. The stress distribution in metal–ceramic specimens for the ISO crack initiation test (three-point flexure bond test) is calculated. Stress concentration at the labial cervical margin of the porcelain-fused-to-metal restoration model can be calculated. One discovers that the shear stress, σ_{xy}, concentration and normal stress, σ_{yy}, concentration along the interfaces occur at the edge of the metal–opaque-ceramic interface, and near the edge of the opaque-body ceramic interface. The former is the most important factor in the failure of bonding, but the latter could also cause fracture within the ceramic layer. Some important conclusions can be drawn. With increase in the thickness of the opaque layer, the stress along the metal–opaque-ceramic interface increases. Therefore the layer of opaque ceramic should be sufficient for an aesthetic feeling of restoration, yet be kept as thin as possible to improve the bond strengths of the metal–ceramic system. Based on stress analysis, the model with unit of porcelain to metal at the labiocervical margin is better than models with porcelain or metal only at the labiocervical margin, for upper central incisor restoration. Stress magnitudes along the metal–ceramic interface enhance with the angle of loading force with axle of central incisor growing. The normal occlusal force at the angle of 45° loading cannot lead the fracture of porcelain from metal. However with increasing angle, for example with deep overbite, the same occlusal force may cause the failure of porcelain-fused-to-metal restoration.

Bibliography

[1] A. K. Aziz, R. B. Kellogg; *On homeomorphisms for an elliptic equation in domains with corners*, Diff. Int. eqs., 8, 2, pp. 333–352, 1995.

[2] I. Babuska, A. Miller; *The post-processing in the finite element method, Parts I & II*, Inter. J. Numer. Math. Engng., 20, pp. 1085–1109, pp. 1110–1129. 1984.

[3] I. Babuska, A. Miller; *A feedback finite element method with a posteriori error estimation: Part I. The finite element method and some basic properties of a posteriori error estimator*, Comput. Meth. Appl. Mech. Engng., 61, pp. 1–40, 1987.

[4] I. Babuska, W. Rheinboldt; *Error estimates for adaptive finite element computations*, SIAM J. Numer. Anal., 15, 4, pp. 736–754, 1978.

[5] I. Babuska, T. von Petersdorff, B. Andersson; *Numerical treatment of vertex singularities and intensity factors for mixed boundary value problems for the Laplace equation in R^3*, SIAM J. Numer. Anal., 31, 5, pp. 1265–1288, 1994.

[6] M. Blumenfeld; *The regularity of interface problems on corner regions*, Lecture Notes in Mathematics 1121, Springer-Verlag, pp. 38–54, 1985.

[7] D. B. Bogy; *Edge-Bonded Dissimilar Orthogonal Elastic Wedges Under Normal and Shear Loading*, J. Appl. Mech., 35, pp. 460–466, 1968.

[8] D. B. Bogy; *Two Edge-Bonded Elastic Wedges on Different Materials and Wedge Angles Under Surface Tractions*, J. Appl. Mech., 38, pp. 377–386, 1971.

[9] D. B. Bogy; *On the Plane Elastostatic Problem of a Loaded Crack Terminating at a Material Interface*, J. Appl. Mech., 38, pp. 911–918, 1971.

[10] C. Chen; *Superconvergence in finite element in domain with reentrant corner*, J. Xiangtan Univ., 12, pp. 134–141, 1990.

[11] Z. Chen; *Multiplicative Adaptive Algorithms for User Preference Retrieval*, Lecture Notes in Computer Science, Vol. 2108, pp. 540–549, Springer-Verlag, 2001.

[12] J. P. Dempsey, G. B. Sinclair; *On the stress singularities in the plane elasticity of the composite wedge*, J. Elasticity, 9, 4, pp. 373–391, 1979.

[13] J. P. Dempsey, G. B. Sinclair; *On the singular behavior at the vertex of a bi-material wedge*, J. Elasticity, 11, 3, pp. 317–327, 1981.

[14] W. Dörfler; *A convergent adaptive algorithm for Poisson's equation*, SIAM J. Numer. Anal., 33, pp. 1106–1124, 1996.

[15] P. Grisvard; *Elliptic Problems in Nonsmooth Domains*, Monographs and Studies in Mathematics 21, Pitman, Boston 1985.

[16] H. D. Han; *The numerical solutions of interface problems by infinite element method*, Numer. Math., 39, pp. 39–50, 1982.

[17] R. B. Kellogg, *Singularities in interface problems*, SYNSPADE II, ed. B. Hubbard, pp. 351–400, 1971.

[18] R. B. Kellogg; *High order regularity for interface problems*, in The Mathematical Foundation of the Finite Element Method, ed. A. K. Aziz, Academic Press, 1972.

[19] V. A. Kondrat'ev; *Boundary value problems for elliptic equations in domains with conical or angular points*, Tr. Mosk. Mat. Obs., 16, pp. 209–292, 1967; English Transl. in Trans. Moscow Math. Soc., 16, pp. 227–313, 1967.

[20] V. A. Kondrat'ev, O. A. Oleinik; *Boundary value problems for partial differential elliptic equations in nonsmooth domains*, Uspekhi Mat. Nauk., 38, 2, pp. 3–76, 1983; English Transl. in Russian Mathematical Surveys, 38, 2, pp. 1–86, 1983.

[21] V. A. Kozlov, V. G. Maz'ya, J. Rossmann; *Elliptic Boundary Value Problems in Domains with Point Singularities*, Mathematical Surveys and Monographs 52, AMS, 1997.

[22] V. A. Kozlov, V. G. Maz'ya, J. Rossmann; *Spectral Problems Associated with Corner Singularities of Solutions to Elliptic Equations*, Mathematical Surveys and Monographs 85, AMS, 2001.

[23] Q. Lin; *Superconvergence of FEM for singular solutions*, J. Comput. Math., 9, pp. 111–114, 1991.

[24] Q. Lin; *Fourth order eigenvalue approximation by extrapolation on domains with reentrant corners*, Numer. Math., 58, pp. 631–640, 1991.

[25] Q. Lin, N. Yan; *A rectangle test for singular solutions with irregular meshes*, Proc. Sys. Sci. Engng., Great Wall Culture Publ. Co., Hong Kong, pp. 230–237, 1991.

[26] X. Lin, P.K. McKinley, and A-H. Esfahanian. *Adaptive multicast wormhole routing in 2D mesh multicomputers.* In 1993 Parallel Architectures and Languages Europe Conference (PARLE '93), pp. 228–241, Munich, Germany, June 1993.

[27] M.C. Lu, F. Erdogan; *Stress intensity factors in two bonded elastic layers containing cracks perpendicular to and on the interface – II. Solution and results*, Eng. Fract. Mech., 18, 3, pp. 507–528, 1983.

[28] M.J. Pitkethly, J.P. Favre, U. Gaur; Sci. Tech., 11, pp. 745–766, 1993.

[29] J.R. Rice; *Elastic Fracture Mechanics Concepts for Interfacial Cracks*, J. Appl. Mech., 55, pp. 98–103, 1988.

[30] J.R. Rice, Z. Suo, J.-S. Wang; *Mechanics and Thermodynamics of Brittle Interfacial Failure in Bimaterial Systems*, in Metal–Ceramic Interfaces (eds. M. Rühle, A. G. Evans, M. F. Ashby and J. P. Hirth), Acta Scripta Metallurgica, Proceedings Series 4, pp. 269–294, Pergamon, 1990.

[31] Z. Suo, J.W. Hutchinson; *Interface crack between two elastic layers*, Int. J. Fracture, 43, 1–18 (1990); Harvard University Report, Mech–118, (1988).

[32] R.W. Thatcher; *Singularities in the solution of Laplace's equation in two dimensions*, J. Inst. Math. Appl., 16, pp. 303–319, 1975.

[33] R.W. Thatcher; *The use of infinite grid refinements at singularities in the solution of Laplace's equation*, Numer. Math., 15, pp. 163–178, 1976.

[34] M.L. Williams; *The Stresses Around a Fault of Crack in Dissimilar Media*, Bulletin of the Seismological Society of America, 49, pp. 199–204, 1959.

[35] J. Wu; *Interface problems for elliptic systems*, J. Partial Diff. Eqs., 12, pp. 313–323, 1999.

[36] J. Wu; *Interface problems for quasilinear elliptic equations*, J. Diff. Eqs., 157, pp. 102–119, 1999.

[37] L.-a. Ying; *The infinite similar element method for calculating stress intensity factors*, Scientia Sinica, 21, 1, pp. 19–43, 1978.

[38] L.-a. Ying; *Infinite element method for elliptic problems*, Science in China (Series A), 34, 12, pp. 1438–1447, 1991.

[39] L.-a. Ying; *Infinite Element Methods*, Peking University Press, Vieweg Publishing, 1995.

[40] L.-a. Ying; *Interface problems for elliptic differential equations*, Chin. Ann. of Math., 18B, pp. 139–152, 1997.

[41] L.-a. Ying; *High order regularity for interface problems*, Northeast. Math. J., 13, pp. 459–476, 1997.

[42] L.-a. Ying; *A decomposition theorem for the solutions to the interface problems of quasi-linear elliptic equations*, Research Report No.31, School of Mathematical Sciences and Institute of Marthematics, Peking University, 2001.

[43] L.-a. Ying; *Two dimensional interface problems for elliptic equations*, J. Partial Diff. Eqs., 16, 1, pp. 37–48, 2003.

[44] O. C. Zienkiewicz, J. Z. Zhu; *The superconvergent patch recovery and a posteriori error estimates, Parts 1 & 2*, Int. J. Numer. Meth. Engng., 33, pp. 1331–1364, pp. 1365–1382, 1992.

[45] O. C. Zienkiewicz, J. Z. Zhu; *A simple error estimator and the adaptive procedure for practical engineering analysis*, Int. J. Numer. Meth. Engng., 24, pp. 337–357, 1987.

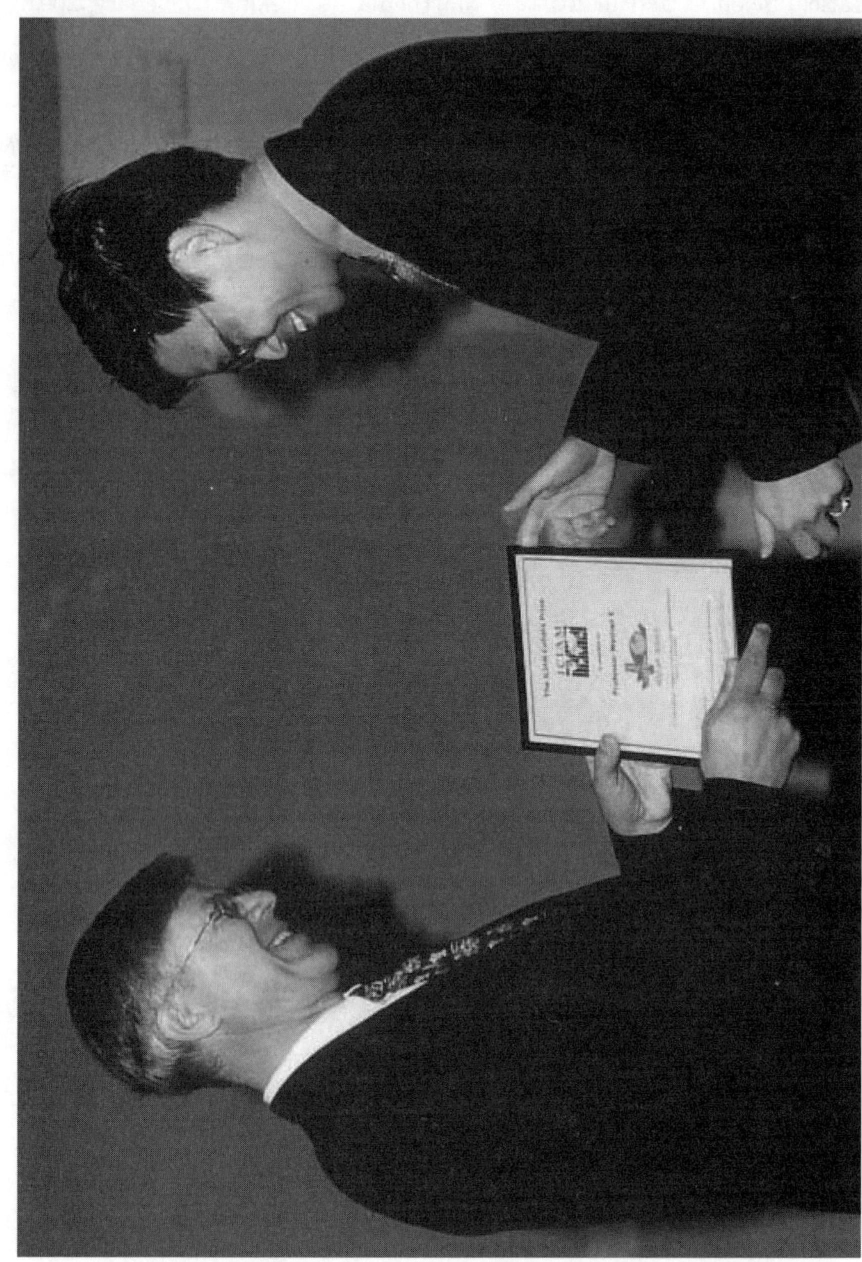

Weinan E accepts the ICIAM Collatz Prize ICIAM President Olavi Nevanlinna. photography by *Happy Medium Photo Co.*

ICIAM

Chapter 20

ICIAM Prizes for 2003

Four prizes are awarded, in connection with the ICIAM Congresses[1]. Prize winners are selected by the ICIAM Prize Committee. For 2003 this committee was chaired by Olavi Nevanlinna (Helsinki University of Technology, Finland), who was President of ICIAM at that time. Other members of the committee were Franco Brezzi (Università di Pavia, Italy), Robert Kohn (Courant Institute, New York University), Reinhard Mennicken (Universität Regensberg, Germany) and Hilary Ockendon (Oxford University, UK).

Each prize has its own subcommittee, chaired by one member of the Prize Committee. These subcommittees work independently, but the final decision is made by the Prize Committee as a whole. Members of subcommittees are made public at the time the prize winners are announced; these are listed below, along with the prize specifications and recipient.

ICIAM Lagrange Prize

This prize is funded by Société de Mathématiques Appliquées et Industrielles (SMAI), Sociedad Española de Matemática Aplicada (SEMA) and Società Italiana di Matematica Applicata e Industriale (SIMAI). It has been established to provide international recognition to individual mathematicians who have made an exceptional contribution to applied mathematics throughout their careers.

The ICIAM Lagrange Prize for 2003 is awarded to Professor Enrico Magenes (Università di Pavia), for his contributions to the development of Applied Mathematics at the world-wide level.

> In a remarkable series of papers, followed and made complete in a three-volume book "Nonhomogeneous Boundary Value Problems and Applications" written in cooperation with J. L. Lions, he set the foundations

[1]The prizes were awarded for the first time at ICIAM'99, held in Edinburgh. At that time they were called the CICIAM Prizes.

for the modern treatment of partial differential equations, and in particular the ones mostly used in applications. This includes the systematic treatment of variational formulations, as well as the paradigm "regularity results–transposition–interpolation", and allows a fully detailed use of the properties of trace spaces. The book has been *the* reference book for more than thirty years, for the completeness of the results reported there, but even more for the strategy of approach to problems. After that, the scientific activity of Magenes moved even further in the direction of application. In the early 1970s he founded the Institute of Numerical Analysis in Pavia, which he directed for more than twenty years, keeping it in close contact with the top-level scientific institutions all over the world, and making it the source of a number of highly successful scientists and of several pioneering results.

Apart from his continuous inspirational influence, he contributed personally to the development of a totally new technique for treating free boundary problems by means of variational inequalities, with remarkable applications to several important problems such as the flow of fluids through porous media or the phase-change phenomena. But even if his own results have been of paramount importance, his major merit is surely in the impulse he gave, and the influence he had in starting, encouraging and sustaining a way of doing mathematics that joined the rigour, the elegance and the deepness of so-called pure mathematics with the real-life problems that have to be faced in applications. If the combination of pure mathematics and applications is what Applied Mathematics is nowadays, Magenes is surely among the ones that deserve most credit.

The subcommittee for ICIAM Lagrange Prize was:

- Franco Brezzi (Chair; Università di Pavia, Italy)
- Martin Bendsøe (Technical University of Denmark)
- Gérard Iooss (Institut Non Linéaire de Nice, CNRS, France)
- Rupert Klein (Potsdam Institut für Klimafolgenforschung, Germany)
- Eitan Tadmor (University of California Los Angeles, USA).

Professor Franco Brezzi accepted the prize on behalf of Enrico Magenes.

ICIAM Collatz Prize

This prize is funded by Gesellschaft für Angewandte Mathematik und Mechanik (GAMM). It has been established to provide international recognition to individual scientists under 42 years of age for outstanding work on industrial and applied mathematics.

The ICIAM Collatz Prize for 2003 is awarded to Professor Weinan E (Princeton University), as a scientist under 42 years of age having already an outstanding scientific reputation in the field of industrial and applied mathematics.

Weinan E was born in 1963 in China where he also finished his bachelor and master degrees. He received his PhD from the University of California at Los Angeles in 1989 (under Björn Engquist). He was a long-term member of the Institute of Advanced Studies in Princeton from 1992 to 1994 and became a professor at the Courant Institute at New York University in 1994. In 1999, he moved to Princeton University where he holds a professorship in the Department of Mathematics and in the Program in Applied and Computational Mathematics. In 1996 he received the US Presidential Early Career Award for Scientists and Engineers, and in 1999 he was awarded the Feng Kang Prize for Scientific Computing.

The scientific work of Weinan E covers many areas of applied mathematics ranging from fluid dynamics to condensed matter physics, including incompressible flows, turbulence, statistical physics, superconductivity, liquid crystals and polymers, epitaxial growth, and micromagnetics. His early contributions were in the field of homogenization of fully nonlinear wave equations. Multiscale problems has remained one of his major fields until today. In his subsequent work on liquid crystals he provided a geometrically nonlinear continuum model, which allowed for a first explanation of the formation of filaments in the smectic-isotropic transition. In micromagnetics he devised, partially together with Garcia, Wang and Gimbutas, new numerical algorithms for finding solutions to the Landau–Lifshitz–Gilbert equation. Thus, for the first time fast-switching processes and the hysteresis effect in ferromagnetic materials could be simulated reliably and efficiently.

Weinan E is a scientist of exceptional vision and scope. His work is a sophisticated combination of modelling, mathematical analysis, and numerics. It is always devoted to providing new insights into real-world processes.

The subcommittee for the ICIAM Collatz Prize was:

- Olavi Nevanlinna (Chair; Helsinki University of Technology, Finland)
- Grigory Barenblatt (University of California at Berkeley, USA)
- Leah Keshet (University of British Columbia, Canada)
- Alexander Mielke (Universität Stuttgart, Germany)
- Etienne Pardoux (Université de Provence, France).

ICIAM Pioneer Prize

This prize is funded by SIAM, is for pioneering work introducing applied mathematical methods and scientific computing techniques to an industrial problem area or a new scientific field of applications. The prize commemorates the spirit and impact of the American pioneers.

The ICIAM Pioneer Prize for 2003 is awarded to Professor Stanley Osher (University of California, Los Angeles) in recognition of his outstanding contributions to applied mathematics and computational science; particularly for his work on shock-capturing schemes, PDE-based image processing, and the level-set method.

Professor Osher's work on shock-capturing schemes for conservation laws has been extremely influential in computational fluid dynamics (CFD). In the late 1970s and early 1980s he developed, with various collaborators, monotone and total-variation-decreasing (TVD) schemes which quickly became very popular. Later, with collaborators, he introduced essentially-non-oscillatory (ENO) schemes, which have found widespread use in compressible CFD. Further developments include WENO schemes, and shock-capturing methods for solving Hamilton–Jacobi equations. Osher's work with L. Rudin on total-variation-based image restoration was among the first applications of PDE methods to image processing. This work has been very influential, stimulating mathematical research on PDE-based image analysis, and leading to the development of related methods for various inverse problems. It has also had commercial success through the activities of Cognitech, a company founded by Osher and Rudin.

His work on level-set methods represents a fresh, very powerful approach to the numerical solution of evolutionary free-boundary problems. In the late 1980s with J. Sethian, Osher addressed the propagation of codimension-one fronts with curvature-dependent speed. Since then, with various collaborators, he has addressed a wide variety of related problems, developing techniques for handling nonlocal velocity laws, triple junctions, and higher-codimension sets. He has, moreover, demonstrated the value of these techniques by applying them to problems from materials science, geometry, and fluid dynamics.

This Pioneer Prize recognizes Professor Osher for his many deep and novel mathematical contributions, which have had remarkable impact on computational science.

The subcommittee for ICIAM Pioneer Prize was:

- Robert Kohn (Chair; Courant Institute, New York)
- Ronald Coifman (Yale University, USA)
- Doina Cioranescu (Université Pierre et Marie Curie, Paris)
- Linda Petzold (University of California, Santa Barbara)
- Gennadi Vainikko (Helsinki University of Technology, Finland)
- Shmuel Winograd (IBM TJ Watson Research Center, USA)
- Yoshikazu Giga (Hokkaido University, Japan).

ICIAM Maxwell Prize

This prize is funded jointly by the Institute of Mathematics and its Applications (IMA) and the James Clerk Maxwell Foundation, to provide international recognition to a mathematician who has demonstrated originality in applied mathematics.

The ICIAM Maxwell Prize for 2003 is awarded to Professor Martin D. Kruskal (Rutgers University), for discovering the particle-like behaviour of solitary waves, which he named 'solitons'; for introducing the inverse scattering transform method of solving the initial-value problem for the KdV equation; and for many other contributions to applied mathematics.

Martin David Kruskal was born in New York in 1925. He did his first degree at the University of Chicago and obtained his PhD from New York University in 1952. He then moved to Princeton, first to Physics and later to Mathematics. Since 1989 he has held the David Hilbert Chair of Mathematics at Rutgers University.

Martin Kruskal is most famous for the invention of the inverse scattering transform method. The key discovery was the particle-like behaviour of solitary wave solutions of the Korteweg de Vries equation; Kruskal named these waves 'solitons' and showed how they could be used to solve initial value problems for a whole class of nonlinear partial differential equations. This work has led to a host of further developments by Kruskal and others and has transformed the theory of nonlinear partial differential equations. Kruskal has also done seminal work in plasma physics and astrophysics; in particular he has shown that the singularity of the Schwarzschild solution of Einstein's equations of general relativity is not an actual singularity of the geometry but is an apparent singularity due to the coordinate system. More recently he has returned to pure mathematics and the study of surreal numbers.

Kruskal's work has already been recognized; he is a member of the National Academy of Sciences, foreign member of the Royal Society of London and the Russian Academy of Natural Sciences and has been awarded a number of prizes including the President's National Medal of Science in 1993. It is very appropriate that the international applied mathematics community should now acknowledge Martin Kruskal's achievements by the award of the James Clerk Maxwell Prize.

The subcommittee for ICIAM Maxwell Prize was:

- Hilary Ockendon (Chair; Oxford University, UK)
- Karl-Heinz Hoffmann (Munich, Germany)
- Joseph Keller (Stanford University, USA)
- Philip Saffman (California Institute of Technology, USA)
- Ian Sloan (University of NSW, Australia)
- Yongji Tan (Fudan University, China)

Professor Nalini Joshi accepted the prize on behalf of Martin Kruskal.